Equilibrium Problems and Variational Models

Nonconvex Optimization and Its Applications

Volume 68

Equilibrium Problems and Variational Models

Edited by

Patrizia Daniele
Department of Mathematics
University of Catania
95125 Catania
Italy

Franco Giannessi
Department of Mathematics
University of Pisa
56127 Pisa
Italy

Antonino Maugeri
Department of Mathematics
University of Catania
95125 Catania
Italy

Distributors for North, Central and South America:
Kluwer Academic Publishers
101 Philip Drive
Assinippi Park
Norwell, Massachusetts 02061 USA
Telephone (781) 871-6600
Fax (781) 871-6528
E-Mail <kluwer@wkap.com>

Distributors for all other countries:
Kluwer Academic Publishers Group
Post Office Box 322
3300 AH Dordrecht, THE NETHERLANDS
Telephone 31 78 6576 000
Fax 31 78 6576 474
E-Mail <orderdept@wkap.nl>

Electronic Services <http://www.wkap.nl>

Library of Congress Cataloging-in-Publication

CIP info or:
Title: Equilibrium Problems and Variational Models
Author: Patrizia Daniele, Franco Giannessi, Antonino Maugeri
ISBN-13:978-1-4613-7955-3 e-ISBN-13:978-1-4613-0239-1
DOI: 10.1007/978-1-4613-0239-1

Contents

Preface

The volume, devoted to variational analysis and its applications, collects selected and refereed contributions, which provide an outline of the field.

The meeting of the title "Equilibrium Problems and Variational Models", which was held in Erice (Sicily) in the period June 23 – July 2 2000, was the occasion of the presentation of some of these papers; other results are a consequence of a fruitful and constructive atmosphere created during the meeting.

New results, which enlarge the field of application of variational analysis, are presented in the book; they deal with the vectorial analysis, time dependent variational analysis, exact penalization, high order derivatives, geometric aspects, distance functions and log–quadratic proximal methodology.

The new theoretical results allow one to improve in a remarkable way the study of significant problems arising from the applied sciences, as continuum model of transportation, unilateral problems, multicriteria spatial price models, network equilibrium problems and many others.

As noted in the previous book "Equilibrium Problems: Nonsmooth Optimization and Variational Inequality Models", edited by F. Giannessi, A. Maugeri and P.M. Pardalos, Kluwer Academic Publishers, Vol. **58** (2001), the progress obtained by variational analysis has permitted to handle problems whose equilibrium conditions are not obtained by the minimization of a functional. These problems obey a more realistic equilibrium condition expressed by a generalized orthogonality (complementarity) condition, which enriches our knowledge of the equilibrium behaviour. Also this volume presents important examples of this formulation.

Finally, recent numerical methods complete this volume that, like the previous one, aims to treat problems from their statement up to the computation of their solutions.

Patrizia Daniele, University of Catania

Franco Giannessi, University of Pisa

Antonino Maugeri, University of Catania

ON VECTOR QUASI–EQUILIBRIUM PROBLEMS

Qamrul Hasan Ansari
Department of Mathematics, Aligarh Muslim University
Aligarh 202 002, India

Jen-Chih Yao
Department of Applied Mathematics, National Sun Yat-sen University
Kaohsiung 804, Taiwan, ROC

Abstract In this paper, we consider the vector quasi-equilibrium problem and prove some existence results for its solution with or without generalized pseudomonotonicity assumption. As consequences of our results, we also derive some existence results for a solution to the vector quasi-optimization problem, vector quasi-saddle point problem and vector quasi-variational inequality problem.

Keywords: Vector quasi-equilibrium problem, vector quasi-variational inequality problem, vector quasi-optimization problem, vector quasi-saddle point problem, pseudomonotonicity, fixed point theorem, one-person game theorem.

2000 Mathematics Subject Classification. 49J, 90C, 65K.

1 Introduction

Let K be a nonempty convex subset of a topological vector space X and let $f : K \times K \to \mathbb{R}$ be a bifunction such that $f(x, x) \geq 0$ for all $x \in K$. The *equilibrium problem* (in short, EP) is to find $\bar{x} \in K$ such that

$$f(\bar{x}, y) \geq 0, \quad \text{for all } y \in K,$$

which includes optimization, saddle point, fixed point, complementarity and variational inequality problems as special cases. The (EP) has been studied by many authors, see for example [3, 5, 6, 9, 10, 11, 13, 14, 15, 20, 21, 23, 27, 32, 34] because of its general form and it includes many fundamental mathematical problems from optimization, operations research, economics, mechanics and engineering.

Let $A : K \to 2^K$ be a multivalued map with nonempty values, where 2^K denotes the family of all subsets of K, then the *quasi-equilibrium problem* (in short, QEP) considered and studied by Ding [17] and Lin and Park [26] which is to find $\bar{x} \in K$ such that

$$\bar{x} \in A(\bar{x}) \quad \text{and} \quad f(\bar{x}, y) \geq 0, \qquad \text{for all } y \in A(\bar{x}).$$

Inspired by the extension of variational inequality problems for vector-valued functions by Giannessi [18], in the recent past (EP) has been extended for vector-valued bifunctions in [1, 2, 8, 19, 20, 24, 27, 29, 31] with applications in vector optimization and vector saddle point problems and Nash equilibrium problem.

Let Y be a topological vector space and C be a pointed, proper, closed and convex cone with int $C \neq \emptyset$, where int C denotes the interior of C. This cone induces the partial ordering on Y by $x \preccurlyeq y$ if and only if $y - x \in C$. Since int $C \neq \emptyset$, we also have $x \prec y$ if and only if $y - x \in$ int C and $x \not\prec y$ if and only if $y - x \notin$ int C. For a given vector-valued bifunction $f : K \times K \to Y$, the *vector equilibrium problem* (in short, VEP), studied in [1, 2, 8, 19, 20, 24, 27, 29, 31], is to find $\bar{x} \in K$ such that

$$f(\bar{x}, y) \not\prec 0, \qquad \text{for all } y \in K.$$

We consider the following *vector quasi-equilibrium problem* (in short, VQEP) of finding $\bar{x} \in K$ such that

$$\bar{x} \in A(\bar{x}) \quad \text{and} \quad f(\bar{x}, y) \not\prec 0, \qquad \text{for all } y \in A(\bar{x}).$$

Let us see some special cases of (VQEP). We denote by $L(X, Y)$ the space of all continuous linear maps from X into Y and $\langle u, x \rangle$ denotes the evaluation of $u \in L(X, Y)$ at $x \in X$.

Special Cases: (I) Let $p : K \to L(X, Y)$ and $\eta : K \times K \to X$ be two maps and let us define $f(x, y) = \langle p(x), \eta(y, x) \rangle$. Then the (VQEP) reduces to the problem of finding $\bar{x} \in K$ such that

$$\bar{x} \in A(\bar{x}) \quad \text{and} \quad \langle p(\bar{x}), \eta(y, \bar{x}) \rangle \not\prec 0, \qquad \text{for all } y \in A(\bar{x}).$$

It is called *vector quasi-variational-like inequality problem* (in short, VQ-VLIP) and considered by Lee et al [25].

When $\eta(y, x) = y - g(x)$, where $g : K \to K$ is a map, then the (VQVLIP) is called *general vector quasi-variational inequality problem* (in short, GVQVIP) considered and studied by Kim and Tan [22].

When $\eta(y, x) = y - x$, then the (VQVLIP) is called *vector quasi-variational inequality problem* (in short, VQVIP).

When $A(x) = K$ for all $x \in K$, then (VQVLIP), (GVQVIP) and (VQVIP) become vector variational-like inequality, general vector variational inequality [30] and vector variational inequality problems, respectively. In a recent volume [19] edited by F. Giannessi, a large number of papers on vector variational and variational-like inequality problems can be found. At the end of this volume, a more comprehensive references on different aspects of vector variational inequalities and their generalizations are also given.

(II) Let $\varphi : K \to Y$ be a vector-valued map and $f(x, y) = \varphi(y) - \varphi(x)$ for all $x, y \in K$, then the (VQEP) is equivalent to the *vector quasi-optimization problem* (in short, VQOP) which is to find $\bar{x} \in K$ such that

$$\bar{x} \in A(\bar{x}) \text{ and } \varphi(y) - \varphi(\bar{x}) \not< 0, \quad \text{for all } y \in A(\bar{x}).$$

When $Y = \mathbb{R}$ and $C = \mathbb{R}_+$, then the (VQEP), (VQVLIP), (VQVIP) and (VQOP) reduce to the quasi-equilibrium problem [17, 26], quasi-variational-like inequality problem [16], quasi-variational inequality problem [7] and quasi-optimization problem, respectively.

From the above special cases, it is clear that our (VQEP) is a more general format of several known problems, for example, vector quasi-variational and quasi-variational-like inequality problems, vector quasi-optimization problems and vector quasi-saddle point problems. Hence by using the existence results for a solution to (VQEP), we can derive some deeper results for the existence of a solution of the problems which are particular cases of (VQEP).

In this paper, we establish some existence results for a solution to the (VQEP) with or without generalized pseudomonotonicity assumption by using a fixed point theorem due to the authors [4] and a one-person game theorem due to Yuan et al [33]. We do not assume that the set K is compact. As consequences of our results, we also derive the existence results for solutions to the (VQOP), vector quasi-saddle point problem and (VQVIP). At the end of last section, it is proved that any solution to the (VQVLIP) is a solution of (VQOP) under certain conditions. By using this fact, we also establish some existence results for a solution to the (VQOP).

2 Preliminaries

Let M be a nonempty subset of a topological vector space X. A function $\varphi : M \to Y$ is called *C-lower semicontinuous* (respectively, *C-upper semicontinuous*) *at* $x_0 \in M$ [8, 28, 31] if, for any neighborhood V of

$\varphi(x_0)$ in Y, there exists a neighborhood U of x_0 in X such that

$$\varphi(x) \in V + C, \quad \text{for all } x \in U \cap M$$

$$(\text{respectively}, \quad \varphi(x) \in V - C, \quad \text{for all } x \in U \cap M).$$

Furthermore, φ is *C-lower semicontinuous* (respectively, *upper semicontinuous*) *on* M if, it is C-lower semicontinuous (respectively, C-upper semicontinuous) at each $x \in M$.

φ is *C-continuous on* M if, it is C-lower semicontinuous as well as C-upper semicontinuous on M.

In [8] it is shown that a function $\varphi : M \to Y$ is C-lower semicontinuous if, and only if, for all $\alpha \in Y$, the set

$$L(\alpha) = \{x \in M : \varphi(x) \not\succ \alpha\}$$

is closed in M.

Similarly we can show that φ is C-upper semicontinuous if, and only if, for all $\alpha \in Y$, the set

$$U(\alpha) = \{x \in M : \varphi(x) \not\prec \alpha\}$$

is closed in M.

Let M be a nonempty convex subset of a topological vector space X. The function $\varphi : M \to Y$ is called *C-quasiconvex* [8, 28] if, for all $\alpha \in Y$, the set $\{x \in M : \varphi(x) \preccurlyeq \alpha\}$ is convex. It is called *C-quasiconcave* if $-\varphi$ is C-quasiconvex.

If φ is C-quasiconvex, then the set $\{x \in K : \varphi(x) \prec \alpha\}$ is also convex; see, for example [8].

For a nonempty subset B of a topological space X, we denote by $\text{int}_X(B)$ and $\text{cl}_X(B)$ the interior of B in X and the closure of B in X, respectively. Also, the subset B is said to be *compactly open* in X if, for each nonempty compact subset D of X, $B \cap D$ is open in D. If B is a subset of a vector space, we shall denote by $\text{co}(B)$ the convex hull of B.

If X and Y are topological spaces and $T : X \to 2^Y$ is a multivalued map, the *graph of* T, denoted by $\mathcal{G}(T)$, is the set

$$\{(x, y) \in X \times Y : y \in T(x)\}.$$

The *inverse* T^{-1} of T is the multivalued map from the range of T to X defined by

$$x \in T^{-1}(y) \quad \text{if and only if} \quad y \in T(x).$$

The multivalued map $\overline{T} : X \to 2^Y$ is defined as

$$\overline{T}(x) = \{y \in Y : (x, y) \in \text{cl}_{X \times Y}\mathcal{G}(T)\},$$

for each $x \in X$. The set $\text{cl}_{X\times Y}\mathcal{G}(T)$ is called the *adherence of the graph of* T.

Let X be a topological vector space and K be a subset of X such that $K = \bigcup_{n=1}^{\infty} K_n$, where $\{K_n\}_{n=1}^{\infty}$ is an increasing (in the sense that $K_n \subseteq K_{n+1}$) sequence of nonempty compact sets. According to Border ([12], pp. 34), a sequence $\{x_n\}_{n=1}^{\infty}$ in K is said to be *escaping from* K (relative to $\{K_n\}_{n=1}^{\infty}$) if, for each $n = 1, 2, \ldots$, there exists $m > 0$ such that $x_k \notin K_n$ for all $k \geq m$.

We shall use the following fixed point and one-person game theorems due to the authors [4] and Yuan et al [33], respectively.

Theorem A. [4] *Let K be a nonempty convex subset of a Hausdorff topological vector space X, and let $S, T : K \to 2^K$ be two multivalued maps. Assume that the following conditions hold.*

(i) *For each $x \in K$, $\text{co}(S(x)) \subseteq T(x)$ and $S(x)$ is nonempty.*
(ii) $K = \bigcup\{\text{int}_K S^{-1}(y) : y \in K\}$.
(iii) *If K is not compact, assume that there exist a nonempty compact convex subset D of K and a nonempty compact (not necessarily convex) subset B of K such that for each $x \in K \setminus B$, there exists $\tilde{y} \in D$ such that $x \in \text{int}_K S^{-1}(\tilde{y})$.*

Then there exists $x_0 \in K$ such that $x_0 \in T(x_0)$.

The following result is the particular form of a one-person game theorem (Theorem 3.1) due to Yuan et al [33].

Theorem B. *Let $(K; A, P)$ be a one-person game such that $K = \bigcup_{i=1}^{\infty} K_n$, where $\{K_n\}_{i=1}^{\infty}$ is an increasing sequence of nonempty compact convex subsets of a topological vector space X and $A, P : K \to 2^K$ are multivalued maps. Assume that the following conditions are satisfied.*

(i) *For each $x \in K$, $A(x)$ is nonempty and convex;*
(ii) *For each $y \in K$, $A^{-1}(y)$ is compactly open in K;*
(iii) *For each $x \in K$, $x \notin \text{co}P(x)$;*
(iv) *For each $y \in K$, $P^{-1}(y)$ is compactly open in K;*
(v) *For each sequence $\{x_n\}_{n=1}^{\infty}$ in K with $x_n \in K_n$ for each $i = 1, 2, \ldots$, which is escaping from K relative to $\{K_n\}_{n=1}^{\infty}$, there exist $n \in \mathbb{N}$ and $y_n \in K_n$ such that $y_n \in \text{co}(A(x_n) \cap P(x_n))$.*

Then there exists $x_0 \in K$ such that $x_0 \in \overline{A}(x_0)$ and $A(x_0) \cap P(x_0) = \emptyset$.

3 Existence Results

Throughout this and next section, unless otherwise specified, we assume that K is a nonempty convex subset of a Hausdorff topological vector space X and Y is a topological vector space with a pointed, proper, closed and convex cone such that int $C \neq \emptyset$. We also assume that $A : K \to 2^K$ is a multivalued map with nonempty convex values such that for each $y \in K$, $A^{-1}(y)$ is open in K, and the set $\mathscr{F} := \{x \in K : x \in A(x)\}$ is closed in K.

We establish some existence results for a solution to the (VQEP) without generalized pseudo- monotonicity assumption.

Theorem 3.1. *Let $f, g : K \times K \to Y$ be vector-valued bifunctions such that*

(i) *for each $x \in K$, $g(x, x) \not\succ 0$;*

(ii) *for each $x \in K$ and for each finite subset $\{y_1, \ldots, y_n\}$ of K, $f(x, y_i) \prec 0$ for all $i = 1, \ldots, n$ imply $g(\hat{y}, x) \succ 0$, where $\hat{y} = \sum_{i=1}^{n} \alpha_i y_i$, $\alpha_i \geq 0$ for all $i = 1, \ldots, n$ and $\sum_{i=1}^{n} \alpha_i = 1$;*

(iii) *for each $y \in K$, the function $x \mapsto f(x, y)$ is C-upper semicontinuous on K;*

(iv) *if K is not compact, there exist a nonempty compact convex subset D of K and a nonempty compact (not necessarily convex) subset B of K such that for each $x \in K \setminus B$, there exists $\tilde{y} \in D$ such that $x \in A^{-1}(\tilde{y})$ and $f(x, \tilde{y}) \prec 0$.*

Then there exists a solution $\bar{x} \in K$ to the (VQEP).

Proof. For each $x \in K$, we define two multivalued maps $P, Q : K \to 2^K$ by

$$P(x) = \{y \in K : f(x, y) \prec 0\}$$

and

$$Q(x) = \{y \in K : g(y, x) \succ 0\}.$$

Then for each $y \in K$, the complement of $P^{-1}(y)$ in K is $[P^{-1}(y)]^c = \{x \in K : f(x, y) \not\prec 0\}$ is closed in K by condition (iii). Therefore $P^{-1}(y)$ is open in K, for each $y \in K$. Now we define another multivalued maps $S, T : K \to 2^K$ by

$$S(x) = \begin{cases} A(x) \cap P(x) & \text{if } x \in \mathscr{F} \\ A(x) & \text{if } x \in K \setminus \mathscr{F} \end{cases}$$

and

$$T(x) = \begin{cases} A(x) \cap Q(x) & \text{if } x \in \mathscr{F} \\ A(x) & \text{if } x \in K \setminus \mathscr{F}, \end{cases}$$

for each $x \in K$.

Let $\{y_1, \ldots, y_n\}$ be any finite subset of $S(x)$ for an arbitrary fixed $x \in K$, and let $\hat{y} = \sum_{i=1}^{n} \alpha_i y_i$ with $\alpha_i \geq 0$, $i = 1, \ldots, n$ and $\sum_{i=1}^{n} \alpha_i = 1$. If $x \in \mathscr{F}$, then $y_i \in A(x)$ and $y_i \in P(x)$ for each $i = 1, \ldots, n$. Since for each $x \in K$, $A(x)$ is convex, we have $\hat{y} \in A(x)$. Also, since each $y_i \in P(x)$, we have

$$f(x, y_i) \prec 0, \quad \text{for all } i = 1, \ldots, n.$$

By condition (ii), $g(\hat{y}, x) \succ 0$ and therefore $\hat{y} \in Q(x)$. Thus for all $x \in \mathscr{F}$, we have $\hat{y} \in A(x) \cap Q(x)$ and so $\mathrm{co}(S(x)) \subseteq T(x)$ for all $x \in \mathscr{F}$. For $x \in K \setminus \mathscr{F}$, $\mathrm{co}(S(x)) \subseteq T(x)$ because for each $x \in K$, $A(x)$ is convex. Since for each $y \in K$, $A^{-1}(y)$ and $P^{-1}(y)$ are open in K, $(A \cap P)^{-1}(y) = A^{-1}(y) \cap P^{-1}(y)$ is open in K. Since for each $y \in K$,

$$S^{-1}(y) = \big(A^{-1}(y) \cap P^{-1}(y)\big) \cup \big((K \setminus \mathscr{F}) \cap A^{-1}(y)\big)$$

(see, for example, the proof of Lemma 2.3 in [17]) and $K \setminus \mathscr{F}$ is open in K, we have $S^{-1}(y)$ is open in K. Now assume that for each $x \in \mathscr{F}$, $A(x) \cap P(x) \neq \emptyset$, then for each $x \in K$, $S(x) \neq \emptyset$ and therefore

$$K = \bigcup_{y \in K} S^{-1}(y) = \bigcup_{y \in K} \mathrm{int}_K S^{-1}(y).$$

Condition (iv) implies condition (iii) of Theorem A. Hence by Theorem A, there exists $x_0 \in K$ such that $x_0 \in T(x_0)$.

From the definition of $\mathscr{F}$ and T, we have $\{x \in K : x \in T(x)\} \subset \mathscr{F}$. Therefore, $x_0 \in \mathscr{F}$ and $x_0 \in A(x_0) \cap Q(x_0)$ and, in particular, $g(x_0, x_0) \succ 0$, a contradiction of (i). Hence there exists $\bar{x} \in \mathscr{F}$ such that $A(\bar{x}) \cap P(\bar{x}) = \emptyset$, that is,

$$\bar{x} \in A(\bar{x}) \quad \text{and} \quad f(\bar{x}, y) \not\prec 0, \quad \text{for all } y \in A(\bar{x}),$$

and the result is proved. □

Remark 3.1. The condition (ii) of Theorem 3.1 holds if,

(0) for each $x, y \in K$, $f(x, y) \prec 0$ implies $g(y, x) \succ 0$; and

(00) for each $x \in K$, the function $y \mapsto g(y, x)$ is C-quasiconcave.

Corollary 3.1. *Let $f : K \times K \to Y$ be a vector-valued bifunction such that*

(i) *for each $x \in K$, $f(x, x) \not\prec 0$;*

(ii) *for each $x \in K$, the function $y \mapsto f(x, y)$ is C-quasiconvex;*

(iii) *for each $y \in K$, the function $x \mapsto f(x,y)$ is C-upper semicontinuous on K;*

(iv) *if K is not compact, there exist a nonempty compact convex subset D of K and a nonempty compact (not necessarily convex) subset B of K such that for each $x \in K \setminus B$, there exists $\tilde{y} \in D$ such that $x \in A^{-1}(\tilde{y})$ and $f(x,\tilde{y}) \prec 0$.*

Then there exists a solution $\bar{x} \in K$ to the (VQEP).

Theorem 3.2. *Let $\{K_n\}_{n=1}^{\infty}$ be an increasing sequence of nonempty compact convex subsets of a topological vector space E such that $K = \bigcup_{n=1}^{\infty} K_n$. Let $A : K \to 2^K$ be a multivalued map with nonempty convex values and for each $y \in Y$, $A^{-1}(y)$ is compactly open in K. Let $f : K \times K \to Y$ be a vector-valued bifunction such that*

(i) *for each $x \in K$, $f(x,x) \not\prec 0$;*

(ii) *for each $x \in K$, the function $y \mapsto f(x,y)$ is C-quasiconvex;*

(iii) *for each $y \in K$, the function $x \mapsto f(x,y)$ is C-upper semicontinuous on each compact subset of K;*

(v) *for each sequence $\{x_n\}_{n=1}^{\infty}$ in K with $x_n \in K_n$ for each $i = 1,2,\ldots$, which is escaping from K relative to $\{K_n\}_{n=1}^{\infty}$, there exist $n \in \mathbb{N}$ and $y_n \in K_n$ such that $y_n \in A(x_n)$ and $f(x_n,y_n) \prec 0$.*

Then there exists $\bar{x} \in K$ such that

$$\bar{x} \in \overline{A}(\bar{x}) \quad \text{and} \quad f(\bar{x},y) \not\prec 0, \quad \textit{for all } y \in A(\bar{x}).$$

Proof. For each $x \in K$, we define a multivalued map $P : K \to 2^K$ by

$$P(x) = \{y \in K : f(x,y) \prec 0\}, \quad \text{for all } x \in K.$$

Then by condition (ii) and (iii), $P(x)$ is convex and $P^{-1}(y)$ is compactly open in K, respectively. Condition (v) implies condition (v) of Theorem B. The result follows from the proof of Theorem 3.1 by using Theorem B. □

Remark 3.2. In Theorem 3.2, E is not necessarily Hausdorff.

Definition 3.1. Let $f, g : K \times K \to Y$ be vector-valued bifunctions. Then f is called

(i) *g-pseudomonotone* if, for all $x, y \in K$,

$$f(x,y) \not\prec 0 \quad \text{implies} \quad g(y,x) \not\succ 0;$$

(ii) *g-pseudodissipative* if, for all $x, y \in K$,

$$g(y,x) \not\succ 0 \quad \text{implies} \quad f(x,y) \not\prec 0.$$

For $f \equiv g$, we simply write pseudomonotone and pseudodissipative, instead of f-pseudomonotone and f-pseudodissipative, respectively.

Now we establish some existence results for a solution to the (VQEP) under g- pseudomonotonicity assumption.

Theorem 3.3. *Let $f, g : K \times K \to Y$ be vector-valued bifunctions such that*

(i) *for each $x \in K$, $f(x,x) \not\prec 0$;*
(ii) *f is g-pseudomonotone and g-pseudodissipative;*
(iii) *for each $x \in K$, the function $y \mapsto f(x,y)$ is C-quasiconvex;*
(iv) *for each $y \in K$, the set $Q(y) := \{x \in K : g(y,x) \not\succ 0\}$ is closed in K;*
(v) *if K is not compact, there exist a nonempty compact convex subset D of K and a nonempty compact (not necessarily convex) subset B of K such that for each $x \in K \setminus B$, there exists $\tilde{y} \in D$ such that $x \in A^{-1}(\tilde{y})$ and $f(x,\tilde{y}) \prec 0$.*

Then there exists a solution $\bar{x} \in K$ to the (VQEP).

Proof. Let P be the same as defined in the proof of Theorem 3.1. For each $x \in K$, we define a multivalued map $S : K \to 2^K$ by

$$S(x) = \begin{cases} A(x) \cap P(x) & \text{if } x \in \mathscr{F} \\ A(x) & \text{if } x \in K \setminus \mathscr{F}. \end{cases}$$

Then for each $y \in K$,

$$\begin{aligned} S^{-1}(y) &= (A^{-1}(y) \cap P^{-1}(y)) \cup (A^{-1}(y) \cap (K \setminus \mathscr{F})) \\ &= A^{-1}(y) \cap (P^{-1}(y) \cup (K \setminus \mathscr{F})) \\ &= A^{-1}(y) \cap (\{x \in K : f(x,y) \prec 0\} \cup (K \setminus \mathscr{F})). \end{aligned}$$

By condition (iii), for each $x \in K$, $P(x) = \{y \in K : f(x,y) \prec 0\}$ is convex. Since $A(x)$ is convex, we have for each $x \in K$, $S(x)$ is convex. For each $y \in K$, let $\Psi(y) = (K \setminus A^{-1}(y)) \cup (\{x \in K : g(y,x) \not\succ 0\} \cap \mathscr{F})$, then $\Psi(y)$ is closed in K for each $y \in K$. By g-pseudomonotonicity of f, the complement of $\Psi(y)$ in K

$$\begin{aligned} [\Psi(y)]^c &= [(K \setminus A^{-1}(y)) \cup (\{x \in K : g(y,x) \not\succ 0\} \cap \mathscr{F})]^c \\ &= A^{-1}(y) \cap (\{x \in K : g(y,x) \not\succ 0\} \cap \mathscr{F})^c \\ &= A^{-1}(y) \cap (\{x \in K : g(y,x) \succ 0\} \cup (K \setminus \mathscr{F})) \\ &\subseteq A^{-1}(y) \cap (\{x \in K : f(x,y) \prec 0\} \cup (K \setminus \mathscr{F})) \\ &= S^{-1}(y). \end{aligned}$$

Similarly, by using g-pseudodissipativeness of f, we can easily prove that $S^{-1}(y) \subseteq [\Psi(y)]^c$ for all $y \in K$. Hence $S^{-1}(y) = [\Psi(y)]^c$ for all $y \in K$. Since for each $y \in K$, $\Psi(y)$ is closed in K, we have $S^{-1}(y)$ is open in K. Assume that for each $x \in K$, there exists $y \in K$ such that $x \in A^{-1}(y)$ and $g(y, x) \succ 0$. Then

$$K \subseteq \bigcup_{y \in K} [\Psi(y)]^c = \bigcup_{y \in K} S^{-1}(y) = \bigcup_{y \in K} \text{int}_K S^{-1}(y) \subseteq K.$$

Therefore $K = \bigcup_{y \in K} \text{int}_K S^{-1}(y) = \bigcup_{y \in K} S^{-1}(y)$ and thus $S(x) \neq \emptyset$ for each $x \in K$. Hence all the conditions of Theorem A are satisfied for $S \equiv T$ and therefore the conclusion follows from the proof of Theorem 3.1 and by using g-pseudodissipativeness of f. □

Remark 3.2. The condition (iv) of Theorem 3.3 holds if, for for each $y \in K$, the function $x \mapsto g(y, x)$ is C-lower semicontinuous on K.

Remark 3.3. (1) In Theorem 3.3, we have not assumed any kind of continuity assumption.
(2) In Theorems 3.1-3.3 and Corollary 3.1, if K is compact subset of X then all these results hold without their last condition.

4 Some Applications

Let $\varphi : K \to Y$ be a vector-valued function and let us define $f(x, y) = \varphi(y) - \varphi(x)$ for all $x, y \in K$, then we have the following existence results for a solution to the (VQOP) by using Corollary 3.1 and Theorem 3.2, respectively.

Theorem 4.1. *Let $\varphi : K \to Y$ be a vector-valued function such that*

(i) *φ is C-quasiconvex and C-lower semicontinuous on K;*

(ii) *if K is not compact, there exist a nonempty compact convex subset D of K and a nonempty compact (not necessarily convex) subset B of K such that for each $x \in K \setminus B$, there exists $\tilde{y} \in D$ such that $x \in A^{-1}(\tilde{y})$ and $\varphi(\tilde{y}) - \varphi(x) \prec 0$.*

Then the (VQOP) has a solution.

Theorem 4.2. *Let $\{K_n\}_{n=1}^{\infty}$ be an increasing sequence of nonempty compact convex subsets of a topological vector space E such that $K = \bigcup_{n=1}^{\infty} K_n$. Let $A : K \to 2^K$ be a multivalued map with nonempty convex values and for each $y \in Y$, $A^{-1}(y)$ is compactly open in K. Let $\varphi : K \to Y$ be a vector-valued function such that*

(i) *it is C-lower semicontinuous on each compact subset of K and C-quasiconvex;*

(ii) *for each sequence* $\{x_n\}_{n=1}^{\infty}$ *in* K *with* $x_n \in K_n$ *for each* $i = 1, 2, \ldots,$ *which is escaping from* K *relative to* $\{K_n\}_{n=1}^{\infty}$, *there exist* $n \in \mathbb{N}$ *and* $y_n \in K_n$ *such that* $y_n \in A(x_n)$ *and* $\varphi(y_n) - \varphi(x_n) \prec 0$.

Then there exists $\bar{x} \in K$ *such that*

$$\bar{x} \in \overline{A}(\bar{x}) \quad \text{and} \quad \varphi(y) - \varphi(\bar{x}) \not\prec 0, \quad \textit{for all } y \in A(\bar{x}).$$

Definition 4.1. Let $X_i, i = 1, 2$, be topological vector spaces and $K_i \subseteq X_i, i = 1, 2$, nonempty convex sets. Let $A : K = K_1 \times K_2 \to 2^K$ be a multivalued map defined as $A(x_1, x_2) = A_1(x_1) \times A_2(x_2)$ for all $x_1 \in K_1$ and $x_2 \in K_2$, where $A_i : K_i \to 2^{K_i}$, $i = 1, 2$, are multivalued maps, $\phi : K \to Y$ a vector-valued function, and Y and C be the same as above. Then $\bar{x} = (\bar{x}_1, \bar{x}_2) \in K_1 \times K_2$ is called

(i) *regular C-saddle point of* ϕ [31] if,

$$\phi(y_1, \bar{x}_2) - \phi(\bar{x}_1, y_2) \not\prec 0, \quad \text{for all } (y_1, y_2) \in K_1 \times K_2;$$

(ii) *regular C-quasi-saddle point of* ϕ if,

$$\bar{x} \in A(\bar{x}) \quad \text{and} \quad \phi(y_1, \bar{x}_2) - \phi(\bar{x}_1, y_2) \not\prec 0, \quad \text{for all } (y_1, y_2) \in A(\bar{x}).$$

From Corollary 3.1, we derive the following existence result of a regular C-quasi-saddle point for ϕ.

Theorem 4.3. *Let* $X_i, i = 1, 2$ *be Hausdorff topological vector spaces,* $K_i \subseteq X_i, i = 1, 2$ *nonempty convex sets and* $A_i : K_i \to 2^{K_i}$, $i = 1, 2$, *multivalued maps. Let* $A : K = K_1 \times K_2 \to 2^K$ *be a multivalued map defined as* $A(x_1, x_2) = A_1(x_1) \times A_2(x_2)$ *for all* $x_1 \in K_1$ *and* $x_2 \in K_2$ *such that for each* $x \in K$, $A(x)$ *is nonempty and convex and for all* $y \in K$, $A^{-1}(y)$ *is open in* K. *Let* $\mathscr{F} := \{x \in K : x \in A(x)\}$ *be closed in* K *and* Y *and* C *be the same as above. Let* $\phi : K_1 \times K_2 \to Y$ *be a vector-valued function such that*

(i) *for each* $x_2 \in K_2$, *the function* $y_1 \mapsto \phi(y_1, x_2)$ *is C-quasiconvex and C-lower semicontinuous on* K;

(ii) *for each* $x_1 \in K_1$, *the function* $y_2 \mapsto \phi(x_1, y_2)$ *is C-quasiconcave and C-upper semicontinuous on* K;

(iii) *if* K *is not compact, there exist a nonempty compact convex subset* D *of* K *and a nonempty compact (not necessarily convex) subset* B *of* K *such that for each* $x \in K \setminus B$, *there exists* $\tilde{y} \in D$ *such that* $x \in A^{-1}(\tilde{y})$ *and* $\phi(\tilde{y}_1, x_2) - \phi(x_1, \tilde{y}_2) \prec 0$.

Then there exists a regular C-quasi-saddle point for ϕ.

Proof. We define $f : K \times K \to Y$ by

$$f(x,y) = \phi(y_1, x_2) - \phi(x_1, y_2)$$

for all $x = (x_1, x_2), y = (y_1, y_2) \in K_1 \times K_2$. Then from Corollary 3.1, we get the conclusion. □

Remark 4.1. When $A(x) = K$ for all $x \in K$, then Theorem 4.3 generalizes Corollary 3.15 in [31] in the following ways:

(a) $X_i, i = 1,2$, need not be locally convex;
(b) $K_i, i = 1,2$, need not be closed.

Now we have the following existence results for a solution to the (VQVLIP) from which existence results for solutions to the (GVQVIP) and (VQVIP) can be easily derived.

Theorem 4.4. *Let $\eta : K \times K \to X$ be a given function with $\eta(x,x) = 0$ for all $x \in K$, and let $p, q : K \to L(X,Y)$ be two maps such that*

(i) *for each $x, y \in K$, $\langle p(x), \eta(y,x)\rangle \prec 0$ implies $\langle q(y), \eta(y,x)\rangle \succ 0$;*
(ii) *for each $x \in K$, the function $y \mapsto \langle q(y), \eta(y,x)\rangle$ is C-quasiconcave;*
(iii) *for each $y \in K$, the function $x \mapsto \langle p(x), \eta(y,x)\rangle$ is C-upper semicontinuous on K;*
(iv) *if K is not compact, there exist a nonempty compact convex subset D of K and a nonempty compact (not necessarily convex) subset B of K such that for each $x \in K \setminus B$, there exists $\tilde{y} \in D$ such that $x \in A^{-1}(\tilde{y})$ and $\langle p(x), \eta(\tilde{y}, x)\rangle \prec 0$.*

Then there exists a solution to the (VQVIP).

Proof. Taking $f(x,y) = \langle p(x), \eta(y,x)\rangle$ and $g(y,x) = \langle q(y), \eta(y,x)\rangle$ for all $x, y \in K$ and applying Theorem 3.1, we get the conclusion. □

Following result can be easily derived from Theorem 3.2 by taking $f(x,y) = \langle p(x), \eta(y,x)\rangle$ for all $x, y \in K$.

Theorem 4.5. *Let $\{K_n\}_{n=1}^{\infty}$ be an increasing sequence of nonempty compact convex subsets of a topological vector space E such that $K = \bigcup_{n=1}^{\infty} K_n$. Let $A : K \to 2^K$ be a multivalued map with nonempty convex values and for each $y \in Y$, $A^{-1}(y)$ is compactly open in K. Let $\eta : K \times K \to X$ be a given function with $\eta(x,x) = 0$ for all $x \in K$, and let $p : K \to L(X,Y)$ be a vector-valued function such that*

(i) *for each* $x \in K$*, the function* $y \mapsto \langle p(x), \eta(y,x)\rangle$ *is* C*-quasiconvex;*

(iii) *for each* $y \in K$*, the function* $x \mapsto \langle p(x), \eta(y,x)\rangle$ *is* C*-upper semi-continuous on each compact subset of* K*;*

(v) *for each sequence* $\{x_n\}_{n=1}^{\infty}$ *in* K *with* $x_n \in K_n$ *for each* $i = 1, 2, \ldots,$ *which is escaping from* K *relative to* $\{K_n\}_{n=1}^{\infty}$*, there exist* $n \in \mathbb{N}$ *and* $y_n \in K_n$ *such that* $y_n \in A(x_n)$ *and* $\langle p(x_n), \eta(y_n, x_n) \prec 0$.

Then there exists $\bar{x} \in K$ *such that*

$$\bar{x} \in \overline{A}(\bar{x}) \quad \text{and} \quad \langle p(\bar{x}), \eta(y, \bar{x})\rangle \not\prec 0, \quad \text{for all } y \in A(\bar{x}).$$

Definition 4.2. Let $\eta : K \times K \to X$ be a given function and let $p, q : K \to L(X, Y)$ be two maps. Then p is called

(i) *q-pseudomonotone w.r.t.* η if, for all $x, y \in K$,

$$\langle p(x), \eta(y,x)\rangle \not\prec 0 \quad \text{implies} \quad \langle q(y), \eta(y,x)\rangle \not\prec 0;$$

(ii) *q-pseudodissipative w.r.t.* η if, for all $x, y \in K$,

$$\langle q(y), \eta(y,x)\rangle \not\prec 0 \quad \text{implies} \quad \langle p(x), \eta(y,x)\rangle \not\prec 0.$$

For $p \equiv q$, we write pseudomonotone w.r.t. η and pseudodissipative w.r.t. η instead of p-pseudomonotone w.r.t. η and p-pseudodissipative w.r.t. η, respectively.

Theorem 4.6. *Let* $\eta : K \times K \to X$ *be a given function with* $\eta(x,y) + \eta(y,x) = 0$ *for all* $x, y \in K$*, and let* $p, q : K \to L(X,Y)$ *be two maps such that*

(i) p *is q-pseudomonotone w.r.t.* η *and q-pseudodissipative w.r.t.* η*;*

(ii) *for each* $x \in K$*, the function* $y \mapsto \langle p(x), \eta(y,x)\rangle$ *is* C*-quasiconvex;*

(iii) *for each* $y \in K$*, the function* $x \mapsto \langle q(y), \eta(y,x)\rangle$ *is* C*-upper semi-continuous;*

(v) *if* K *is not compact, there exist a nonempty compact convex subset* D *of* K *and a nonempty compact (not necessarily convex) subset* B *of* K *such that for each* $x \in K \setminus B$*, there exists* $\tilde{y} \in D$ *such that* $x \in A^{-1}(\tilde{y})$ *and* $\langle p(x), \eta(\tilde{y}, x)\rangle \prec 0$.

Then there exists a solution $\bar{x} \in K$ *to the (VQVIP).*

Proof. Define $f(x,y) = \langle p(x), \eta(y,x)\rangle$ and $g(y,x) = -\langle q(y), \eta(y,x)\rangle$ for all $x, y \in K$ and apply Theorem 3.3, we get the conclusion. □

Now we provide an example of a function which satisfies condition (i) of Theorem 4.6.

Example 4.1. Let $X = Y = \mathbb{R}$ and $K = C = \mathbb{R}_+$ and we define $p, q : K \to \mathbb{R}$ and $\eta : K \times K \to \mathbb{R}$ by

$$p(x) = 2x, \quad q(y) = 3y + 1 \quad \text{and} \quad \eta(y, x) = y^2 - x^2,$$

for all $x, y \in K$. Then p is q-pseudomonotone w.r.t. η and also q-pseudissipative w.r.t. η.

Remark 4.2. When $Y = \mathbb{R}$, $C = \mathbb{R}_+$ and $\eta(y, x) = y - x$, then Theorem 4.6 provides the existence of a solution to the quasivariational inequality problem under generalized pseudomonotonicity assumption, and hence Theorem 4.6 is a new result in this direction.

To prove the existence of a solution to the (VQOP) by using Theorem 4.5, we need to introduce the following concept.

Let $\eta : K \times K \to X$ be a given function. A vector-valued function $\varphi : K \to Y$ is called *η-differentiable on K* if, the set

$$\partial_\eta \varphi(x) = \{u \in L(X, Y) : \varphi(y) - \varphi(x) \not\geq \langle u, \eta(y, x) \rangle, \quad \text{for all } y \in K\}$$

is a singleton denoted by $D_\eta \varphi(x)$. For $\eta(y, x) = y - x$, this definition is used by Tan and Tinh [31]. It is easy to check that $D_\eta \varphi : X \to L(X, Y)$ is pseudomonotone w.r.t. η if, $\eta(x, y) + \eta(y, x) = 0$ for all $x, y \in X$.

Proposition 4.1. *Let $\eta : K \times K \to X$ be a given function and let $\varphi : K \to Y$ be η-differentiable on K. Then any solution of (VQVLIP) is a solution of (VQOP) with $p(x) = D_\eta \varphi(x)$, for all $x \in K$.*

Proof. Suppose that $\bar{x} \in K$ is a solution of (VQVLIP), then

$$\bar{x} \in A(\bar{x}) \quad \text{and} \quad \langle D_\eta \varphi(\bar{x}), \eta(y, \bar{x}) \rangle \notin -\text{int}\, C, \quad \text{for all } y \in A(\bar{x}).$$

Since φ is differentiable on K, we have

$$\varphi(y) - \varphi(\bar{x}) - \langle D_\eta \varphi(\bar{x}), \eta(y, \bar{x}) \rangle \in C.$$

Since $a \notin -\text{int}\, C$ and $b - a \in C$ then $b \notin -\text{int}\, C$, we have

$$\varphi(y) - \varphi(\bar{x}) \notin -\text{int}\, C \quad \text{for all } y \in A(\bar{x})$$

and the result is proved. □

By using Proposition 4.1 and Theorem 4.5, we have the following existence result for a solution to the (VQOP).

Theorem 4.7. *Let $\{K_n\}_{n=1}^{\infty}$ be an increasing sequence of nonempty compact convex subsets of a topological vector space E such that $K = \bigcup_{n=1}^{\infty} K_n$. Let $A : K \to 2^K$ be a multivalued map with nonempty convex values and for each $y \in Y$, $A^{-1}(y)$ is compactly open in K. Let $\eta : K \times K \to X$ be a given function with $\eta(x,x) = 0$ for all $x \in K$, and let $\varphi : K \to Y$ be η-differentiable function such that*

(i) *for each $x \in K$, the function $y \mapsto \langle D_\eta\varphi(x), \eta(y,x)\rangle$ is C-quasiconvex;*

(iii) *for each $y \in K$, the function $x \mapsto \langle D_\eta\varphi(x), \eta(y,x)\rangle$ is C-upper semicontinuous on each compact subset of K;*

(v) *for each sequence $\{x_n\}_{n=1}^{\infty}$ in K with $x_n \in K_n$ for each $i = 1, 2, \ldots$, which is escaping from K relative to $\{K_n\}_{n=1}^{\infty}$, there exist $n \in \mathbb{N}$ and $y_n \in K_n$ such that $y_n \in A(x_n)$ and $\langle D_\eta\varphi(x_n), \eta(y_n, x_n) \prec 0$.*

Then there exists $\bar{x} \in K$ such that

$$\bar{x} \in \overline{A}(\bar{x}) \quad \text{and} \quad \varphi(y) - \varphi(\bar{x}) \not\prec 0, \quad \textit{for all } y \in A(\bar{x}).$$

Acknowledgement. This research was supported by the National Science Council of the Republic of China.

References

[1] Q.H. Ansari, *Vector Equilibrium Problems and Vector Variational Inequalities*, In [19], pp. 1-16, 2000.

[2] Q.H. Ansari, W. Oettli and D. Schläger, *A Generalization of Vectorial Equilibria*, Mathematical Methods of Operations Research, Vol. 46, pp. 147-152, 1997.

[3] Q.H. Ansari, N.-C. Wong and J.-C. Yao, *The Existence of Nonlinear Inequalities*, Applied Mathematics Letters, Vol. 12, pp. 89-92, 1999.

[4] Q.H. Ansari and J.-C. Yao, *A Fixed Point Theorem and Its Applications to the System of Variational Inequalities*, Bulletin of the Australian Mathematical Society, Vol. 59, pp. 433-442, 1999.

[5] A.S. Antipin, *On Convergence of Proximal Methods to Fixed Point of Extremal Mappings and Estimates of Their Rate of Convergence*, Computational Mathematics and Mathematical Physics, Vol. 35, pp. 539-551, 1995.

[6] J.-P. Aubin, *L'Analyse Non Linéaire et Ses Motivations Économiques*, Masson, Paris, 1984.

[7] C. Baiocchi and A. Capelo, *Variational and Quasivariational Inequalities, Applications to Free Boundary Problems*, John Wiley & Sons, New York, 1984.

[8] M. Bianchi, N. Hadjisavvas and S. Schaible, *Vector Equilibrium Problems with Generalized Monotone Bifunctions*, Journal of Optimization Theory and Applications, Vol. 92, pp. 527-542, 1997.

[9] M. Bianchi and S. Schaible, *Generalized Monotone Bifunctions and Equilibrium Problems*, Journal of Optimization Theory and Applications, Vol. 90, pp. 31-43, 1996.

[10] E. Blum and W. Oettli, *From Optimization and Variational Inequalities to Equilibrium Problems*, The Mathematics Student, Vol. 63, pp. 123-145, 1994.

[11] H. Brézis, L. Nirenberg and G. Stampacchia, *A Remark on Ky Fan's Minimax Principle*, Bolletin Uni. Mat. Italiana, Vol. 6(4), pp. 293-300, 1972.

[12] K.C. Border, *Fixed Point Theorems with Applications to Economics and Game Theory*, Cambridge University Press, Cambridge, U.K., 1985.

[13] O. Chadli, Z. Chbani and H. Riahi, *Recession Methods for Equilibrium Problems and Applications to Variational and Hemivariational Inequalities*, Discret and Continuous Dynamical Systems, Vol. 5, pp. 185-195, 1999.

[14] O. Chadli, Z. Chbani and H. Riahi, *Equilibrium Problems and Noncoercive Variational Inequalities*, Optimization, Vol. 49, pp. 1-12, 1999.

[15] O. Chadli, Z. Chbani and H. Riahi, *Equilibrium problems with Generalized Monotone Bifunctions and Applications to Variational Inequalities*, Journal of Optimization Theory and Applications, Vol. 105, pp.299-323, 2000.

[16] N.H. Dien, *Some Remarks on Variational-like and Quasivariational-like Inequalities*, Bulletin of the Australian Mathematical Society, Vol. 46, pp. 335-342, 1992.

[17] X.P. Ding, *Existence of Solutions for Quasi-Equilibrium Problems in Noncompact Topological Spaces*, Computers and Mathematics with Applications, Vol. 39, pp. 13-21, 2000.

[18] F. Giannessi, *Theorems of the Alternative, Quadratic Programs and Complementarity Problems*, Variational Inequalities and Complementarity Problems, Edited by R.W. Cottle, F. Giannessi, and J.L. Lions, John Wiley and Sons, New York, pp.151-186, 1980.

[19] F. Giannessi, (Ed.), *Vector Variational Inequalities and Vector Equilibria. Mathematical Theories*, Kluwer Academic Publishers, Dordrecht-Boston-London, 2000.

[20] N. Hadjisavvas and S. Schaible, *From Scalar to Vector Equilibrium Problems in the Quasimonotone Case*, Journal of Optimization Theory and Applications, Vol. 96, pp. 297-309, 1998.

[21] T. Husain and E. Tarafdar, *Simultaneous Variational Inequalities, Minimization Problems and Related Results*, Mathematica Japonica, Vol. 39, pp. 221-231, 1994.

[22] W.K. Kim and K.-K. Tan, *On generalized Vector Quasi-Variational Inequalities*, Optimization, Vol. 46, pp. 185-198, 1999.

[23] I.V. Konnov, *A General Approach to Finding Stationary Point and the Solution of Related Problems*, Computational Mathematics and Mathematical Physics, Vol. 36, pp. 585-593, 1996.

[24] G.M. Lee, D.S. Kim and B.S. Lee, *On Noncooperative Vector Equilibrium*, Indian Journal of Pure and Applied Mathematics, Vol. 27, pp. 735-739, 1996.

[25] G.M. Lee, D.S. Kim and B.S. Lee, *On Vector Quasivariational-like Inequality*, Bulletin of the Korean Mathematical Society, Vol. 33, pp. 45-55, 1996.

[26] L.-J. Lin and S. Park, *On Some Generalized Quasi-Equilibrium Problems*, Journal of Mathematical Analysis and Applications, Vol. 224, pp. 167-181, 1998.

[27] L.-J. Lin and Z.-T. Yu, *Fixed-point Theorems and Equilibrium Problems*, To appear in Nonlinear Analysis, Theory, Methods and Applications, 2000.

[28] D.T. Luc, *Theory of Vector Optimization*, Lecture Notes in Economics and Mathematical Systems, Springer-Verlag, Berlin, Vol. 319, 1989.

[29] W. Oettli, *A Remark on Vector-Valued Equilibria and Generalized Monotonicity*, Acta Mathematica Vietnamica, Vol. 22, pp. 213-221, 1997.

[30] A.H. Siddiqi, Q.H. Ansari and A. Khaliq, *On Vector Variational Inequalities*, Journal of Optimization Theory and Applications, Vol. 84, pp. 171-180, 1995.

[31] N.X. Tan and P.N. Tinh, *On the Existence of Equilibrium Points of Vector Functions*, Numerical Functional Analysis and Optimization, Vol. 19, pp. 141-156, 1998.

[32] E. Tarafdar and G. X.-Z. Yuan, *Generalized Variational Inequalities and Its Applications*, Nonlinear Analysis, Theory, Methods and Applications, Vol. 30, pp. 4171-4181, 1997.

[33] G. X.-Z. Yuan, G. Isac, K. K. Tan, and J. Yu, *The Study of Minimax Inequalities, Abstract Economics and Applications to Variational Inequalities and Nash Equilibra*, Acta Applicandae Mathematicae, Vol. 54, pp. 135-166, 1998.

[34] G. X.-Z. Yuan, *KKM Theory and Applications in Nonlinear Analysis*, Marcel Dekker, Inc., New York, Basel, 1999.

THE LOG–QUADRATIC PROXIMAL METHODOLOGY IN CONVEX OPTIMIZATION ALGORITHMS AND VARIATIONAL INEQUALITIES

Alfred Auslender

Departement de Mathematiques, Universite Lyon I

Lyon, France

e-mail: `auslen@poly.polytechnique.fr`

Marc Teboulle

School of Mathematical Sciences

Tel-Aviv University,

Ramat-Aviv 69978, Israel

e-mail: `teboulle@math.tau.ac.il`

Abstract The logarithmic-quadratic proximal map recently introduced by the authors allows for deriving several efficient algorithms in convex optimization and variational inequalities. This brief survey outlines the power and usefulness of the resulting logarithmic-quadratic methodology.

Keywords: Proximal algorithms, Logarithmic-quadratic proximal methods, Convex programming Decomposition schemes, Large scale separable convex problems. Lagrangian multiplier methods, Variational inequalities, Convex Feasibility problems, Linear Programming, Nondifferentiable convex optimization, Bundle methods.

2000 Mathematics Subject Classification. 49J, 90C, 65K.

1 Introduction

The classical quadratic proximal algorithm is at the root for the development of many important algorithms to solve constrained minimization problems and variational inequalities. This include:

- Augmented Lagrangian methods
- Primal/dual methods for Variational inequalities and maximal monotone inclusions
- Decomposition methods
- Block nonlinear Gauss-Seidel methods for linearly constrained problems
- Primal iterative scheme for Linearly constrained convex problems
- Projection techniques for convex feasibility problems
- Bundle methods in nondifferentiable optimization.

Recently, there has been an increased interest in analyzing proximal-methods based on replacing the standard quadratic term by a more general (not necessarily quadratic) term playing the role of a distance-like function . This paper will focus on the *logarithmic quadratic proximal* (LQP) map recently introduced by the authors. We give a short survey of results recently derived through the LQP methodology.

The topics itemized above will be analyzed through the LQP framework. Our emphasize will be on explaining the usefulness and benefits of using the logarithmic-quadratic function and its conjugate within the proximal framework, over both the classical quadratic and other nonquadratic proximal methods recently suggested in the literature. The contents of the paper is well summarized by the items above. We start in the next section with the motivation for using nonquadratic proximal methods, outlining the difficulties associated with the quadratic as well as other nonquadratic proximal and Lagrangian methods for the itemized areas. This is followed by Section 3 in which we summarize the basic properties of the LQ-function and the resulting LQ proximal algorithm. Section 4 then surveys its applications to the itemized topics. The definitions and notation used in this paper in relation to convex analysis are standard and can be found in [58]. The finite dimensional Euclidean space of dimension p is denoted by $\mathbb{R}^p$ with nonnegative and positive orthant denoted by $\mathbb{R}^p_+$ and $\mathbb{R}^p_{++}$ respectively. We use $\langle \cdot, \cdot \rangle$ to denote the inner product of two vectors in $\mathbb{R}^p$ with $|| \cdot ||$ for the corresponding norm.

2 Lagrangians and Proximal Methods

The basic idea of augmented Lagrangian methods for solving nonlinear programming problems, also called *multiplier methods*, is to transform a constrained problem into a sequence of unconstrained problems. The approach differs from the penalty-barrier methods, from the fact that in the functional defining the unconstrained problem to be solved, in addition to a penalty parameter, there are also multipliers associated with the constraints. Multiplier methods can be seen as a combination of penalty and dual methods. The motivation for these methods came from the desire of avoiding ill-conditioning associated with the usual penalty-barrier methods. Indeed, in contrast to penalty methods, the penalty parameter need not to go to infinity to achieve convergence of the multiplier methods. As a consequence, the augmented Lagrangian has a "good" conditioning, and the methods are robust for solving nonlinear programs. Augmented Lagrangians methods were proposed independently by Hestenes [39] and Powell [57] for the case of equality constraints, and extended for the case of inequality constraints by Rockafellar [59]. Many other authors have contributed to the development of augmented Lagrangian methods, and for an excellent treatment and comprehensive study of multiplier methods, we refer the reader to the monograph of Bertsekas [13] and references therein. For a more recent summary on Lagrangian multipliers in convex programming, see Teboulle [66].

2.1 The quadratic augmented Lagrangian

We start by briefly describing the basic steps involved in the classical quadratic multiplier method for solving convex program,

$$(P)\quad \min\{f(x) : g_i(x) \leq 0, i = 1, \dots, m,\ x \in \mathbb{R}^n\}.$$

where f, g_i are given convex functions on $\mathbb{R}^n$.
The quadratic method of multipliers (QMM) (see e.g., [61]), then generates the sequences $\{u^k\} \subset \mathbb{R}^m$, $\{x^k\} \subset \mathbb{R}^n$ as follows. Given a sequence of nondecreasing numbers $c_k > 0$, compute

$$\begin{aligned} x^{k+1} &= \operatorname{argmin}\{L_{c_k}(x, u^k) :\ x \in \mathbb{R}^n\} \\ u^{k+1} &= \max\{0, u^k + c_k g(x^{k+1})\}. \end{aligned}$$

where

$$L_c(x, u) = f(x) + \frac{1}{2c}\sum_{i=1}^{m}[\max{}^2\{0, u_i + c g_i(x)\} - u_i^2]. \tag{2.1}$$

is the so-called quadratic augmented Lagrangian associated to problem (P).

This method can be derived by using slack variables in the inequality constraints and then apply the multiplier method which was originally devised in [39], [57], for problems having only equality constraints. An alternative and interesting way for deriving an augmented Lagrangian method is via the proximal framework.

2.2 Proximal Minimization Algorithms

Consider the convex optimization problem

$$(C)\qquad \min\{F(x) : x \in \mathbb{R}^n\}$$

where $F : \mathbb{R}^n \to (-\infty, +\infty]$ is a proper, lower semi-continuous convex function. One method to solve (C) is to *regularize* the objective function using the proximal mapping of Moreau [52]. Given a real positive number λ, a proximal approximation of F is defined by:

$$F_\lambda(x) = \inf_u\{F(u) + 1/2\lambda\|x - u\|^2\}. \tag{2.2}$$

The resulting function F_λ enjoys several important properties: it is convex and differentiable and when minimized possesses the same set of minimizers and the same optimal value than problem (C), namely

$$\inf\{F(u) :\ u \in \mathbb{R}^n\} = \inf\{F_\lambda(x) :\ x \in \mathbb{R}^n\}. \tag{2.3}$$

The later property is in fact at the root of an iterative procedure for solving problem (C), called the *proximal minimization algorithm* [51], [60]. The method is as follows: given an initial point $x_0 \in \mathbb{R}^n$ a sequence $\{x_k\}$ is generated by solving:

$$x^{k+1} = \operatorname{argmin}\{F(x) + (1/2\lambda_k)\|x - x^k\|^2\} \tag{2.4}$$

where $\{\lambda_k\}_{k=1}^{\infty}$ is a sequence of positive numbers. This algorithm converges to an optimal solution of (C) under the sole assumption that such a point exists, see e.g.,[38], [48],[60]. In view of (2.3), it can be seen that the proximal minimization algorithm is nothing else but a Gauss-Seidel method (see Section 4) applied to the right-hand side of (2.3). Solving (2.4) might in fact be as difficult as solving the original problem (C) and at this stage, it is legitimate to question on the usefulness of such an algorithm.

It turns out that a powerful application of the proximal minimization algorithm is when applied to the *dual* (D) of the convex optimization problem (P) defined by

$$(D) \qquad \sup\{p(u) : u \in \mathbb{R}^p_+\}$$

where $p(u) = \inf_x l(x,u)$, and $l(x,u) = f(x) + \sum_{i=1}^m u_i g_i(x)$ is the usual Lagrangian associated to (P). Indeed, as shown by Rockafellar [61, p. 107], the quadratic multiplier method (QMM) is nothing else but applying the proximal algorithm on the dual problem (D). Under Slater's condition for problem (P), and assuming that the solution set of the primal problem is nonempty and compact, it can be shown that the dual sequence of multipliers $\{u^k\}$ converges to some Lagrangian multiplier u^* and that any limit point of the primal sequence $\{x^k\}$ solves (P).

The interplay between primal proximal methods and dual multiplier methods offers the possibility of designing and analyzing the convergence properties of the later from the former. Furthermore, this interplay provides the motivation to consider useful potential extensions to eliminate the difficulties associated with the QMM and which are discussed next.

2.3 Entropic Proximal Methods and Modified Lagrangians

Suppose that problem (P) is given with smooth data, f, g_i, say $C^2(\mathbb{R}^n)$ or even $C^\infty(\mathbb{R}^n)$. The main computational task in the QMM is the unconstrained minimization of L_c given in (2.1). Two fundamental difficulties arise with this functional:

(a) The functional L_c is only $C^{1,1}(\mathbb{R}^n)$.

(b) Furthermore, if the original problem is assumed also with *separable* data, the augmented Lagrangian losts the separability structure and is not amenable to decomposability.

The resulting disadvantage of (a) is clear: efficient Newton type methods cannot be used to solve the unconstrained minimization of the augmented Lagrangian L_c. Indeed, the lack of continuity in the second derivative of L_c can seriously slow down the rate of convergence of (QMM) and cause algorithmic failure. The disadvantage (b) is not only inherent to the structure of L_c but also even when separability can be achieved via some special devices, one still encounters the first difficulty for the resulting simpler decomposable problems. This will be discussed in Section 4. For now, we concentrate on the origin of the difficulty arising in (a) and on a possible cure to it.

One way of coping with difficulty (a) is in view of the results of the previous section, to try alternative proximal regularization terms, which could lead to better differentiability properties of the corresponding augmented Lagrangian functional. Indeed, the main problem emerges when applying the Moreau proximal regularization on the dual, which does not take into account the nonnegativity constraints and thus leads to the augmented Lagrangian L_c which is only $C^{1,1}$. Thus, what is needed is a "smarter" proximal term, which when applied to a constrained problem (such as for example the dual) will *automatically absorbs* the constraints within the minimization process. This was the motivation of the work of Teboulle [64], who introduced and studied new classes of *Entropic* proximal approximation of a function given by

$$F_\lambda(x) := \inf_u\{f(u) + \lambda^{-1}D(u,x)\},\ (\lambda > 0). \tag{2.5}$$

Here, $D(\cdot,\cdot)$ which replaces the quadratic proximal term $||u-x||^2$ is a measure of "closeness" between x, y satisfying $D(x,y) \geq 0$ with equality if and only if $x = y$. Specifically, D are chosen to be *entropy-like distances*: the Csiszar's φ-divergence functional proposed in the context of probability and information theory problems or a Bregman's type distance proposed for solving linearly constrained convex programs.

Here, we will concentrate only on the former. The φ-divergence is defined by

$$D(x,y) := d_{\varphi}(x,y) := \sum_{i=1}^{n} y_i \varphi(\frac{x_i}{y_i}) \tag{2.6}$$

where φ is a given convex function defined on the positive real line and satisfying some other technical conditions, [64],[65]. A typical example for the function φ is the choice $\varphi(t) = t\log t - t + 1,\ t \geq 0$, which leads to the so-called Kullback-Liebler relative entropy type functional, and justifies the proposed entropic terminology. Many other examples can be found in [64],[65].

Thus, nonnegativity constraints are now "built-in" in d_{φ}. This allows for eliminating such constraints, which are naturally present in dual problems, and will also lead to "smoother" Lagrangians. In similarity with (2.4), the iterative procedure emerging from using this entropy-like proximal regularization takes the form:

$$x^{k+1} = \text{argmin}\{f(x) + \lambda_k^{-1} d_{\varphi}(x, x^k)\} \tag{2.7}$$

On the dual side, applying (2.7) on the dual problem (D), one can then design a wide variety of Entropic Lagrangians/Multiplier methods. The basic steps of the Entropic Multipliers Method (EMM) then emerging can be described as follows:

Entropic Multipliers Method-EMM

Given a sequence of positive numbers $\{\lambda_k\}$, and initial points $\{x^k, u^k\}$, generate iteratively the next points by solving

$$x^{k+1} \in \text{argmin}\{L_{\lambda_k}(x, u^k) : x \in \mathbb{R}^n\} \tag{2.8}$$

$$y^{k+1} \in \text{argmax}\{\langle u, g(x^{k+1})\rangle - \lambda_k^{-1} d_{\varphi}(y, y^k) :\ u \in \mathbb{R}^m_+\} \tag{2.9}$$

where L_λ is the Entropic Lagrangian defined by

$$L_\lambda(x,u) = \sup_{\mu \in \mathbb{R}^m_+} \{l(x,\mu) - \lambda^{-1} d_{\varphi}(\mu, u)\} \tag{2.10}$$

i.e., the entropic regularization of the usual Lagrangian $l(x,\mu)$ associated with problem (P). The method is viable since both (2.9) and (2.10) can be solved *analytically*, and the computational analysis and effort should concentrate on (2.8). As shown in [64], the (EMM) gives rise to a rich family of new numerical methods, which includes, (with an appropriate choice of d_φ), the classical quadratic method of multipliers, [61], the class $\hat{P}_I$ of [13] and the modified barrier functions of Polyak [54]. The analysis of algorithms of the form (2.7) and the related entropic Lagrangians methods can be found in several recent studies with respect to convex programs, variational inequalities and semi-definite programming, see e.g., [3], [12], [24], [29], [40], [41], [55], [56], [62], [65], [67] and references therein. For iterative schemes of the form (2.7) but using Bregman-based functional

$$B(x,y) := \psi(x) - \psi(y) - \langle x - y, \nabla\psi(y)\rangle,$$

instead of (2.7), we refer the reader to the recent work of Kiwiel [43] which includes up to date convergence results for these methods and a comprehensive list of references.

2.4 Difficulties with Entropic Proximal Methods

The use of entropic proximal maps opens the way of treating the difficulty outlined in (a), but does not yet provide with the complete desired cure to the problem. Let us briefly explain why the main difficulty (a) essentially remains, for both convex programs and variational inequalities, when solved within the entropic framework which uses d_φ or B.

Convex problems

Primal entropic proximal methods defined through (2.7) can be shown to be globally convergent under a variety of assumptions on the problem data (usually mild) and on the kernel φ, (in some cases restrictive ones). Yet, when applied within the dual framework, the situation is not as powerful. For example, with d_φ as defined in (2.6), depending on the choice of the kernel φ, the corresponding augmented Lagrangians *are not always*:

- defined on the whole space,

- C^∞, assuming the data possess similar smoothness assumptions,
- with bounded Hessians (given the problem's data with similar assumptions),
- Self-concordant (see [53]) for quadratically constrained problems.

Moreover, as an additional serious drawback, the corresponding primal sequence can be shown to converge only in an ergodic sense to an optimal solution of the primal problem, see e.g., [41]. We also note that similar difficulties are encounter with methods based on Bregman functionals, see e.g., [43] and references therein.

Variational Inequalities
Consider now the class of variational inequality problems. Let C be a polyhedral set on $\mathbb{R}^m$ defined by

$$C := \{x \in \mathbb{R}^m : Ax \leq b\}$$

where A is an (p, m) matrix, $b \in \mathbb{R}^p$, $p \geq m$. We suppose that the matrix A is of maximal rank, i.e., $\text{rank}A = m$ and that $\text{int}C = \{x : Ax < b\}$ is nonempty. Let T be a maximal monotone set valued map such that $\text{dom}T \cap \text{int}C \neq \emptyset$ ($\text{dom}T := \{x : T(x) \neq \emptyset\}$).
Consider the variational inequality

$$(VI) \quad \text{“ Find a point } x^* \in C \text{ and } g^* \in T(x^*) \text{ satisfying}$$

$$(g^*, x - x^*) \geq 0 \quad \forall x \in C\text{”},$$

which includes as a special case the nonlinear complementarity problem by choosing $C = \mathbb{R}^m_+$.

A number of recent papers have concentrated on generalization of the proximal point algorithm for solving (VI) and based on entropic proximal terms d_φ or B, see, e.g., [3], [16], [27], [65]. In all these papers, convergence was proved under restrictive assumptions on the problem data. In [65] only ergodic convergence is proved and in [3], [16], [19], [27], it is assumed that the map T is paramonotone that is: (x, y), $(x', y') \in G(T)$ and $(x - x', y - y') = 0$ collectively imply that $(x, y') \in G(T)$. This notion, introduced by Bruck [15] to prove convergence of an iterative method, is very close to strict monotonicity and quite restrictive. For example, it is not even satisfied by the maximal monotone operator associated with the saddle function of a convex program. Here, a suitable result would be to derive an algorithm which will be globally convergent to a solution of (VI), under the only assumption that the set of solutions of (VI) is nonempty, a result which is not possible within the above entropic proximal schemes.

2.5 Toward Solutions to Difficulties

The resolution of all the drawbacks outlined above is by concentrating on the entropic proximal framework, but with a proximal term of the form

$$d_\varphi(x, y) := \sum_{j=1}^{p} y_j^2 \varphi(y_j^{-1} x_j). \tag{2.11}$$

The specific form of this functional can in fact be is found in the work of Tseng-Bertsekas [67], which uses the particular function $\varphi(t) = t \log t - t + 1$ leading to the so-called exponential multiplier methods, see e.g., [13]. However, global convergence of the method was not obtained in [67], but interesting quadratic rate of convergence were established. More recently, in Ben-Tal and Zibulevsky [12], this difficulty has been overcome within a more general proximal scheme where it was proven that the corresponding dual multiplier methods, produce primal-dual sequences which are bounded and each of their limit points solve the primal-dual problems. More importantly, a new type of penalty-barrier function having a logarithmic branch *glued* to a quadratic one was also introduced and used to construct an efficient algorithm for large scale problems, with computational implementation reporting good numerical results, (see [12, Section 8]). However, *global* convergence of the proximal scheme proposed in [12] of the primal or/and dual sequences was not obtained. In addition, the proposed penalty-barrier Lagrangian functional is not C^3, and this can cause numerical difficulties when implementing a Newton-type method. Furthermore, in both [12],[67], no extension of these methods to variational inequality problems were analyzed, ruling out the possibility of deriving primal-dual type algorithms.

An attractive solution to difficulty (a), which will eliminate all the drawbacks above, and leads to a Lagrangian sharing the smoothness properties of the problem's data (and thus opening the possibility of using efficiently Newton's type methods) is through the use of a specific choice for φ called the logarithmic-quadratic function, within the definition of d_φ given in (2.11) and which is introduced next.

3 The Logarithmic-Quadratic Proximal Framework

The Logarithmic-Quadratic function (LQ for short) and the associated proximal scheme has been introduced by Auslender-Teboulle-Ben-Tiba [4] to originally solve variational inequalities associated with maximal monotone maps over a polyhedra. In this section we survey the fundamental properties of the LQ function and the resulting LQ proximal method (LQP for short).

3.1 The LQ-Function and its Conjugate: Basic Properties

Let $\nu > \mu > 0$ be given fixed parameters, and define

$$\varphi(t) = \begin{cases} \frac{\nu}{2}(t-1)^2 + \mu(t - \log t - 1) & \text{if } t > 0 \\ +\infty & \text{otherwise} \end{cases} \tag{3.12}$$

Associated with φ we define for any $v \in \mathbb{R}^p_{++}$:

$$d(u,v) = \begin{cases} \sum_{i=1}^p \frac{\nu}{2}(u_i - v_i)^2 + \mu(v_i^2 \log \frac{v_i}{u_i} + u_i v_i - v_i^2) & \text{if } u \in \mathbb{R}^p_{++} \\ +\infty & \text{otherwise} \end{cases} \tag{3.13}$$

It will also be convenient later on to use the parameter

$$\theta := (\nu + \mu)/2., \tag{3.14}$$

and to note that $\nu > \theta$.
The functional d defined in (3.13) can be rewritten as

$$d(u,v) = \sum_{j=1}^p v_j^2 \varphi(u_j v_j^{-1}), \ \forall u, v > 0. \tag{3.15}$$

Second-order homogeneous functionals of the form (3.15) with other choices of the kernel φ, are possible and the reader is refer to [5] for more details. In the rest of this paper we remove the index φ in the notation d_φ, since all the described results will be based on the fixed choice of the logarithmic-quadratic function φ as defined in (3.12).

The next Theorem recalls some important properties of φ, its conjugate φ^* and d, see [4], [5] for proofs.

Theorem 3.1 *Let φ be given in (3.12) and d as defined in (3.13). Then,*
(i) φ is a differentiable strongly convex function on $\mathbb{R}_{++}$ with modulus $\nu > 0$ and $\varphi_\infty(d) = +\infty$, $\forall d \neq 0$, where φ_∞ denotes the recession function of φ.
(ii) $\lim_{t\to 0} \varphi'(t) = -\infty$.
(iii) For any $u, v > 0$ we have $d(u,v) \geq \mu||u-v||^2$ and $d(u,v) = 0$ if and only if $u = v$.
(iv) The conjugate of φ is given by

$$\varphi^*(s) = \frac{\nu}{2}t^2(s) + \mu \log t(s) - \frac{\nu}{2}, \tag{3.16}$$

$$\begin{aligned} t(s) &:= (2\nu)^{-1}\{(\nu-\mu) + s + \sqrt{((\nu-\mu)+s)^2 + 4\mu\nu}\} \\ &= (\varphi^*)'(s). \end{aligned} \tag{3.17}$$

(v) $dom\varphi^* = \mathbb{R}$, and $\varphi^* \in C^\infty(\mathbb{R})$.
(vi)$(\varphi^*)'(s) = (\varphi')^{-1}(s)$ is Lipschitz for all $s \in \mathbb{R}$, with constant ν^{-1}.
(vii) φ^* is strictly convex and increasing on $\mathbb{R}$.
(viii) $(\varphi^*)''(s) < \nu^{-1}$, $\forall s \in \mathbb{R}$.
(ix) $(\varphi^*)_\infty(-1) = 0$ and $(\varphi^*)_\infty(1) = +\infty$ where $(\varphi^*)_\infty$ is the recession function of φ^*.
(x) The conjugate φ^* is self-concordant with parameter 2, i.e.,

$$(\varphi^*)'''(s) \leq 2(\varphi^*)''^{3/2}(s), \; \forall s \in \mathbb{R}.$$

An interesting implication of (x) is to the composite function $c(x) := \varphi^*(q(x))$, where $q : \mathbb{R}^n \to \mathbb{R}$ is a given convex quadratic function. Then, $c(x)$ is also self-concordant.

3.2 The Logarithmic-Quadratic Proximal Minimization

Let $F : \mathbb{R}^n \to (-\infty, +\infty]$ be a closed proper convex function. Thus to solve the convex program

$$(C) \quad F_* = \inf\{F(x) : \ x \in \mathbb{R}^p_+\}$$

we may consider now the following Logarithmic-Quadratic Proximal method (LQP), which starting with $x^0 > 0$ generates the sequence $\{x^k\}$ via

$$x^k = \operatorname{argmin}\{F(x) + \lambda_k^{-1} d(x, x^{k-1}) : \ x \in \mathbb{R}^p_+\}$$

where $\lambda_k > 0$ and d is defined in (3.13).

In fact, we will consider the more general scheme proposed and analyzed in [5] which also allows for approximate proximal computations. Let $\varepsilon - \operatorname{argmin} F(x) := \{z : F(z) \leq \inf F + \varepsilon\}$, with $\varepsilon \geq 0$, and define

$$F_k(x) = F(x) + \lambda_k^{-1} d(x, x^{k-1}). \tag{3.18}$$

Logarithmic-Quadratic-Proximal Method (LQP)
Given φ defined in (3.12), $x^0 \in \mathbb{R}^p_{++}, \varepsilon_k \geq 0, \lambda_k > 0$, generate the sequence $\{x^k\} \subset \mathbb{R}^p_{++}$ satisfying

$$g^k \ \in \ \partial_{\varepsilon_k} F(x^k) \tag{3.19}$$

$$\lambda_k g^k \ + \ \Phi'(x^{k-1}, x^k) = 0 \tag{3.20}$$

where

$$\Phi'(a, b) := (a_1 \varphi'(b_1/a_1), \ldots, a_p \varphi'(b_p/a_p))^T \ \ \forall a, b \in \mathbb{R}^p_{++} \tag{3.21}$$

with $\varphi'(t) = \nu(t-1) + \mu(1 - t^{-1}), \ t > 0$ and $\partial_\varepsilon F$ denotes the ε-subdifferential of F.

It can be verified that the above iteration scheme is nothing else but an approximate version of the proximal method in the sense that

$$x^k \in \varepsilon_k - \operatorname{argmin} F_k(x).$$

We denote by X_* the set of optimal solutions of problem (C) and throughout it is assumed that $\operatorname{dom} F \cap \mathbb{R}^p_{++} \neq \emptyset$. The next result proven in [5, Lemma 3.3], guaranteed that the LQP algorithm is well defined.

Lemma 3.1 *Let $x^{k-1} \in \mathbb{R}^p_{++}, \lambda_k > 0$ and F_k as defined in (3.18). Then, there exists a unique $y^k \in \mathbb{R}^p_{++}$ such that*

$$y^k = argmin_x F_k(x), \tag{3.22}$$

$$\lambda_k g^k + \Phi'(x^{k-1}, y^k) = 0, \qquad g^k \in \partial F(y^k). \tag{3.23}$$

Remark 3.1 The LQP can be similarly extended to handle the more general case of linearly constrained problems, i.e., problem (C) with constraints in the form $Ax \leq b$, instead of $\mathbb{R}^p_+$, see [4, Theorem 4.1] for details.

The main convergence analysis result for the (LQP) method given below was established in [5, Theorem 3.2].

Theorem 3.2 *Let $\{\lambda_k\}$ be an arbitrary sequence of positive numbers and $\sigma_n := \sum_{k=1}^n \lambda_k$. Let $\{x^k\}$ be the sequence generated by* **LQP***. Then,*

(i) there exists a unique $y^k \in \mathbb{R}^p_{++}$ such that (3.22) and (3.23) hold.

(ii) $F(x^n) - F(x) \leq \theta\sigma_n^{-1}||x - x^0||^2 + \sigma_n^{-1}\sum_{k=1}^n \sigma_k\varepsilon_k, \ \forall x \in \mathbb{R}^p_+$.

(iii) If $\sigma_n \to \infty$ and $\varepsilon_k \to 0$, then $\liminf_{n\to\infty} F(x^n) = F_*$. Moreover, if $\sum_{k=1}^\infty \varepsilon_k < \infty$ then, $F(x^n) \to F_* = \inf\{F(x) : x \in \mathbb{R}^p_+\}$ if F_* is finite.

(iv) Furthermore, if $X_* \neq \emptyset$, $\sigma_n \to \infty$ and $\sum_{k=1}^\infty \lambda_k\varepsilon_k < \infty$, then the sequence $\{x^n\}$ converges to an optimal solution of (C). In addition, if $\lambda_k \geq \lambda > 0$, then the following relations hold:

$$\liminf_{k\to\infty} g_i^k \geq 0, \qquad \lim_{k\to\infty} g_i^k x_i^k = 0 \quad \forall i = 1, \ldots, p.$$

The fundamental aspect of LQM, as opposed to *all* other known entropic proximal methods, is that it behaves like the classical quadratic proximal algorithm, namely one can measure/estimate progress of the method still in terms of *quadratic* norms, yet with the key advantage of automatically producing iterates that stay in the interior of the feasible set. In fact, the convergence analysis and properties of the LQP method relies on the following key result given in [5, Lemma 3.4], which clearly explains the later point just mentioned.

Lemma 3.2 *Let φ be given* (3.12) *with $\nu > \mu > 0$ and θ as defined in (3.14). For any $a, b \in \mathbb{R}^p_{++}$ and $c \in \mathbb{R}^p_+$, we have*

$$(c - b, \Phi'(a,b)) \leq \theta(||c-a||^2 - ||c-b||^2) - \frac{(\nu-\theta)}{2}||b-a||^2,$$

where $\Phi'(a,b)$ is defined in (3.21).

We would like to emphasize that the second order homogeneity property of the functional d combined with the special choice of the logarithmic-quadratic function, is the key to obtain such a type of result, which is not available through the use of first order homogeneous functional (2.6). In the next section, we revisit all the problems encountered and described in Section 2, and demonstrate the usefulness and applications of the LQP framework.

4 The LQP in Action

In this section, the function φ is as defined in (3.12).

4.1 Primal LQP for Variational Inequalities over Polyhedra

Recall quickly the problem set up in Section 2. Let $C = \{x : Ax \leq b\}$ and let T be a maximal monotone set valued map such that $\text{dom}T \cap \text{int}C \neq \emptyset$ ($\text{dom}T := \{x : T(x) \neq \emptyset\}$).
Consider the variational inequality

(VI) " Find a point $x^* \in C$ and $g^* \in T(x^*)$ satisfying

$$(g^*, x - x^*) \geq 0 \quad \forall x \in C\text{''},$$

The associated LQP method for this problem was studied in [4], where it is proved that the whole sequence produced by the proposed method converges to a single point essentially under the sole assumption that the solution set of (VI) is nonempty. Thus, the LQP framework, allows to eliminate all the restrictions discussed in Section 2 and imposed by other generalized proximal methods. In particular the restrictive paramonoonicity hypothesis is not needed anymore. Recently, this method has been further analyzed in two directions: In [17], where inexact stopping rules has been proposed, and in [70], where the method has been modified in its primal form, in order to apply it on problems with linear constraints, whose feasible set may not have a nonempty interior and can be started from an arbitrary initial point.

4.2 Lagrangian Methods for convex optimization and variational inequalities

Consider the standard convex program

$$(P) \quad \inf\{f_0(x):\ f_i(x) \le 0,\ i = 1, \dots, m\}$$

where the functions f_i are convex on $\mathbb{R}^n$.

Given $\lambda > 0$, we define for $u > 0$

$$H(x, u, \lambda) = f_0(x) + \lambda^{-1} \sum_{i=1}^{m} u_i^2 \varphi^*(\lambda f_i(x)/u_i) \tag{4.24}$$

To solve the convex problem (P) one uses the following Log-Quadratic-Multiplier Method (**LQ-MM**):

LQ-MM: Given $u^0 \in \mathbb{R}^m_{++}$ and $\lambda_k \ge \lambda > 0$, $\forall k \ge 1$, generate the sequences $\{x^k, u^k\}$ according to

$$x^k \in \arg\min\{H(x, u^{k-1}, \lambda_k) :\ x \in \mathbb{R}^n\} \tag{4.25}$$

$$u_i^k = u_i^{k-1}(\varphi^*)'(\lambda_k f_i(x^k)/u_i^{k-1}),\ i = 1, \dots, m \tag{4.26}$$

It can be proved that under Slater's condition and under the assumption that the optimal primal set is nonempty and compact, that the whole dual sequence u^k converges to an optimal dual solution and that the sequence x^k is bounded with limit points optimal primal solutions. Such a result cannot be obtained by other entropic multiplier methods, see [5] for proof and more details.

To obtain convergence of the whole sequence $\{x^k\}$ to a single primal optimal point Auslender and Teboulle [7] have developed a Lagrangian duality scheme for variational inequalities with maximal monotone set-valued maps over closed convex sets described by a finite number of convex inequalities. Applied within the primal-dual framework, the LQP produces a sequence $\{x^k, u^k\}$ converging to *a single point* (x, u) which is optimal for (P)-(D) under the sole assumption that such a point exists and Slater's holds for the constraint set described by convex inequalities. The method is exactly the same as (LQ-MM), except that the Lagrangian in that case is of the form:

$$H(x, u, \lambda) = f_0(x) + \lambda^{-1} \sum_{i=1}^{m} u_i^2 \varphi^*(\lambda f_i(x)/u_i) + (2\lambda_k)^{-1}||x - x^{k-1}||^2.$$

This results in a C^∞ *strongly convex Lagrangian which is also self-concordant whenever the constraints are convex quadratic.* Such properties are not shared by other (EMM) methods.

4.3 Dual and Primal-Dual Decomposition schemes

It is well known that the development and analysis of decomposition schemes relies fundamentally on using the proximal point algorithm ([51], [60]) applied to the primal, dual and primal-dual formulation of the given problem. The resulting algorithms involve typically the minimization of a specific decomposable Augmented Lagrangian followed or/and preceded by some dual multiplier explicit updating dual multipliers formula. Well know methods include for example: alternating direction of multipliers [35], [50], [34], [30], [37], [25], [68], the method of partial inverse of Spingarn [63], and more recently the proximal-based method of Chen and Teboulle [21] and its extension given by Tseng [69]. For more details and references on decomposition methods, we refer the reader to the book of Bertsekas-Tsitsiklis [14]. A key advantage of these methods over the usual augmented Lagrangians techniques ([13]), is that they can readily exploit the particular structure of the given problem and decompose according to that structure. However, the main difficulty (b) explained in Section 2 remains with all these types of decomposition algorithms when applied to convex separable problems, since the resulting Lagrangians to be minimized are only $C^{1,1}$ and thus cannot be solved efficiently via Newton type methods.

In [9], Auslender and Teboulle present a new decomposition method which overcomes this difficulty and results in appropriately decomposable Modified Lagrangians which are C^∞ (provided the problem's data are such) and also possess several other interesting properties. The proposed algorithm is in fact an extension of Chen and Teboulle method's [21] which was developed for solving convex programs with a particular separable structure. The extension is in three directions. Firstly they consider the more general framework of variational inequalities, secondly they use the LQ proximal theory in place of the usual quadratic proximal theory and thirdly they consider non-negativity constraints explicitly, the latter being fundamentally important in the context of the primal-dual framework for usual convex programs. The resulting algorithm is fully decomposable according to the given problem's structure, produces *positive* dual sequences thanks to the logarithmic proximal term, and is accordingly called the *Entropic Proximal Decomposition Method.* In particular, using a novel way of decomposing a saddle function, they derive within the primal-dual framework, a decomposition method for solving standard separable convex programs with Cartesian product structures. The resulting algorithm leads to the minimization of smooth separable Lagrangians which are automatically strongly convex , and produces a primal-dual sequence which globally converges to an optimal primal-dual solution, provided such a point exists. As a consequence these Lagrangians possess all the properties needed to apply Newton methods on the reduced dimensional spaces with efficiency, in contrast to classical splitting methods derived from the quadratic proximal methods. We briefly outline the main results described above. For the more general decomposition scheme to solve variational inequalities, we refer to [9] for more details and proofs.

Convex optimization with linear coupling constraints

Consider the convex program with the particular separable structure involving linear coupling and non negativity constraints:

$$(PL) \qquad \min\{f(x) + g(z) : \; Ax + Bz = b, x \in \mathbb{R}^n_+, z \in \mathbb{R}^m\}.$$

where $f : \mathbb{R}^n \to (-\infty, +\infty]$ and $g : \mathbb{R}^m \to (-\infty, +\infty]$ are given closed proper convex functions, and A is a given $m \times n$ matrix, B an $m \times p$ matrix, and $b \in \mathbb{R}^m$.

Algorithm I: Start with an initial arbitrary triple $(x^0, z^0, y^0) \in \mathbb{R}^n_{++} \times$

$\mathbb{R}^p \times \mathbb{R}^m$, and given positive parameters λ_k, generate the sequence $(x^k, z^k, y^k) \in \mathbb{R}^n_{++} \times \mathbb{R}^p \times \mathbb{R}^m$, by the following steps:
Step 1: Compute $p^{k+1} = y^k + (2\theta)^{-1}\lambda_k(Ax^k + Bz^k - b)$.
Step 2: Solve

$$\begin{aligned} x^{k+1} &= \operatorname*{argmin}_{x \geq 0}\{f(x) + \langle p^{k+1}, Ax\rangle + \lambda_k^{-1} d(x, x^k)\}, \\ z^{k+1} &= \operatorname*{argmin}_{z}\{g(z) + \langle p^{k+1}, Bz\rangle + \theta\lambda_k^{-1} \parallel z - z^k \parallel^2\}. \end{aligned}$$

Step 3: Compute $y^{k+1} = y^k + (2\theta)^{-1}\lambda_k(Ax^{k+1} + Bz^{k+1} - b)$.

Under the following Assumption for problem (PL):

$$\exists x \in \mathbb{R}^n_{++} \cap \text{ri }(\text{dom} f),\ z \in \text{ri }(\text{dom} g) \text{ such that} Ax + Bz = b,$$

the existence of an optimal multiplier y^* is guaranteed, and it is proved in [9] that the sequence (x^k, z^k, y^k) converges to (x^*, z^*, y^*), with (x^*, z^*) optimal for (PL) and y^* optimal for the associated dual problem of (PL).

We emphasize that in both minimization steps, one has the advantage of having *built-in strongly convex* separable functions for any convex functions f, g. Furthermore, it is important to notice that the minimizations in Step 2 can be performed efficiently if $f, g \in C^3$ are assumed self-concordant functions ([53]). Indeed, since both $d(\cdot, x^k)$ and $||\cdot - z^k||^2$ are self-concordant as well, then the computation of (x^{k+1}, z^{k+1}) can be performed efficiently using Newton's method for which the theory of self concordance applied, see [53]. Finally, our method can be used in particular to solve linear programs with block structures i.e., with $f(x) = c^T x, g(z) = d^T z$ in problem (PL). In that particular case, even Step 2 can be solved *analytically* for (x^{k+1}, z^{k+1}).

Separable Convex optimization problems

Consider the separable convex program (SP):

$$(SP) \quad \min\Big\{\sum_{j=1}^{n} f_j(v_j) : \sum_{j=1}^{n} c_{ij}(v_j) \leq 0,\ i = 1, \ldots, m,$$
$$v_j \in \mathbb{R}^{d_j},\ j = 1, \ldots n\Big\}.$$

where f_j, c_{ij} are finite valued convex functions on $\mathbb{R}^{d_j}$. In the following all the vectors are assumed to be column vectors, but to simplify notation, we will often write $v = (v_1, \dots, v_n) \in \Pi_{j=1}^n \mathbb{R}^{d_j} := \mathbb{R}^d$, with $d = d_1 + \dots d_n$.

The decomposition algorithm for (SP) is obtained by applying EPDM on an appropriate saddle point formulation of (SP).
Let L be the associated usual Lagrangian for (SP):

$$L(w, v) = \begin{cases} \sum_{j=1}^n f_j(v_j) + \langle w, c_j(x_j) \rangle & \text{if } w \geq 0 \\ -\infty & \text{if } w \not\geq 0, \end{cases}$$

where $c_j(x_j) := (c_{1j}(x_j), \dots, c_{mj}(x_j))$, $j = 1, \dots n$, and $\langle \cdot, \cdot \rangle$ is the inner product in $\mathbb{R}^m$. The dual problem associated with (SP) is then equivalent to solve:

$$(DSP) \qquad \inf_{w \in \mathbb{R}_+^m} \sup_{v \in \mathbb{R}^d} -L(w, v)$$

We suppose that there exists (w^*, v^*) saddle point of $-L$. As a consequence it follows that $(w^*, v^*) \in \mathbb{R}_+^m \times \mathbb{R}^d$ are optimal solutions for (DSP) and (SP) respectively.

Now, fix $u^k \in \mathbb{R}_{++}^{nm}$ and define for each $j = 1, \dots, n$ the following separable modified Lagrangian:

$$H_j^k(u^k, v)$$

$$:= f_j(v_j) + \lambda_k^{-1} \sum_{i=1}^m (u_{ij}^k)^2 \varphi^*(\lambda_k(p_{ij}^{k+1} + c_{ij}(v_j))/u_{ij}^k) + \theta \lambda_k^{-1} ||v_j - v_j^k||^2 \tag{4.27}$$

Algorithm II.
Given φ defined in (3.12), $(v_j^0, p_j^0, y_j^0) \in \mathbb{R}^{d_j} \times \mathbb{R}^m \times \mathbb{R}^m$ for $j = 1, \dots n$, $(u^0, w^0) \in \mathbb{R}_{++}^{nm} \times \mathbb{R}_{++}^m$ and $\lambda_k > 0$, generate for $j = 1, \dots, n$ the sequences $\{p_j^k, v_j^k, y_j^k\}$ and the positive sequences $\{w^k, u_j^k\}$ according to
Step 1. Compute

$$p_j^{k+1} = y_j^k + (2\theta)^{-1} \lambda_k (w^k - u_j^k), \; j = 1, \dots, n.$$

Step 2. For each $j = 1, \dots, n$ find the unique minimizer v_j^{k+1} solution of

$$v_j^{k+1} = \operatorname{argmin}\{H_j^k(u^k, v_j) : \; v_j \in \mathbb{R}^{d_j}\}.$$

Step 4. Compute

$$
\begin{aligned}
w_i^{k+1} &= w_i^k(\varphi^*)'(-\lambda_k \sum_{j=1}^{n} p_{ij}^{k+1}/w_i^k),\ i = 1, \ldots, m, \\
u_{ij}^{k+1} &= u_{ij}^k(\varphi^*)'(\lambda_k(c_{ij}(v^{k+1}) + p_{ij}^{k+1})/u_{ij}^k),\ i = 1, \ldots, m, \\
y_j^{k+1} &= y_j^k + (2\theta)^{-1}\lambda_k(w^{k+1} - u_j^{k+1}),\ j = 1, \ldots, n.
\end{aligned}
$$

We have thus obtained a fully decomposable method for solving (SP). Note that the main computational part is step 2 which requires to minimize the strongly convex function H_j^k. Assuming f_j, c_{ij} smooth enough, the minimization can be performed efficiently via Newton's method, due to the additional nice properties shared by the function φ^*, (see Theorem 2.1, properties (v)-(viii)).

Under suitable assumptions, (see [9, Theorem 3] for details), it can be proved that the sequence $\{u^k, v^k, w^k, p^k, y^k\}$ generated by algorithm II converges to a point $(u^*, v^*, w^*, p^*, y^*)$ such that (v^*, w^*) is a saddle point of $-L(v, w)$, i.e., v^* is optimal for (SP) and $w^* = u_j^* \in \mathbb{R}_+^m$, $\forall j = 1, \ldots, n$ is an optimal solution of its associated dual problem (DSP).

4.4 Primal Decomposition: Block Gauss-Seidel Schemes for Linearly constrained Problems

Recent advances in computer technology such as parallel and distributed computing systems have considerably increased computing power and motivated the search for decomposition techniques and related algorithms that can exploit it. Here we will concentrate on optimization problems with block structures and analyze a new variant of the block nonlinear Gauss-Seidel method under linear constraints which was recently suggested in [6].

Let $X \subset \mathbb{R}^n$, be the Cartesian product of m closed nonempty and convex subsets $X_i \subset \mathbb{R}^{n_i}$, i.e, $X = \prod_{i=1}^{m} X_i$, with $n_1 + n_2 + \ldots + n_m = n$. Any vector $x \in X$ is accordingly decomposed as $x = (x_1, x_2, \ldots, x_m)$ with, $x_i \in X_i$. Let $f : \mathbb{R}^n \to \mathbb{R}$ be a given function and consider the problem:

$$\min\{f(x) : x \in X\}. \tag{4.28}$$

Given the current iterate $x^k = (x_1^k, x_2^k, \ldots, x_m^k)$, the block-nonlinear Gauss-Seidel (**GS**) method generates sequentially the next iterate $x^{k+1} = (x_1^{k+1}, x_2^{k+1}, \ldots, x_m^{k+1})$ via the iteration

$$x_i^{k+1} = \operatorname*{argmin}_{x \in X_i}\{f(x_1^{k+1}, \ldots, x_{i-1}^{k+1}, x, x_{i+1}^k, \ldots, x_m^k)\},\ i = 1, \ldots, m. \quad (4.29)$$

At each iteration, the function f is minimized with respect to each of the block-coordinate vectors x_i^k, taken in cyclic order. Thus, the **GS** method naturally exploits the Cartesian product structure and is attractive when the minimization in (4.29) can be solved efficiently. Here we concentrate on the important class of convex minimization problems over polyhedral constraints with block structure, namely when the sets X_i are given in the form

$$X_i = \{x \in \mathbb{R}^{n_i} : A^i x \leq b^i\}, \quad (4.30)$$

where A^i is a $p_i \times n_i$ matrix and $b^i \in \mathbb{R}^{p_i}$.

In [6], a new Gauss-Seidel type algorithm to solve (4.28) under the constraints (4.30) was introduced. The main objective there to eliminate the constraints in each minimization step of the form (4.29), replacing it by a suitable *unconstrained* minimization step. Indeed, as mentioned earlier while the **GS** method is useful for decomposition, yet one still have to solve in (4.29) for each block, a constrained minimization problem which can be time consuming and quite inefficient, as opposed to an unconstrained problem, which could be solved via an efficient Newton type algorithm (given a smooth enough objective function).

Let a_j^i denotes the rows of the matrix A^i and define for $i = 1, \ldots, m$; $j = 1, \ldots, p_i$,

$$\begin{aligned} l_j^i(x_i) &= b_j^i - (a_j^i, x_i), \\ L^i(x_i) &= (l_1^i(x_i), \ldots, l_{p_i}^i(x_i))^T, \\ D_i(x_i, y_i) &= d(L^i(x_i), L^i(y_i)). \end{aligned}$$

Here $(.,.)$ denotes the inner product between vectors in $\mathbb{R}^{n_i}$ and $\lambda_i^k \geq \lambda > 0$.

To achieve their task, namely to eliminate the constraints, the authors propose to couple the Gauss-Seidel algorithm with the LQP algorithm, which generates a sequence $\{x^k\}$ via the scheme:

$$x_i^{k+1} = \operatorname*{argmin}_{x_i \in \mathbb{R}^{n_i}}\{f(x_1^{k+1}, \ldots, x_{i-1}^{k+1}, x_i, x_{i+1}^k, \ldots, x_m^k) + \lambda_i^k D(x_i, x_i^k)\}, \quad (4.31)$$

where $D_{(.,.)}$ is the proximal logarithmic-quadratic term which forces the iterates $\{x_i^k\}$ to stay in the interior of the feasible set $\text{int}X_i$. The new algorithm (4.31) seems promising from the numerical point of view since the logarithmic-quadratic proximal term satisfies the self-concordance properties, which appear fundamental to solve problems of the form (4.31) via efficient Newton methods. Another advantage of such a method is that convergence of the method is proved without the usual strong convexity assumptions made of $v_i \rightarrow f(v_i) = f(x_1, \ldots x_{i-1}, v_i, x_{i+1}, \ldots, x_n)$.

4.5 Convex Feasibility Problems

The convex feasibility (CF) problem consists of finding a point in the intersection of convex sets. It occurs in a wide variety of contexts within mathematics as well as in many applied sciences problems and particularly in image reconstruction problems. For nice surveys on CF problems, projections algorithms toward their solutions and an extensive bibliography we refer for example to [11] and [22]. Let $C_i,\ i = 1, \ldots, m$ be given closed convex set of $\mathbb{R}^p$ with a nonempty intersection $C := C_1 \cap \ldots \cap C_m$. The CF problem then consists of finding some point $x \in C$.

One the basic approach to solve the CF problem is simply to use projections onto each set C_i and to generate a sequence of points converging a solution of the CF problem. There are many ways of using projections to generate a solution to the CF problem, (see e.g., [11]. In a recent work Auslender and Teboulle [8], focus on the simple idea of averaging projections (barycenter method), which goes back at least to Cimmino [20] who has considered the special case when each C_i is a half-space. Denoting by P_{C_i} the projection onto the set C_i the barycenter projection method consists of two main steps.

Barycenter Projection Method. Start with $x_0 \in \mathbb{R}^n$ and generate sequences
$\{x_k^i : i = 1, \ldots, m\}$ via:
(i) Project: $p_k^i = P_{C_i}(x_k),\ \forall i = 1, \ldots, m,$
(ii) Average: $x_{k+1} = m^{-1} \sum_{i=1}^m p_k^i$.

More generally, in the averaging step, one can assign nonnegative weights such that $w_k^i \geq w > 0$ and replace step (ii) above by

$$x_{k+1} = (\sum_{i=1}^{m} w_k^i)^{-1} p_k^i)/(\sum_{i=1}^{m} w_k^i)^{-1}).$$

Clearly step above (ii) is then recovered with the choice $w_k^i = m^{-1}$, $\forall k$.

This basic algorithm produces a sequence $\{x_k\}$ which converges to a solution of the CF problem, provided the intersection C is assumed nonempty. In many applications, the convex feasibility problem occurs in the nonnegative orthant, namely we search for a point $x \in C \cap \mathbb{R}_+^n$. In such cases, even when the C_i are simple hyperplanes we do not have an explicit formula for the projection $P_{C_i \cap \mathbb{R}_+^n}$, i.e., when intersecting the hyperplanes with the nonnegative orthant. In fact, in order to take into account the geometry of $C \cap \mathbb{R}_+^n$, it appears more natural to use projections not necessarily based on Euclidean quadratic squared distances, but rather on some other type of projections-like operators that which will be able to ensure automatically the positiveness of the required point solving CF in these cases. This has lead several authors to consider nonquadratic distance-like functions. Most common has been the use of Bregman type distances, see for example the work of Censor and Elfving [18] and references therein. Yet, when using non quadratic distance-like functions, the projections cannot be computed exactly or analytically. It is therefore required to build a nonorthogonal (nonlinear) projection algorithm which first allows to control the errors with inexact computations, and secondly allows for computing the projections efficiently, e.g., via Newton's method. Previous works using nonquadratic projections do not address these issues. Motivated by these recent approaches based on nonorthogonal Bregman type projections, see e.g., [11], [18], Auslender and Teboulle [8] suggested a nonorthogonal projection algorithm for CF, based on the Logarithmic-Quadratic (LQ) functional. The LQ-projection onto $C \cap \mathbb{R}_+^n$, based on d is defined as follows: for each $y \in \mathbb{R}_{++}^p$,

$$E_C(y) := \text{argmin}\{d(x, y) : \; x \in C \cap \mathbb{R}_+^p\}$$

As already seen, the LQ functional and its associated conjugate which also plays a key role in nonlinear projections, enjoys nice and very important properties not shared by any other nonquadratic distance-like functions previously used and proposed in the literature. In [8], the authors derive a convergent projection algorithm, under very mild assumptions on the problems data and which address the two difficulties alluded above. We outline below their main result.

The LQ Projection Method Let φ be defined in (3.12).

Let $\{\epsilon_k\}$ be a nonnegative sequence such that $\sum_{k=1}^{\infty} \epsilon_k < +\infty$. Start with $x_0 \in \mathbb{R}^p_{++}$ and generate iteratively the sequence $\{x_k\}$ by :

Step 1. *Compute approximate LQ projections.* For each $i = 1, \ldots, m$ compute x_k^i such that

$$x_k^i \in \mathbb{R}^p_{++}, \quad \epsilon_k^i = x_k^i - E_{C_i}(x_k), \quad \text{with } ||\epsilon_k^i|| \leq \epsilon_k.$$

Step 2. *Average:* $x_{k+1} = m^{-1} \sum_{i=1}^{m} x_k^i$.

Assuming

$$C \cap \mathbb{R}^p_{++} \neq \emptyset$$

it is proved in [8] that the sequence $\{x_k\}$ produced by the LQ-projection method converges to a point $x \in C \cap \mathbb{R}^p_+$.

4.6 Bundle Methods in Nonsmooth Optimization

Standard bundle methods for nonsmooth convex minimization have been considered in several papers, e.g., [1], [23], [31], [36], [42], [49]. These methods are directly related to the quadratic proximal algorithm. More recently, Kabbadj [45] and Kiwiel [44] have proposed to use generalize bundle type methods by replacing the quadratic proximal term with a Bregman's type distance. When applied to the dual (D) of a standard convex program (P), these modifications lead to a new class of interior bundle methods. Motivated by the nice properties of the log-quad function φ and its conjugate, and by these recent modifications, Auslender and Teboulle [10] recently proposed to analyze bundle methods through the LQ framework. A short summary of these results is now presented, and the reader is referred to [10] for more results and proofs, and which also includes the analysis for linearly constrained nonsmooth problem.

Consider nondifferentiable convex program over the nonnegative orthant,

$$(P) \quad f_* = \inf\{f(x) : \; x \in \mathbb{R}^p_+\},$$

where $f : \mathbb{R}^p \to \mathbb{R} \cup +\infty$ is a proper lsc convex function with the subdifferential map ∂f assumed bounded on the bounded subsets of $\mathbb{R}^p_{++}$. Note that the later always holds when f is real-valued, an assumption which is commonly used for most bundle methods.

To minimize f on $\mathbb{R}^p_+$ we have to first solve approximately the following problem. Let $\lambda > 0$, $\bar{x} \in \mathbb{R}^p_{++}$ and consider the minimization problem:

$$P_\lambda(\bar{x}) \qquad m_\lambda(\bar{x}) = \inf\{f(x) + \lambda^{-1} d(x, \bar{x}) : \ x \in \mathbb{R}^p\}.$$

As a consequence of Lemma 3.1, problem $P_\lambda(\bar{x})$ admits a unique solution denoted by $\operatorname{prox}^d_{\lambda f}(\bar{x}) \in \mathbb{R}^p_{++}$.

We now define an algorithm for solving $P_\lambda(\bar{x})$. The starting data are $x_1 \in \mathbb{R}^p_{++}$, $g_1 \in \partial f(x_1)$ and $\psi^0(x) = (g_1, x - x_1) + f(x_1)$.

Algorithm A Starting with ψ^0, generate the sequence $\{x_k, \psi^k\}$ as follows: suppose at iteration k that ψ^k is known, then compute
(1) $x^k = \operatorname{prox}^d_{\lambda\psi^k}(\bar{x})$,
Set $d_k = -\lambda^{-1}\nabla_x d(x_k, \bar{x})$; $l^k(x) = \psi^k(x_k) + (d_k, x - x_k)$.
(2) Let $\psi^{k+1} : \mathbb{R}^p \to \mathbb{R} \cup +\infty$, be a proper lsc convex function satisfying:
$H_1 \quad \psi^{k+1} \leq f$,
$H_2 \quad l^k \leq \psi^{k+1}$,
$H_3 \quad f(x_k) + (g_k, x - x_k) \leq \psi^{k+1}(x), \ \forall x \ \text{ with } \ g_k \in \partial f(x_k)$.

Note that using the necessary optimality conditions for (1) in Algorithm A, it follows that

$$d_k \in \partial \psi^k(x_k) \tag{4.32}$$

Two commonly used examples for the choice of ψ^k satisfying the above are,

$$\begin{aligned} \psi^{k+1}(x) &= \max\{f(x_i) + (g_i, x - x_i) : \ i \leq k\}, \ \ g_k \in \partial f(x_k), \\ \psi^{k+1}(x) &= \max\{l^k(x), f(x_k) + (g_k, x - x_k)\} \end{aligned}$$

To solve the initial problem (P), one then use the following algorithm: Let $\eta \in (0, 1)$ be a given fixed parameter.

LQ-Bundle Algorithm (LQ-BA).

Start with $y_0 \in \mathbb{R}^p_{++}$ and generate inductively the sequence $\{y_k\}$, $k = 0, 1, \ldots$, as follows: (a) If y_k is optimal for (P), then Stop. Else, (b)

Compute via Algorithm A in a finite number of steps $n(k)$, a point $x_{n(k)}$ satisfying

$$\begin{aligned} x_{n(k)} &= \operatorname{prox}^{d}_{\lambda_{n(k)}\psi^{n(k)}}(y^k), && (4.33) \\ f(y_k) - f(x_{n(k)}) &\geq \eta(f(y_k) - \psi^k(x_{n(k)})). && (4.34) \end{aligned}$$

(c) Set $y_{k+1} = x_{x_{n(k)}}$ and goto (a).

It has been proved in [10], that if problem (P) admits an optimal solution, then the sequence $\{y_k\}$ generated by LQ-BA is well defined and converges to an optimal solution of (P), for any $0 < \lambda \leq \lambda_k \leq \bar{\lambda}$.

When applied to the dual problem of an ordinary convex program, it has been shown in [10] that the LQ-BA with the choice

$$\psi^{k+1}(x) = \max\{l^k(x), f(x_k) + (g_k, x - x_k)\},$$

produces a decomposition scheme for which the only dual computation at each step k will consist of minimizing a scalar function on the interval $[0, 1]$, of the form

$$\Phi_k(\rho) = \sum_{i=1}^{p} c_i \varphi^*(L_i(\rho)) + L_{p+1}(\rho)$$

where $L_i(\cdot)$ is linear, $c_i > 0$, and φ^*, is the conjugate of the log-quadratic function (cf. (3.16)). Thus, the resulting Φ is self-concordant, (a property not shared by entropic bundle based on Bregman distance) and then the minimization can be done efficiently.

References

[1] A. Auslender, "Numerical methods for nondifferentiable convex optimization", *Mathematical Programming Study*, **30**, 1987, 102-126.

[2] A. Auslender, R. Cominetti and M. Haddou, "Asymptotic analysis of penalty and barrier methods in convex and linear programming",*Mathematics of Operations Research*, **22**, 1997, 43-62.

[3] A. Auslender and M. Haddou, "An interior proximal method for convex linearly constrained problems and its extension to variational inequalities", *Mathematical Programming*, **71**, 1995, 77-100.

[4] A. Auslender, M. Teboulle and S. Ben-Tiba,A Logarithmic-Quadratic Proximal Method for Variational Inequalities, *Computational Optimization and Applications* **12** (1998) 31–40.

[5] A. Auslender, M. Teboulle and S. Ben-Tiba, Interior Proximal and Multiplier Methods based on Second Order Homogeneous Kernels, *Mathematics of Operations Research* **24** (1999) 645–668.

[6] A. Auslender, M. Teboulle and S. Ben-Tiba, "Coupling the logarithmic-quadratic proximal method and the block nonlinear Gauss-Seidel Algorithm for linearly constrained convex minimization", *Ill-Posed Variational Problems and Regularization Techniques*, Editors, M. Thera and R. Tichatschke. Lecture notes in ecomomics and Mathematical Systems. **Vol. 477**, 1999, Springer Verlag, 35-47.

[7] A. Auslender and M. Teboulle, "Lagrangian Duality and Related Multiplier Methods for Variational Inequalities", *SIAM J. Optimization*, **10**, 2000, 1097-1115.

[8] A. Auslender and M. Teboulle, "A Log-quadratic projection method for convex feasibility problems", in D. Butanariu, Y. Censor and S. Reich Eds. Inherently parallel algorithms in fesibility and optimiztaiom and their aplications. *Studies in Computational Mathematics*,**Vol. 8**, 2001, 1-10. Elsevier, Amsterdam

[9] A. Auslender and M. Teboulle, "Entropic proximal decomposition methods for convex programs and variational inequalities", *Mathematical Programming*, In press.

[10] A. Auslender and M. Teboulle, " An entropic proximal bundle method for nonsmooth convex minimization", Working paper, September 2000.

[11] H. H. Bauschke and J. M. Borwein, On projection algorithms for solving convex feasibility problems, *SIAM Review* **38** (1996) 367–426.

[12] A. Ben-Tal and M. Zibulevsky, "Penalty-Barrier Methods for convex programming problems", *SIAM J. Optimization*, **7**, 1997, 347-366.

[13] D. Bertsekas, *Constrained Optimization and Lagrange Multiplier Methods*, Academic Press, NY, 1982.

[14] D. P. Bertsekas and J. N. Tsitsiklis, *Parallel and Distributed Computation: Numerical Methods* (Prentice-Hall, New Jersey, 1989).

[15] R. D. Bruck, An iterative solution of a variational inequality for certain monotone operators in Hilbert space, *Bulletin of the American Math. Soc.*, **81**, 1975, 890-892. (With corrigendum, in **82**, 1976, p. 353).

[16] R. S. Burachik and A. N. Iusem, "A generalized proximal point algorithm for the variational inequality problem in a Hilbert space", *SIAM Journal on Optimization*, **8**, 1998, 197-216.

[17] R. S. Burachick and B. Svaiter, "A relative error tolerance for a family of generalized proximal point methods". To appear in *Mathematics of Operations Research.*

[18] Y. Censor and T. Elfving, A multiprojection algorithm using Bregman projections in a product space, *Numerical Algorithms* **8** (1994) 221-239.

[19] Y. Censor, A. N. Iusem and S. A. Zenios, An interior-point method with Bregman functions for the variational inequality problem with paramonotone operators. Working paper, University of Haifa, 1994.

[20] G. Cimmino, Calcolo appproximato per le soluzioni dei sistemi di equazioni lineari, *La Ricerc Scientifica Roma* **1** (1938) 326–333.

[21] G. Chen and M. Teboulle, "A proximal based decomposition method for convex minimization problems", *Mathematical Programming* 64, (1994) 81–101.

[22] P. L. Combettes, Hilbertian convex feasibility problem: convergence and projection methods, *Applied Mathematics and Optimization*, **35** (1997) 311–330.

[23] R. Correa and C. Lemarechal, "Convergence of some algorithm for convex programming", *Mathematical Programming*, **62**, 1993, 261-275.

[24] M. Doljansky and M. Teboulle, "An Interior proximal algorithm and the exponential multiplier method for semi-definite programming", **9**, 1998, 1-13.

[25] J. Eckstein and D. P. Bertsekas, "On the Douglas-Rachford splitting method and the proximal point algorithm for maximal monotone operators," *Mathematical Programming* 55, (1992) 293–318.

[26] J. Eckstein, " Non linear proximal point algorithms using Bregman functions, with applications to convex programming", *Mathematics of Operations Research*, **18**, 1993, 202-226.

[27] J. Eckstein, Approximate iterations in Bregman-function-based proximal algorithms, *RRR*, 12-96, January 97, Rutgers University.

[28] J. Eckstein and M. Ferris, " Smooth methods of multipliers for complementarity problems" Research report RRR 27-96 (revised February 1997), RUTCOR, Rutgers University, New Brunswick, NJ 08903.

[29] P. P. B. Eggermont, "Multiplicatively iterative algorithms for convex programming", *Linear Algebra and Its Applications*, 130, 1990, 25-42.

[30] M. Fortin and R. Glowinski, *Augmented Lagrangian Methods: Applications to the Solution of Boundary-Valued Problems* (North-Holland, Amsterdam, 1983).

[31] M. Fukushima, "A descent algorithm for nonsmooth convex programming", *Mathematical Programming*, **30**, 1984, 163-175.

[32] M. Fukushima, "Application of the alternating direction method of multipliers to separable convex programming problems," *Computational Optimization and Applications* 1, (1992) 93–111.

[33] A. V. Fiacco and G. P. McCormick, "*Nonlinear Programming: Sequential Unconstrained Minimization Techniques*", Classics in Applied Mathematics, SIAM , Philadelphia, 1990.

[34] D. Gabay, "Applications of the method of multipliers to variational inequalities," in: M. Fortin and R. Glowinski, ed., *Augmented Lagrangian Methods: Applications to the Solution of Boundary-Valued Problems* (North-Holland, Amsterdam, 1983) pp. 299–331.

[35] D. Gabay and B. Mercier, "A dual algorithm for the solution of nonlinear variational problems via finite-element approximations," *Comp. Math. Appl.* 2, (1976) 17–40.

[36] M. Gaudioso and M. F. Monaco, "A bundle type approach to unconstrained minimization of convex nonsmooth functions", *Mathematical Programming*, **23**, 1982, 216-226.

[37] R. Glowinski and P. Le Tallec, "Augmented lagrangian and operator-splitting methods in nonlinear mechanics," in *SIAM Studies in Applied Mathematics* (SIAM, Philadelphia, 1989).

[38] O. Güler, "On the convergence of the proximal point algorithm for convex minimization", *SIAM J. of Control and Optimization*, **29**, 1991, 403–419.

[39] M. R. Hestenes, "Multiplier and gradient methods", *J. Optimization Theory and Applications*, **4**, 1969, 303-320.

[40] A. N. Iusem, B. Svaiter and M. Teboulle, "Entropy-Like proximal methods in convex programming", *Mathematics of Operations Research*, **19**, 1994, 790-814.

[41] A. Iusem and M. Teboulle, "Convergence rate analysis of nonquadratic proximal and augmented Lagrangian methods for convex and linear programming", *Mathematics of Operations Research*, **20**, 1995, 657-677.

[42] K. C. Kiwiel, "An aggregate subgradient method for nonsmooth convex optimization", *Mathematical Programming*, **27**, 1983, 320-341.

[43] K. C. Kiwiel, "Proximal minimization methods with generalized Bregman functions", *SIAM J. Control and Optimization*, **35**, 1997, 1142-1168.

[44] K. C. Kiwiel, "A bundle Bregman proximal method for nondifferentiable convex minimization", *Mathematical Programming*, **85**, 1999, Ser. A, 241–258. .

[45] M. Kabbadj, *Methodes proximale entropiques*, Ph.D.Thesis, Universite de Montpellier II, France, 1984.

[46] M. Kyono and M. Fukushima, "Nonlinear proximal decomposition methods for convex programming ". Preprint, July 1999.

[47] L. S. Lasdon, *Optimization Theory for Large Systems* (Macmillan, New York, 1970).

[48] B. Lemaire, "The proximal algorithm". In International Series of Numerical Mathematics (J.P. Penot, ed.) Birkhauser Verlag, Basel, 87, 1989, 73-87.

[49] C. Lemarechal, "Bundle methods in nonsmooth optimization", In, *Nonsmooth Optimization*, C. Lemarechal and R. Mifflin Eds., Pergamon Press Oxford, 1978.

[50] P. L. Lions and B. Mercier, "Splitting algorithms for the sum of two nonlinear operators," *SIAM J. Numerical Analysis* 16, (1979) 964–979.

[51] B. Martinet, "Regularisation d'inéquations variationnelles par approximations successive", *Revue Francaise d'Automatique et Informatique Recherche Opérationnelle*, **4**, 1970, 154-159.

[52] J. J. Moreau, "Proximité et dualité dans un espace Hilbertien", *Bull. Soc. Math. France* **93**, 1965, 273–299.

[53] Y. Nesterov, A. Nemirovski, "Interior point polynomial algorithms in convex programming", *SIAM Publications*, Philadelphia, PA, 1994.

[54] R. A. Polyak, "Modified barrier functions (theory and methods)", *Mathematical Programming* , **54**, 1992, 177–222.

[55] R. A. Polyak, "Nonlinear rescaling vs. smoothing technique in constrained optimization". Research report, George Mason University, September 2000.

[56] R. A. Polyak and M. Teboulle, “Nonlinear rescaling and proximal-like methods in convex optimization”, *Mathematical Programming*, **76**, (1997), 265-284.

[57] M. J. D. Powell, "A method for nonlinear constraints in minimization problems", in *Optimization*, R. Fletcher Editor, Academic Press, New-York, 1969, 283-298.

[58] R. T. Rockafellar, *Convex Analysis*, Princėton University Press, Princeton, NJ, 1970.

[59] R. T. Rockafellar, “A dual approach to solving nonlinear prpgramming problems by unconstrained minimization",*Mathematical Programming*, **5**, 1973, 354-373.

[60] R.T. Rockafellar, “Monotone operators and the proximal point algorithm", *SIAM J. of Control and Optimization*, **14**, 1976, 877- 898.

[61] R.T. Rockafellar, "Augmented Lagrangians and applications of the proximal point algorithm in convex programming", *Mathematics of Operations Research*, **1**, 1976, 97- 116.

[62] P. J. Silva, J. Eckstein, and C. Humes, "Rescaling and stepsize selection in proximal methods using separable generalized distances". Rutcor Research Report, November 2000.

[63] J. E. Spingarn, “Applications of the method of partial inverses to convex programming: decomposition," *Mathematical Programming* 32, (1985) 199–223.

[64] M. Teboulle, “Entropic proximal mappings with application to nonlinear programming", *Mathematics of Operations Research*, **17**, 1992, 670–690.

[65] M. Teboulle, “Convergence of Proximal-like Algorithms", *SIAM J. of Optimization*, **7**, 1997, 1069-1083.

[66] M. Teboulle, “Lagrangian multiplier methods for convex programs", In *Encyclopedia of Optimization*, Eds. Floudas and Pardalos, 2001.

[67] P. Tseng, D. Bertsekas, "On the convergence of the exponential multiplier method for convex programming", *Mathematical Programming*, **60**, 1993, 1-19.

[68] P. Tseng, "Applications of a splitting algorithm to decomposition in convex programming and variational inequalities," *SIAM J. of Control and Optimization* 29, (1991) 119–138.

[69] P. Tseng, "Alternating projection-proximal methods for convex programming and variational inequalities," *SIAM J. Optimization* **7**, (1997) 951–967.

[70] N. Yamashuta, C. Kanzow, T. Morimoto and M. Fukushima, "An infeasible interior proximal methods for convex programming problems with linear constraints", Preprint 2000.

THE CONTINUUM MODEL OF TRANSPORTATION PROBLEM

Patrizia Daniele
Dipartimento di Matematica - Università di Catania,
Viale A. Doria, 6 - 95125 CATANIA (ITALIA)
e-mail: daniele@dmi.unict.it

Giovanna Idone
D.I.M.E.T. – Università di reggio Calabria,
Via Graziella, Loc. Feo di Vito - 89100 REGGIO CALABRIA (ITALIA)
e-mail: idone@ing.unirc.it

Antonino Maugeri
Dipartimento di Matematica - Università di Catania,
Viale A. Doria, 6 - 95125 CATANIA (ITALIA)
e-mail: maugeri@dmi.unict.it

Abstract A continuum model of transportation network is considered in presence of capacity constraints on the flow. The equilibrium conditions are expressed in terms of a Variational Inequality for which an existence theorem and a computational procedure are provided.

Keywords: Variational Inequality, continuum traffic equilibrium problems, capacity constraints, quasi-relative interior, Lagrangean theory, subgradient method.

1 Introduction

In the paper [10] the author considers a continuum model of transportation network and characterizes the equilibrium conditions by means of the following Variational Inequality:

$$\text{Find } u \in \mathbb{K} \text{ such that } \int_\Omega c(x, u(x))(v(x) - u(x))\, dx \geq 0 \quad \forall v \in \mathbb{K}, \tag{1}$$

where

$$\mathbb{K} = \{v = \mathcal{E}(\Omega, \mathbb{R}^2) : v_1(x) \geq 0,\ v_2(x) \geq 0,\ div\, v\, +\, t(x) = 0,$$
$$u_1|_{\partial\Omega} = \varphi_1(x),\ u_2|_{\partial\Omega} = \varphi_2(x)\}.$$

Ω is a simply connected bounded domain in $\mathbb{R}^2$ of generic point $x = (x_1, x_2)$, with Lipschitz boundary $\partial\Omega$. $v(x) = (v_1(x), v_2(x))$ represents the unknown flow at each point $x \in \Omega$ and the components $v_1(x)$, $v_2(x)$ are the traffic density through a neighbourhood of x in the directions of the increasing axes x_1 and x_2. $\varphi = (\varphi_1, \varphi_2) \in L^2(\partial\Omega, \mathbb{R}^2)$ is the fixed flow on the boundary $\partial\Omega$ (or on a part of $\partial\Omega$), $c(x, u(x)) = (c_1(x, u(x)), c_2(x, u(x)))$ is the "personal cost" whose components $c_1(x, u(x))$, $c_2(x,\ u(x))$ represent the travel cost along the axes x_1 and x_2 respectively. It is assumed

i) $c(x, u) : \Omega \times \mathbb{R}^2 \to \mathbb{R}^2$ is a Carathéodory function such that

$$\|c(x, u)\|_{\mathbb{R}^2} \leq \alpha(x) + \|u\|_{\mathbb{R}^2} \text{ a. e. in } \Omega, \ \ u \in \mathbb{R}^2 \tag{2}$$

with $\alpha(x) \in L^2(\Omega)$.

$\mathcal{E}(\Omega, \mathbb{R}^2)$ is the functional space defined in the following way:

$$\mathcal{E}(\Omega, \mathbb{R}^2) = \{v \in L^2(\Omega, \mathbb{R}^2) : \ \text{div } v \in L^2(\Omega)\}$$

endowed with the norm:

$$\|u\|^2 = \int_\Omega \|u\|^2 \, dx + \int_\Omega |\ \text{div } v|^2 \, dx.$$

The equilibrium condition is the following:

Definition 1 $u(x) \in \mathbb{K}$ *is an equilibrium distribution flow if there exists a potential* $\mu \in H^1(\Omega)$ *such that*

$$\left(c_i(x, u(x)) - \frac{\partial \mu}{\partial x_i}\right) u_i(x) = 0 \quad i = 1, 2, \ \ a.\,e.\,in\,\Omega \tag{3}$$

$$c_i(x, u(x)) - \frac{\partial \mu}{\partial x_i} \geq 0 \quad i = 1, 2, \ \ a.\,e.\,in\,\Omega. \tag{4}$$

μ *measures the cost occurred when a network user travels from the point* x *to the boundary* $\partial\Omega$ *using the cheapest possible path.*

In [10] it is proved the following equivalence result:

Theorem 1 $u \in \mathbb{K}$ *is an equilibrium distribution according to definition 1 if and only if*

$$\int_\Omega c(x, u(x))(v(x) - u(x)) \, dx \geq 0 \quad \forall v \in \mathbb{K}.$$

For better represent the reality, it is convenient to to consider capacity constraints on the flow:

$$0 \leq s(x) \leq v(x) \leq z(x) \text{ a. e. in } \Omega \tag{5}$$

with $s(x) = (s_1(x), s_2(x))$, $z(x) = (z_1(x), z_2(x))$ such that

$$(s_1(x), z_2(x)), (z_1(x), s_2(x)) \in \mathbb{K}. \tag{6}$$

Now the set of feasible flows becomes

$$\tilde{\mathbb{K}} = \{v(x) = \mathcal{E}(\Omega, \mathbb{R}^2) : s(x) \leq v(x) \leq z(x) \text{ a. e. in } \Omega,$$
$$\text{div } v(x) + t(x) = 0 \quad v_1|_{\partial\Omega} = \varphi_1(x),\ v_2|_{\partial\Omega} = \varphi_2(x)\}.$$

In virtue of (6) $\tilde{\mathbb{K}}$ is nonempty and the new formulation of the equilibrium conditions is the following:

Definition 2 *$u(x) \in \tilde{\mathbb{K}}$ is an equilibrium distribution flow if there exists a potential $\mu \in H^1(\Omega)$ such that*

$$if\ s_i(x) < u_i(x) \leq z_i(x),\ then\ c_i(x, u(x)) = \frac{\partial \mu}{\partial x_i}; \tag{7}$$

$$if\ u_i(x) = s_i(x),\ then\ c_i(x, u(x)) \geq \frac{\partial \mu}{\partial x_i}; \tag{8}$$

$$if\ u_i(x) = z_i(x),\ then\ c_i(x, u(x)) \leq \frac{\partial \mu}{\partial x_i}. \tag{9}$$

Under the assumption *i*) in [6] it is proved the following:

Theorem 2 *$u \in \tilde{\mathbb{K}}$ is an equilibrium distribution according with definition 2 if and only if*

$$\int_\Omega c(x, u(x))(v(x) - u(x))\, dx \geq 0 \quad \forall v(x) \in \tilde{\mathbb{K}}. \tag{10}$$

Moreover in [6] the following existence result for the Variational Inequality (10) has been provided.

Theorem 3 *Assume that condition i) holds and that the following monotonicity condition holds:*

$$ii) \int_\Omega (c(x, u) - c(x, v))\,(u - v)\, dx \geq 0\ \forall u, v \in \tilde{\mathbb{K}},\ a.\ e.\ in\ \Omega.$$

Then the Variational Inequality (10) admits solutions.

Theorem 3 ensures the existence of a solution fulfilling the equilibrium conditions (7)–(9). As it is well known, these conditions provide an equilibrium flow that follows the so–called "user's optimization" approach. This equilibrium flow is different from the equilibrium flow obtained minimizing a cost functional. For more details about these questions we refer to references [4], [5], [8], [9].

The aim of this paper is to improve paper [6], adding a calculation procedure based on the subgradient method.

2 Calculus of the solution

The computational procedure we present is based on the adaptation of the subgradient method. In addition to the assumptions *i*) and *ii*), let us assume that:

iii) $c(x,u(x)) = c(x)u(x) + b(x)$ where $c(x)$ is a symmetric 2×2 matrix of class $C^1(\bar{\Omega})$ and $b(x)$ is a vector of $\mathbb{R}^2$ of class $L^\infty(\Omega)$ and, for sake of simplicity, $t(x) = 0$.

In virtue of the assumption *ii*) the solution to the variational inequality (10) is also solution to the Minty variational inequality:

$$\int_\Omega c(x,v(x))(u(x)-v(x))\,dx \le 0 \quad \forall v(x) \in \tilde{\mathbb{K}}. \tag{11}$$

Let us set

$B = \{u \in \mathcal{E}(\Omega) : s(x) \le u(x) \le z(x),\ u_1|_{\partial\Omega} = \varphi_1(x)\ u_2|_{\partial\Omega} = \varphi_2(x),$
$\|\text{ grad div } u\|_{\mathcal{E}^*(\Omega)} \le R,\ \text{with } R \text{ constant large enough}\}$ *

and $\forall u \in B$

$$\Psi_1(u) = \max_{v\in\tilde{\mathbb{K}}} \int_\Omega c(x,v(x))(u(x)-v(x))\,dx. \tag{12}$$

Being the operator

$$v \to \int_\Omega c(x,v(x))(u(x)-v(x))\,dx$$

weakly upper semicontinuous and $\tilde{\mathbb{K}}$ a bounded subset of $\mathcal{E}(\Omega)$ and hence a weakly compact set, $\Psi_1(u)$ is well defined. Moreover, $\Psi_1(u)$, being the

*Taking into account the natural trace theorem related to the space $\mathcal{E}(\Omega)$, we assume that

$$\varphi_1 \cdot X_1 + \varphi_2 \cdot X_2 \in H^{-\frac{1}{2}}(\partial\Omega),$$

where X_1 and X_2 are the components of the outer normal to $\partial\Omega$.

maximum of a family of continuous and affine functions, is convex and weakly lower semicontinuous.

Now let us set, for all $u \in B$:

$$\Psi_2(u) = \int_\Omega [\text{ div } u]^2 \, dx$$

and

$$\Psi(u) = \max\{\Psi_1(u), \Psi_2(u)\}.$$

Ψ turns out to be convex and weakly lower semicontinuous; it is easily seen taht $\Psi(u) \geq 0$ and $\Psi(u) = 0 \Leftrightarrow u \in \tilde{\mathbb{K}}$ and verifies (11) and (10).

Now let us show that the subdifferential

$$\partial\Psi(u) = \{\tau \in \mathcal{E}^*(\Omega) : \ \Psi(v) - \Psi(u) \geq \langle \tau, v - u \rangle \ \forall v \in B\}$$

is nonempty for all $u \in B$. Let us define:

$$\tau_1 = c(x, \bar{v}(x)),$$

where $\bar{v}$ is a solution of

$$\max_{w \in \tilde{\mathbb{K}}} \int_\Omega c(x, w(x))(u(x) - w(x)) \, dx = \int_\Omega c(x, \bar{v}(x))(u(x) - \bar{v}(x)) \, dx.$$

Let us prove that $\tau_1 \in \partial\Psi_1(u)$, namely that

$$\max_{w \in \tilde{\mathbb{K}}} \int_\Omega c(x, w(x))(v(x) - w(x)) \, dx - \max_{w \in \tilde{\mathbb{K}}} \int_\Omega c(x, w(x))(u(x) - w(x)) \, dx$$
$$\geq \int_\Omega c(x, \bar{v}(x))(v(x) - u(x)) \, dx \quad \forall v \in B.$$

We get:

$$\int_\Omega c(x, \bar{v}(x))(v(x) - u(x)) \, dx = \int_\Omega c(x, \bar{v}(x))(v(x) - \bar{v}(x) + \bar{v}(x) - u(x)) \, dx$$
$$= \int_\Omega c(x, \bar{v}(x))(v(x) - \bar{v}(x)) \, dx + \int_\Omega c(x, \bar{v}(x))(\bar{v}(x) - u(x)) \, dx$$
$$= \int_\Omega c(x, \bar{v}(x))(v(x) - \bar{v}(x)) \, dx - \max_{w \in \tilde{\mathbb{K}}} \int_\Omega c(x, w(x))(u(x) - w(x)) \, dx$$
$$\leq \max_{w \in \tilde{\mathbb{K}}} \int_\Omega c(x, w(x))(v(x) - w(x)) \, dx - \max_{w \in \tilde{\mathbb{K}}} \int_\Omega c(x, w(x))(u(x) - w(x)) \, dx$$

and hence the assert.

Now let us define:

$$\tau_2 = -2 \text{ grad div } u$$

and let us prove

$$\Psi_2(v) - \Psi_2(u) \geq \langle \Psi'(u), v - u \rangle.$$

$\forall v \in B$, we have:

$$\Psi_2(v) - \Psi_2(u) = \int_\Omega \left[\frac{\partial v_1}{\partial x_1} + \frac{\partial v_2}{\partial x_2}\right]^2 dx - \int_\Omega \left[\frac{\partial u_1}{\partial x_1} + \frac{\partial u_2}{\partial x_2}\right]^2 dx$$
$$= \int_\Omega \left[\frac{\partial v_1}{\partial x_1} + \frac{\partial v_2}{\partial x_2} - \frac{\partial u_1}{\partial x_1} - \frac{\partial u_2}{\partial x_2}\right] \cdot \left[\frac{\partial v_1}{\partial x_1} + \frac{\partial v_2}{\partial x_2} + \frac{\partial u_1}{\partial x_1} + \frac{\partial u_2}{\partial x_2}\right] dx$$
$$= \int_\Omega \left[\frac{\partial (v_1 - u_1)}{\partial x_1} + \frac{\partial (v_2 - u_2)}{\partial x_2}\right] \cdot \left[\frac{\partial (v_1 + u_1)}{\partial x_1} + \frac{\partial (v_2 + u_2)}{\partial x_2}\right] dx$$
$$= \int_\Omega \left[\frac{\partial (v_1 - u_1)}{\partial x_1} + \frac{\partial (v_2 - u_2)}{\partial x_2}\right]$$
$$\cdot \left\{2\left[\frac{\partial u_1}{\partial x_1} + \frac{\partial u_2}{\partial x_2}\right] + \left[\frac{\partial v_1}{\partial x_1} + \frac{\partial v_2}{\partial x_2}\right] - \left[\frac{\partial u_1}{\partial x_1} + \frac{\partial u_2}{\partial x_2}\right]\right\} dx$$
$$= \int_\Omega 2\left[\frac{\partial (v_1 - u_1)}{\partial x_1} + \frac{\partial (v_2 - u_2)}{\partial x_2}\right] \cdot \left[\frac{\partial u_1}{\partial x_1} + \frac{\partial u_2}{\partial x_2}\right] dx$$
$$+ \int_\Omega \left[\frac{\partial (v_1 - u_1)}{\partial x_1} + \frac{\partial (v_2 - u_2)}{\partial x_2}\right] \cdot \left[\frac{\partial (v_1 - u_1)}{\partial x_1} + \frac{\partial (v_2 - u_2)}{\partial x_2}\right] dx.$$

Then, we obtain:

$$\Psi_2(v) - \Psi_2(u) \geq \int_\Omega 2\left[\frac{\partial (v_1 - u_1)}{\partial x_1} + \frac{\partial (v_2 - u_2)}{\partial x_2}\right] \cdot \left[\frac{\partial u_1}{\partial x_1} + \frac{\partial u_2}{\partial x_2}\right] dx$$
$$= \langle -2 \text{ grad div } u, v - u \rangle.$$

As a consequence, the subdifferential of Ψ is nonempty, because if

$$\begin{matrix} \tau_1 \in \partial\Psi_1(u) \\ \\ \tau_2 \in \partial\Psi_2(u) \end{matrix} \Rightarrow \tau = \begin{cases} \tau_1 \text{ if } \Psi_1(u) > \Psi_2(u) \\ \\ \tau_2 \text{ if } \Psi_2(u) > \Psi_1(u) \end{cases} \in \partial\Psi(u)$$

The subgradient method runs as follows.
Let

$$\Gamma = \{u \in B : \ \Psi(u) = 0\}.$$

Then $u \in B$ is a solution if and only if $u \in \Gamma$. Choose $u^0 \in B$. Given $u^n \in B$ and $u^n \notin \Gamma$, let

$$u^{(n+1)} = \text{Proj}_B(u^n - \rho_n \tau_n)$$

where

$$\tau_n \in \partial\Psi(u^n) \quad \text{and} \quad \rho_n = \frac{\Psi(u^n)}{\|\tau_n\|^2_{\mathcal{E}^*(\Omega)}}.$$

If $u^n \notin \Gamma$, τ_n is different from zero (if $\tau_n = 0$ it follows $\Psi(v) \geq \Psi(u^n)$ $\forall v \in B$ and in particular $\Psi(u^n) = 0$, namely $u^n \in \Gamma$).

Moreover, in virtue of assumption *iii)* and of the definition of B, $\|\tau_n\|_{\mathcal{E}^*(\Omega)}$ remains bounded.

If $\{\tau_n\}$ remains bounded and $u^n \notin \Gamma$ for all n, we have the following result:

Theorem 4 *There holds:*

$$\Psi(u^n) \to 0.$$

The sequence $\{u^n\}$ *has weak cluster points and every weak cluster point is in* Γ.

Proof. For arbitrary $\gamma \in \Gamma$, using the nonexapansivity of the projection mapping and the support inequality

$$\langle \tau_n, \gamma - u^n \rangle \leq \Psi(\gamma) - \Psi(u^n) = -\Psi(u^n),$$

we obtain

$$\|u^{n+1} - \gamma\|^2 \leq \|u^n - \rho_n \tau_n - \gamma\|^2$$

$$= \|u^n - \gamma\|^2 + 2\rho_n \langle \tau_n, \gamma - u^n \rangle + \rho_n^2 \|\tau_n\|^2_{\mathcal{E}^*(\Omega)}$$

$$\leq \|u^n - \gamma\|^2 - 2\rho_n \Psi(u^n) + \rho^2 \|\tau_n\|^2_{\mathcal{E}^*(\Omega)} = \|u^n - \gamma\|^2 - \frac{\Psi(u^n)}{\|\tau_n\|^2_{\mathcal{E}^*(\Omega)}}.$$

Since the sequence $\{\|u^n - \gamma\|^2\}$ is decreasing and bounded from below and since $\|\tau_n\|_{\mathcal{E}^*(\Omega)}$ are bounded from above, it follows that

$$\Psi^2(u^n) = \|\tau_n\|^2 \left(\|u^n - \gamma\|^2 - \|u^{n+1} - \gamma\|^2\right) \to 0.$$

In particular

$$\|u^n - \gamma\| \leq \|u^0 - \gamma\| \quad \forall n.$$

This shows that $\{u^n\}$ is bounded and therefore has a weak cluster point. Since $\Psi(u^n) \to 0$, and since Ψ is weakly lower semicontinuous, it follows

$$0 = \liminf \Psi(u^n) \geq \Psi(u) \geq 0,$$

namely $\Psi(u) = 0$ and u is the solution to the Variational Inequality.

References

[1] J.M. Borwein - A.S. Lewis, *Practical conditions for Fenchel duality in Infinite Dimensions*, Pitman Research Notes in Mathematics Series **252**, M.A. Théra - J.B. Baillon Editors, 83–89.

[2] S. Dafermos, *Continuum Modelling of Transportation Networks*, Transportation Res. **14** B (1980), 295–301.

[3] P. Daniele, *Lagrangean Function for Dynamic Variational Inequalities*, Rendiconti del Circolo Matematico di Palermo, , Serie II, **58** (1999), 101–119.

[4] P. Daniele - A. Maugeri, *Vector Variational Inequalities and a Continuum Modelling of Traffic Equilibrium Problem*, in Vector Variational Inequalities and Vector Equilibria, F. Giannessi Ed., Kluwer Academic Publishers (2000), 97–111.

[5] J. Gwinner, *On continuum Modelling of Large Dense Networks in Urban Road Traffic*, in Mathematics in Transport Planning and Control (J.D. Griffiths Ed.), IMA Conference, Cardiff (1988).

[6] G. Idone, *Variational Inequalities and Applications to a Continuum Model of Transportation Network with Capacity Constraints*, to appear in Journal of Global Optimization, 2002.

[7] A. Maugeri, *Dynamic Models and Generalized Equilibrium Problems*, in New Trends in Mathematical Programming, F. Giannessi et al. Eds., Kluwer Academis Publishers (1998), 191–202.

[8] A. Maugeri, *New Classes of Variational Inequalities anf Applications to Equilibrium Problems*, Methods of Operation Research **53** (1985), 129–131.

[9] A. Maugeri, *New Classes of Variational Inequalities anf Applications to Equilibrium Problems*, Rendiconti Accademia Nazionale delle Scienze deta dei XL **11** (1987), 224–285.

[10] A. Maugeri, *Equilibrium Problems and Variational Inequalities*, in Equilibrium Problems: Nonsmooth Optimization and Variational Inequalities Models, A. Maugeri, F. Giannessi, P. Pardalos Eds., Kluwer Academic Publishers (2001), 187–205.

THE ECONOMIC MODEL FOR DEMAND–SUPPLY MARKETS

Patrizia Daniele
Dipartimento di Matematica e Informatica- Università di Catania,
Viale A. Doria, 6 - 95125 CATANIA (ITALIA)
e-mail: daniele@dmi.unict.it

Antonino Maugeri
Dipartimento di Matematica e Informatica- Università di Catania,
Viale A. Doria, 6 - 95125 CATANIA (ITALIA)
e-mail: maugeri@dmi.unict.it

Abstract In this paper we aim to present a model of spatially distributed economic markets under the assumptions that each demand market becomes a supply market after the time Δ.

Keywords: Variational inequalities, demand and supply excesses.

1 Introduction

In this paper we consider the spatial price equilibrium problem in the more general case when supply and demand excesses are present and the data depend on the time. This time dependence of the data allows us to study the situation in which the markets are demand and supply markets in two different times.

In [10] the authors first consider this realistic model when it is possible that not all the supply commodity is sold and not all the demand commodity is satisfied, and they show that this model, which can be called "disequilibrium model", is expressed in terms of a variational inequality studied by many authors (see [2], [6], [10], [11], [12] and in particular [8]).

An improvement of the model can be obtained when we consider the influence of the evolution in the time of the economic reality and time dependent models. Also in this case we are lead to a time – dependent variational inequality which describes the trajectories of the flows representing the equilibrium prices during the evolution of the time (see [3], [4], [5], [7], [9], [14]).

Another improvement studied in this paper is obtained considering the case in which the demand markets use the commodity offered at the time t by the first supply markets in order to produce another commodity at the time $t + \Delta$ which will be sold to the former supply markets. Also this frame can be expressed by a variational inequality, namely:

$$\ll v(u), \tilde{u} - u \gg + \ll \nu(\omega), \tilde{\omega} - \omega \gg \geq 0 \quad \forall (\tilde{u}, \tilde{\omega}) \in \mathbb{K} \times (\Xi \cap D),$$

where, roughly speaking, v is the operator which expresses the quantities of the first commodity produced by the first supply markets as function of the prices u; ν represents the quantities of the second commodity produced by the first demand markets as function of the prices ω; $\mathbb{K}$ is the set of feasible prices of the first commodity; Ξ is the set of feasible prices of the second commodity; D is the set of constraints that express how the availability of the first commodity influences the production of the second commodity.

For this model we provide an existence theorem and also illustrate this interactive pair of markets by means of an example which is solved by the direct computational procedure.

2 The first phase: formalization of the equilibrium.

The supply markets $P_1, P_2, \ldots, P_n$ and the demand markets $Q_1, Q_2, \ldots,$ Q_m of the commodity m_1, whose geometry remains fixed, are considered at all time $t \in \mathcal{T} = [0, T]$. For each time $t \in \mathcal{T}$ we have:

- the total supply vector $g(t) \in \mathbb{R}^n$,

- the supply price vector $p(t) \in \mathbb{R}^n$,

· the total demand vector $f(t) \in \mathbb{R}^m$,

· the demand price vector $q(t) \in \mathbb{R}^m$,

+ the flow vector $x(t) \in \mathbb{R}^{nm}$,

+ the unit cost vector $c(t) \in \mathbb{R}^{nm}$.

The feasible vectors $u(t) = (p(t), q(t), x(t))$ have to satisfy the time-dependent constraints on prices and transportation flows, namely that:

$$u(t) \in \prod_{i=1}^{n} [\underline{p}_i(t), \overline{p}_i(t)] \times \prod_{j=1}^{m} [\underline{q}_j(t), \overline{q}_j(t)] \times \prod_{i=1}^{n} \prod_{j=1}^{m} [\underline{x}_{ij}(t), \overline{x}_{ij}(t)]$$

where $\underline{p}_i(t)$, $\overline{p}_i(t)$, $\underline{q}_j(t)$, $\overline{q}_j(t)$, $\underline{x}_{ij}(t)$, $\overline{x}_{ij}(t)$ are given.

For technical reasons, the functional setting for the trajectories $u(t)$ is the reflexive Banach space: L

$$L = L^2(\mathcal{T}, \mathbb{R}^n) \times L^2(\mathcal{T}, \mathbb{R}^m) \times L^2(\mathcal{T}, \mathbb{R}^{nm}).$$

The set of feasible vectors $u(t) = (p(t), q(t), x(t))$ is given by

$$\begin{aligned} &\mathbb{K} = \mathbb{K}_1 \times \mathbb{K}_2 \times \mathbb{K}_3 = \\ &= \{p \in L^2(\mathcal{T}, \mathbb{R}^n) : \underline{p}(t) \le p(t) \le \overline{p}(t) \text{ a. e. on } \mathcal{T}\}\times \\ &\times\{q \in L^2(\mathcal{T}, \mathbb{R}^m) : \underline{q}(t) \le q(t) \le \overline{q}(t) \text{ a. e. on } \mathcal{T}\}\times \\ &\times\{x \in L^2(\mathcal{T}, \mathbb{R}^{nm}) : \underline{x}(t) \le x(t) \le \overline{x}(t) \text{ a. e. on } \mathcal{T}\}. \end{aligned}$$

$\mathbb{K}$ is a convex, closed and bounded set. Furthermore we are giving the mappings:

$$\begin{aligned} g &= g(p(t)) : \mathbb{K}_1 \to L^2(\mathcal{T}, \mathbb{R}^n), \\ f &= f(q(t)) : \mathbb{K}_2 \to L^2(\mathcal{T}, \mathbb{R}^m), \\ c &= c(x(t)) : \mathbb{K}_3 \to L^2(\mathcal{T}, \mathbb{R}^{nm}), \end{aligned}$$

which assign to each price trajectory $p \in \mathbb{K}_1$ and $q \in \mathbb{K}_2$ the supply $g \in L^2(\mathcal{T}, \mathbb{R}^n)$ and the demand $f \in L^2(\mathcal{T}, \mathbb{R}^m)$ respectively and to the flow trajectory $x \in \mathbb{K}_3$ the cost $c \in L^2(\mathcal{T}, \mathbb{R}^{nm})$.

Introducing the supply excesses $s_i(t)$ and the demand excesses $t_j(t)$ we must have:

$$g_i(p(t)) = \sum_{j=1}^{m} x_{ij}(t) + s_i(t) \quad i = 1, 2, \ldots, n \tag{1}$$

$$f_j(q(t)) = \sum_{i=1}^{n} x_{ij}(t) + t_j(t) \quad j = 1, 2, \ldots, m. \tag{2}$$

Obviously $s(t) \in L^2(\mathcal{T}, \mathbb{R}^n)$ and $t(t) \in L^2(\mathcal{T}, \mathbb{R}^m)$. "Dynamic" market equilibrium takes the following form:

Definition 1 *$u(t) = (p(t), q(t), x(t)) \in L$ is a dynamic market equilibrium if and only if for each $i = 1, 2, \ldots, n$ and $j = 1, 2, \ldots, m$ and a. e. on $\mathcal{T}$ there hold:*

$$\left.\begin{aligned} s_i(t) > 0 \qquad &\Rightarrow p_i(t) = \underline{p}_i(t), \\ \underline{p}_i(t) < p_i(t) \le \overline{p}_i(t) \quad &\Rightarrow s_i(t) = 0 \end{aligned}\right\} \quad i = 1, \ldots, n; \tag{3}$$

$$\begin{array}{lll} t_j(t) > 0 & \Rightarrow q_j(t) = \overline{q}_j(t), & \\ & & j = 1, \ldots, m; \\ \underline{q}_j(t) \leq q_j(t) < \overline{q}_j(t) & \Rightarrow t_j(t) = 0 & \end{array} \tag{4}$$

$$p_i(t) + c_{ij}(t) \begin{cases} > q_j(t) & if \quad x_{ij}(t) = \underline{x}_{ij}(t) \\ = q_j(t) & if \quad \underline{x}_{ij}(t) \leq x_{ij}(t) \leq \overline{x}_{ij}(t) \\ < q_j(t) & if \quad x_{ij}(t) = \overline{x}_{ij}(t). \end{cases} \tag{5}$$

Let $v : \mathbb{K} \to L$ be the operator defined setting

$$\begin{aligned} v &= v(p(t), q(t), x(t)) = \\ &= \Big((g_i(p(t)) - \sum_{j=1}^{m} x_{ij}(t))_{i=1,\ldots,n}, (f_j(q(t)) - \sum_{i=1}^{n} x_{ij}(t))_{j=1,\ldots,m}, \\ & \quad (p_i(t) + c_{ij}(x(t)) - q_j(t))_{\substack{i=1,\ldots,n \\ j=1,\ldots,m}} \Big). \end{aligned}$$

The following characterization holds.

Theorem 1 *Suppose that for each $i = 1, 2, \ldots, n$ and $j = 1, 2, \ldots, m$ there hold*

1. $q_j(t) = 0$ *on a set $E \subseteq \mathcal{T}$ having positive measure*

$$\Rightarrow f_j(q(t)) \geq 0 \text{ in } E; \tag{6}$$

2. $x_{ij}(t) > 0$ *on a set $E \subseteq \mathcal{T}$ having positive measure* $\Rightarrow c_{ij}(x(t)) > 0 \,\text{in}\, E$.

Then $u(t) = (p(t), q(t), x(t)) \in \mathbb{K}$ is a dynamic market equilibrium if and only if $u(t)$ is a solution to

$$\begin{aligned} \ll v(u), \tilde{u} - u \gg = \int_0^T \Big\{ \sum_{i=1}^{n} \Big(g_i(p(t)) - \sum_{j=1}^{m} x_{ij}(t) \Big) (\tilde{p}_i(t) - p_i(t)) + \\ - \sum_{j=1}^{m} \Big(f_j(q(t)) - \sum_{i=1}^{n} x_{ij}(t) \Big) (\tilde{q}_j(t) - q_j(t)) + \\ + \sum_{i=1}^{n} \sum_{j=1}^{m} \Big(p_i(t) + c_{ij}(x(t)) - q_j(t) \Big) (\tilde{x}_{ij}(t) - x_{ij}(t)) \Big\} \, dt \geq 0 \end{aligned} \tag{7}$$

$\forall \tilde{u} = (\tilde{p}, \tilde{q}, \tilde{x}) \in K.$

Proof. Assume that (1)-(5) hold. Let $\tilde{u} \in \mathbb{K}$. It follows:

$$\sum_{i=1}^{n} \Big(g_i(p(t)) - \sum_{j=1}^{m} x_{ij}(t)\Big)(\tilde{p}_i(t) - p_i(t))$$
$$-\sum_{j=1}^{m} \Big(f_j(q(t)) - \sum_{i=1}^{n} x_{ij}(t)\Big)(\tilde{q}_j(t) - q_j(t))$$
$$+\sum_{i=1}^{n}\sum_{j=1}^{m} \Big(p_i(t) + c_{ij}(x(t)) - q_j(t)\Big)(\tilde{x}_{ij}(t) - x_{ij}(t)) \geq 0$$
$$\text{a. e. on} \quad \mathcal{T}.$$

Hence (7) follows.

Viceversa, let us assume that (7) holds. As a consequence we obtain, assuming in turns $\tilde{q} = q$ and $\tilde{x} = x$, $\tilde{p} = p$ and $\tilde{x} = x$, $\tilde{p} = p$ and $\tilde{q} = q$:

$$\int_0^T \sum_{i=1}^{n} \Big(g_i(p(t)) - \sum_{j=1}^{m} x_{ij}(t)\Big)(\tilde{p}_i(t) - p_i(t))\, dt \geq 0 \quad \forall p \in \mathbb{K}_1,$$

$$\int_0^T \sum_{j=1}^{m} \Big(f_j(q(t)) - \sum_{i=1}^{n} x_{ij}(t)\Big)(\tilde{q}_j(t) - q_j(t))\, dt \geq 0 \quad \forall q \in \mathbb{K}_2,$$

$$\int_0^T \sum_{i=1}^{n}\sum_{j=1}^{m} \Big(p_i(t) + c_{ij}(x(t)) - q_j(t)\Big)(\tilde{x}_{ij}(t) - x_{ij}(t))\, dt \geq 0 \quad \forall x \in \mathbb{K}_3$$

from which (1) and (3), (2) and (4), (5) follow respectively.

In fact, let us assume that (1) does not hold. Then there exists an index i^* together with a set $E \subseteq \mathcal{T}$ having positive measure such that

$$s_{i^*}(t) = g_{i^*}(t) - \sum_{j=1}^{m} x_{i^*j}(t) < 0 \quad \text{on} \quad \text{E}.$$

Then the choice

$$\tilde{p}_i(t) = p_i(t) \quad \text{for} \quad \text{i} \neq \text{i}^*$$
$$\tilde{p}_i^*(t) \begin{cases} = & p_i(t) \quad \text{on} \quad \mathcal{T} - \text{E} \\ > & p_i(t) \quad \text{on} \quad \text{E} \end{cases}$$

leads to a contradiction:

$$\int_0^T \sum_{i=1}^{n} \Big(g_i(p(t)) - \sum_{j=1}^{m} x_{ij}(t)\Big)(\tilde{p}_i(t) - p_i(t))\, dt =$$
$$\int_E \Big(g_i^*(p(t)) - \sum_{j=1}^{m} x_{i^*j}(t)\Big)(\tilde{p}_i(t) - p_i(t))\, dt < 0.$$

Suppose now that (3) is not verified. Then there exists an index i^* together with a set $E \subset \mathcal{T}$ having positive measure such that either $s_{i^*}(t) > 0$ and $p_{i^*}(t) > \underline{p}_{i^*}(t)$ or $p_{i^*}(t) > \underline{p}_{i^*}(t)$ and $s_{i^*}(t) > 0$ on E.

In both the cases the choice

$$\tilde{p}_i(t) = p_i(t) \quad \text{for} \quad \mathrm{i} \neq \mathrm{i}^*$$

$$\tilde{p}_i^*(t) = \begin{cases} p_i^*(t) & \text{on} \quad \mathcal{T} - \mathrm{E} \\ \underline{p}_i^*(t) & \text{on} \quad \mathrm{E} \end{cases}$$

leads to a contradiction. Similarly one can proceed in other cases and the equivalence is achieved. □

For what concerns the existence, we may provide some existence theorems under general assumptions. Let us recall the following general definitions. Let E be a real topological vector space, $\mathbb{K} \subseteq E$ convex. Then $C : \mathbb{K} \to E^*$ is said to be

- *pseudomonotone* iff, for all $x, y \in \mathbb{K}$,

$$\langle C(x), y - x \rangle \geq 0 \Rightarrow \langle C(y), x - y \rangle \leq 0;$$

- *hemicontinuous* iff, for all $y \in \mathbb{K}$, the function $\xi \mapsto \langle C(\xi), y - \xi \rangle$ is upper semicontinuous on $\mathbb{K}$;

- *hemicontinuous along line segments* iff, for all $x, y \in \mathbb{K}$, the function $\xi \mapsto \langle C(\xi), y - x \rangle$ is upper semicontinuous on the line segment $[x, y]$.

The general theorem related to the existence of the solutions of the variational inequalities is the following:

Theorem 2 *Let E be a real topological vector space and $\mathbb{K} \subseteq E$ be convex and nonempty. Let $C : \mathbb{K} \to E^*$ be given such that*

1. *there exist $A \subseteq \mathbb{K}$ nonempty, compact, and $B \subseteq \mathbb{K}$ compact, convex such that, for every $x \in \mathbb{K} \setminus A$, there exists $y \in B$ with $\langle C(x), y - x \rangle < 0$;*

and either

2. *C is hemicontinuous*

or

3. *C is pseudomonotone and hemicontinuous along line segments.*

Then there exists $\overline{x} \in A$ such that $\langle C(\overline{x}), y - \overline{x} \rangle \geq 0$ for all $y \in \mathbb{K}$.

Adapting this theorem to our case, we have the following:

Theorem 3 *Each of the following conditions is sufficient for the existence of a solution to the V.I.:*

1. $v(u)$ *is hemicontinuous with respect to the strong topology and there exist* $A_1 \subseteq \mathbb{K}_1$, $A_2 \subseteq \mathbb{K}_2$, $A_3 \subseteq \mathbb{K}_3$ *compact and* $B_1 \subseteq \mathbb{K}_1$, $B_2 \subseteq \mathbb{K}_2$, $B_3 \subseteq \mathbb{K}_3$ *convex and compact such that* $\forall u_1 = (p_1, q_1, x_1) \in (\mathbb{K}_1 \setminus A_1) \times (\mathbb{K}_2 \setminus A_2) \times (\mathbb{K}_3 \setminus A_3)$ $\exists u_2 = (p_2, q_2, x_2) \in B_1 \times B_2 \times B_3$ *such that* $\ll v(u_1), u_2 - u_1 \gg < 0$;

2. $v(u)$ *is pseudomonotone and hemicontinuous along line segments;*

3. $v(u)$ *is hemicontinuous with respect to the weak topology on* $\mathbb{K}$.

3 The second phase: formalization of the equilibrium.

Let us assume that, using the commodity m_1, the markets $Q_1, Q_2, \ldots, Q_m$ produce another commodity m_2 which is required by the markets $P_1, P_2, \ldots, P_n$. Now the markets will be considered at all time $t \in \overline{\mathcal{T}} = [\Delta, T + \Delta]$. For each time $t \in \overline{\mathcal{T}}$ we have a total supply vector $\gamma(t + \Delta) \in \mathbb{R}^n$, a supply price vector $\pi(t+\Delta) \in \mathbb{R}^n$, the total demand vector $\varphi(t+\Delta) \in \mathbb{R}^m$, the demand price vector $\rho(t + \Delta) \in \mathbb{R}^m$, the flow vector $y(t + \Delta) \in \mathbb{R}^{nm}$ and the unit cost vector $c(t + \Delta) \in \mathbb{R}^{nm}$.

The feasible vectors $\omega(t + \Delta) = (\pi(t + \Delta), \rho(t + \Delta), y(t + \Delta))$ have to satisfy the time-dependent bounds on prices and transportation flows, namely that, almost everywhere on $\overline{\mathcal{T}}$

$$\omega(t+\Delta) \in \prod_{j=1}^{m} [\underline{\pi}_j(t+\Delta), \overline{\pi}_j(t+\Delta)] \times \prod_{i=1}^{n} [\underline{\rho}_i(t+\Delta), \overline{\rho}_i(t+\Delta)] \times$$
$$\times \prod_{i=1}^{n} \prod_{j=1}^{m} [\underline{y}_{ij}(t+\Delta), \overline{y}_{ij}(t+\Delta)]$$

where $\underline{\pi}_j(t + \Delta)$, $\overline{\pi}_j(t + \Delta)$, $\underline{\rho}_i(t + \Delta)$, $\overline{\rho}_i(t + \Delta)$, $\underline{y}_{ij}(t + \Delta)$, $\overline{y}_{ij}(t + \Delta)$ are given.

For technical reasons, the functional setting for the trajectories $\omega(t+\Delta)$ is the reflexive Banach space: L

$$L^2(\overline{\mathcal{T}}, \mathbb{R}^m) \times L^2(\overline{\mathcal{T}}, \mathbb{R}^n) \times L^2(\overline{\mathcal{T}}, \mathbb{R}^{nm})$$

which we abbreviate by

$$\overline{L} = \overline{L}_1 \times \overline{L}_2 \times \overline{L}_3.$$

The set of feasible vectors $\omega(t+\Delta) = (\pi(t+\Delta), \rho(t+\Delta), y(t+\Delta))$ is given by

$$\begin{aligned}\Xi &= \Xi_1 \times \Xi_2 \times \Xi_3 = \\ &= \{\pi \in \overline{L}_1 : \underline{\pi}(t+\Delta) \leq \pi(t+\Delta) \leq \overline{\pi}(t+\Delta) \text{ a. e. on } \overline{\mathcal{T}}\}\times \\ &\times\{\rho \in \overline{L}_2 : \underline{\rho}(t+\Delta) \leq \rho(t+\Delta) \leq \overline{\rho}(t+\Delta) \text{ a. e. on } \overline{\mathcal{T}}\}\times \\ &\times\{y \in \overline{L}_3 : \underline{y}(t+\Delta) \leq y(t+\Delta) \leq \overline{y}(t+\Delta) \text{ a. e. on } \overline{\mathcal{T}}\}.\end{aligned}$$

It is easily seen that Ξ is a convex, closed and bounded set. Furthermore we are giving the mappings:

$$\gamma : \Xi_1 \to \overline{L}_1, \quad \varphi : \Xi_2 \to \overline{L}_2, \quad c : \Xi_3 \to \overline{L}_3$$

which assign to each price trajectories $\pi \in \Xi_1$ and $\rho \in \Xi_2$ the supply $\gamma \in \overline{L}_1$ and the demand $\varphi \in \overline{L}_2$ respectively and to the flow trajectory $y \in \Xi_3$ the cost $c \in \overline{L}_3$.

Introducing the supply excesses $\sigma_j(t+\Delta)$ and the demand excesses $\tau_i(t+\Delta)$ we must have:

$$\gamma_j(\pi(t+\Delta)) = \sum_{i=1}^{n} y_{ij}(t+\Delta) + \sigma_j(t+\Delta) \quad j = 1, 2, \dots, m \tag{8}$$

$$\varphi_i(\rho(t+\Delta)) = \sum_{j=1}^{m} y_{ij}(t+\Delta) + \tau_i(t+\Delta) \quad i = 1, 2, \dots, n. \tag{9}$$

Obviously $\sigma \in \overline{L}_1$ and $\tau \in \overline{L}_2$. Dynamic market equilibrium takes the following form:

Definition 2 $\omega(t+\Delta) = (\pi(t+\Delta), \rho(t+\Delta), y(t+\Delta)) \in \overline{L}$ *is a dynamic market equilibrium if and only if for each* $j = 1, 2, \dots, m$ *and* $i = 1, 2, \dots, n$ *and a. e. on* $\overline{\mathcal{T}}$ *there hold:*

$$\left.\begin{aligned}&\sigma_j(t+\Delta) > 0 \Rightarrow \pi_j(t+\Delta) = \underline{\pi}_j(t+\Delta),\\ &\underline{\pi}_j(t+\Delta) < \pi_j(t+\Delta) < \overline{\pi}_j(t+\Delta) \Rightarrow \sigma_j(t+\Delta) = 0;\end{aligned}\right\} \quad j = 1, \dots, m \tag{10}$$

$$\left.\begin{aligned}&\tau_i(t+\Delta) > 0 \Rightarrow \rho_i(t+\Delta) = \overline{\rho}_i(t+\Delta),\\ &\underline{\rho}_i(t+\Delta) < \rho_i(t+\Delta) < \overline{\rho}_i(t+\Delta) \Rightarrow \tau_i(t+\Delta) = 0;\end{aligned}\right\} \quad i = 1, \dots, n \tag{11}$$

$$\pi_j(t+\Delta)+c_{ij}(t+\Delta)\begin{cases} \geq \rho_i(t+\Delta) & if\ y_{ij}(t+\Delta)=\underline{y}_{ij}(t+\Delta) \\ = \rho_i(t+\Delta) & if\ \underline{y}_{ij}(t+\Delta)\leq y_{ij}(t+\Delta) \\ & \qquad \leq \overline{y}_{ij}(t+\Delta) \\ \leq \rho_i(t+\Delta) & if\ y_{ij}(t+\Delta)=\overline{y}_{ij}(t+\Delta). \end{cases} \tag{12}$$
$$j=1,\dots,m \quad i=1,\dots,n.$$

Let $\nu : \Xi \to \overline{L}$ be the operator defined setting

$$\nu=\nu(\pi(t+\Delta),\rho(t+\Delta),y(t+\Delta))=$$
$$\Big((\gamma_j(t+\Delta)-\sum_{i=1}^{n}y_{ij}(t+\Delta))_{j=1,\dots,m},(\varphi_i(t+\Delta)-\sum_{j=1}^{m}y_{ij}(t+\Delta))_{i=1,\dots,n},$$
$$(\pi_j(t+\Delta)+c_{ij}(y(t+\Delta))-\rho_i(t+\Delta))_{\substack{j=1,\dots,m\\ i=1,\dots,n}}\Big).$$

The following characterization holds.

Theorem 4 *Suppose that for each $i=1,2,\dots,n$ and $j=1,2,\dots,m$ there hold*

1. $\rho_i(t+\Delta)=0$ *on a set* $E\subseteq\overline{\mathcal{T}}$ *having positive measure*
$$\Rightarrow \varphi_i(\rho(t+\Delta))\geq 0 \text{ in } E; \tag{13}$$
2. $y_{ij}(t+\Delta)>0$ *on a set* $E\subseteq\overline{\mathcal{T}}$ *having positive measure* $\Rightarrow c_{ij}(y(t+\Delta))>0$ in E.

Then $\omega(t+\Delta)=(\pi(t+\Delta),\rho(t+\Delta),y(t+\Delta))\in\Xi$ is a dynamic market equilibrium if and only if $\omega(t+\Delta)$ is a solution to

$$\begin{aligned} &\ll \nu(\omega),\tilde{\omega}-\omega \gg \\ = &\int_{\Delta}^{T+\Delta}\Big\{\sum_{j=1}^{m}\Big(\gamma_j(\pi(t+\Delta))-\sum_{i=1}^{n}y_{ij}(t+\Delta)\Big)(\tilde{\pi}_j(t+\Delta)-\pi_j(t+\Delta)) \\ - &\sum_{i=1}^{n}\Big(\varphi_i(\rho(t+\Delta))-\sum_{j=1}^{m}y_{ij}(t+\Delta)\Big)(\tilde{\rho}_i(t+\Delta)-\rho_i(t+\Delta)) \\ + &\sum_{j=1}^{m}\sum_{i=1}^{n}\Big(\pi_j(t+\Delta)+c_{ij}(y(t+\Delta))-\rho_i(t+\Delta)\Big) \\ &(\tilde{y}_{ij}(t+\Delta)-y_{ij}(t+\Delta))\Big\}\,dt\geq 0 \quad \forall\tilde{\omega}=(\tilde{\pi},\tilde{\rho},\tilde{y})\in\Xi. \end{aligned} \tag{14}$$

Proof. See [3].

The existence theorems for the previous variational inequality are analogous to the ones related to the first phase.

4 The dependence of the second phase on the first one.

Now we introduce additional constraints on $\omega(t+\Delta)$, because the production $\omega(t+\Delta)$ depends on the availability of the commodity m_1, namely by $u(t)$. We assume that these constraints can be described collectively by the requirement $\omega \in D$, where $D \subseteq L$. We assume that D is convex and

$$\Xi \cap ri\, D \neq \emptyset. \tag{15}$$

An example of set D could be

$$D = \{\omega(t+\Delta) \in \Xi : \omega(t+\Delta) \leq v(u(t)) \text{ a. e. on } \overline{\mathcal{T}}\},$$

where g are continuous functions and Ξ remains the same as in the previous section. The presence of these additional constraints leads to an equilibrium condition expressed by the following variational inequality:

$$\omega \in \Xi \cap D \text{ and } \ll \nu(\omega), \tilde{\omega} - \omega \gg \geq 0 \quad \forall \tilde{\omega} \in \Xi \cap D. \tag{16}$$

We want to prove the following characterization which shows how the constraints depending on the first phase quantities of m_1 influence the produced quantities of m_2.

Theorem 5 *$\omega \in \Xi \cap D$ is a solution of (16) if, and only if, there exists $\mu \in L^*$ such that*

$$\ll \mu, \tilde{\omega} - \omega \gg \geq 0 \qquad \forall \tilde{\omega} \in D, \tag{17}$$

$$\ll \nu(\omega) - \mu, \tilde{\omega} - \omega \gg \geq 0 \quad \forall \tilde{\omega} \in \Xi. \tag{18}$$

Proof. If (17) and (18) are satisfied, then it is clear that

$$\ll \nu(\omega), \tilde{\omega} - \omega, \gg \geq \ll \nu(\omega) - \mu, \tilde{\omega} - \omega \gg \geq 0 \quad \forall \tilde{\omega} \in \Xi \cap D,$$

and ω is an equilibrium in the sense of (16).

Conversely, let ω satisfy (16). We apply the separation theorem to the sets

$$\begin{aligned} A &:= \{(\tilde{\omega}, \gamma) \in D \times \mathbb{R} \,|\, \gamma < 0\}, \\ B &:= \{(\tilde{\omega}, \gamma) \in \Xi \times \mathbb{R} \,|\, \ll \nu(\omega), \tilde{\omega} - \omega \gg \leq \gamma\}. \end{aligned}$$

Then A and B are nonempty, convex subsets of $L \times \mathbb{R}$; they are disjoint because of (16). Moreover A has nonempty relative interior, since D has nonempty relative interior. So, from the separation theorem for convex sets there exist $(\mu, k) \in L^* \times \mathbb{R}$, $(\mu, k) \neq (0,0)$, and $\alpha \in \mathbb{R}$ such that

$$\ll \mu, \tilde{\omega} \gg + k\gamma \leq \alpha \quad \forall\, (\tilde{\omega}, \gamma) \in A, \tag{19}$$

$$\ll \mu, \tilde{\omega} \gg + k\gamma \geq \alpha \quad \forall\, (\tilde{\omega}, \gamma) \in B. \tag{20}$$

It follows from (19) or (20) that $k \geq 0$. Assume, for contradiction, that $k = 0$. Choose $\omega_0 \in \Xi \cap ri\, D$, which is possible from (15). Then from (19) and (20) follows $\ll \mu, \omega_0 \gg = \alpha$, and since now $\mu \neq 0$, there exists $\tilde{\omega} \in ri\, D$ with $\ll \mu, \tilde{\omega} \gg > \alpha$, which contradicts (19). Thus $k > 0$, and setting $\mu = -\overline{\mu}\, k$ and dividing by k, we obtain:

$$\ll -\overline{\mu}, \tilde{\omega} \gg + \gamma \leq \frac{\alpha}{k}.$$

Then we get, choosing γ arbitrarily close to zero in (19), that

$$- \ll \overline{\mu}, \tilde{\omega} \gg \leq \frac{\alpha}{k} \quad \forall \tilde{\omega} \in D, \tag{21}$$

and, choosing $\gamma := \ll \nu(\omega), \tilde{\omega} - \omega \gg$ in (20), that

$$- \ll \overline{\mu}\, k, \tilde{\omega} \gg + k \ll \nu(\omega), \tilde{\omega} - \omega \gg \geq \alpha \quad \forall \tilde{\omega} \in \Xi,$$

from which:

$$- \ll \overline{\mu}, \tilde{\omega} \gg + \ll \nu(\omega), \tilde{\omega} - \omega \gg \geq \frac{\alpha}{k} \quad \forall \tilde{\omega} \in \Xi, \tag{22}$$

In particular, choosing $\tilde{\omega} = \omega$ in (21) and (22), we obtain $- \ll \overline{\mu}, \omega \gg = \alpha$. Substituting for α in (21) and (22) gives (17) and (18). □

5 General model.

Reassuming, the model that describes the supply–demand market problem can be synthesized in the following way:

Theorem 6 *The first and second phase equilibrium problems are equivalent to the problem:*

$$\text{“ Find } (u, \omega) \in \mathbb{K} \times (\Xi \cap D) :$$

$$\ll v(u), \tilde{u} - u \gg + \ll \nu(\omega), \tilde{\omega} - \omega \gg \geq 0 \quad \forall (\tilde{u}, \tilde{\omega}) \in \mathbb{K} \times (\Xi \cap D)\text{”}. \tag{23}$$

Proof. Let us assume that u is a solution to (7) and ω is a solution to (14), then

$$\langle v(u), \tilde{u} - u \rangle \geq 0 \quad \forall \tilde{u} \in \mathbb{K}$$
$$\ll \nu(\omega), \tilde{\omega} - \omega \gg \geq 0 \quad \forall \tilde{\omega} \in \Xi \cap D,$$

form which we get

$$\ll v(u), \tilde{u} - u \gg + \ll \nu(\omega), \tilde{\omega} - \omega \gg \geq 0 \quad \forall (\tilde{u}, \tilde{\omega}) \in \mathbb{K} \times (\Xi \cap D)$$

Viceversa, let us choose $\tilde{\omega} = \omega$ and $\tilde{u} = u$ in (23) respectively, then we obtain:

$$\langle v, \tilde{u} - u \rangle \geq 0 \quad \forall \tilde{u} \in \mathbb{K}$$

and

$$\ll \nu(\omega), \tilde{\omega} - \omega \gg \geq 0 \quad \forall \tilde{\omega} \in \Xi \cap D$$

and the assertion holds. □

6 Example

Let us consider an example. In the first phase the problem consists of two supply markets P_1, P_2 and two demand markets Q_1, Q_2. P_1 and P_2 produce the commodity m_1 which is required by the markets Q_1 and Q_2 at the time t. The supply functions and the demand functions depend on two non negative parameters $h(t)$ and $k(t)$, by means of which it is possible to control the production and the demand.

Let us assume that the supply functions are:

$$g_1(p_1(t), p_2(t)) = p_1(t) + h(t), \quad g_2(p_1(t), p_2(t)) = -p_1(t) + p_2(t) + k(t);$$

let us assume that the demand functions are:

$$f_1(q_1(t), q_2(t)) = q_2(t), \quad f_2(q_1(t), q_2(t)) = q_1(t) - q_2(t) + h(t);$$

let us assume that the transportation cost functions are:

$$\begin{array}{ll} c_{11}(x(t)) = x_{11}(t) + 1, & c_{12}(x(t)) = x_{12}(t) + 1, \\ c_{21}(x(t)) = x_{21}(t) + x_{12}(t), & c_{22}(x(t)) = x_{22}(t) + 1, \end{array}$$

where $x(t) = (x_{11}(t), x_{12}(t), x_{21}(t), x_{22}(t))$. As the feasible set $\mathbb{K}$ of the vectors $u(t) = (p_1(t), p_2(t), q_1(t), q_2(t), x_{11}(t), x_{12}(t), x_{21}(t), x_{22}(t))$, we assume the positive orthant $\mathbb{K} = \mathbb{R}^8_+$ and the components of the operator $v = v(u(t))$ are:

$$\begin{cases} g_1(p(t)) - x_{11}(t) - x_{12}(t) = p_1(t) - x_{11}(t) - x_{12}(t) + h(t), \\ g_2(p(t)) - x_{21}(t) - x_{22}(t) = -p_1(t) - x_{21}(t) - x_{22}(t) + p_2(t) + k(t), \\ f_1(q(t)) - x_{11}(t) - x_{21}(t) = q_2(t) - x_{11}(t) - x_{21}(t), \\ f_2(q(t)) - x_{12}(t) - x_{22}(t) = q_1(t) - q_2(t) - x_{12}(t) - x_{22}(t) + h(t), \\ p_1(t) + c_{11}(x(t)) - q_1(t) = p_1(t) - q_1(t) + x_{11}(t) + 1, \\ p_1(t) + c_{12}(x(t)) - q_2(t) = p_1(t) - q_2(t) + x_{12}(t) + 1, \\ p_2(t) + c_{21}(x(t)) - q_1(t) = p_2(t) - q_1(t) + x_{21}(t) + x_{12}(t), \\ p_2(t) + c_{22}(x(t)) - q_2(t) = p_2(t) - q_2(t) + x_{22}(t) + 1. \end{cases}$$

The supply and demand excesses are:

$$\begin{aligned}
s_1(t) &= p_1(t) + h(t) - x_{11}(t) - x_{12}(t)\\
s_2(t) &= -p_1(t) + p_2(t) + k(t) - x_{21}(t) - x_{22}(t)\\
t_1(t) &= q_2(t) - x_{11}(t) - x_{21}(t)\\
t_2(t) &= q_1(t) - q_2(t) + h(t) - x_{12}(t) - x_{22}(t)
\end{aligned}$$

Solving the system

$$\begin{cases} v(u) = 0\\ u \in \mathbb{K}, \end{cases}$$

we find the following equilibrium supply prices, demand prices and shipments:

$$\begin{cases}
p_1(t) = \dfrac{-2h(t) + 4k(t) - 4}{3}, & p_2(t) = \dfrac{-2h(t) + 4k(t) - 4}{3},\\[2ex]
q_1(t) = \dfrac{-2h(t) + 7k(t) - 4}{3}, & q_2(t) = \dfrac{-h(t) + 5k(t) - 2}{3},\\[2ex]
x_{11}(t) = k(t) - 1, & x_{12}(t) = \dfrac{h(t) + k(t) - 1}{3},\\[2ex]
x_{21}(t) = \dfrac{-h(t) + 2k(t) + 1}{3}, & x_{22}(t) = \dfrac{h(t) + k(t) - 1}{3},
\end{cases}$$

and

$$g_1(p_1(t), p_2(t)) = \frac{h(t) + 4k(t) - 4}{3},$$

$$g_2(p_1(t), p_2(t)) = k(t),$$

$$f_1(q_1(t), q_2(t)) = \frac{-h(t) + 5k(t) - 2}{3},$$

$$f_2(q_1(t), q_2(t)) = \frac{2h(t) + 2k(t) - 2}{3}.$$

If

$$\begin{cases}
h(t) - 2k(t) \le -2, & 2h(t) - 7k(t) \le -4,\\[1ex]
h(t) - 5k(t) \le -2, & k(t) \ge 1,\\[1ex]
h(t) + k(t) \ge 1, & h(t) - 2k(t) \le 1,
\end{cases}$$

that is

$$\text{if } h(t) - 2k(t) \le -2,$$

then the solution lies in $\mathbb{K}$, otherwise we have to search the solution in the boundary of $\mathbb{K}$.

In the second phase the problem consists of two supply markets Q_1, Q_2 and two demand markets P_1, P_2. Q_1 and Q_2 produce the commodity m_2 which is required by the markets P_1 and P_2 at the time $t+\Delta$. The supply functions and the demand functions depend on two non negative parameters $\alpha(t+\Delta)$ and $\beta(t+\Delta)$, by means of which it is possible to control the production and the demand.

Let us assume that the supply functions are:

$$\gamma_1(\pi_1(t+\Delta), \pi_2(t+\Delta)) = \pi_1(t+\Delta) + \alpha(t+\Delta),$$
$$\gamma_2(\pi_1(t+\Delta), \pi_2(t+\Delta)) = -\pi_1(t+\Delta) + \pi_2(t+\Delta) + \beta(t+\Delta);$$

let assume that the demand functions are:

$$\varphi_1(\rho_1(t+\Delta), \rho_2(t+\Delta)) = \rho_2(t+\Delta),$$
$$\varphi_2(\rho_1(t+\Delta), \rho_2(t+\Delta)) = \rho_1(t+\Delta) - \rho_2(t+\Delta) + \alpha(t+\Delta);$$

let us assume that the transportation cost functions are:

$$c_{11}(y(t+\Delta)) = y_{11}(t+\Delta) + 1,$$

$$c_{12}(y(t+\Delta)) = y_{12}(t+\Delta) + 1,$$

$$c_{21}(y(t+\Delta)) = y_{21}(t+\Delta) + y_{12}(t+\Delta),$$

$$c_{22}(y(t+\Delta)) = y_{22}(t+\Delta) + 1,$$

where $y(t+\Delta) = (y_{11}(t+\Delta), y_{12}(t+\Delta), y_{21}(t+\Delta), y_{22}(t+\Delta))$.

The components of the operator $\nu = \nu(\omega(t+\Delta))$ are:

$$\begin{cases}
\gamma_1 - y_{11} - y_{21} = \pi_1(t+\Delta) - y_{11}(t+\Delta) - y_{21}(t+\Delta) + \alpha(t+\Delta), \\[2ex]
\gamma_2 - y_{12} - y_{22} = -\pi_1(t+\Delta) - y_{12}(t+\Delta) - y_{22}(t+\Delta) \\
\qquad\qquad\qquad +\pi_2(t+\Delta) + \beta(t+\Delta), \\[2ex]
\varphi_1 - y_{11} - y_{12} = \rho_2(t+\Delta) - y_{11}(t+\Delta) - y_{12}(t+\Delta), \\[2ex]
\varphi_2 - y_{21} - y_{22} = \rho_1(t+\Delta) - \rho_2(t+\Delta) - y_{21}(t+\Delta) \\
\qquad\qquad\qquad -y_{22}(t+\Delta) + \alpha(t+\Delta), \\[2ex]
\pi_1 + c_{11} - \rho_1 = \pi_1(t+\Delta) - \rho_1(t+\Delta) + y_{11}(t+\Delta) + 1, \\[2ex]
\pi_1 + c_{21} - \rho_2 = \pi_1(t+\Delta) - \rho_2(t+\Delta) + y_{21}(t+\Delta) + y_{12}(t+\Delta), \\[2ex]
\pi_2 + c_{12} - \rho_1 = \pi_2(t+\Delta) - \rho_1(t+\Delta) + y_{12}(t+\Delta) + 1, \\[2ex]
\pi_2 + c_{22} - \rho_2 = \pi_2(t+\Delta) - \rho_2(t+\Delta) + y_{22}(t+\Delta) + 1.
\end{cases}$$

The supply and demand excesses are:

$$\sigma_1(t+\Delta) = \pi_1(t+\Delta) + \alpha(t+\Delta) - y_{11}(t+\Delta) - y_{21}(t+\Delta)$$

$$\begin{aligned}\sigma_2(t+\Delta) = {} & -\pi_1(t+\Delta) + \pi_2(t+\Delta) + \beta(t+\Delta) \\ & -y_{12}(t+\Delta) - y_{22}(t+\Delta)\end{aligned}$$

$$\tau_1(t+\Delta) = \rho_2(t+\Delta) - y_{11}(t+\Delta) - y_{12}(t+\Delta)$$

$$\begin{aligned}\tau_2(t+\Delta) = {} & \rho_1(t+\Delta) - \rho_2(t+\Delta) + \alpha(t+\Delta) \\ & -y_{21}(t+\Delta) - y_{22}(t+\Delta)\end{aligned}$$

Solving the system

$$\begin{cases} \nu(\omega) = 0 \\ \omega \in \Xi, \end{cases}$$

we find the following equilibrium supply prices, demand prices and shipments:

$$\begin{cases} \pi_1(t+\Delta) = \dfrac{-\alpha(t+\Delta) + 6\beta(t+\Delta) - 10}{4}, \\ \pi_2(t+\Delta) = 2\beta(t+\Delta) - 4, \\ \rho_1(t+\Delta) = 3\beta(t+\Delta) - 4, \\ \rho_2(t+\Delta) = \dfrac{\alpha(t+\Delta) + 10\beta(t+\Delta) - 14}{4}, \\ y_{11}(t+\Delta) = \dfrac{\alpha(t+\Delta) + 6\beta(t+\Delta) - 10}{4}, \\ y_{12}(t+\Delta) = \beta(t+\Delta) - 1, \\ y_{21}(t+\Delta) = \dfrac{\alpha(t+\Delta)}{2}, \\ y_{22}(t+\Delta) = \dfrac{\alpha(t+\Delta) + 2\beta(t+\Delta) - 2}{4}, \end{cases}$$

and

$$\gamma_1(\pi_1(t+\Delta),\pi_2(t+\Delta)) = \frac{3\alpha(t+\Delta)+6\beta(t+\Delta)-10}{4},$$

$$\gamma_2(\pi_1(t+\Delta),\pi_2(t+\Delta)) = \frac{\alpha(t+\Delta)+6\beta(t+\Delta)-6}{4},$$

$$\varphi_1(\rho_1(t+\Delta),\rho_2(t+\Delta)) = \frac{\alpha(t+\Delta)+10\beta(t+\Delta)-14}{4},$$

$$\varphi_2(\rho_1(t+\Delta),\rho_2(t+\Delta)) = \frac{3\alpha(t+\Delta)+2\beta(t+\Delta)-2}{4}.$$

If

$$\begin{cases} \alpha(t+\Delta)-6\beta(t+\Delta)\le -10, & \alpha(t+\Delta)+10\beta(t+\Delta)\ge 14, \\ \alpha(t+\Delta)+6\beta(t+\Delta)\ge 10, & \alpha(t+\Delta)+2\beta(t+\Delta)\ge 2, \\ \beta(t+\Delta)\ge 2, \end{cases}$$

that is

$$\text{if } \beta(t+\Delta)\ge 2 \text{ and } \alpha(t+\Delta)-6\beta(t+\Delta)\le -10,$$

then the solution lies in Ξ, otherwise we have to search the solution in the boundary of Ξ.

If we introduce as additional constraint the following one:

$$\sum_{j=1}^{2}\pi_j(t+\Delta)\le\sum_{i=1}^{2}p_i(t),$$

then it must be:

$$\pi_1(t+\Delta)+\pi_2(t+\Delta)\le p_1(t)+p_2(t),$$

which means:

$$\alpha(t+\Delta)\ge\frac{42\beta(t+\Delta)+16h(t)-32k(t)-46}{3}.$$

So the following conditions must be satisfied:

$$\gamma_1(\pi_1(t+\Delta),\pi_2(t+\Delta))\ge 12\beta(t+\Delta)+4h(t)-8k(t)-14,$$

$$\gamma_2(\pi_1(t+\Delta),\pi_2(t+\Delta))\ge\frac{15\beta(t+\Delta)+4h(t)-8k(t)-16}{3},$$

$$\varphi_1(\rho_1(t+\Delta),\rho_2(t+\Delta))\ge\frac{18\beta(t+\Delta)+4h(t)-8k(t)-22}{3},$$

$$\varphi_2(\rho_1(t+\Delta),\rho_2(t+\Delta))\ge 11\beta(t+\Delta)+4h(t)-8k(t)-12.$$

References

[1] S. Dafermos, *"Traffic Equilibrium and Variational Inequalities"*, Transportation Sc. 14 (1980), 42-54;

[2] S. Dafermos, *"Exchange price equilibrium and Variational Inequalities"*, Math. Programming 46 (1990), 391-402;

[3] P. Daniele, A. Maugeri, *"On Dynamical Equilibrium Problems and Variational Inequalities"*, Equilibrium Problems: Nonsmooth Optimization and Variational Inequality Models, Kluwer Academic Publishers, F. Giannessi - A. Maugeri - P. Pardalos Eds. (2001);

[4] P. Daniele, A. Maugeri, W. Oettli, *"Variational Inequalities and time-dependent traffic equilibria"*, C. R. Acad. Sci. Paris t. 326, serie I (1998);

[5] P. Daniele, A. Maugeri, W. Oettli, *"Time Dependent Traffic Equilibria"*, Jou. Optim. Th. Appl., Vol. 103, No. 3, 1999, pp. 543-555;

[6] P. Daniele, *"Variational Inequalities for Static Equilibrium Market. Lagrangean Function and Duality"*, Equilibrium Problems: Nonsmooth Optimization and Variational Inequality Models, Kluwer Academic Publishers, F. Giannessi - A. Maugeri - P. Pardalos Eds. (2001);

[7] M. De Luca, *Existence of Solutions for a Time–dependent Quasi–Variational Inequality*, Rend. Circ. Matem. di Palermo, Serie II, Suppl. **48**, 1997, pp. 101–106;

[8] J. Gwinner, *"Stability and Monotone Variational Inequalities with Various Applications"*, Variational Inequalities and Network Equilibrium Problems, F. Giannessi - A. Maugeri Eds. Plenum Press (1995), 123-142;

[9] J. Gwinner, *"Time Dependent Variational Inequalities - Some Recent Trends"*, Equilibrium Problems and Variational Models, Kluwer Academic Publishers, P. Daniele - F. Giannessi - A. Maugeri Eds., to appear;

[10] A. Nagurney, L. Zhao, *"Disequilibrium and Variational Inequalities"*, I Comput. Appl. Math. 33 (1990), 181-198;

[11] A. Nagurney, *"Network Economics - A Variational Inequality Approach"*, Kluwer Academic Publishers, 1993;

[12] A. Nagurney, *"Multicriteria Spatial Price Networks: Statics and Dynamics"*, Equilibrium Problems and Variational Models, Kluwer Academic Publishers, P. Daniele - F. Giannessi - A. Maugeri Eds., to appear;

[13] A.C. Pigou, *"The Economics of Welfare"*, MacMillan, London, England, 1920;

[14] M. J. Smith, *"A New Dynamic Traffic Model and the Existence and calculation of Dynamic user Equilibria on Congested Capacity-constrained Road Network"*, Transpn. Res. B 27B (1993), 49-63.

CONSTRAINED PROBLEMS OF CALCULUS OF VARIATIONS VIA PENALIZATION TECHNIQUE

Vladimir F. Demyanov *
Applied Mathematics Dept., St.Petersburg State University
Staryi Peterhof, St.Petersburg, 198904 RUSSIA

Abstract In the paper the classical constrained problems of Calculus of Variations are treated by making use of the penalty function approach. The initial constrained problem is reduced to an unconstrained one. It looks attractive from numerical considerations. It is shown that the most known classical results related to optimality conditions can be obtained from the formulated conditions. However, the related unconstrained problem is essentially nonsmooth. Nevertheless, the present state of Nonsmooth Analysis makes it possible to get proper analytical tools to overcome the arising difficulties. The main advantage of the proposed technique is its possible numerical applications for computing solutions of constrained variational problems since the conditions stated are in a form applicable for immediate implementation and the available algorithms for solving unconstrained optimization problems can effectively be employed.

Keywords: Constrained problems of Calculus of Variations, Necessary optimality conditions, Penalty Function, Exact Penalty Function, Nonsmooth Analysis.

2000 Mathematics Subject Classification. 90C30, 52A20.

1 Introduction

In the paper the classical constrained problems of Calculus of Variations ([1, 16]) are treated by making use of the penalty function approach. The idea to reduce the initial constrained problem to an unconstrained one is very old and productive: the Lagrange multipliers rule is effectively used to solve constrained variational problems (see ([1, 16, 17])). However, trying to use this rule for solving practical problems we need exact

*The research was supported by the Russian Foundation for Fundamental Studies (grant RFFI No. 00-01-00454)

values of the multipliers. This creates a serious computational problem. Another approach is to use penalty functions, especially exact penalty functions. Such an approach was first proposed by [13] and later by [21] for solving convex programming problems (see [2, 14, 15] and bibliography therein). Some new conditions for application of exact penalties were stated in [12, 6, 8]. These conditions turns out to be useful for solving problems of Optimal Control and Calculus of Variations (see [9]).

In the present paper we consider a general problem of Calculus of Variations with constraints. These constraints can be "deleted" by introducing a proper exact penalty function. The resulting new functional is essentially nonsmooth, however, it has some nice differential properties, and modern methods of Nondifferentiable Optimization and Nonsmooth Analysis (see [3, 10, 17, 19, 20]) allow to attack this problem numerically. It is demonstrated that the classical necessary optimality conditions can be derived by making use of penalty functions. However, the main advantage of the proposed technique is its possible numerical applications for computing solutions of constrained variational problems since the conditions stated are in a form applicable for immediate implementation. Note also that this approach is based on the theory of Nonsmooth Analysis which is now available.

The paper is organized as follows. In section 2 the problem is formulated. In Section 3 an equivalent statement of the problem is given. Local minima and penalty functions are discussed in Sections 4 and 5. An Exact penalty function is introduced in Section 6. Differential properties of the penalty function are studied. Necessary optimality conditions are established in Section 7. These conditions are of the Euler type since only classical variations of arcs are used. To get Weierstrass type conditions it is required to employ more sophisticated needle type variations (see [16, 18, 4, 5]).

2 Statement of the problem

Let $T > 0$ be fixed. By $P^1[0,T]$ we denote the class of functions which are continuous on $[0,T]$ and whose derivatives are piecewise continuous on $[0,T]$. If $x \in P^1[0,T]$ and $t_0 \in [0,T)$ is a point of discontinuity of the function $x'(t)$, then for the sake of definiteness it will be assumed that

$$x'(t_0) = x'_+(t_0) := \lim_{\alpha \downarrow 0} \frac{x(t_0+\alpha) - x(t)}{\alpha}. \tag{1}$$

At the point T we put

$$x'(T) = \lim_{\alpha \uparrow 0} \frac{x(T+\alpha) - x(T)}{\alpha}.$$

Let the functions $l(x, z, t)$, $\dfrac{\partial l(x, z, t)}{\partial x}$ and $\dfrac{\partial l(x, z, t)}{\partial z}$ be continuous in all their arguments on $\mathbb{R} \times \mathbb{R} \times [0, T]$.

Fix $x_0 \in \mathbb{R}$ and $x_1 \in \mathbb{R}$. Put

$$\Omega = \{x \in P^1[0, T] \mid x(0) = x_0, \ x(T) = x_1; \\ l(x(t), x'(t), t) \leq 0 \ \forall \ t \in [0, T]\}. \tag{2}$$

Assume that

$$\Omega \neq \emptyset. \tag{3}$$

Consider the functional

$$\mathcal{I}(x) = \int_0^T F(x(t), x'(t), t)\, dt, \tag{4}$$

where the functions $F(x, z, t)$, $\frac{\partial F}{\partial x}$ and $\frac{\partial F}{\partial z}$ are continuous in all their arguments on $\mathbb{R} \times \mathbb{R} \times [0, T]$.

The set of points $\{(t, x(t)) | t \in [0, T]\}$ represents an arc on the plane (t, x). Sometimes the function $x(t)$ $(t \in [0, T])$ will also be called an *arc.* An arc $x \in \Omega$ is called *admissible.*

An admissible arc $x_0 \in \Omega$ is called a *strong extremal* of the functional (4) if there exists $\varepsilon > 0$ such that

$$\mathcal{I}(x_0) \leq \mathcal{I}(x) \qquad \forall x \in \Omega \cap B_\varepsilon(x_0), \tag{5}$$

where

$$B_\varepsilon(x_0) = \{x \in P^1[0, T] \mid \max_{0 \leq t \leq T} |x(t) - x_0(t)| < \varepsilon\}. \tag{6}$$

If

$$\mathcal{I}(x_0) \leq \mathcal{I}(x) \quad \forall x \in \Omega \cap \widetilde{B}_\varepsilon(x_0),$$

where

$$\widetilde{B}_\varepsilon(x_0) = \{x \in B_\varepsilon(x_0) \mid \sup_{0 \leq t \leq T} |x'(t) - x_0'(t)| < \varepsilon\}, \tag{7}$$

then the arc $x_0 \in \Omega$ is called a *weak extremal* of the functional (4).

Every strong extremal is a weak one as well (the converse is, generally speaking, not true). *Therefore all necessary conditions for a weak extremum are necessary conditions for a strong extremum, while sufficient conditions for a strong extremum are sufficient for a weak extremum.*

3 An equivalent statement of the problem

Let us reformulate the above stated problem. Let $P[0, T]$ be the set of piecewise continuous and bounded on $[0, T]$ functions. If $t_0 \in [0, T)$ is a discontinuity point of the function $z(t)$ then for the definiteness we put

$$z(t_0) = \lim_{t \downarrow t_0} z(t). \tag{8}$$

At the point T it is assumed that

$$z(T) = \lim_{t \uparrow T} z(t). \tag{9}$$

Let us introduce the functional

$$f(z) = \int_0^T F(x_0 + \int_0^t z(\tau)\, d\tau, z(t), t)\, dt. \tag{10}$$

Put

$$h_0(T, z) := x_0 - x_1 + \int_0^T z(t) dt, \tag{11}$$

$$l(t, z) := l(x_0 + \int_0^t z(\tau)\, d\tau, z(t), t), \tag{12}$$

$$h(t, z) := \max\{0, l(t, z)\}. \tag{13}$$

Since the function $l(x, z, t)$ is continuous in its variables, then the functions $l(t, z)$ and $h(t, z)$ are piecewise continuous in t for every $z \in P[0, T]$.

Let us consider the set

$$Z = \left\{ z \in P[0, T] \;\middle|\; x_0 + \int_0^T z(t)\, dt = x_1\ ,\ l(t, z) \le 0\ \forall t \in [0, T] \right\}. \tag{14}$$

The set Z can be represented in the form

$$Z = \left\{ z \in P[0, T] \;\middle|\; h_0(T, z) = 0,\ \ h(t, z) = 0\ \forall t \in [0, T] \right\}. \tag{15}$$

Lemma 3.1 *The problem*

$$\mathcal{I}(x) \longrightarrow \min_{x \in \Omega} \tag{16}$$

is equivalent to the problem

$$f(z) \longrightarrow \min_{z \in Z} \tag{17}$$

in the following sense: if $x^* \in \Omega$ is a solution of the problem (16) then the function $z^*(t) = x^{*\prime}(t)$ is a solution of the problem (17); and vice versa, if $z^* \in Z$ is a minimizer of the function f on the set Z, then the function $x^*(t) = x_0 + \int_0^t z^*(\tau)\, d\tau$ is a solution of the problem (16).

Proof. Let $x^* \in \Omega$ and

$$\mathcal{I}(x^*) \leq \mathcal{I}(x) \quad \forall x \in \Omega.$$

Put $z^*(t) = x^{*\prime}(t)$ and take an arbitrary $z \in Z$. Then the function

$$x(t) = x_0 + \int_0^t z(\tau)\, d\tau$$

is piecewise continuously differentiable and

$$x(T) = x_1, \quad l(x(t), x'(t), t) \leq 0 \ \forall t \in [0, T],$$

i.e. $x \in \Omega$. Hence,

$$f(z^*) = \mathcal{I}(x^*) \leq \mathcal{I}(x) = f(z).$$

Since $z \in Z$ is arbitrary then $f(z^*) = \min_{z \in Z} f(z)$, i.e. z^* is a solution of the problem (17).

Now let $z^* \in Z$ and

$$f(z^*) \leq f(z) \quad \forall z \in Z.$$

Put $x^*(t) = x_0 + \int_0^t z^*(\tau)\, d\tau$. Take any $x \in \Omega$ and put $z(t) = x'(t)$. It follows from (2) that $\int_0^T z(t)\, dt = x_1 - x_0$ and $h(t, z) = 0 \ \forall t \in [0, T]$, i.e. $z \in Z$, therefore

$$\mathcal{I}(x^*) = f(z^*) \leq f(z) = \mathcal{I}(x).$$

Since $x \in \Omega$ is arbitrary then

$$\mathcal{I}(x^*) = \min_{x \in \Omega} \mathcal{I}(x),$$

i.e. x^* is a solution of the problem (16).

Thus, if $z^* \in Z$ is a global minimizer of the functional f on the set Z, then the function $x^*(t) = x_0 + \int_0^t z^*(\tau)\, d\tau \in \Omega$ and is a global minimizer of the functional $\mathcal{I}$ on the set Ω, and, conversely, if $x^* \in \Omega$ is a global minimizer of the functional $\mathcal{I}$ on the set Ω, then $z^* = x^{*\prime} \in Z$ and is a global minimizer of the function f on the set Z.

4 Local minima

Consider the problem (17). On the set $P[0, T]$ let us introduce a metric $\rho(z_1, z_2)$. As ρ it is possible to choose, for example, one of the following functions:

$$\rho_1(z_1, z_2) = \max_{t \in [0,T]} \left| \int_0^t (z_1(\tau) - z_2(\tau))\, d\tau \right|, \tag{18}$$

$$\rho_2(z_1, z_2) = \int_0^T |z_1(t) - z_2(t)|\, dt, \tag{19}$$

$$\rho_3(z_1, z_2) = \sup_{t\in[0,T]} |z_1(t) - z_2(t)|, \tag{20}$$

$$\rho_4(z_1, z_2) = \left(\int_0^T (z_1(t) - z_2(t))^2 dt\right)^{1/2}. \tag{21}$$

In (18) the function $\int_0^t (z_1(\tau) - z_2(\tau))\, d\tau$ is continuous, therefore sup is attained (therefore one can write max), while in (20) sup may not be attained.

It is not difficult to show that the functions ρ_1, ρ_2, ρ_3 and ρ_4 are metrics.

Remark 4.1 *Since the functions $\bar{\rho}_i(z) = \rho_i(0, z)$ are positively homogeneous then on the set $P[0,T]$ one can introduce a norm using one of the functions $\bar{\rho}_i(z) = \rho_i(0, z)$ $(i \in 1:4)$, and the set $P[0,T]$ becomes a normed space.*

The following relations exist between the metrics ρ_i $(i \in 1:4)$:

$$\rho_1(z_1, z_2) = \max_{t\in[0,T]} \left| \int_0^t (z_1(\tau) - z_2(\tau)\, d\tau \right| \leq \int_0^T |z_1(\tau) - z_2(\tau)|\, d\tau =$$

$$= \rho_2(z_1, z_2) \leq T \sup_{t\in[0,T]} |z_1(t) - z_2(t)| = T\rho_3(z_1, z_2),$$

$$\rho_4(z_1, z_2) \leq \left(T \sup_{t\in[0,T]} (z_1(t) - z_2(t))^2\right)^{1/2} =$$

$$= \sqrt{T} \sup_{t\in[0,T]} |z_1(t) - z_2(t)| = \sqrt{T}\rho_3(z_1, z_2),$$

i.e.

$$\rho_1(z_1, z_2) \leq \rho_2(z_1, z_2) \leq T\rho_3(z_1, z_2), \tag{22}$$

$$\rho_4(z_1, z_2) \leq \sqrt{T}\rho_3(z_1, z_2). \tag{23}$$

Thus, the metric ρ_3 majorizes the metrics ρ_4, ρ_2 and ρ_1, and the metric ρ_2 majorizes the metric ρ_1.

The inequalities (22) and (23) imply the inclusions

$$\{z \mid \rho_3(z, z_0) < \frac{\varepsilon}{T}\} \subset \{z \mid \rho_2(z, z_0) < \varepsilon\} \subset \{z \mid \rho_1(z, z_0) < \varepsilon\},$$

$$\{z \mid \rho_3(z, z_0) < \frac{\varepsilon}{\sqrt{T}}\} \subset \{z \mid \rho_4(z, z_0) < \varepsilon\}.$$

Therefore one concludes that a local minimizer of the function f on the set Z in the metric ρ_1 is a local minimizer of f in both the metrics ρ_2 and ρ_3, while a local minimizer of f in the metric ρ_2 is a local minimizer of f in the metric ρ_3 and a local minimizer of f in the metric ρ_4 is a local minimizer of f in the metric ρ_3.

It is not difficult to see that if $z^* \in Z$ is a local minimizer of the functional f on the set Z with the metric ρ_1, then the function $x^*(t) = x_0 + \int_0^t z^*(\tau)\,d\tau$ belongs to the set Ω and is a strong extremal of the functional $\mathcal{I}$ on the set Ω.

If $z^* \in Z$ is a local minimizer of the functional f on the set Z with the metric ρ_4, ρ_3 or ρ_2, then the function $x^*(t) = x_0 + \int_0^t z^*(\tau)\,d\tau$ is a weak extremal of the functional $\mathcal{I}$ on Ω (but not necessarily a strong extremal).

Conversely, if $x^* \in \Omega$ is a weak extremal of the problem (16), then the function $z^*(t) = x^{*\prime}(t)$ is a local minimizer of the function f on the set Z in the metric ρ_3.

If $x^*(t) \in \Omega$ is a strong extremal of the problem (16), then the function $z^*(t) = x^{*\prime}(t)$ is a local minimizer of the function f on Z in the metrics ρ_1, ρ_2 and ρ_3 (but not necessarily in the metric ρ_4).

Remark 4.2 *It has already been observed that the metric ρ_3 majorizes the metrics ρ_4, ρ_2 and ρ_1, while the metric ρ_2 majorizes the metric ρ_1. It is not difficult to show that neither pair of metrics (ρ_1, ρ_2), (ρ_1, ρ_3), (ρ_2, ρ_3) and (ρ_4, ρ_3) is a pair of equivalent metrics.*

Remark 4.3 *Let*

$$Z = \{z \in P[0,T] \mid \varphi(z) = 0\},$$

where $\varphi(z) \geq 0 \quad \forall z \in P[0,T]$. The inequalities (22) and (23) imply that if

$$\varphi(z) \geq a\rho_3(z, Z),$$

where $a > 0$, then

$$\varphi(z) \geq \frac{a}{\sqrt{T}}\rho_4(z, Z),$$

$$\varphi(z) \geq \frac{a}{T}\rho_2(z, Z) \geq \frac{a}{T}\rho_1(z, Z).$$

If

$$\varphi(z) \geq a\rho_2(z, Z),$$

then

$$\varphi(z) \geq a\rho_1(z, Z).$$

5 Penalty functions

Let $f : X \to I\!R$ and $\varphi : X \to I\!R_+ = \{x \in I\!R \mid x \geq 0\}$ be given functions, X be a given set. Consider the problem of minimizing the function f on the set

$$\Omega = \{x \in X \mid \varphi(x) = 0\}.$$

Assume that $\Omega \neq \emptyset$. Let us fix $\lambda \geq 0$ and introduce the function

$$F_\lambda(x) = f(x) + \lambda\varphi(x). \tag{24}$$

The function $F_\lambda(x)$ is called a *penalty function* (for the given f and φ), the value λ is called a *penalty parameter*. Put

$$F_\lambda^* = \inf_{x \in X} F_\lambda(x). \tag{25}$$

For any $\lambda \geq 0$ there exists a sequence $\{x_{\lambda k}\}$, which is minimizing for the function $F_\lambda(x)$ on X, i.e.

$$F_\lambda(x_{\lambda k}) \longrightarrow \inf_{x \in X} F_\lambda(x). \tag{26}$$

Lemma 5.1 *If*

$$\inf_{x \in X} f(x) = f^* > -\infty, \tag{27}$$

then

$$d(\lambda) = \limsup_{k \to \infty} \varphi(x_{\lambda k}) \underset{\lambda \to \infty}{\longrightarrow} 0. \tag{28}$$

Proof. By the assumption $\Omega \neq \emptyset$, therefore there exists $\bar{x} \in \Omega$. Then $F_\lambda(\bar{x}) = f(\bar{x})$, therefore

$$F_\lambda^* \leq F_\lambda(\bar{x}) = f(\bar{x}) \quad \forall \lambda \geq 0. \tag{29}$$

Assume that (28) does not hold, then a sequence $\{\lambda_s\}$ and a value $a > 0$ exist such that

$$\lambda_s \to +\infty, \quad d(\lambda_s) \geq a. \tag{30}$$

For every λ_s let us find a k_s such that the following inequalities hold:

$$\varphi(x_{\lambda_s k_s}) \geq \frac{a}{2}, \quad F_{\lambda_s}(x_{\lambda_s k_s}) = F_{\lambda_s}^* + \varepsilon_s, \tag{31}$$

where $\varepsilon_s \downarrow 0$. Due to (30) and (31)

$$F_\lambda(x_{\lambda_s k_s}) = f(x_{\lambda_s k_s}) + \lambda_s \varphi(x_{\lambda_s k_s}) \geq f^* + \lambda_s \frac{a}{2} \underset{s \to \infty}{\longrightarrow} +\infty. \tag{32}$$

The relations (29) and (31) imply

$$F_\lambda(x_{\lambda_s k_s}) = F^*_{\lambda_s} + \varepsilon_s \le f(\bar{x}) + \varepsilon_s \longrightarrow f(\bar{x}).$$

The left-hand side in (32) tends to $+\infty$. This contradiction completes the proof.

Corollary 5.1 *If for every $\lambda \ge 0$ there exists $x_\lambda \in X$ such that $F_\lambda(x_\lambda) = F^*_\lambda$, then*

$$\varphi(x_\lambda) \underset{\lambda \to \infty}{\longrightarrow} 0. \tag{33}$$

Remark 5.1 *If for some $\lambda \ge 0$ there exists $x_\lambda \in X$ such that $F_\lambda(x_\lambda) = F^*_\lambda$ and $\varphi(x_\lambda) = 0$ (i.e. $x_\lambda \in \Omega$) then x_λ is a minimizer of f on Ω.*

We have

$$F^*_\lambda = \inf_{x \in X} F_\lambda(x) \le \inf_{x \in \Omega} F_\lambda(x) = \inf_{x \in \Omega} f(x) = f^*_\Omega \quad \forall \lambda \ge 0. \tag{34}$$

Let X be a metric space with a metric ρ. Put

$$\Omega_\delta = \{x \in X \mid \varphi(x) < \delta\}, \quad \rho(y, \Omega) = \inf_{x \in \Omega}(y, x), \; r_\delta = \sup_{y \in \Omega_\delta} \rho(y, \Omega).$$

Lemma 5.2 *Let X be a metric space,*

$$\inf_{x \in X} f(x) = f^* > -\infty, \tag{35}$$

$$r_\delta \underset{\delta \downarrow 0}{\longrightarrow} 0. \tag{36}$$

and the function f be uniformly continuous on Ω_{δ_0} for some $\delta_0 > 0$. Then

$$\lim_{\lambda \to \infty} F^*_\lambda = f^*_\Omega. \tag{37}$$

Proof. Assume the contrary. Then $a > 0$ and sequences $\{\lambda_k\}$ and $\{x_k\}$ exist such that $\lambda_k \longrightarrow \infty$, $x_k \in X$,

$$f(x_k) + \lambda_k \varphi(x_k) \le f^*_\Omega - a. \tag{38}$$

It follows from (37) that

$$\varphi(x_k) \to 0. \tag{39}$$

Let us choose the sequence $\{\varepsilon_k\}$ such that $\varepsilon_k \downarrow 0$. For each x_k let us find $y_k \in \Omega$ such that

$$\rho(y_k, x_k) \leq \rho(x_k, \Omega) + \varepsilon_k \leq r_{\delta_k} + \varepsilon_k, \tag{40}$$

where $\delta_k = \varphi(x_k) + \varepsilon_k$. The relation (39) implies $\delta_k \to 0$, and due to (37) $r_{\delta_k} \longrightarrow 0$. Therefore (40) and (39) yield $\rho(y_k, x_k) \longrightarrow 0$. ¿From (38) and the uniform continuity of f on Ω_{δ_0} one concludes that for k sufficiently large

$$f(y_k) = f(x_k) + [f(y_k) - f(x_k)] < f_\Omega^* - \frac{a}{2},$$

which contradicts the definition of F_Ω^*, since $y_k \in \Omega$.

6 Exact penalty functions

Thus, we consider the problem of minimizing the functional

$$f(z) = \int_0^T F(x_0 + \int_0^t z(\tau)\, d\tau, z(t), t)\, dt \tag{41}$$

on the set $Z \subset P[0,T]$ given by the relation (14)(see also (15)). The set Z can be represented in the form

$$Z = \{z \in P[0,T] \mid \varphi(z) = 0\}, \tag{42}$$

where

$$\varphi(z) = \left[\int_0^T h^2(t,z)\, dt + h_0^2(T,z)\right]^{1/2}. \tag{43}$$

6.1 Properties of the function φ.

Let us study the differential properties of the function φ defined by (43). Let $z \in P[0,T]$ be fixed. Choose and fix an arbitrary $v \in P[0,T]$ and for $\varepsilon \geq 0$ put

$$z_\varepsilon(t) = z(t) + \varepsilon v(t), \tag{44}$$

$$x_\varepsilon(t) = x_0 + \int_0^t z_\varepsilon(\tau)\, d\tau = x(t) + \varepsilon \int_0^t v(\tau)\, d\tau. \tag{45}$$

Clearly, $x_\varepsilon \in P[0,T]$ for all $\varepsilon > 0$. Moreover, x_ε is piecewise continuously differentiable, i.e. $x_\varepsilon \in P^1[0,T]$ for all $\varepsilon > 0$

The function z_ε is piecewise continuous on $[0,T]$, i.e. $z_\varepsilon \in P[0,T]$ for all $\varepsilon > 0$. The function $\Delta z(t) = z_\varepsilon(t) - z(t) = \varepsilon v(t)$ is called the *classical variation of the arc* z, and the function $\Delta x(t) = x_\varepsilon(t) - x(t) = \varepsilon \int_0^t v(\tau)\, d\tau$ will be referred to as the *classical variation of the arc* x. We have

$$l(t, z_\varepsilon) = l(x_0 + \int_0^t z_\varepsilon(\tau)\, d\tau, z_\varepsilon(t), t) =$$

$$= l(x_0 + \int_0^t z(\tau)d\tau + \varepsilon \int_0^t v(\tau)\, d\tau, z(t) + \varepsilon v(t), t) =$$

$$= l(x_0 + \textstyle\int_0^t z(\tau)d\tau, z(t), t) + \varepsilon H_1(t, z, v) + o(\varepsilon, t) =$$

$$= l(t, z) + \varepsilon H_1(t, z, v) + o(\varepsilon, t), \tag{46}$$

where

$$H_1(t, z, v) = \frac{\partial l(t, z)}{\partial x} \int_0^t v(\tau)d\tau \ + \ \frac{\partial l(t, z)}{\partial z} v(t), \tag{47}$$

$$l(t, z) = l(x_0 + \int_0^t z(\tau)d\tau, z(t), t) = l(x, z, t),$$

$$\frac{o(\varepsilon, t)}{\varepsilon} \underset{\varepsilon \downarrow 0}{\longrightarrow} 0; \tag{48}$$

$$\frac{\partial l(t, z)}{\partial x} := \frac{\partial l(x, z, t)}{\partial x}, \quad \frac{\partial l(t, z)}{\partial z} := \frac{\partial l(x, z, t)}{\partial z},$$

calculated at the point $x = x_0 + \int_0^t z(\tau)d\tau, z = z(t)$;

$$h(t, z_\varepsilon) = \max\{0, l(t, z_\varepsilon)\} = \max\{0, l(t, z) + \varepsilon H_1(t, z, v) + o(\varepsilon, t)\} =$$

$$= h(t, z) + \varepsilon H(t, z, v) + o(\varepsilon, t), \tag{49}$$

$$H(t, z, v) = \begin{cases} H_1(t, z, v), & if\ l(t, z) > 0, \\ 0, & if\ l(t, z) < 0, \\ \max\{0, H_1(t, z, v)\}, & if\ l(t, z) = 0. \end{cases} \tag{50}$$

Let us introduce the sets

$$\sigma_+(z) = \{t \in [0, T] \mid l(t, z) > 0\ \},$$

$$\sigma_-(z) = \{t \in [0, T] \mid l(t, z) < 0\ \}, \tag{51}$$

$$\sigma_0(z) = \{t \in [0, T] \mid l(t, z) = 0\ \}.$$

It is clear that

$$\sigma_+(z) \cup \sigma_-(z) \cup \sigma_0(z) \ = \ [0, T]. \tag{52}$$

First let us consider the case $\varphi(z) > 0$ (i.e. $z \notin Z$).

Theorem 6.1 *If $z \notin Z$ then the function φ is Gâteaux differentiable.*

Proof. We have

$$\varphi(z_\varepsilon) = \left[\int_0^T h^2(t, z_\varepsilon)\, dt + h_0^2(T, z_\varepsilon) \right]^{1/2} =$$

$$= \left[\int_0^T \Big(h(t,z) + \varepsilon H(t,z,v) + o(\varepsilon, t) \Big)^2 dt + \right.$$

$$\left. + \left(h_0(T,z) + \varepsilon \int_0^T v(\tau) d\tau \right)^2 \right]^{1/2} =$$

$$= \left\{ \int_0^T h^2(t,z)\, dt + h_0^2(T,z) + \right.$$

$$\left. +2\varepsilon \left[\int_0^T h(t,z) H(t,z,v) dt + h_0(T,z) \int_0^T v(\tau) d\tau \right] + o(\varepsilon) \right\}^{1/2} =$$

$$= \varphi(z) + \varepsilon \left[\int_0^T \frac{h(t,z)}{\varphi(z)} H(t,z,v) dt + \frac{h_0(T,z)}{\varphi(z)} \int_0^T v(\tau) d\tau \right] + o(\varepsilon),$$

where

$$\frac{o(\varepsilon)}{\varepsilon} \underset{\varepsilon \downarrow 0}{\longrightarrow} 0.$$

It follows from (50) and the definition of the function h (see (13)) that

$$\varphi(z_\varepsilon) = \varphi(z) +$$
$$\varepsilon \left[\int_0^T \frac{h(t,z)}{\varphi(z)} H_1(t,z,v) dt + \frac{h_0(T,z)}{\varphi(z)} \int_0^T v(\tau) d\tau \right] + o(\varepsilon). \quad (53)$$

The relation (53) yields

$$\varphi(z_\varepsilon) = \varphi(z) + \varepsilon \varphi'(z,v) + o(\varepsilon), \quad (54)$$

where

$$\varphi'(z,v) = \lim_{\varepsilon \downarrow 0} \frac{\varphi(z_\varepsilon) - \varphi(z)}{\varepsilon} =$$
$$= \int_0^T \frac{h(t,z)}{\varphi(z)} H_1(t,z,v) dt + \frac{h_0(T,z)}{\varphi(z)} \int_0^T v(t) dt. \quad (55)$$

Taking into account (47) one gets

$$\varphi'(z,v) =$$

$$= \int_0^T \frac{h(t,z)}{\varphi(z)} \left(\frac{\partial l(t,z)}{\partial x} \int_0^t v(\tau) d\tau \; + \; \frac{\partial l(t,z)}{\partial z} v(t) \right) dt + \qquad (56)$$
$$+ \frac{h_0(T,z)}{\varphi(z)} \int_0^T v(\tau) d\tau.$$

Since

$$\int_0^T a(t) \int_0^t v(\tau) d\tau \; dt = \int_0^T \left(\int_t^T a(\tau) d\tau \right) v(t) dt \,, \qquad (57)$$

then (56) implies

$$\varphi'(z, v) = \int_0^T G(t,z) v(t) dt \,, \qquad (58)$$

where

$$G(t,z) = \frac{h_0(T,z)}{\varphi(z)} + \int_t^T \frac{h(t,z)}{\varphi(z)} \frac{\partial l(\tau,z)}{\partial x} d\tau \; + \frac{h(t,z)}{\varphi(z)} \frac{\partial l(t,z)}{\partial z}. \qquad (59)$$

The function $G(t,z)$ is piecewise continuous on $[0,T]$. The relation (58) means that at the point $z \notin Z$ the function φ is Gâteaux differentiable with the "gradient" $G \in P[0,T]$ given by the formula (59). Put

$$\frac{h_0(T,z)}{\varphi(z)} = w_0(z) \in I\!R, \quad \frac{h(t,z)}{\varphi(z)} = w_1(t,z) \in P[0,T]. \qquad (60)$$

Clearly, $w_1(t,z) \geq 0 \quad \forall t \in [0,T]$,

$$||w(z)|| = \left[\int_0^T w_1^2(t,z) \; dt + w_0^2(z) \right]^{1/2} = 1. \qquad (61)$$

Then

$$G(t,z) = w_0(z) + \frac{\partial l(t,z)}{\partial z} w_1(t,z) + \int_t^T \frac{\partial l(\tau,z)}{\partial x} w_1(\tau,z) d\tau. \qquad (62)$$

Now let us consider the case $\varphi(z) = 0$ (i.e. $z \in Z$).

Theorem 6.2 *If $z \in Z$ then the function φ is Dini directionally differentiable.*

Proof. We have

$$\varphi(z_\varepsilon) = \left[\int_0^T h^2(t, z_\varepsilon) \, dt + h_0^2(T, z_\varepsilon) \right]^{1/2} =$$

$$= \left[\int_0^T \Big(h(t,z) + \varepsilon H(t,z,v) + o(\varepsilon,t) \Big)^2 dt + \left(h_0(T,z) + \varepsilon \int_0^T v(\tau) d\tau \right)^2 \right]^{1/2}$$

Since $\varphi(z)=0$ then

$$h(t,z)=0 \ \forall t\in[0,T], \ h_0(T,z)=0.$$

Therefore

$$\varphi(z_\varepsilon)=\varepsilon||H(z,v);v|| \ + \ o(\varepsilon),$$

where

$$||H(z,v);v||=\left[\int_0^T H^2(t,z,v)dt \ + \ \left(\int_0^T v(t)dt\right)^2\right]^{1/2}.$$

This implies

$$\varphi'(z,v)=\lim_{\varepsilon\downarrow 0}\frac{\varphi(z_\varepsilon)-\varphi(z)}{\varepsilon}=||H(z,v);v||=$$

$$=\max_{||w||\le 1}\left[\int_0^T H(t,z,v)w_1(t)dt+\int_0^T v(\tau)d\tau \ w_0\right]. \tag{63}$$

Strictly speaking, in (63) it should be written sup instead of max, however, in our case (due to the piecewise continuity of H w.r. to t) the supremum is attained.

Here

$$w=[w_0,w_1], \ w_0\in I\!R, \ w_1\in P[0,T],$$

$$||w||=\left[\int_0^T w_1^2(t)\ dt+w_0^2\right]^{1/2}=1. \tag{64}$$

Since $z\in Z$, then (see (51)) $\sigma_+(z)=\emptyset$, and (50) and (51) yield

$$\varphi'(z,v)=\max_{||w||\le 1}\left[\int_{\sigma_0} w_1(t)\max_{\bar{w}(t)\in[0,1]}\Big(\bar{w}(t)H_1(t,z,v)\Big)dt+\right.$$

$$\left.+\int_0^T v(\tau)d\tau \ w_0\right]. \tag{65}$$

This expression can be rewritten as

$$\varphi'(z,v)=$$

$$=\max_{\substack{||w||\le 1\\ w_1(t)\ge 0\ \forall t\in\sigma_0(z)\\ w_1(t)=0\ \forall t\in\sigma_-(z)}}\left[\int_0^T H_1(t,z,v)w_1(t)dt \ + \ \int_0^T v(\tau)d\tau \ w_0\right].$$

The relation (47) implies

$$\varphi'(z,v)=$$

$$= \max_{w \in W(z)} \left\{ \int_0^T \left[\frac{\partial l(t,z)}{\partial x} \int_0^t v(\tau)d\tau \ + \ \frac{\partial l(t,z)}{\partial z} v(t) \right] w_1(t)dt \ + \ \int_0^T v(t)dt \ w_0 \right\}, \tag{66}$$

where

$$W(z) \ = \ \Big\{ w = [w_0, w_1] \ \Big| \ ||w|| \leq 1, \ w_0 \in I\!R,$$

$$w_1 \in P[0,T], \ w_1(t) \geq 0 \ \forall t \in \sigma_0(z), \ w_1(t) = 0 \ \forall t \in \sigma_-(z) \Big\}. \tag{67}$$

Note that the set $W(z)$ is convex. Integrating by parts in (66), we get

$$\varphi'(z,v) =$$

$$= \max_{w \in W} \int_0^T \left[\int_t^T \frac{\partial l(\tau,z)}{\partial x} w_1(\tau)d\tau + \ \frac{\partial l(t,z)}{\partial z} w_1(t) + w_0 \right] v(t)dt. \tag{68}$$

We conclude from (68) that at a point $z \in Z$ the function φ is Dini directionally differentiable, and the formula (68) shows that the function φ is subdifferentiable at z, i.e.

$$\varphi'(z,v) = \max_{A \in \partial\varphi(z)} \int_0^T A(t)v(t)dt, \tag{69}$$

where

$$\partial\varphi(z) = \Big\{ A \in P[0,T] \ \Big| \ A(t) =$$

$$= \int_t^T \frac{\partial l(\tau,z)}{\partial x} w_1(\tau)d\tau + \ \frac{\partial l(t,z)}{\partial z} w_1(t) + w_0, \tag{70}$$

$$w = [w_0, w_1] \in W(z) \Big\},$$

and the set $W(z)$ is defined in (67). Observe that the set $\partial\varphi(z)$ is a convex one.

Remark 6.1 *For the functional $f(z)$ the following expansion holds:*

$$f(z_\varepsilon) = f(z) + \varepsilon \int_0^T \left[\frac{\partial F(t,z)}{\partial x} \int_0^t v(\tau)d\tau \ + \ \frac{\partial F(t,z)}{\partial z} v(t) \right] dt \ + o(\varepsilon) =$$

$$= f(z) + \varepsilon \int_0^T \left[\int_t^T \frac{\partial F(\tau,z)}{\partial x} d\tau + \ \frac{\partial F(t,z)}{\partial z} \right] v(t)dt \ + \ o(\varepsilon), \tag{71}$$

where

$$\frac{o(\varepsilon)}{\varepsilon} \underset{\varepsilon \downarrow 0}{\longrightarrow} 0, \ F(t,z) = F(x_0 + \int_0^t z(\tau)d\tau, z(t), t).$$

6.2 Properties of the function G.

Consider again the case $z \notin Z$. It has been shown in 6.1 that the function φ is Gâteaux diffrentiable at the point $z \notin Z$, and its "gradient" $G(t,z)$ takes the form (62). Let us study the behaviour of the function $G(t,z)$ under different assumptions.

Case 2a. Assume that in (2)

$$l(x(t), x'(t), t) \leq 0 \ \forall \ t \in [0, T_1), \tag{72}$$

where $T_1 < T$ (i.e. the constraint in (72) takes place only on the interval $[0, T_1)$). Then (56) is of the form

$$\varphi'(z,v) =$$

$$= \int_0^{T_1} \frac{h(t,z)}{\varphi(z)} \left(\frac{\partial l(t,z)}{\partial x} \int_0^t v(\tau) d\tau \ + \ \frac{\partial l(t,z)}{\partial z} v(t) \right) dt + \tag{73}$$
$$+ \frac{h_0(T,z)}{\varphi(z)} \int_0^T v(\tau) d\tau.$$

As above, we get

$$\varphi'(z,v) = \int_0^T G(t,z) v(t) dt \ , \tag{74}$$

where

$$G(t,z) =$$

$$= \begin{cases} \dfrac{h_0(T,z)}{\varphi(z)} + \displaystyle\int_t^{T_1} \dfrac{h(t,z)}{\varphi(z)} \dfrac{\partial l(\tau,z)}{\partial x} d\tau \ + \dfrac{h(t,z)}{\varphi(z)} \ \dfrac{\partial l(t,z)}{\partial z}, & t \in [0, T_1), \\[2ex] \dfrac{h_0(T,z)}{\varphi(z)}, & t \in [T_1, T]. \end{cases} \tag{75}$$

Putting

$$\frac{h_0(T,z)}{\varphi(z)} = w_0(z) \in I\!R, \quad \frac{h(t,z)}{\varphi(z)} = w_1(t) \quad \forall t \in [0, T_1),$$

we get

$$G(t,z) =$$

$$= \begin{cases} w_0(z) + \displaystyle\int_t^{T_1} \dfrac{\partial l(\tau,z)}{\partial x} w_1(\tau,z) d\tau \ + \dfrac{\partial l(t,z)}{\partial z} \ w_1(t,z), & t \in [0, T_1), \\[2ex] w_0(z), & t \in [T_1, T], \end{cases} \tag{76}$$

where

$$||w(z)|| = \left[\int_0^{T_1} w_1^2(t,z)\ dt + w_0^2(z) \right]^{1/2} = 1. \tag{77}$$

Theorem 6.3 *Let $S_i \subset P[0,T]$ be a bounded open set (in the metric ρ_i). If condition (72) with $T_1 < T$ holds and*

$$\left| \frac{\partial l(t,z)}{\partial z} \right| \geq b > 0 \quad \forall t \in [0,T_1],\ \forall z \in S_i \setminus Z \tag{78}$$

then there exist $a_1 > 0$ and $a_2 > 0$ such that

$$||G(z)|| = \left[\int_0^{T} G^2(t,z)\ dt \right]^{1/2} \geq a_1 \quad \forall z \in S_i \setminus Z, \tag{79}$$

$$\sup_{t\in[0,T]} |G(t,z)| \geq a_2. \tag{80}$$

Proof. First let us prove (79). Assume that (79) does not hold. Then there exists a sequence $\{z_k\}$ such that

$$z_k \in S_i \setminus Z \quad \forall k,\ ||G(z_k)|| \underset{k\to\infty}{\longrightarrow} 0. \tag{81}$$

Since

$$||G(z_k)||^2 = |w_0(z_k)|^2(T - T_1) + \int_0^{T_1} G^2(t,z_k)dt$$

then it follows from (81) that

$$w_0(z_k) \underset{k\to\infty}{\longrightarrow} 0. \tag{82}$$

Taking into account (78) and the boundedness of $\dfrac{\partial l(t,z)}{\partial x}$, we conclude from the properties of solutions of integral equations that

$$||w_1(t,z_k)|| \underset{k\to\infty}{\longrightarrow} 0. \tag{83}$$

However, (82) and (83) contradict (61). This contradiction completes the proof of (79).

Now let us prove (80). Assume that (80) does not hold. Then there exists a sequence $\{z_k\}$ such that

$$z_k \in S_i \setminus Z \quad \forall k,\ \sup_{t\in[0,T]} |G(t,z_k)| \underset{k\to\infty}{\longrightarrow} 0. \tag{84}$$

It follows from (81) and (77) that

$$w_0(z_k) \underset{k\to\infty}{\longrightarrow} 0. \tag{85}$$

Taking into account (78) and the boundedness of $\dfrac{\partial l(t,z)}{\partial x}$, we conclude from the properties of solutions of integral equations that

$$||w_1(t,z_k)|| \underset{k\to\infty}{\longrightarrow} 0. \tag{86}$$

However, (85) and (86) contradict (61). This contradiction completes the proof of (80) and that of the theorem.

Remark 6.2 *Theorem 6.3 is valid also in the case where in (2)*

$$l(x(t), x'(t), t) \leq 0 \ \forall \ t \in [0, T_1) \cup [T_2, T], \tag{87}$$

where $0 \leq T_1 < T_2 \leq T$ *(i.e. on the interval* $[T_1, T_2)$ *there is no constraint).*

Case 2b. Assume that the function $l(x,z,t)$ in (2) does not depend on z.

Theorem 6.4 *Let* $S_i \subset P[0,T]$ *be a bounded open set (in the metric* ρ_i*). If the function* $l(x,z,t)$ *in (2) does not depend on* z *and the condition*

$$\left|\frac{\partial l(t,z)}{\partial x}\right| \geq b > 0 \quad \forall t \in [0,T], \ \forall z \in S_i \setminus Z \tag{88}$$

holds then there exists $a_2 > 0$ *such that*

$$\sup_{t\in[0,T]} |G(t,z)| \ \geq a_2. \tag{89}$$

Proof. Since $l(x,z,t)$ does not depend on z then it follows from (62) that

$$G(t,z) = w_0(z) + \int_t^T \frac{\partial l(\tau,z)}{\partial x} w_1(\tau,z)d\tau. \tag{90}$$

Assume that (89)does not hold. As in the proof of (6.3) then there exists a sequence $\{z_k\}$ such that

$$z_k \in S_i \setminus Z \quad \forall k, \ \sup_{t\in[0,T]} |G(t,z_k)| \underset{k\to\infty}{\longrightarrow} 0. \tag{91}$$

The relation (90) yields $G(T, z_k) = w_0(z_k)$. Therefore (see (91))

$$w_0(z_k) \underset{k\to\infty}{\longrightarrow} 0. \tag{92}$$

Taking into account (88),(90) and (91), we again conclude that

$$||w_1(t, z_k)|| \underset{k\to\infty}{\longrightarrow} 0. \tag{93}$$

However, (92) and (93) contradict (61). Thus, (89) is proved.
The relation (89) can be proved in an analogous way.

Case 2c. Assume that there exist t_1 and t_2 such that

$$0 \le t_1 < t_2 \le T, \quad l(t, z_0) < 0 \quad \forall t \in [t_1, t_2].$$

Then for every $i \in 1:4$ there exists $\delta_i > 0$ such that

$$l(t, z) < 0 \quad \forall t \ \in [t_1, t_2], \quad \forall z \in S_i = \{z \in P[0, T] \mid \rho_i(z, z_0) < \delta_i\}.$$

Theorem 6.5 *If*

$$\left|\frac{\partial l(t, z)}{\partial z}\right| \ge b > 0 \quad \forall t \in [0, t_1] \cup [t_1, T], \ \forall z \in S_i \setminus Z \tag{94}$$

then there exist $a_1 > 0$ and $a_2 > 0$ such that

$$||G(z)|| = \left[\int_0^T G^2(t, z)\, dt\right]^{1/2} \ge a_1 \quad \forall z \in S_i \setminus Z, \tag{95}$$

$$\sup_{t\in[0,T]} |G(t, z)| \ge a_2. \tag{96}$$

Proof is analogous to that of Theorems 6.3 and 6.4.

6.3 The rate of descent of the function φ.

Thus, Theorems 6.3, 6.4 and 6.5 imply that in the cases 2a and 2c there exist $a_1 > 0$ and $a_2 > 0$ such that

$$||G(z)|| \ge a_1 \quad \forall z \in S_i \setminus Z, \tag{97}$$

$$b(z) = \sup_{t\in[0,T]} |G(t, z)| \ge a_2 \quad \forall z \in S_i \setminus Z. \tag{98}$$

In the case 2b there exists a_2 such that (98) holds. Fix $z \in S_i \setminus Z$. Due to (98), one can find $t_1(z) \in [0,T]$ and $t_2(z) \in [0,T]$ such that

$$t_1(z) < t_2(z),$$

$$|G(t,z)| \geq \frac{1}{2}b(z) \quad \forall t \in \sigma_z = [t_1(z), t_2(z)]. \tag{99}$$

Put $\gamma(z) = t_2(z) - t_1(z)$. For every $\alpha \geq 0$ take $z_\alpha(t) = z(t) + \alpha v_z(t)$, where

$$v_z(t) = \begin{cases} -G(t,z), & t \in \sigma_z, \\ 0, & t \in [0,T] \setminus \sigma_z. \end{cases} \tag{100}$$

Then $z_\alpha \in P[0,T]$ and (see (54) and (58))

$$\varphi(z_\alpha) =$$

$$= \varphi(z) + \alpha \int_0^T G(t,z)v_z(t)dt + o(\alpha) \leq \varphi(z) - \alpha\gamma(z)\frac{b^2(z)}{4} + o(\alpha). \tag{101}$$

We have

$$\rho_1(z_\alpha, z) = \alpha \max_{t\in[0,T]} \left| \int_0^t v_z(\tau)d\tau \right| \leq$$

$$\leq \alpha \max_{t\in[0,T]} \int_0^t |v_z(\tau)|d\tau = \alpha \int_0^T |v_z(\tau)|d\tau \leq \alpha\gamma(z)b(z), \tag{102}$$

$$\rho_2(z_\alpha, z) = \alpha \int_0^T |v_z(\tau)|d\tau \leq \alpha\gamma(z)b(z), \tag{103}$$

$$\rho_3(z_\alpha, z) = \alpha \sup_{t\in[0,T]} |v_z(t)| = \alpha b(z). \tag{104}$$

The relations (101), (102) and (98) yield

$$\varphi_1^{\downarrow}(z) = \liminf_{z'\to z} \frac{\varphi(z') - \varphi(z)}{\rho_1(z', z)} \leq$$

$$\leq \liminf_{\alpha\downarrow 0} \frac{\varphi(z_\alpha) - \varphi(z)}{\rho_1(z_\alpha, z)} \leq -\frac{b(z)}{4} \leq -\frac{a_2}{4} \qquad \forall z \in S_1 \setminus Z. \tag{105}$$

In a similar way it follows from (101), (103) and (98) that

$$\varphi_2^{\downarrow}(z) = \liminf_{z'\to z} \frac{\varphi(z') - \varphi(z)}{\rho_2(z', z)} \leq$$

$$\leq \liminf_{\alpha\downarrow 0} \frac{\varphi(z_\alpha) - \varphi(z)}{\rho_2(z_\alpha, z)} \leq -\frac{b(z)}{4} \leq -\frac{a_2}{4} \qquad \forall z \in S_2 \setminus Z. \tag{106}$$

To get an analogous inequality for the metric

$$\rho_4(z', z) = \left(\int_0^T (z'(t) - z(t))^2 dt \right)^{1/2}$$

(if (97) holds), let us take $z_\alpha(t) = z(t) - \alpha G(t, z)$. Then

$$\varphi(z_\alpha) = \varphi(z) - \alpha \int_0^T G^2(t, z) dt + o(\alpha) = \varphi(z) - \alpha ||G(z)||^2 + o(\alpha), \quad (107)$$

$$\rho_4(z_\alpha, z) = \left(\int_0^T (z_\alpha(t) - z(t))^2 dt \right)^{1/2} = \alpha ||G(z)||. \quad (108)$$

Taking into account (97), one concludes that

$$\varphi_4^\downarrow(z) = \liminf_{z' \to z} \frac{\varphi(z') - \varphi(z)}{\rho_4(z', z)} \le$$

$$\le \liminf_{\alpha \downarrow 0} \frac{\varphi(z_\alpha) - \varphi(z)}{\rho_4(z_\alpha, z)} = -||G(z)|| \le -a_1 \qquad \forall z \in S_4 \setminus Z. \quad (109)$$

Remark 6.3 *It is impossible to get an inequality similar to (105), (106) and (109) for the metric ρ_3 since $\gamma(z)$ may take any small (though positive) values. It is due to the fact that the function $G(t, z)$ in (62) is bounded for every specific z but this boundedness is not uniform in $z \in S_3$ (see the definition (60) of the function $w(z)$).*

6.4 An Exact Penalty function.

Thus, we consider the problem of minimizing the functional

$$f(z) = \int_0^T F(x_0 + \int_0^t z(\tau)\, d\tau, z(t), t)\, dt$$

on the set $Z \subset P[0, T]$, described by the relation (42). Let $z_0 \in Z$ be a local minimizer of the function f on the set Z in the metric ρ_i. Assume that in some neighbourhood $B_{i\delta}(z_0) = \{z \in P[0, T] \mid \rho_i(z, z_0) < \delta\}$ of the point z_0 the function φ satisfies the condition

$$\varphi_i^\downarrow(z) \le -a < 0 \qquad \forall z \in B_{i\delta}(z_0) \setminus Z. \quad (110)$$

Some of such conditions are stated in Theorems 6.3 and 6.4.

If follows from [7], Lemma III.4.3, that (110) implies (under some additional conditions)

$$\varphi(z) \ge a \rho_i(z, Z) \qquad \forall z \in B_{i\delta}(z_0) \setminus Z \quad (111)$$

where

$$\rho_i(z, Z) = \inf_{y \in Z} \rho_i(z, y). \tag{112}$$

Now it follows from [7], Theorem III.4.6, that if the function f is Lipschitz on $B_{i\delta}(z_0)$ in the metric ρ_i then there exists $\lambda^* < \infty$ such that for every $\lambda > \lambda^*$ the point z_0 is a local (unconstrained) minimizer of the function $\Phi_\lambda(z) = f(z) + \lambda\varphi(z)$ in the metric ρ_i.

7 Necessary conditions for an Extremum

7.1 Necessary conditions generated by classical variations.

Let $z_0 \in Z$ be a local minimizer of the function f on the set Z in the metric ρ_i. Assume that in some neighbourhood $B_{i\delta}(z_0)$ of the point z_0 the function φ satisfies the condition (110) and assume also that f is Lipschitz on $B_{i\delta}(z_0)$ in the metric ρ_i. Then, as it has just been observed, for $\lambda > \lambda^*$ the point z_0 is a local unconstrained minimizer of the function $\Phi_\lambda(z)$ in the metric ρ_i. Fix $v \in P[0, T]$ and for $\varepsilon \geq 0$ put

$$z_\varepsilon(t) = z_0(t) + \varepsilon v(t). \tag{113}$$

Note that

$$\rho_i(z_\varepsilon, z_0) \underset{\varepsilon \downarrow 0}{\longrightarrow} 0 \quad \forall i \in 1:4. \tag{114}$$

Then (69) and (71) yield

$$\Phi_\lambda(z_\varepsilon) = \Phi_\lambda(z_0) + \varepsilon\Big\{ \int_0^T \left[\int_t^T \frac{\partial F(\tau, z_0)}{\partial x} d\tau + \frac{\partial F(t, z_0)}{\partial z} \right] v(t)dt +$$

$$+\lambda \max_{A \in \partial\varphi(z_0)} \int_0^T A(t)v(t)dt \Big\} + o(\varepsilon, v), \tag{115}$$

where

$$\frac{o(\varepsilon, v)}{\varepsilon} \underset{\varepsilon \downarrow 0}{\longrightarrow} 0 \quad \forall v \in P[0, T],$$

$$\partial\varphi(z_0) = \Big\{ A \in P[0, T] \ \Big| \ A(t) =$$

$$= \int_t^T \frac{\partial l(\tau, z_0)}{\partial x} w_1(\tau)d\tau + \frac{\partial l(t, z_0)}{\partial z} w_1(t) + w_0, \tag{116}$$

$$w = [w_0, w_1] \in W(z_0) \Big\},$$

$$W(z_0) = \Big\{ w = [w_0, w_1] \;\Big|\; ||w|| \le 1,\ w_0 \in I\!R,$$

$$w_1 \in P[0,T],\ w_1(t) \ge 0\ \forall t \in \sigma_0(z_0), \tag{117}$$
$$w_1(t) = 0\ \forall t \in \sigma_-(z_0) \Big\},$$

$$\sigma_-(z_0) = \{ t \in [0,T] \mid l(t, z_0) < 0 \}, \tag{118}$$

$$\sigma_0(z_0) = \{ t \in [0,T] \mid l(t, z_0) = 0 \}, \tag{119}$$

$$l(t, z_0) = l(x_0 + \int_0^t z_0(\tau)\, d\tau, z_0(t), t). \tag{120}$$

Since $z_0 \in Z$ then

$$\sigma_-(z_0) \cup \sigma_0(z_0) = [0, T]. \tag{121}$$

It follows from (115) that

$$\Phi_\lambda(z_\varepsilon) = \Phi_\lambda(z_0) + \varepsilon \Phi'_\lambda(z_0, v) + o(\varepsilon, v), \tag{122}$$

where

$$\Phi'_\lambda(z_0, v) = \max_{C \in \partial\Phi_\lambda(z_0)} \int_0^T C(t) v(t) dt, \tag{123}$$

$$\partial\Phi_\lambda(z_0) = \Big\{ C \in P[0,T] \;\Big|\; C(t) = \int_t^T \frac{\partial F(\tau, z_0)}{\partial x} d\tau + \frac{\partial F(t, z_0)}{\partial z} +$$

$$+ \int_t^T \frac{\partial l(\tau, z_0)}{\partial x} \lambda w_1(\tau) d\tau + \frac{\partial l(t, z_0)}{\partial z} \lambda w_1(t) + \lambda w_0, \tag{124}$$
$$w = [w_0, w_1] \in W(z_0) \Big\}.$$

Since the point z_0 is a local unconstrained minimizer of the function $\Phi_\lambda(z)$ in the metric ρ_i and since (see (114)) $\rho_i(z_\varepsilon, z_0) \underset{\varepsilon \downarrow 0}{\longrightarrow} 0 \quad \forall i \in 1:4$, then (see [10, 20]) the following necessary condition for a minimum should hold:

$$\Phi'_\lambda(z_0, v) \ge 0 \quad \forall v \in P[0, T]. \tag{125}$$

This condition is equivalent to

$$\inf_{A \in \partial\Phi_\lambda(z_0)} ||A|| = 0. \tag{126}$$

It follows from (126) that there exists a sequence $\{A_k\}$, such that

$$A_k \in \partial\Phi_\lambda(z_0), \quad ||A_k|| \to 0.$$

It means that a sequence $\{w_k\}$ exists such that $w_k = [w_{0k}, w_{1k}] \in W$,

$$\int_t^T \frac{\partial F(\tau, z_0)}{\partial x} d\tau + \frac{\partial F(t, z_0)}{\partial z} +$$

$$+ \int_t^T \frac{\partial l(\tau, z_0)}{\partial x} \lambda w_{1k}(\tau) d\tau + \frac{\partial l(t, z_0)}{\partial z} \lambda w_{1k}(t) + \lambda w_{0k} \tag{127}$$
$$= A_k(t) \quad \forall t \in P[0, T],$$

$$||A_k|| \to 0. \tag{128}$$

Put

$$\Psi_k(t) = \lambda w_{1k}(t), \quad C_{0k} = \lambda w_{0k}. \tag{129}$$

Then (127) implies

$$\int_t^T \frac{\partial F(\tau, z_0)}{\partial x} d\tau + \frac{\partial F(t, z_0)}{\partial z} +$$

$$+ \int_t^T \frac{\partial l(\tau, z_0)}{\partial x} \Psi_k(\tau) d\tau + \frac{\partial l(t, z_0)}{\partial z} \Psi_k(t) + C_{0k} \tag{130}$$
$$= A_k(t) \quad \forall t \in P[0, T].$$

Since $\frac{\partial F(t,z_0)}{\partial x}, \frac{\partial F(t,z_0)}{\partial z}, \frac{\partial l(t,z_0)}{\partial x}, \frac{\partial l(t,z_0)}{\partial z}$ are piecewise continuous and bounded on $[0, T]$, then it follows from (128) and the properties of solutions of integral equations that a nonnegative function $\Psi(t)$ and a constant C_0 exist such that

$$\int_t^T \frac{\partial F(\tau, z_0)}{\partial x} d\tau + \frac{\partial F(t, z_0)}{\partial z} +$$

$$+ \int_t^T \frac{\partial l(\tau, z_0)}{\partial x} \Psi(\tau) d\tau + \frac{\partial l(t, z_0)}{\partial z} \Psi(t) + C_0 = 0 \quad \forall t \in P[0, T]. \tag{131}$$

Remark 7.1 *The sequence $\{w_k\}$ is bounded, and formally it is easy to get condition (131) choosing a convergent subsequence of the sequence $\{w_k\}$. However, in general, this fact (the existence of a convergent subsequence) does not hold in the infinite-dimensional spaces. Nevertheless, we got this result by making use of the properties of solutions of integral equations.*

Thus, condition (126) can be rewritten in the form

$$\mathbf{0} \in \partial \Phi_\lambda(z_0), \tag{132}$$

where $\mathbf{0} \in P[0, T]$ is the zero element of the space $P[0, T]$.

We conclude from (117) and (131) that

$$\int_t^T \frac{\partial F(\tau, z_0)}{\partial x} d\tau + \frac{\partial F(t, z_0)}{\partial z} = C \quad \forall t \in \sigma_-(z_0). \tag{133}$$

Here C depends on the interval of $\sigma_-(z_0)$ where t is located.

The condition (133) is the *Euler integral condition.* If $t \in \sigma_-(z_0)$ and the function z_0 is continuous at t then, differentiating (133), we get the *Euler condition*:

$$\frac{d}{dt}\frac{\partial F(t, z_0)}{\partial z} - \frac{\partial F(t, z_0)}{\partial x} = 0 \quad \forall t \in \sigma_-(z_0). \tag{134}$$

If $t \in int\ \sigma_0(z_0)$ and the function z_0 is continuous at t then, differentiating (133), we get

$$\frac{\partial l(t, z_0)}{\partial z}\dot{\Psi}(t) + \left[\frac{d}{dt}\frac{\partial l(t, z_0)}{\partial z} - \frac{\partial l(t, z_0)}{\partial x}\right]\Psi(t) +$$

$$+ \frac{d}{dt}\frac{\partial F(t, z_0)}{\partial z} - \frac{\partial F(t, z_0)}{\partial x} = 0 \quad \forall t \in int\ \sigma_0(z_0), \tag{135}$$

$$\frac{\partial l(T, z_0)}{\partial z}\Psi(T) = -C_0 - \frac{\partial F(T, z_0)}{\partial z}. \tag{136}$$

The above results can be formulated as the following statement.

Theorem 7.1 *Assume that in some neighbourhood $B_{i\delta}(z_0)$ of a point $z_0 \in Z$ the function φ satisfies the condition (110) and let the function f be Lipschitz on $B_{i\delta}(z_0)$ in the metric ρ_i. Then for the point z_0 to be a local minimizer of f on Z in the metric ρ_i, it is necessary that a constant $C_0 \in I\!R$ and a nonnegative function $\Psi(t) \in P[0, T]$ exist such that*

$$\Psi(t) = 0 \quad \forall t \in \sigma_-(z_0), \tag{137}$$

$$\int_t^T \frac{\partial F(\tau, z_0)}{\partial x} d\tau + \frac{\partial F(t, z_0)}{\partial z} +$$

$$+ \int_t^T \frac{\partial l(\tau, z_0)}{\partial x}\Psi(\tau) d\tau + \frac{\partial l(t, z_0)}{\partial z}\Psi(t) + C_0 = 0 \quad \forall t \in P[0, T]. \tag{138}$$

7.2 Discussion and Remarks

It has been shown in Corollary 5.1 that if for every $\lambda \geq 0$ there exists $z_\lambda \in P[0, T]$ such that

$$F_\lambda(z_\lambda) = F_\lambda^* = \inf_{z \in P[0,T]} F_\lambda(z)$$

and if

$$f^* = \inf_{z\in P[0,T]} f(z) > -\infty$$

then

$$\varphi(z_\lambda) \underset{\lambda\to\infty}{\longrightarrow} 0, \quad \lim_{\lambda\to\infty} F_\lambda^* = \inf_{z\in Z} f(z) = f_Z^*. \tag{139}$$

Take a $\lambda \geq 0$. If $\varphi(z) = 0$ then (see Remark 5.1) z_λ is a minimizer of f on Z (and then condition (138) holds). Otherwise, if $\varphi(z_\lambda) > 0$, then the function φ is Gâteaux dufferentiable at the point z_λ, and, hence, the function $F_\lambda(z)$ is Gâteaux dufferentiable at z_λ with the "gradient"

$$F'_\lambda(t,z_\lambda) = \int_t^T \frac{\partial F(\tau,z_\lambda)}{\partial x}d\tau + \frac{\partial F(t,z_\lambda)}{\partial z} + \lambda G(t,z_\lambda),$$

where $G(t,z_\lambda)$ is given by formula (62). At the point z_λ the necessary condition

$$F'_\lambda(t,z_\lambda) = 0 \quad \forall t\in[0,T]$$

must be satisfied, i.e.

$$\int_t^T \frac{\partial F(\tau,z_\lambda)}{\partial x}d\tau + \frac{\partial F(t,z_\lambda)}{\partial z} +$$

$$+\int_t^T \frac{\partial l(\tau,z_\lambda)}{\partial x}\lambda w_1(\tau,z_\lambda)d\tau + \frac{\partial l(t,z_\lambda)}{\partial z}\lambda w_1(t,z_\lambda) + \tag{140}$$

$$+\lambda w_0(z_\lambda) = 0 \quad \forall t\in[0,T],$$

where

$$w_0(z_\lambda) = \frac{h_0(T,z_\lambda)}{\varphi(z_\lambda)} \in I\!R, \quad w_1(t,z_\lambda) = \frac{h(t,z_\lambda)}{\varphi(z_\lambda)}. \tag{141}$$

Putting

$$\Psi_\lambda(t) = \lambda w_1(t,z_\lambda), \quad C_{0\lambda} = \lambda w_0(z_\lambda), \tag{142}$$

we get from (140)

$$\int_t^T \frac{\partial F(\tau,z_\lambda)}{\partial x}d\tau + \frac{\partial F(t,z_\lambda)}{\partial z} +$$

$$+\int_t^T \frac{\partial l(\tau,z_\lambda)}{\partial x}\Psi_\lambda(\tau)d\tau + \frac{\partial l(t,z_\lambda)}{\partial z}\Psi_\lambda(t) + C_{0\lambda} = 0 \quad \forall t\in[0,T]. \tag{143}$$

Formally condition (143) is of the same type as condition (138). The difference is in the way of definition of the pairs $[C_{0\lambda},\Psi_\lambda]$ and $[C_0,\Psi]$.

The pair $[C_{0\lambda}, \Psi_\lambda]$ is defined by formulas (141) and (142) while the pair $[C_0, \Psi]$ is defined by relation (131). Observe that

$$||C_{0\lambda}, \Psi_\lambda|| = \left[C_{0\lambda}^2 + \int_0^T \Psi_\lambda^2(t)dt\right]^{1/2} = \lambda,$$

while $||C_0, \Psi|| \in [0, \lambda]$. Note also, that $\Psi(t) \geq 0$, $\Psi_\lambda(t) \geq 0$ and the function $\Psi_\lambda(t)$ is well-defined by (141) and (142).

Remark 7.2 *It has already been observed that on the set $\sigma_-(z_0)$ the condition (138) becomes the classical Euler condition (see (133) and (134)).*

Remark 7.3 *In the cases 2a, 2b and 2c the condition (110) holds for the metrics ρ_1, ρ_2 and ρ_4. It means, in particular, that the condition (138) is a necessary condition for a point z_0 to be a strong extremal. If condition (110) holds for the metric ρ_3 then the condition (138) is a necessary condition for a point z_0 to be a weak extremal.*

Remark 7.4 *If the condition 2b holds (i.e. l does not depend on z) then (90) and (138) imply*

$$\int_t^T \frac{\partial F(\tau, z_0)}{\partial x} d\tau + \frac{\partial F(t, z_0)}{\partial z} +$$

$$+ \int_t^T \frac{\partial l(\tau, z_0)}{\partial x} \Psi(\tau) d\tau + C_0 = 0 \quad \forall t \in P[0, T], \tag{144}$$

where

$$\Psi(t) \geq 0 \quad \forall t \in [0, T], \tag{145}$$

$$\Psi(t) = 0 \quad \forall t \in \sigma_-(z_0). \tag{146}$$

It follows from (144) that

$$\frac{\partial l(t, z_0)}{\partial x} \Psi(t) = -\frac{\partial F(t, z_0)}{\partial x} + \frac{d}{dt} F(t, z_0) \quad \forall t \in int\ \sigma_0, \tag{147}$$

$$C_0 = -\frac{\partial F(T, z_0)}{\partial z}. \tag{148}$$

For $t \in \sigma_-(z_0)$ condition (134) is satisfied.

Remark 7.5 *In the present paper we have employed the classical variations of z (see (44)). To get more sophisticated optimality conditions it is required to use the so-called needle variations of different types (see [16, 18, 4, 5]).*

Remark 7.6 *The described approach can be extended to the case of a vector-valued function f and several constraints of the equality– or inequality–type.*

Remark 7.7 *Consider the functional*

$$f_1(z) = f(z) + \int_0^T \Psi(t)l(t,z)dt + C_0 h_0(T,z) =$$

$$= \int_0^T F(x_0 + \int_0^t z(\tau)\,d\tau, z(t), t)\,dt +$$

$$+ \int_0^T \Psi(t)l(x_0 + \int_0^t z(\tau)\,d\tau, z(t), t)\,dt \; + \; C_0\Big(x_0 - x_1 + \int_0^T z(t)dt\Big),$$

where $\Psi(t)$ and C_0 are from Theorem 7.1. The condition (138) means that z_0 is a stationary point of the functional f_1, i.e. $\Psi(t)$ and C_0 can be viewed as the Lagrange multipliers of the problem:

Minimize f subject to

$$h_0(T,z) = 0, \qquad l(t,z) \le 0 \quad \forall t \in [0,T].$$

Remark 7.8 *We have demonstrated that the classical necessary optimality conditions can be derived by making use of the penalty function approach. However, the main advantage of the proposed technique is its possible numerical applications for computing solutions of constrained variational problems since the conditions stated are in a form applicable for numerical implementation. Note also that this approach is based on the theory of Nonsmooth Analysis which is now available (see, e.g., [3, 10, 17, 19, 20]).*

Acknowledgment The author is thankful to Prof. F. Giannessi and G. Tamasyan for useful comments.

References

[1] G. A. Bliss, *Lectures on the Calculus of Variations*, Chicago, University of Chicago Press, 1946.

[2] D. Boukary, A.V. Fiacco, Survey of penalty, exact-penalty and multiplier methods from 1968 to 1993. *Optimization*, Vol. 32, No. 4, pp. 301–334, 1995.

[3] F.H. Clarke, *Optimization and Nonsmooth Analysis.* New York, Wiley Interscience, 1983.

[4] V.F. Demyanov, Nonsmooth problems in Calculus of variations. In *Lecture Notes in Economics and Math. Systems. v. 382. Advances in Optimization*, Eds. W.Oettli, D.Pallaschke, pp. 227–238, Berlin, Springer Verlag, 1992.

[5] V.F. Demyanov, Calculus of variations in nonsmooth presentation. In *Nonsmooth Optimization: Methods and Applications.* Ed. F.Giannessi, pp. 76–91. Singapore, Gordon and Breach, 1992a.

[6] V.F. Demyanov, Exact penalty functions in nonsmooth optimization problems. *Vestnik of St. Petersburg University*, ser. 1, issue 4 (No. 22), pp. 21-27, 1994.

[7] V.F. Demyanov, *Conditions for an extremum and variational problems*, St. Petersburg, St. Petersburg University Press, 2000.

[8] V.F. Demyanov, G. Di Pillo, F. Facchinei, Exact penalization via Dini and Hadamard conditional derivatives, *Optimization Methods and Software*, Vol. 9, pp. 19–36, 1998.

[9] V.F. Demyanov, F. Giannessi, V.V. Karelin, Optimal Control Problems via Exact Penalty Functions, *Journal of Global Optimization*, Vol. 12, No. 3, pp.215–223, 1998.

[10] V.F. Demyanov, A.M. Rubinov, *Constructive Nonsmooth Analysis*, Frankfurt a/M., Peter Lang Verlag, 1995.

[11] G. Di Pillo, L. Grippo, On the Exactness of a class of Nondifferentiable Penalty functions. *J. Optim. Theory Appl.*, Vol. 57, pp. 397–408, 1988.

[12] G. Di Pillo, F. Facchinei, Exact penalty functions for nondifferentiable programming problems. In *Nonsmooth Optimization and Related Topics*, Eds. F.H. Clarke, V.F. Demyanov and F. Giannessi, pp. 89-107. New York, Plenum, 1989.

[13] I.I. Eremin, The method of penalties in convex programming. *Doklady Acad. Nauk SSSR*, Vol. 173, pp. 748 - 751, 1967.

[14] R. Fletcher, Penalty functions, in *Mathematical programming: the state of the art*, (Eds. A. Bachen, M. Grötschel, B. Korte), Springer–Verlag, Berlin, pp. 87–114, 1983.

[15] S. Han, O. Mangasarian, Exact penalty functions in nonlinear programming, *Mathematical Programming*, Vol.17, pp. 251–269, 1979.

[16] M.R. Hestenes, *Calculus of Variations and Optimal Control Theory*, New York, John Wiley & Sons, 1966.

[17] A.D. Ioffe, V.M. Tichomirov, *Theory of Extremal Problems*, North-Holland, 1979.

[18] L.S. Pontryagin, V.G. Boltyanskii, R.V. Gamkrelidze, E.F. Mishchenko, *The Mathematical theory of Optimal Processes*, New York, Interscience Publishers, 1962.

[19] B.N. Pschenichnyi, *Convex analysis and Extremal Problems*, (In Russian) Moscow, Nauka Publishers, 1980.

[20] R.T. Rockafellar, *Convex Analysis*, Princeton, N.J., Princeton University Press, 1970.

[21] W.L. Zangwill, Nonlinear programming via penalty functions. *Management Science*, Vol. 13, pp. 344–358, 1967.

VARIATIONAL PROBLEMS WITH CONSTRAINTS INVOLVING HIGHER–ORDER DERIVATIVES

Vladimir F.Demyanov *
Applied Mathematics Dept., St.Petersburg State University
Staryi Peterhof, St.Petersburg, 198904 RUSSIA

Franco Giannessi
Dipartimento di Matematica, Università di Pisa
Via Buonarroti, 2, Pisa – 56127 ITALY

Abstract The Exact Penalization approach to solving constrained problems of Calculus of Variations described in [7] is extended to the case of variational problems where the functional and the constraints contain higher-order derivatives. The constraints are of the inequality type. The initial constrained problem is reduced to an unconstrained one. The related unconstrained problem is essentially nonsmooth. Nevertheless, the present state of Nonsmooth Analysis makes it possible to get proper analytical tools to overcome the arising difficulties. The main advantage of the proposed technique is its possible numerical applications for computing solutions of constrained variational problems.

Keywords: Calculus of Variations, Inequality-type Constraints, Higher-order derivatives, Necessary optimality conditions, Penalty Function, Exact Penalty Function, Nonsmooth Analysis.

2000 Mathematics Subject Classification. 90C30, 52A20.

1 Introduction

In the paper the inequality-type constrained problems of Calculus of Variations ([1, 16]) are treated by making use of the penalty function approach. The paper is a continuation and generalization of the paper [7] where classical variational problems were studied by means of the

*The research was supported by the Russian Foundation for Fundamental Studies (grant RFFI No. 00-01-00454)

exact penalization technique. The idea is to reduce the initial constrained problem to an unconstrained one using exact penalty functions. Such an approach was first proposed by [13] and later by [19] for solving convex programming problems (see [2, 14, 15] and bibliography therein). Some new conditions for application of exact penalties were stated in [12, 5, 8]. These conditions turns out to be useful for solving problems of Optimal Control and Calculus of Variations (see [9]).

In the present paper we consider a general problem of Calculus of Variations with inequality-type constraints involving higher-order derivatives. The case where only the first-order derivatives are present was studied in [7]. It was demonstrated that the classical necessary optimality conditions can be derived by making use of the penalty function approach. However, the main advantage of the proposed technique is its possible numerical applications for computing solutions of constrained variational problems since the conditions stated are in a form applicable for immediate implementation. In the present paper we extend these results to the case where second-order derivatives are present in both the functional and the constraints. The proposed approach can be applied to the case of higher-order derivatives.

The paper is organized as follows. In section 2 the problem is formulated. In Section 3 an equivalent statement of the problem is given. Local minima and penalty functions are discussed in Section 4. Properties of the function describing the constraints are studied in Section 5. An Exact penalty function is introduced and studied in Section 6. Necessary optimality conditions are established in Section 7. These conditions are of the Euler-Poisson type since only classical variations of arcs are used. To get Weierstrass-type conditions it is required to employ more sophisticated needle type variations (see [16, 17, 3, 4]).

2 Statement of the problem

Let $T > 0$ be fixed. By $P^2[0,T]$ we denote the class of functions which are continuous on $[0,T]$ with continuous first-order derivatives and whose second-order derivatives are piecewise continuous on $[0,T]$. If $x \in P^2[0,T]$ and $t_0 \in [0,T)$ is a point of discontinuity of the function $x''(t)$, then for the sake of definiteness it will be assumed that

$$x''(t_0) = x''_+(t_0) = \lim_{\alpha \downarrow 0} \frac{x'(t_0+\alpha) - x'(t)}{\alpha}. \tag{1}$$

At the point T we put

$$x''(T) = \lim_{\alpha \uparrow 0} \frac{x'(T+\alpha) - x'(T)}{\alpha}.$$

Let the function $l(x, y, z, t)$ be continuous with $\frac{\partial l}{\partial x}$, $\frac{\partial l}{\partial y}$ and $\frac{\partial l}{\partial z}$ in all their arguments on $\mathbb{R} \times \mathbb{R} \times \mathbb{R} \times [0, T]$.

Fix $x_0 \in \mathbb{R}$ and $x_1 \in \mathbb{R}$. Put

$$\begin{aligned}\Omega = \{x \in P^2[0,T] \mid x(0) = x_0,\ x(T) = x_1;\\ l(x(t), x'(t), x''(t), t) \leq 0\ \forall\, t \in [0,T]\}.\end{aligned} \tag{2}$$

Assume that

$$\Omega \neq \emptyset. \tag{3}$$

Consider the functional

$$\mathcal{I}(x) = \int_0^T F(x(t), x'(t), x'', t)\, dt, \tag{4}$$

where the functions $F(x, y, z, t)$, $\dfrac{\partial F}{\partial x}$, $\dfrac{\partial l}{\partial y}$ and $\dfrac{\partial F}{\partial z}$ are continuous in all their arguments on $\mathbb{R} \times \mathbb{R} \times \mathbb{R} \times [0, T]$.

3 An equivalent statement of the problem

Let $P[0, T]$ be the class of piecewise continuous and bounded on $[0, T]$ functions. If $t_0 \in [0, T)$ is a discontinuity point of the function $q(t) \in P[0, T]$ then for the definiteness we put

$$q(t_0) = \lim_{t \downarrow t_0} q(t). \tag{5}$$

At the point T it is assumed that

$$q(T) = \lim_{t \uparrow T} q(t). \tag{6}$$

Let $z = [z_1, z_2, C] \in P[0, T] \times P[0, T] \times \mathbb{R} = U$. Let us introduce the functional

$$f(z) = \int_0^T F\left(x_0 + \int_0^t z_1(\tau)\, d\tau, C + \int_0^t z_2(\tau)\, d\tau, z_2(t), t\right) dt. \tag{7}$$

Put

$$h_0(T, z) := x_0 - x_1 + \int_0^T z_1(t) dt, \tag{8}$$

$$l(t, z) := l(x_0 + \int_0^t z_1(\tau)\, d\tau, C + \int_0^t z_2(\tau)\, d\tau, z_2(t), t), \tag{9}$$

$$h_2(t, z) = C + \int_0^t z_2(\tau)\, d\tau - z_1(t), \tag{10}$$

$$h(t,z) = \max\{0, l(t,z)\}. \tag{11}$$

Since the function l in continuous in its variables, then the functions $l(t,z)$, $h_2(t,z)$ and $h(t,z)$ are piecewise continuous in t for every $z \in U$.

Let us consider the set

$$Z = \left\{ z \in U \;\middle|\; h_0(T,z) = 0, \quad h(t,z) = 0 \;\forall t \in [0,T], \right.$$
$$\left. h_2(t,z) = 0 \;\forall t \in [0,T] \right\}. \tag{12}$$

Lemma 3.1 *The problem*

$$\mathcal{I}(x) \longrightarrow \min_{x \in \Omega} \tag{13}$$

is equivalent to the problem

$$f(z) \longrightarrow \min_{z \in Z} \tag{14}$$

in the following sense: if $x^* \in \Omega$ is a solution of the problem (13) then the point $z^* = [x^{*'}(t), x^{*''}(t), x^{*'}(0)]$ is a solution of the problem (14); and vice versa, if $z^* = [z_1^*(t), z_2^*(t), C^*] \in Z$ is a minimizer of the function f on the set Z, then the function $x^*(t) = x_0 + \int_0^t z_1^*(\tau)\, d\tau$ is a solution of the problem (13).

Proof. Let $x^* \in \Omega$ and

$$\mathcal{I}(x^*) \leq \mathcal{I}(x) \quad \forall x \in \Omega. \tag{15}$$

Put $z^*(t) = [z_1^*(t), z_2^*(t), C^*]$ where

$$z_1^*(t) = x^{*'}(t), \;\; z_2^*(t) = x^{*''}(t), \;\; C^* = x^{*'}(0).$$

Take an arbitrary $z = [z_1(t), z_2(t), C] \in Z$. Since $z \in Z$ then for the function

$$x(t) = x_0 + \int_0^t z_1(\tau)\, d\tau$$

(due to (10) and (12)) we have

$$z_1(t) = C + \int_0^t z_2(\tau)\, d\tau, \quad x_0 + \int_0^T z_1(t)\, dt = x_1,$$

i.e.

$$x(t) = x_0 + \int_0^t z_1(\tau)\, d\tau = x_0 + Ct + \int_0^t \int_0^\tau z_2(\tau)\, d\tau dt.$$

Therefore x is twice differentiable and

$$x'(t) = C + \int_0^t z_2(\tau)\, d\tau, \;\; x''(t) = z_2(t), \;\; x(T) = \\ = x_0 + \int_0^T z_1(t)\, dt = x_1. \tag{16}$$

Since $h(t,z) = \max\{0, l(t,z)\}$ then (see (9))

$$\max\{0, l(x(t), x'(t), x''(t), t)\} = 0,$$

i.e.

$$l(x(t), x'(t), x''(t), t) \leq 0 \quad \forall t \in [0,T].$$

Taking into account (16), we conclude that $x \in \Omega$. The relation (15) yields

$$\mathcal{I}(x^*) \leq \mathcal{I}(x) \quad \forall x \in \Omega,$$

i.e.

$$\int_0^T F(x_0 + \int_0^t z_1^*(\tau)\, d\tau, C^* + \int_0^t z_2^*(\tau)\, d\tau, z_2^*(t), t)\, dt = f(z^*) \leq \\ \leq \int_0^T F(x_0 + \int_0^t z_1(\tau)\, d\tau, C + \int_0^t z_2(\tau)\, d\tau, z_2(t), t)\, dt = f(z). \tag{17}$$

Since $z \in Z$ is arbitrary then $f(z^*) = \min_{z \in Z} f(z)$, i.e. z^* is a solution of the problem (14).

Let now $z^* = [z_1^*(t), z_2^*(t), C^*] \in Z$ be a minimizer of f on Z, i.e.

$$f(z^*) \leq f(z) \quad \forall z \in Z. \tag{18}$$

We have

$$h_0(T, z^*) = x_0 - x_1 + \int_0^T z_1^*(t)\, dt = 0,$$

$$h_2(t, z^*) = C^* + \int_0^t z_2^*(\tau)\, d\tau \; - \; z_1^*(t) = 0 \; \forall t \in [0,T],$$

$$h(t, z^*) = \max\{0, l(t, z^*)\} = \\ = \max\{0, l(x_0 + \int_0^t z_1^*(\tau)\, d\tau, C^* + \int_0^t z_2^*(\tau)\, d\tau, z_2^*(t), t)\} = 0 \; \forall t \in [0,T].$$

Put $x^*(t) = x_0 + \int_0^t z_1^*(\tau)\, d\tau$. Take any $x \in \Omega$ and put $z = [z_1(t), z_2(t), C] \in Z$ where $z_1(t) = x'(t), z_2(t) = x''(t), C = x'(0)$. It follows from (2) that

$$x(T) = x_0 + \int_0^T x'(t) dt = x_0 + \int_0^T z_1(t) dt = x_1,$$

i.e.

$$h_0(T,z)=0, \tag{19}$$

$$h_2(t,z)=C+\int_0^t z_2(\tau)d\tau \;-\; z_1(t)=$$

$$=x'(0)+\int_0^t x''(\tau)d\tau - x'(t)=0\ \forall t\in[0,T], \tag{20}$$

$$l(x(t),x'(t),x''(t),t) \;=\; l(x_0+\int_0^t x'(\tau)\,d\tau, x'(0)+\int_0^t x''(\tau)\,d\tau, x''(t),t)=$$

$$=l(x_0+\int_0^t z_1(\tau)\,d\tau, C+\int_0^t z_2(\tau)\,d\tau, z_2(t),t)\,dt=l(t,z)\le 0. \tag{21}$$

Then

$$h(t,z)=\max\{0,l(t,z)\}=0 \quad \forall t\in[0,T]. \tag{22}$$

If follows from (19),(20) and (22) that $z\in Z$. therefore (18) yields

$$\mathcal{I}(x^*)=f(z^*)\le f(z)=\mathcal{I}(x).$$

Since $x\in\Omega$ is arbitrary then

$$\mathcal{I}(x^*)=\min_{x\in\Omega}\mathcal{I}(x),$$

i.e. x^* is a solution of the problem (13).

Thus, if $z^*=[z_1^*(t),z_2^*(t),C^*]\in Z$ is a global minimizer of the functional f on the set Z, then the function $x^*(t)=x_0+\int_0^t z_1^*(\tau)\,d\tau\in\Omega$ and is a global minimizer of the functional $\mathcal{I}$ on the set Ω, and, conversely, if $x^*\in\Omega$ is a global minimizer of the functional $\mathcal{I}$ on the set Ω, then $z^*=[x^{*'}(t),x^{*''}(t),x^{*'}(0)]\in Z$ and is a global minimizer of the function f on the set Z.

4 Local minima

Consider the problem (14). On the set $U=P[0,T]\times P[0,T]\times I\!R$ let us introduce a metric $\rho(z',z'')$ where $z'=[z_1'(t),z_2'(t),C']$, $z''=[z_1''(t),z_2''(t),C'']$. As ρ it is possible to choose, for example, one of the following functions:

$$\rho_1(z',z'')=\max_{t\in[0,T]}\left|\int_0^t (z_1'(\tau)-z_1''(\tau))\,d\tau\right| +$$

$$+\max_{t\in[0,T]}\left|\int_0^t (z_2'(\tau)-z_2''(\tau))\,d\tau\right|+|C'-C''|, \tag{23}$$

$$\rho_2(z',z'') = \int_0^T |z_1'(t) - z_1''(t)|dt + \int_0^T |z_2'(t) - z_2''(t)|dt + |C' - C''|, \quad (24)$$

$$\rho_3(z',z'') = \sup_{t\in[0,T]} \left|z_1'(t) - z_1''(t)\right| + \sup_{t\in[0,T]} \left|z_2'(t) - z_2''(t)\right| + \frac{1}{T}|C' - C''|, \quad (25)$$

$$\rho_3(z',z'') = \left[\int_0^T (z_1'(t) - z_1''(t))^2 dt + \right.$$
$$\left. + \int_0^T (z_2'(t) - z_2''(t))dt + (C' - C'')^2\right]^{\frac{1}{2}}. \quad (26)$$

The following relationship exists between the metrics ρ_i $(i \in 1:4)$:

$$\rho_1(z',z'') \leq \rho_2(z',z'') \leq T\rho_3(z',z''),$$

$$\rho_4(z',z'') \leq \sqrt{T}\rho_3(z',z'').$$

These inequalities imply the inclusions

$$\{z|\ \rho_3(z,z_0) < \frac{\varepsilon}{T}\} \subset \{z|\ \rho_2(z,z_0) < \varepsilon\} \subset \{z|\ \rho_1(z,z_0) < \varepsilon\},$$

$$\{z|\ \rho_3(z,z_0) < \frac{\varepsilon}{\sqrt{T}}\} \subset \{z|\ \rho_4(z,z_0) < \varepsilon\}.$$

Therefore one concludes that a local minimizer of the functional f on the set Z in the metric ρ_1 is a local minimizer of f in both the metrics ρ_2 and ρ_3, while a local minimizer of f in the metric ρ_2 is a local minimizer of f in the metric ρ_3, and a local minimizer of f in the metric ρ_4 is a local minimizer of f in the metric ρ_3.

5 Properties of the function φ

5.1 A classical variation of z.

Thus, we consider the problem of minimizing the functional

$$f(z) = \int_0^T F(x_0 + \int_0^t z_1(\tau)\,d\tau, C + \int_0^t z_2(\tau)\,d\tau, z_2(t), t)\,dt \quad (27)$$

on the set $Z \subset P[0,T] \times P[0,T] \times I\!R = U$ described by (12). The set Z can be represented in the form

$$Z = \{z = [z_1, z_2, C] \in U \mid \varphi(z) = 0\}, \quad (28)$$

where

$$\varphi(z) = \left[\int_0^T h^2(t,z)dt + \int_0^T h_2^2(t,z)dt + h_0^2(T,z)\right]^{1/2}, \quad (29)$$

where h_0 is given by (8), h – by (11), h_2 – by (10).

Let us study the differential properties of the function φ. Let $z \in U$ be fixed. Choose and fix an arbitrary $v = [v_1, v_2, v_0] \in U$ and for any $\varepsilon \geq 0$ put

$$z_\varepsilon(t) = z(t) + \varepsilon v(t) = [z_{1\varepsilon}, z_{2\varepsilon}, C_\varepsilon] \tag{30}$$

where

$$z_{1\varepsilon} = z_1(t) + \varepsilon v_1(t), \ z_{2\varepsilon} = z_2(t) + \varepsilon v_2(t), \ C_\varepsilon = C + \varepsilon v_0.$$

We have

$$l(t, z_\varepsilon) = l(x_0 + \int_0^t z_{1\varepsilon}(\tau)\, d\tau, C_\varepsilon + \int_0^t z_{2\varepsilon}(\tau)\, d\tau, z_{2\varepsilon}(t), t) =$$

$$= l\Bigg(x_0 + \int_0^t z_1(\tau) d\tau + \varepsilon \int_0^t v_1(\tau)\, d\tau,$$

$$C + \int_0^t z_2(\tau)\, d\tau + \varepsilon(v_0 + \int_0^t v_2(\tau)\, d\tau), z_2(t) + \varepsilon v_2(t), t\Bigg) =$$

$$= l(x_0 + \int_0^t z_1(\tau) d\tau, C + \int_0^t z_2(\tau)\, d\tau, z_2(t), t) + \varepsilon H_1(t, z, v) + o(\varepsilon, t) =$$

$$= l(t, z) + \varepsilon H_1(t, z, v) + o(\varepsilon, t), \tag{31}$$

$$h_2(t, z_\varepsilon) = -z_{1\varepsilon}(t) + C_\varepsilon + \int_0^t z_{2\varepsilon}(\tau)\, d\tau = h_2(t, z) + \varepsilon H_2(t, z, v), \tag{32}$$

$$h_0(T, z_\varepsilon) = h_0(T, z) + \varepsilon \int_0^T v_1(t) dt, \tag{33}$$

where

$$H_1(t, z, v) = \frac{\partial l(t, z)}{\partial x} \int_0^t v_1(\tau) d\tau \ + \frac{\partial l(t, z)}{\partial y} \left(v_0 + \int_0^t v_2(\tau) d\tau \right) \ + $$
$$+ \frac{\partial l(t, z)}{\partial z} v_2(t), \tag{34}$$

$$l(t, z) = l(x_0 + \int_0^t z_1(\tau)\, d\tau, C + \int_0^t z_2(\tau)\, d\tau, z_2(t), t) = l(x, y, z, t),$$

$$\frac{o(\varepsilon, t)}{\varepsilon} \underset{\varepsilon \downarrow 0}{\longrightarrow} 0, \tag{35}$$

$$H_2(t, z, v) = -v_1(t) + v_0 + \int_0^t v_2(\tau) d\tau. \tag{36}$$

It follows from (31), (32) and (11) that

$$h(t, z_\varepsilon) \;=\; h(t,z) + \varepsilon H(t,z,v) + o(\varepsilon, t), \tag{37}$$

where

$$H(t,z,v) = \begin{cases} H_1(t,z,v), & if\ l(t,z) > 0, \\ 0, & if\ l(t,z) < 0, \\ \max\{0, H_1(t,z,v)\}, & if\ \ l(t,z) = 0. \end{cases} \tag{38}$$

Here $o(\varepsilon, t)$ satisfies (35). Let us introduce the sets

$$\begin{aligned} \sigma_+(z) &= \{t \in [0,T] \mid l(t,z) > 0\ \}, \\ \sigma_-(z) &= \{t \in [0,T] \mid l(t,z) < 0\ \}, \\ \sigma_0(z) &= \{t \in [0,T] \mid l(t,z) = 0\ \}. \end{aligned} \tag{39}$$

It is clear that

$$\sigma_+(z) \cup \sigma_-(z) \cup \sigma_0(z) \;=\; [0,T]. \tag{40}$$

5.2 The case $z \notin Z$.

First let us consider the case $\varphi(z) > 0$ (i.e. $z \notin Z$).

Theorem 5.1 *If $z \notin Z$ then the function φ is Gâteaux differentiable at z.*

Proof. We have (see (31), (32), (33))

$$\varphi(z_\varepsilon) = \left[\int_0^T h^2(t, z_\varepsilon)\, dt + \int_0^T h_2^2(t, z_\varepsilon)\, dt + h_0^2(T, z_\varepsilon)\right]^{1/2} =$$

$$= \left[\int_0^T \Big(h(t,z) + \varepsilon H(t,z,v) \;+\; o(\varepsilon, t)\Big)^2 dt + \right.$$

$$\left. + \int_0^T \Big(h_2(t,z) + \varepsilon H_2(t,z,v)\Big)^2 dt \;+\; \left(h_0(T,z) + \varepsilon \int_0^T v_1(\tau) d\tau\right)^2\right]^{1/2} =$$

$$= \left\{ \int_0^T \Big[h^2(t,z)\, dt + 2\varepsilon h(t,z) H(t,z,v) + o(\varepsilon, t)\Big] dt + \right.$$

$$+ \int_0^T \Big[h_2^2(t,z)\, dt + 2\varepsilon h_2(t,z) H_2(t,z,v) + o(\varepsilon, t)\Big] dt + h_0^2(T,z) +$$

$$\left. + 2\varepsilon h_0(T,z) \int_0^T v_1(\tau) d\tau\Big] \;+\; o(\varepsilon)\right\}^{1/2} =$$

$$= \varphi(z) \ + \varepsilon\Big[\int_0^T \frac{h(t,z)}{\varphi(z)} H(t,z,v)dt \ + \ \int_0^T \frac{h_2(t,z)}{\varphi(z)} H_2(t,z,v)dt +$$

$$+ \frac{h_0(T,z)}{\varphi(z)} \int_0^T v(\tau)d\tau \Big] \ + \ o(\varepsilon) = \varphi(z) \ + \varepsilon\varphi'(z,v) \ + \ o(\varepsilon), \tag{41}$$

where

$$\frac{o(\varepsilon)}{\varepsilon} \underset{\varepsilon\downarrow 0}{\longrightarrow} 0, \tag{42}$$

$$\varphi'(z,v) = \lim_{\varepsilon\downarrow 0} \frac{\varphi(z_\varepsilon) - \varphi(z)}{\varepsilon} =$$

$$= \int_0^T \frac{h(t,z)}{\varphi(z)} H(t,z,v)dt + \int_0^T \frac{h_2(t,z)}{\varphi(z)} H_2(t,z,v)dt +$$

$$+ \frac{h_0(T,z)}{\varphi(z)} \int_0^T v_1(t)dt. \tag{43}$$

Taking into account (36), (38) and (11) we get

$$\varphi'(z,v) = \int_0^T \frac{h(t,z)}{\varphi(z)} H_1(t,z,v)dt +$$

$$+ \int_0^T \frac{h_2(t,z)}{\varphi(z)} H_2(t,z,v)dt \ + \ \frac{h_0(T,z)}{\varphi(z)} \int_0^T v_1(t)dt =$$

$$= \int_0^T \left[\frac{h(t,z)}{\varphi(z)} \frac{\partial l(t,z)}{\partial x} \int_0^t v_1(\tau)d\tau \ + \ \frac{h(t,z)}{\varphi(z)} \frac{\partial l(t,z)}{\partial y} \left(v_0 + \int_0^t v_2(\tau)d\tau \right) + \right.$$

$$\left. + \frac{h(t,z)}{\varphi(z)} \frac{\partial l(t,z)}{\partial z} v_2(t) \ + \ \frac{h_2(t,z)}{\varphi(z)} \left(\int_0^t v_2(\tau)d\tau \ + \ v_0 - v_1(t) \right) \right] dt +$$

$$+ \frac{h_0(T,z)}{\varphi(z)} \int_0^T v_1(t)dt \ = I_1 + I_2 + I_3, \tag{44}$$

where

$$I_1 = \int_0^T \left[\frac{h(t,z)}{\varphi(z)} \frac{\partial l(t,z)}{\partial y} + \ \frac{h_2(t,z)}{\varphi(z)} \right] dt \ v_0, \tag{45}$$

$$I_2 = \int_0^T \left[\frac{h(t,z)}{\varphi(z)} \frac{\partial l(t,z)}{\partial x} \int_0^t v_1(\tau)d\tau \ - \ \frac{h_2(t,z)}{\varphi(z)} v_1(t) + \right.$$

$$\left. + \frac{h_0(t,z)}{\varphi(z)} v_1(t) \right] dt, \tag{46}$$

$$I_3 \ = \ \int_0^T \left[\frac{h(t,z)}{\varphi(z)} \frac{\partial l(t,z)}{\partial y} \int_0^t v_2(\tau)d\tau + \right.$$

$$+ \frac{h(t,z)}{\varphi(z)}\frac{\partial l(t,z)}{\partial z}v_2(t) + \frac{h_2(t,z)}{\varphi(z)}\int_0^t v_2(\tau)d\tau\Bigg] dt. \tag{47}$$

Put

$$\frac{h_0(T,z)}{\varphi(z)} := w_0(z) \in I\!R, \tag{48}$$

$$\frac{l(t,z)}{\varphi(z)} := w_1(z) \in P[0,T], \quad \frac{h_2(t,z)}{\varphi(z)} := w_2(t,z) \in P[0,T]. \tag{49}$$

Clearly,

$$||w(z)|| = \left[\int_0^T w_1^2(t,z)\; dt + \int_0^T w_2^2(t,z)\; dt + w_0^2(z)\right]^{1/2} = 1. \tag{50}$$

Integrating by parts in (46) and (47), we have

$$I_1 = \int_0^T \left[w_1(t,z)\frac{\partial l(t,z)}{\partial y} + \; w_2(t,z)\right] dt\; v_0, \tag{51}$$

$$I_2 = \int_0^T \left[\int_t^T w_1(\tau,z)\frac{\partial l(\tau,z)}{\partial x}d\tau \; - w_2(t,z) + w_0(z)\right] v_1(t)dt, \tag{52}$$

$$I_3 = \int_0^T \Bigg\{ w_1(t,z)\frac{\partial l(t,z)}{\partial z} +$$

$$+ \int_t^T \left[w_1(\tau,z)\frac{\partial l(t,z)}{\partial y} + w_2(\tau,z)\right] d\tau \Bigg\} v_2(t)dt. \tag{53}$$

It follows from (44), (51), (52) and (53) that if $z \notin Z$ then the function φ is Gâteaux differentiable at z, i.e.

$$\varphi'(z,v) \;=\; (G(z),v) \;=$$

$$= \int_0^T \Big[G_1(t,z)v_1(t)dt \; + G_2(t,z)v_2(t)\Big] dt \; + G_0(z)v_0, \tag{54}$$

where

$$G(z) = [G_1(t,z),\; G_2(t,z),\; G_0(z)], \tag{55}$$

$$G_1(t,z) = w_0(z) - w_2(t,z) + \int_t^T \frac{\partial l(\tau,z)}{\partial x}w_1(\tau,z)d\tau, \tag{56}$$

$$G_2(t,z) = \frac{\partial l(t,z)}{\partial z}w_1(t,z) + \int_t^T \left[\frac{\partial l(\tau,z)}{\partial y}w_1(\tau,z)d\tau + w_2(\tau,z)\right] d\tau, \tag{57}$$

$$G_0(z) = \int_0^T \left[\frac{\partial l(t,z)}{\partial y}w_1(t,z)dt + w_2(t,z)\right] dt. \tag{58}$$

Here $w(z) = [w_1(t,z), w_2(t,z), w_0(z)]$ is defined by (49) and satisfies (50). The point $G(z) \in U$ will be referred to as the "gradient" of φ at z.

5.3 The case $z \in Z$.

Now let us consider the case $\varphi(z) = 0$ (i.e. $z \in Z$).

Theorem 5.2 *If $z \in Z$ then the function φ is Dini directionally differentiable at z.*

Proof. We have (see (31), (32),(33))

$$\varphi(z_\varepsilon) = \left[\int_0^T h^2(t, z_\varepsilon)\, dt + \int_0^T h_2^2(t, z_\varepsilon)\, dt + h_0^2(T, z_\varepsilon) \right]^{1/2} =$$

$$= \left[\int_0^T \Big(h(t, z) + \varepsilon H(t, z, v) + o(\varepsilon, t) \Big)^2 dt + \right.$$

$$+ \int_0^T \Big(h_2(t, z) + \varepsilon H_2(t, z, v) \Big)^2 dt +$$

$$\left. + \left(h_0(T, z) + \varepsilon \int_0^T v_1(\tau) d\tau \right)^2 \right]^{1/2}. \tag{59}$$

Since $\varphi(z) = 0$ then

$$h(t, z) = 0, \;\; h_2(t, z) = 0 \;\forall t \in [0, T], \;\; h_0(T, z) = 0. \tag{60}$$

Therefore

$$\varphi(z_\varepsilon) = \varepsilon ||H(z, v); H_2(z, v); v||,$$

where

$$||H(z, v); H_2(z, v); v|| = \left[\int_0^T H^2(t, z, v) dt + \right.$$

$$\left. + \int_0^T H_2^2(t, z, v) dt \; + \; \left(\int_0^T v_1(t) dt \right)^2 \right]^{1/2}.$$

This implies

$$\varphi'(z, v) = \lim_{\varepsilon \downarrow 0} \frac{\varphi(z_\varepsilon) - \varphi(z)}{\varepsilon} = ||H(z, v); H_2(z, v); v|| =$$

$$= \max_{||w|| \le 1} \left[\int_0^T H(t, z, v) w_1(t) dt \; + \right.$$

$$\left. + \int_0^T H_2(t, z, v) w_2(t) dt \; + \; \int_0^T v_1(\tau) d\tau \; w_0 \right]. \tag{61}$$

Here

$$w = [w_1, w_2, w_0], \;\; w_1 \in P[0, T], \;\; w_2 \in P[0, T], \;\; w_0 \in I\!R,$$

$$||w|| = \left[\int_0^T (w_1^2(t) \ + \ w_2^2(t)) \ dt + w_0^2 \right]^{1/2} = 1. \tag{62}$$

Since $z \in Z$, then (see (39)) $\sigma_+(z) = \emptyset$, therefore (38) and (39) yield

$$\varphi'(z,v) = \max_{||w|| \le 1} \left[\int_{\sigma_0} w_1(t) \max_{\bar{w}(t) \in [0,1]} \Big(\bar{w}(t) H_1(t,z,v) \Big) dt + \right.$$

$$\left. + \int_0^T H_2(t,z,v) w_2(t) dt \ + \ \int_0^T v_1(\tau) d\tau \ w_0 \right]. \tag{63}$$

This expression can be rewritten as

$$\varphi'(z,v) = \max_{\substack{||w|| \le 1 \\ w_1(t) \ge 0 \ \forall t \in \sigma_0(z) \\ w_1(t) = 0 \ \forall t \in \sigma_-(z)}} \left[\int_0^T H_1(t,z,v) w_1(t) dt + \right.$$

$$\left. + \int_0^T H_2(t,z,v) w_2(t) dt \ + \ \int_0^T v_1(t) dt \ w_0 \right]. \tag{64}$$

Taking into account (34) and (36) one has

$$\varphi'(z,v) \ = \ \max_{w \in W(z)} \left\{ \int_0^T \left[\frac{\partial l(t,z)}{\partial x} \int_0^t v_1(\tau) d\tau \ + \right. \right.$$

$$+ \ \frac{\partial l(t,z)}{\partial y} \left(v_0 \ + \ \int_0^t v_2(\tau) d\tau \right) \ + \ \frac{\partial l(t,z)}{\partial z} v_2(t) \Bigg] w_1(t) dt \ +$$

$$\left. + \int_0^T \left[v_0 \ - \ v_1(t) \ + \ \int_0^t v_2(\tau) d\tau \right] w_2(t) dt \ + \ \int_0^T v_1(t) dt \ w_0 \right\} \ =$$

$$= \max_{w \in W(z)} \left\{ \int_0^T \left[\frac{\partial l(t,z)}{\partial y} w_1(t) \ + \ w_2(t) \right] dt \ v_0 \ + \right.$$

$$+ \int_0^T \left[\frac{\partial l(t,z)}{\partial x} w_1(t) \int_0^t v_1(\tau) d\tau \ + \ (w_0 - w_2(t)) v_1(t) \right] dt \ +$$

$$+ \int_0^T \left[\left(\frac{\partial l(t,z)}{\partial y} w_1(t) \ + \ w_2(t) \right) \int_0^t v_2(\tau) d\tau \ + \right.$$

$$\left. \left. + \ \frac{\partial l(t,z)}{\partial z} w_1(t) v_2(t) \ \right] dt \right\}, \tag{65}$$

where

$$W(z) \ = \ \Big\{ w = [w_1, w_2, w_0] \ \Big| \ ||w|| \le 1, \ w_0 \in \mathbb{R}, \ w_1 \in P[0,T],$$

$$w_2 \in P[0,T],\ w_1(t) \geq 0\ \forall t \in \sigma_0(z),\ w_1(t) = 0\ \forall t \in \sigma_-(z)\Big\}. \tag{66}$$

Note that the set $W(z)$ is convex. Integrating by parts in (65), we have

$$\varphi'(z,v) = \max_{w \in W(z)} \Big\{ \int_0^T \left[\frac{\partial l(t,z)}{\partial y} w_1(t) \ +\ w_2(t) \right] dt\ v_0 \ +$$

$$+ \int_0^T \left[\int_t^T \frac{\partial l(\tau,z)}{\partial x} w_1(\tau) d\tau \ +\ w_0 \ -\ w_2(t) \right] v_1(t) dt \ +$$

$$+ \int_0^T \left[\int_t^T \left(\frac{\partial l(\tau,z)}{\partial y} w_1(\tau) d\tau \ +\ w_2(\tau) \right) d\tau \ + \right.$$

$$\left. + \frac{\partial l(t,z)}{\partial z} w_1(t) \right] v_2(t) dt \Big\}. \tag{67}$$

We conclude from (67) that at a point $z \in Z$ the function φ is Dini directionally differentiable, and, moreover, the function φ is subdifferentiable, i.e.

$$\varphi'(z,v) = \max_{A \in \partial\varphi(z)} (A, v), \tag{68}$$

where

$$(A,v) \ =\ \int_0^T A_1(t) v_1(t) dt \ +\ \int_0^T A_2(t) v_2(t) dt \ +\ A_0 v_0, \tag{69}$$

$$\partial\varphi(z) = \Big\{ A = (A_1, A_2, A_0) \in U \ \Big| \ A_1(t) \ =$$

$$= \int_t^T \frac{\partial l(\tau,z)}{\partial x} w_1(\tau) d\tau \ +\ w_0 \ -\ w_2(t),$$

$$A_2(t) \ =\ \int_t^T \left(\frac{\partial l(\tau,z)}{\partial y} w_1(\tau) d\tau \ +\ w_2(\tau) \right) d\tau \ +\ \frac{\partial l(t,z)}{\partial z} w_1(t),$$

$$A_0 \ =\ \int_0^T \left[\frac{\partial l(t,z)}{\partial y} w_1(t) + w_2(t) \right] dt, \quad w = [w_1, w_2, w_0] \in W(z) \Big\}, \tag{70}$$

and the set $W(z)$ is defined in (66). Note that the set $\partial\varphi(z)$ is a convex one.

Remark 5.1 *For the functional $f(z)$ the following expansion holds:*

$$f(z_\varepsilon) = f(z) \ +\ \varepsilon \int_0^T \left[\frac{\partial F(t,z)}{\partial x} \int_0^t v_1(\tau) d\tau \ + \right.$$

$$\left. + \frac{\partial F(t,z)}{\partial y} \left(v_0 \ +\ \int_0^t v_2(\tau) d\tau \right) + \frac{\partial F(t,z)}{\partial z} v_2(t) \right] dt \ +$$

$$+ \, o(\varepsilon) \; = \; f(z) \, + \, \varepsilon \int_0^T \left[\int_t^T \frac{\partial F(\tau, z)}{\partial x} d\tau \; v_1(t) \; + \right.$$

$$\left. + \left(\frac{\partial F(t,z)}{\partial z} + \int_t^T \frac{\partial F(\tau, z)}{\partial y} d\tau \right) v_2(t) \; + \; \frac{\partial F(t,z)}{\partial y} \; v_0 \right] dt \; + \; o(\varepsilon), \quad (71)$$

where

$$\frac{o(\varepsilon)}{\varepsilon} \underset{\varepsilon \downarrow 0}{\longrightarrow} 0, \;\; F(t,z) = F(x_0 + \int_0^t z_1(\tau) d\tau, C + \int_0^t z_2(\tau) d\tau, z_2(t), t).$$

6 Exact penalty functions

6.1 Properties of the function G.

Theorems analogous to Theorems 6.3, 6.4 and 6.5 from [7] are valid in our case as well. We shall formulate (without proof) only one of them.

Let $z_0 \in Z$. Assume that there exist t_1 and t_2 such that

$$0 \le t_1 < t_2 \le T, \;\; l(t, z_0) < 0 \quad \forall t \in [t_1, t_2]. \qquad (72)$$

Then for every $i \in 1:4$ there exist $\delta_i > 0$ such that

$$l(t,z) < 0 \quad \forall t \in [t_1, t_2], \quad \forall z \in S_i = \{ z \in U \mid \rho_i(z, z_0) < \delta_i \}. \qquad (73)$$

Theorem 6.1 *If*

$$\left| \frac{\partial l(t,z)}{\partial z} \right| \ge b > 0 \quad \forall t \in [0, t_1] \cup [t_2, T], \; \forall z \in S_i \setminus Z \qquad (74)$$

then there exist $a_1 > 0$ and $a_2 > 0$ such that

$$||G(z)|| = \left[\int_0^T G_1^2(t,z) dt \, + \int_0^T G_2^2(t,z) dt \, + \, G_0^2 \right]^{1/2} \ge a_1 \quad \forall z \in S_i \backslash Z, \qquad (75)$$

$$b(z) \; = \; \sup_{t \in [0,T]} |G_1(t,z)| \; + \; \sup_{t \in [0,T]} |G_2(t,z)| \; + \; |G_0(z)| \; \ge a_2 \quad \forall z \in S_i \setminus Z. \qquad (76)$$

Proof is similar to that of Theorem 6.4 from [7].

Fix $z \in S_i \setminus Z$. It follows from (76) that at least one of the following inequalities holds:

$$a) \quad \sup_{t \in [0,T]} |G_1(t,z)| \; \ge \frac{1}{3} b(z), \qquad (77)$$

$$b) \quad \sup_{t\in[0,T]} |G_2(t,z)| \geq \frac{1}{3}b(z), \tag{78}$$

$$c) \quad |G_0(z)| \geq \frac{1}{3}b(z). \tag{79}$$

In the case a) one can find $t_1(z) \in [0,T]$ and $t_2(z) \in [0,T]$ such that

$$t_1(z) < t_2(z), \quad |G_1(t,z)| \geq \frac{1}{6}b(z) \quad \forall t \in \sigma_z = [t_1(z), t_2(z)]. \tag{80}$$

Put $\gamma(z) = t_2(z) - t_1(z)$. For every $\alpha \geq 0$ take $z_\alpha(t) = z(t) + \alpha v_z(t)$, where

$$v_z(t) = [v_{1z}(t), v_{2z}(t), v_{0z}] \in U,$$
$$v_{1z}(t) = \begin{cases} -G_1(t,z), & t \in \sigma_z, \\ 0, & t \in [0,T] \setminus \sigma_z, \end{cases} \tag{81}$$

$$v_{2z}(t) = 0 \quad \forall t \in [0,T], \quad v_{0z} = 0. \tag{82}$$

Then $z_\alpha \in P[0,T]$ and (see (41) and (42))

$$\varphi(z_\alpha) = \varphi(z) + \alpha(G(z), v_z) + o(\alpha) = \varphi(z) +$$
$$+\alpha\left[\int_0^T G_1(t,z)v_{1z}(t)dt + \int_0^T G_2(t,z)v_{2z}(t)dt + G_0(z)v_{0z}\right] + o(\alpha) =$$
$$= \varphi(z) + \alpha\int_0^T G_1(t,z)v_{1z}(t)dt + o(\alpha) \leq \varphi(z) - \alpha\gamma(z)\frac{b^2(z)}{36} + o(\alpha). \tag{83}$$

We have

$$\rho_1(z_\alpha, z) = \alpha\left[\max_{t\in[0,T]}\left|\int_0^t v_{1z}(\tau)d\tau\right| + \max_{t\in[0,T]}\left|\int_0^t v_{2z}(\tau)d\tau\right| + |v_{0z}|\right] =$$
$$= \alpha \max_{t\in[0,T]}\left|\int_0^t v_{1z}(\tau)d\tau\right| \leq$$
$$\leq \alpha \max_{t\in[0,T]}\int_0^t |v_{1z}(\tau)|d\tau = \alpha\int_0^T |v_{1z}(\tau)|d\tau \leq \alpha\gamma(z)b(z), \tag{84}$$

$$\rho_2(z_\alpha, z) = \alpha\int_0^T |v_{1z}(\tau)|d\tau \leq \alpha\gamma(z)b(z), \tag{85}$$

$$\rho_3(z_\alpha, z) = \alpha \sup_{t\in[0,T]} |v_{1z}(t)| = \alpha b(z). \tag{86}$$

In the case b) one can find $t_1(z) \in [0,T]$ and $t_2(z) \in [0,T]$ such that

$$t_1(z) < t_2(z), \quad |G_2(t,z)| \geq \frac{1}{6}b(z) \quad \forall t \in \sigma_z = [t_1(z), t_2(z)]. \tag{87}$$

Put $\gamma(z) = t_2(z) - t_1(z)$. For every $\alpha \geq 0$ take $z_\alpha(t) = z(t) + \alpha v_z(t)$, where

$$v_z(t) = [v_{1z}(t), v_{2z}(t), v_{0z}] \in U, \tag{88}$$

$$v_{1z}(t) \;=\; 0 \quad \forall t \in [0,T], \quad v_{0z} = 0, \tag{89}$$

$$v_{2z}(t) = \begin{cases} -G_2(t,z), & t \in \sigma_z, \\ 0, & t \in [0,T] \setminus \sigma_z. \end{cases} \tag{90}$$

Then $z_\alpha \in P[0,T]$. As above, we get the relations

$$\varphi(z_\alpha) \;\leq \varphi(z) - \alpha\gamma(z)\frac{b^2(z)}{36} + o(\alpha), \tag{91}$$

$$\rho_1(z_\alpha, z) \;\leq\; \alpha\gamma(z)b(z), \tag{92}$$

$$\rho_2(z_\alpha, z) \;\leq\; \alpha\gamma(z)b(z), \tag{93}$$

$$\rho_3(z_\alpha, z) \;=\; \alpha b(z). \tag{94}$$

In the case c) let us take $z_\alpha(t) = z(t) + \alpha v_z(t)$, where

$$v_z(t) = [v_{1z}(t), v_{2z}(t), v_{0z}] \in U, \tag{95}$$

$$v_{1z}(t) \;=\; v_{2z}(t) \;=\; 0 \quad \forall t \in [0,T], \quad v_{0z} \;=\; -G_0(z). \tag{96}$$

Then $z_\alpha \in P[0,T]$. As above, we get the relations

$$\varphi(z_\alpha) \;=\; \varphi(z) + \alpha G_0(z)v_{0z} \;+\; o(\alpha) \leq \varphi(z) - \alpha G_0^2(z) \;+\; o(\alpha), \tag{97}$$

$$\rho_1(z_\alpha, z) \;=\; \alpha|G_0(z)|, \tag{98}$$

$$\rho_2(z_\alpha, z) \;=\; \alpha|G_0(z)|, \tag{99}$$

$$\rho_3(z_\alpha, z) \;=\; \frac{\alpha}{T}|G_0(z)|. \tag{100}$$

In the case a) it follows from (83), (84), (85) and (86) that

$$\varphi_1^{\downarrow}(z) = \liminf_{z' \to z} \frac{\varphi(z') - \varphi(z)}{\rho_1(z', z)} \leq \liminf_{\alpha \downarrow 0} \frac{\varphi(z_\alpha) - \varphi(z)}{\rho_1(z_\alpha, z)} \leq$$

$$\leq \; -\frac{\gamma(z)b^2(z)}{36\gamma(z)b(z)} \;=\; -\frac{b(z)}{36} \;\leq\; -\frac{a_2}{36} \qquad \forall z \in S_1 \setminus Z. \tag{101}$$

In a similar way one gets

$$\varphi_2^{\downarrow}(z) = \liminf_{z' \to z} \frac{\varphi(z') - \varphi(z)}{\rho_2(z', z)} \;\leq\; -\frac{b(z)}{36} \;\leq\; -\frac{a_2}{36} \qquad \forall z \in S_2 \setminus Z. \tag{102}$$

In the case b) it follows from (91), (92), (93) and (94) that

$$\varphi_1^{\downarrow}(z) = \liminf_{z' \to z} \frac{\varphi(z') - \varphi(z)}{\rho_1(z', z)} \leq \liminf_{\alpha \downarrow 0} \frac{\varphi(z_\alpha) - \varphi(z)}{\rho_1(z_\alpha, z)} \leq$$

$$\leq -\frac{\gamma(z)b^2(z)}{36\gamma(z)b(z)} = -\frac{b(z)}{36} \leq -\frac{a_2}{36} \qquad \forall z \in S_1 \setminus Z. \tag{103}$$

In a similar way one gets

$$\varphi_2^{\downarrow}(z) - \frac{b(z)}{36} \leq -\frac{a_2}{36} \qquad \forall z \in S_2 \setminus Z. \tag{104}$$

In the case c) the relations (97), (98), (99) and (100) yield

$$\varphi_1^{\downarrow}(z) \leq -|G_0(z)| \leq -\frac{b(z)}{3} \leq -\frac{b(z)}{36} \leq -\frac{a_2}{36} \qquad \forall z \in S_1 \setminus Z, \tag{105}$$

$$\varphi_2^{\downarrow}(z) \leq -\frac{b(z)}{3} \leq -\frac{b(z)}{36} \leq -\frac{a_2}{36} \qquad \forall z \in S_2 \setminus Z. \tag{106}$$

Finally, from the relations (101), (105), (102) and (106) we conclude that

$$\varphi_1^{\downarrow}(z) \leq -\frac{a_2}{36} \qquad \forall z \in S_1 \setminus Z, \tag{107}$$

$$\varphi_2^{\downarrow}(z) \leq -\frac{a_2}{36} \qquad \forall z \in S_2 \setminus Z. \tag{108}$$

To get an analogous inequality for the metric

$$\rho_4(z', z) = \left[\int_0^T (z_1'(t) - z_1(t))^2 dt + \int_0^T (z_2'(t) - z_2(t))^2 dt + (C' - C'')^2\right]^{1/2},$$

let us take $z_\alpha(t) = z(t) - \alpha G(t, z)$. Then

$$\begin{aligned} \varphi(z_\alpha) &= \varphi(z) - \alpha(G(t,z), G(t,z)) + o(\alpha) = \\ &= \varphi(z) - \alpha||G(z)||^2 + o(\alpha), \end{aligned} \tag{109}$$

$$\rho_4(z_\alpha, z) = (G(t,z), G(t,z))^{1/2} = \alpha||G(z)||. \tag{110}$$

Taking into account (75), one has

$$\varphi_2^{\downarrow}(z) = \liminf_{z' \to z} \frac{\varphi(z') - \varphi(z)}{\rho_4(z', z)} \leq$$

$$\leq \liminf_{\alpha \downarrow 0} \frac{\varphi(z_\alpha) - \varphi(z)}{\rho_4(z_\alpha, z)} = -||G(z)|| \leq -a_1 \qquad \forall z \in S_4 \setminus Z. \tag{111}$$

Remark 6.1 *It is impossible to get an inequality similar to (107), (108) and (111) for the metric ρ_3 since $\gamma(z)$ (in the cases a) and b)) may take any small (though positive) values. It is due to the fact that the function $G(t, z)$ in (45) is bounded for every specific z but this boundedness is not uniform in $z \in S_3$ (see the definitions (55) of the function G and (49) of the function $w(z)$).*

6.2 An Exact Penalty function.

Thus, we consider the problem of minimizing the functional

$$f(z) = \int_0^T F(x_0 + \int_0^t z_1(\tau)\,d\tau, C + \int_0^t z_2(\tau)\,d\tau, z_2(t), t)\,dt \qquad (112)$$

on the set $Z \subset U$, described by the relation (28). Let $z_0 \in Z$ be a local minimizer of the functional f on the set Z in the metric ρ_i. Assume that in some neighbourhood $B_{i\delta}(z_0) = \{z \in P[0,T] \mid \rho_i(z, z_0) < \delta\}$ of the point z_0 the function φ satisfies the condition

$$\varphi_i^{\downarrow}(z) \leq -a < 0 \qquad \forall z \in B_{i\delta}(z_0) \setminus Z. \qquad (113)$$

Some of such conditions are stated in Theorem 6.1.

Now it follows from [7] that if the function f is Lipschitz on $B_{i\delta}(z_0)$ in the metric ρ_i then there exists $\lambda^* < \infty$ such that for every $\lambda > \lambda^*$ the point z_0 is a local unconstrained minimizer of the functional $\Phi_\lambda(z) = f(z) + \lambda\varphi(z)$ in the metric ρ_i.

Remark 6.2 *It follows from Theorem 6.1 that if z_0 is such that $l(t, z_0) \neq$ $\mathbf{0}$ then condition (113) holds. This implies that the points which are not covered by our Theorem are solutions of the implicit differential equation*

$$l(x, x', x'', t) = 0 \quad \forall t \in [0, T].$$

7 Necessary conditions for an Extremum

Let $z_0 \in Z$ be a local minimizer of the functional f on the set Z in the metric ρ_i. Assume that in some neighbourhood $B_{i\delta}(z_0)$ of the point z_0 the function φ satisfies the condition (113) and assume also that f is Lipschitz on $B_{i\delta}(z_0)$ in the metric ρ_i. Then, as it has just been observed, for $\lambda > \lambda^*$ the point z_0 is a local unconstrained minimizer of the functional $\Phi_\lambda(z)$ in the metric ρ_i. Fix $v = [v_1, v_2, v_0] \in U$ and for $\varepsilon \geq 0$ put

$$z_\varepsilon(t) = z_0(t) + \varepsilon v(t). \qquad (114)$$

Note that

$$\rho_i(z_\varepsilon, z_0) \underset{\varepsilon\downarrow 0}{\longrightarrow} 0 \quad \forall i \in 1:4. \qquad (115)$$

Then (68), (69) and (71) yield

$$\Phi_\lambda(z_\varepsilon) = \Phi_\lambda(z_0) +$$

$$+\varepsilon\Big\{\int_0^T\Big[\int_t^T\frac{\partial F(\tau,z_0)}{\partial x}d\tau\, v_1(t)+\Big(\frac{\partial F(t,z_0)}{\partial z}+\int_t^T\frac{\partial F(\tau,z_0)}{\partial y}d\tau\Big)\,v_2(t)+$$

$$+\frac{\partial F(t,z_0)}{\partial y}\,v_0\Big]dt\;+\;\lambda\max_{A\in\partial\varphi(z_0)}(A,v)\Big\}\;+\;o(\varepsilon,v), \qquad (116)$$

where

$$\frac{o(\varepsilon,v)}{\varepsilon}\underset{\varepsilon\downarrow 0}{\longrightarrow}0\quad\forall v\in U,$$

$$\partial\varphi(z)=\Big\{A=(A_1,A_2,A_0)\in U\;\Big|$$

$$A_1(t)\;=\;\int_t^T\frac{\partial l(\tau,z_0)}{\partial x}w_1(\tau)d\tau\;+\;w_0\;-\;w_2(t),$$

$$A_2(t)\;=\;\int_t^T\Big(\frac{\partial l(\tau,z_0)}{\partial y}w_1(\tau)d\tau\;+\;w_2(\tau)\Big)d\tau\;+\;\frac{\partial l(t,z_0)}{\partial z}w_1(t),$$

$$A_0\;=\;\int_0^T\Big[\frac{\partial l(t,z_0)}{\partial y}w_1(t)\;+\;w_2(t)\Big]dt,$$

$$w=[w_1,w_2,w_0]\in W(z_0)\Big\}, \qquad (117)$$

$$W(z_0)\;=\;\Big\{w=[w_1,w_2,w_0]\;\Big|\;||w||\le 1,$$

$$w_0\in I\!R,\quad w_1\in P[0,T],\quad w_2\in P[0,T], \qquad (118)$$

$$w_1(t)\ge 0\ \forall t\in\sigma_0(z_0),\ w_1(t)=0\ \forall t\in\sigma_-(z_0)\Big\},$$

$$\sigma_-(z_0)=\{t\in[0,T]\mid l(t,z_0)<0\ \}, \qquad (119)$$

$$\sigma_0(z_0)=\{t\in[0,T]\mid l(t,z_0)=0\ \}, \qquad (120)$$

$$l(t,z_0)=l(x_0+\int_0^t z_{10}(\tau)\,d\tau,\ C_0+\int_0^t z_{20}(\tau)\,d\tau,z_{20}(t),t). \qquad (121)$$

Since $z_0\in Z$ then

$$\sigma_-(z_0)\cup\sigma_0(z_0)\;=\;[0,T]. \qquad (122)$$

It follows from (116) that

$$\Phi_\lambda(z_\varepsilon)=\Phi_\lambda(z_0)\;+\varepsilon\Phi'_\lambda(z_0,v)\;+\;o(\varepsilon,v), \qquad (123)$$

where

$$\Phi'_\lambda(z_0,v)=\max_{C\in\partial\Phi_\lambda(z_0)}(C,v)\;=$$

$$= \max_{C \in \partial\Phi_\lambda(z_0)} \left[\int_0^T C_1(t) v_1(t) dt \; + \; \int_0^T C_2(t) v_2(t) dt \; + \; C_0 v_0 \right], \tag{124}$$

$$\partial\Phi_\lambda(z_0) = \Bigg\{ C = [C_1, C_2, C_0] \in U \;\Bigg|$$

$$C_1(t) = \int_t^T \frac{\partial F(\tau, z_0)}{\partial x} d\tau \; +$$

$$+ \int_t^T \frac{\partial l(\tau, z_0)}{\partial x} \lambda w_1(\tau) d\tau \; + \; \lambda w_0 \; - \; w_2(t) \quad \forall t \in [0, T],$$

$$C_2 \; = \; \frac{\partial F(t, z_0)}{\partial z} \; + \; \int_t^T \frac{\partial F(\tau, z_0)}{\partial y} d\tau \; +$$

$$+ \int_t^T \left(\frac{\partial l(\tau, z_0)}{\partial y} \lambda w_1(\tau) + \lambda w_2(\tau) \right) d\tau \; + \; \frac{\partial l(t, z_0)}{\partial z} \lambda w_1(t) \quad \forall t \in [0, T],$$

$$C_0 \; = \; \int_0^T \left[\frac{\partial F(t, z_0)}{\partial y} \; + \; \frac{\partial l(t, z_0)}{\partial y} \lambda w_1(t) + \lambda w_2(t) \right] dt,$$

$$w = [w_1, w_2, w_0] \in W(z_0) \Bigg\}. \tag{125}$$

Since the point z_0 is a local unconstrained minimizer of the functional $\Phi_\lambda(z)$ in the metric ρ_i and since (see (115))

$$\rho_i(z_\varepsilon, z_0) \underset{\varepsilon \downarrow 0}{\longrightarrow} 0 \quad \forall i \in 1:4$$

then (see [10, 18]) the following necessary condition for a minimum should hold:

$$\Phi'_\lambda(z_0, v) \; \geq 0 \quad \forall v \in U. \tag{126}$$

Arguing as in [7], it is possible to show that this condition is equivalent to

$$\mathbf{0} \in \partial\Phi_\lambda(z_0), \tag{127}$$

where $\mathbf{0} \in U$ is the zero element of the space U. Again arguing as in [7], we conclude from condition (127) that there exists an element $w^* = [w_1^*, w_2^*, w_0^*] \in W(z_0)$ such that

$$\int_t^T \frac{\partial F(\tau, z_0)}{\partial x} d\tau \; + \; \int_t^T \frac{\partial l(\tau, z_0)}{\partial x} \lambda w_1^*(\tau) d\tau \; +$$
$$\lambda w_0^* \; - \; w_2^*(t) \; = \; 0 \quad \forall t \in [0, T], \tag{128}$$

$$\frac{\partial F(t, z_0)}{\partial z} \; + \; \int_t^T \frac{\partial F(\tau, z_0)}{\partial y} d\tau \; +$$

$$+ \int_t^T \left(\frac{\partial l(\tau, z_0)}{\partial y} \lambda w_1^*(\tau) + \lambda w_2^*(\tau) \right) d\tau \ +$$
$$\frac{\partial l(t, z_0)}{\partial z} \lambda w_1^*(t) \ = \ 0 \quad \forall t \in [0, T], \tag{129}$$

$$\int_0^T \left[\frac{\partial F(t, z_0)}{\partial y} \ + \ \frac{\partial l(t, z_0)}{\partial y} \lambda w_1^*(t) + \lambda w_2^*(t) \right] dt \ = \ 0. \tag{130}$$

It follows from (128) that

$$w_2^*(t) \ = \ \int_t^T \frac{\partial F(\tau, z_0)}{\partial x} d\tau \ + \ \int_t^T \frac{\partial l(\tau, z_0)}{\partial x} \lambda w_1^*(\tau) d\tau \ + \ \lambda w_0^*. \tag{131}$$

Substituting (131) into (129) and (130) one gets

$$\frac{\partial F(t, z_0)}{\partial z} \ + \ \int_t^T \frac{\partial F(\tau, z_0)}{\partial y} d\tau \ +$$
$$+ \int_t^T \int_\tau^T \frac{\partial F(s, z_0)}{\partial x} ds d\tau \ + \ \frac{\partial l(t, z_0)}{\partial z} \lambda w_1^*(t) \ + \ \int_t^T \frac{\partial l(\tau, z_0)}{\partial y} \lambda w_1^*(\tau) d\tau \ +$$
$$+ \int_t^T \int_\tau^T \frac{\partial l(s, z_0)}{\partial x} \lambda w_1^*(s) ds d\tau \ + \ \int_t^T \lambda w_0^* d\tau \ = \ 0 \quad \forall t \in [0, T], \tag{132}$$
$$\int_0^T \left[\frac{\partial F(t, z_0)}{\partial y} \ + \ \int_t^T \frac{\partial F(\tau, z_0)}{\partial x} d\tau \ + \right.$$
$$\left. \frac{\partial l(t, z_0)}{\partial y} \lambda w_1^*(t) \ + \ \int_t^T \frac{\partial l(\tau, z_0)}{\partial x} \lambda w_1^*(\tau) d\tau \ + \ \lambda w_0^* \right] dt \ = \ 0. \tag{133}$$

Put

$$\Psi(t) = \lambda w_1^*(t), \quad C_0 = \lambda w_0^*. \tag{134}$$

We conclude from (132) and (133) that

$$\frac{\partial F(t, z_0)}{\partial z} \ + \ \int_t^T \frac{\partial F(\tau, z_0)}{\partial y} d\tau \ +$$
$$+ \int_t^T \int_\tau^T \frac{\partial F(s, z_0)}{\partial x} ds d\tau \ + \ \frac{\partial l(t, z_0)}{\partial z} \Psi(t) \ + \ \int_t^T \frac{\partial l(\tau, z_0)}{\partial y} \Psi(\tau) d\tau \ +$$
$$+ \int_t^T \int_\tau^T \frac{\partial l(s, z_0)}{\partial x} \Psi(s) ds d\tau \ + \ \int_t^T C_0 d\tau \ = \ 0 \quad \forall t \in [0, T], \tag{135}$$
$$\int_0^T \left[\frac{\partial F(t, z_0)}{\partial y} \ + \ \int_t^T \frac{\partial F(\tau, z_0)}{\partial x} d\tau \ + \right.$$
$$\left. \frac{\partial l(t, z_0)}{\partial y} \Psi(t) \ + \ \int_t^T \frac{\partial l(\tau, z_0)}{\partial x} \Psi(\tau) d\tau \ + \ C_0 \right] dt \ = \ 0. \tag{136}$$

It follows from (135) and (136) that

$$\frac{\partial F(T,z_0)}{\partial z} + \frac{\partial l(T,z_0)}{\partial z}\Psi(T) = 0, \tag{137}$$

$$\frac{\partial F(0,z_0)}{\partial z} + \frac{\partial l(0,z_0)}{\partial z}\Psi(0) = 0. \tag{138}$$

Note that

$$\Psi(t) = 0 \quad \forall t \in \sigma_-(z_0).$$

The above results can be formulated as the following statement.

Theorem 7.1 *Let $z_0 \in Z$. Assume that in some neighbourhood $B_{i\delta}(z_0) = \{z \mid \rho_i(z,z_0) < \delta\}$ of the point z_0 the function φ satisfies the condition (113) and let the function f be Lipschitz on $B_{i\delta}(z_0)$ in the metric ρ_i. Then for the point z_0 to be a local minimizer of f on Z in the metric ρ_i, it is necessary that a constant $C_0 \in I\!R$ and a nonnegative function $\Psi(t) \in P[0,T]$ exist such that*

$$\Psi(t) = 0 \quad \forall t \in \sigma_-(z_0), \tag{139}$$

$$\frac{\partial F(t,z_0)}{\partial z} + \int_t^T \frac{\partial F(\tau,z_0)}{\partial y}d\tau +$$
$$+ \int_t^T\int_\tau^T \frac{\partial F(s,z_0)}{\partial x}dsd\tau + \frac{\partial l(t,z_0)}{\partial z}\Psi(t) + \int_t^T \frac{\partial l(\tau,z_0)}{\partial y}\Psi(\tau)d\tau +$$
$$\int_t^T\int_\tau^T \frac{\partial l(s,z_0)}{\partial x}\Psi(s)dsd\tau + \int_t^T C_0 d\tau = 0 \quad \forall t \in [0,T], \tag{140}$$

$$\frac{\partial F(T,z_0)}{\partial z} + \frac{\partial l(T,z_0)}{\partial z}\Psi(T) = 0, \tag{141}$$

$$\frac{\partial F(0,z_0)}{\partial z} + \frac{\partial l(0,z_0)}{\partial z}\Psi(0) = 0. \tag{142}$$

A point $z_0 \in Z$ satisfying (139), (140), (141) and (142) is called a *stationary point* of f on Z.

Remark 7.1 *Since*

$$\Psi(t) = 0 \quad \forall t \in \sigma_-(z_0),$$

then (140) implies

$$\frac{\partial F(t,z_0)}{\partial z} + \int_t^T \frac{\partial F(\tau,z_0)}{\partial y}d\tau +$$
$$+ \int_t^T\int_\tau^T \frac{\partial F(s,z_0)}{\partial x}dsd\tau + (T-t)C_0 = 0 \quad \forall t \in \sigma\_0(z_0).$$

If the function z_0 is continuous at $t \in int\ \sigma_-(z_0)$ then, differentiating twice this identity, we get

$$\frac{d^2}{dt^2}\frac{\partial F(t,z_0)}{\partial z} - \frac{d}{dt}\frac{\partial F(t,z_0)}{\partial y} + \frac{\partial F(t,z_0)}{\partial x} = 0. \tag{143}$$

The equation (143) is just the classical Euler–Poisson equation.

Remark 7.2 *If the function z_0 is continuous at $t \in int\ \sigma_0(z_0)$ then, differentiating twice the equality (140), we have*

$$\frac{\partial l(t,z_0)}{\partial z}\ddot{\Psi}(t) + \dot{\Psi}\left[2\frac{d}{dt}\frac{\partial l(t,z_0)}{\partial z} - \frac{\partial l(t,z_0)}{\partial y}\right] +$$

$$+ \Psi\left[\frac{d^2}{dt^2}\frac{\partial l(t,z_0)}{\partial z} - \frac{d}{dt}\frac{\partial l(t,z_0)}{\partial y} + \frac{\partial l(t,z_0)}{\partial x}\right] +$$

$$+ \frac{d^2}{dt^2}\frac{\partial F(t,z_0)}{\partial z} - \frac{d}{dt}\frac{\partial F(t,z_0)}{\partial y} + \frac{\partial F(t,z_0)}{\partial x} = 0, \tag{144}$$

where Ψ satisfies the boundary conditions (141) and (142).

Remark 7.3 *To check whether a given function $z_0 \in Z$ satisfies the necessary conditions, one has to solve equation (144) with the boundary conditions (141) and(142) (and taking into acount (139)). Under some rather mild hypothesis there exists a unique solution $\Psi(t)$ of (144). If*

$$\Psi(t) \geq 0 \quad \forall t \in [0,T]$$

then z_0 is a stationary point, otherwise z_0 is not a stationary point.

Remark 7.4 *Variational problems with constraints containing higher-order*
derivatives can be treated by the Exact Penalization Technique as well.

Remark 7.5 *In the present paper we have employed the classical variations of z (see (30)). To get more sophisticated optimality conditions it is required to use the so-called needle variations of different types (see [16, 17, 3, 4]).*

References

[1] G.A. Bliss, *Lectures on the Calculus of Variations*, Chicago, University of Chicago Press, 1946.

[2] D. Boukary, A.V. Fiacco, Survey of penalty, exact-penalty and multiplier methods from 1968 to 1993, *Optimization*, Vol. 32, No. 4, pp. 301–334, 1995.

[3] V.F. Demyanov, Nonsmooth problems in Calculus of variations, in *Lecture Notes in Economics and Math. Systems. v. 382. Advances in Optimization*, Eds. W.Oettli, D.Pallaschke, pp. 227–238. Berlin, Springer Verlag, 1992.

[4] V.F. Demyanov, Calculus of variations in nonsmooth presentation. In *Nonsmooth Optimization: Methods and Applications*, Ed. F.Giannessi, pp. 76–91, Singapore, Gordon and Breach, 1992a.

[5] V.F. Demyanov, Exact penalty functions in nonsmooth optimization problems, *Vestnik of St. Petersburg University*, ser. 1, issue 4 (No. 22), pp. 21-27, 1994.

[6] V.F. Demyanov, *Conditions for an extremum and variational problems*, St. Petersburg, St. Petersburg University Press, 2000.

[7] V.F. Demyanov, *Constrained problems of Calculus of Variations via penalization Technique.* See this volume, pp. **00-00**, 2001.

[8] V.F. Demyanov, G. Di Pillo, F. Facchinei, Exact penalization via Dini and Hadamard conditional derivatives, *Optimization Methods and Software*, Vol. 9, pp. 19–36, 1998.

[9] V.F. Demyanov, F. Giannessi, V.V. Karelin, Optimal Control Problems via Exact Penalty Functions, *Journal of Global Optimization*, Vol. 12, No. 3, pp.215–223, 1998.

[10] V.F. Demyanov, A.M. Rubinov, *Constructive Nonsmooth Analysis*, Frankfurt a/M., Peter Lang Verlag, 1995.

[11] G. Di Pillo, L. Grippo, On the Exactness of a class of Nondifferentiable Penalty functions, *J. Optim. Theory Appl.*, Vol. 57, pp. 397–408, 1988.

[12] G. Di Pillo, F. Facchinei, Exact penalty functions for nondifferentiable programming problems, in *Nonsmooth Optimization and Related Topics*, Eds. F.H. Clarke, V.F. Demyanov and F. Giannessi, pp. 89-107, New York, Plenum, 1989.

[13] I.I. Eremin, The method of penalties in convex programming, *Doklady Acad. Nauk SSSR*, Vol. 173, pp. 748 – 751, 1967.

[14] R. Fletcher, Penalty functions, in *Mathematical programming: the state of the art*, (Eds. A. Bachen, M. Grötschel, B. Korte), Springer–Verlag, Berlin, pp. 87–114, 1983.

[15] S. Han, O. Mangasarian, Exact penalty functions in nonlinear programming, *Mathematical Programming*, Vol.17, pp. 251–269, 1979.

[16] M.R. Hestenes, *Calculus of Variations and Optimal Control Theory*, New York, John Wiley & Sons, 1966.

[17] L.S. Pontryagin, V.G. Boltyanskii, R.V. Gamkrelidze, E.F. Mishchenko, *The Mathematical theory of Optimal Processes*, New York, Interscience Publishers, 1962.

[18] R.T. Rockafellar, *Convex Analysis*, Princeton, N.J., Princeton University Press, 1970.

[19] W.L. Zangwill, Nonlinear programming via penalty functions, *Management Science*, Vol. 13, pp. 344–358, 1967.

ON THE STRONG SOLVABILITY OF A UNILATERAL BOUNDARY VALUE PROBLEM FOR NONLINEAR PARABOLIC OPERATORS IN THE PLANE

Rosalba Di Vincenzo
Dipartimento di Matematica - Università di Catania,
Viale A. Doria, 6 - 95125 CATANIA (ITALIA)
e-mail: divincenzo@dmi.unict.it

1 Introduction

In this paper we study the strong solvability of the following boundary value problem with unilateral boundary conditions

$$(1.1)\qquad \begin{cases} \mathcal{A}(X, H(u)) - \dfrac{\partial u}{\partial t} = f(X) & \text{a.e. in } \mathrm{Q} = \Omega \times]0, \mathrm{T}[\\ u(x,0) = 0 & \text{in } \Omega \\ u \geq 0, \ \dfrac{\partial u}{\partial n} \geq 0, \ u \cdot \dfrac{\partial u}{\partial n} = 0 & \text{on } \partial\Omega \times]0, \mathrm{T}[, \end{cases}$$

where n is the unit outward normal to the boundary of Ω, $\partial\Omega$.
Let us observe that when it results $\mathcal{A}(X, H(u)) = \Delta u$, the problem

$$\begin{cases} \Delta u - \alpha \dfrac{\partial u}{\partial t} = f(X) & \text{a.e. in } \mathrm{Q} \\ u(x,0) = 0 & \text{in } \Omega \\ u \geq 0, \ \dfrac{\partial u}{\partial n} \geq 0, \ u \cdot \dfrac{\partial u}{\partial n} = 0 & \text{on } \partial\Omega \times]0, \mathrm{T}[\end{cases}$$

is uniquely solvable for each $f \in L^2(Q)$ and for each $\alpha > 0$ in the space $H^{2,1}(Q)$ (cfr. [1]).
We assume $\mathcal{A}(X, \xi)$, $(X, \xi) \in Q \times \mathbb{R}^4$, to satisfy Carathéodory's condition, i.e. it is measurable in $X \in Q$ for each $\xi = \{\xi_{ij}\}_{i,j=1,2} \in \mathbb{R}^4$ and continuous in ξ for a.a. $X \in Q$. In addition $\mathcal{A}$ verifies the following condition (A), introduced by S.Campanato (see [2]):
(A) *there exist three positive constants* α, γ, δ, *with* $\gamma + \delta < 1$, *such that,*

for almost all $X \in Q$ and for all ξ, $\tau \in \mathbb{R}^4$, it results

$$\left|\sum_{i=1}^{2} \xi_{ii} - \alpha\left[\mathcal{A}(X, \xi+\tau) - \mathcal{A}(X, \tau)\right]\right| \leq \gamma \|\xi\| + \delta \left|\sum_{i=1}^{2} \xi_{ii}\right|,$$

where $\|\xi\| = \left(\sum_{i,j=1}^{2} \xi_{ij}^2\right)^{\frac{1}{2}}$.

Let us note that Campanato's (A) condition does not imply the continuity of the mapping $\mathcal{A}(X, \xi)$ with respect to X, and, taking into account Carathéodory's condition, we need only measurability of $\mathcal{A}(X, \xi)$ with respect to X.

2 Hypotheses and results

Let $\Omega \subset \mathbb{R}^2$ be an open, bounded convex set with C^2-smooth boundary $\partial\Omega$. We assume that $\partial\Omega$ is a closed curve and let $x_1 = x_1(\varphi), x_2 = x_2(\varphi)$ be the normal parameterization of $\partial\Omega$, with φ being a curvilinear parameter, $\varphi \in [0, L]$.
Let $n = (x_2', -x_1')$ be the unit outward normal to $\partial\Omega$.
Let us set $Q = \Omega \times]0, T[$ $(T > 0)$ and denote by $X = (x, t)$ its points.
We consider a real valued function $\mathcal{A}(X, \xi) : \Omega \times \mathbb{R}^4 \to \mathbb{R}$ verifying Carathéodory's condition, Campanato's (A) condition and such that $\mathcal{A}(X, 0) = 0$. We denote by $H(u) = \{D_{ij}u\}_{i,j=1,2}$ the Hessian matrix of u and by $H^{2,1}(Q)$ the (Hilbert) space of the functions $u \in L^2(0, T; W^{2,2}(\Omega))$, u, $u_t \in L^2(Q)$ with the norm

$$\|u\|_{2,1} = \left(\int_0^T dt \int_\Omega (u^2 + \sum_{i=1}^{2} u_{x_i}^2 + \sum_{i,k=1}^{2} u_{x_i x_k}^2 + u_t^2)\, dx\right)^{\frac{1}{2}}.$$

In this paper we are concerned with the "strong" solvability, namely in the space $H^{2,1}(Q)$ of the following problem

$$\text{(2.1)} \qquad \begin{cases} \mathcal{A}(X, H(u)) - \dfrac{\partial u}{\partial t} = f(X) & \text{a.e. in } \mathrm{Q} = \Omega \times]0, \mathrm{T}[\\ u(x, 0) = 0 & \text{in } \Omega \\ u \geq 0, \ \dfrac{\partial u}{\partial n} \geq 0, \ u \cdot \dfrac{\partial u}{\partial n} = 0 & \text{on } \partial\Omega \times]0, \mathrm{T}[. \end{cases}$$

The main result is the following theorem

Theorem 2.1 *Under assumption (A), for each $f \in L^2(Q)$, problem (2.1) is uniquely solvable in the Sobolev space $H^{2,1}(Q)$.*

3 Preliminary results

In the proof of Theorem 2.1 we will use the nearness theory. Now we recall some definitions and results.

Definition 3.1 *Let $\mathcal{B}$ be a set and $\mathcal{B}_1$ a real Banach space. Consider two mappings A and B defined on $\mathcal{B}$ with values in $\mathcal{B}_1$. The mapping A is said to be near to B if there exist two positive constants α and k, $k \in (0,1)$, such that $\forall u, v \in \mathcal{B}$ it implies*

$$\|B(u) - B(v) - \alpha[A(u) - A(v)]\|_{\mathcal{B}_1} \leq \|B(u) - B(v)\|_{\mathcal{B}_1}.$$

The following theorem is proved by S.Campanato [2].

Theorem 3.1 *The mapping $A : \mathcal{B} \to \mathcal{B}_1$ is injective, or surjective, or bijective if and only if it is near to a mapping $B : \mathcal{B} \to \mathcal{B}_1$, which is injective, or surjective, or bijective respectively.*

We recall the definition of monotone operators:

Definition 3.2 *Let $\mathcal{B}$ be a set, $\mathcal{B}_1$ a real Hilbert space, A, B two operators from $\mathcal{B}$ into $\mathcal{B}_1$. The operator A is said to be monotone with respect to B if, for each $u, v \in \mathcal{B}$, it results*

$$(A(u) - A(v), B(u) - B(v))_{\mathcal{B}_1} \geq 0.$$

We also use the following results:

Theorem 3.2 *Let A, B, C be operators from $\mathcal{B}$ into $\mathcal{B}_1$, with $\mathcal{B}_1$ being a real Hilbert space. If:*
A is near to B with costants α e k,
C is monotone with respect to B,
then the mapping $A + C$ is near to $B + \alpha C$ with the same constants α e k.

In the proof of Theorem 2.1 we will use the following existence and uniqueness theorem.

Theorem 3.3 *For all $f \in L^2(Q)$, for all $\alpha > 0$, the problem*

$$\begin{cases} \Delta u - \alpha \dfrac{\partial u}{\partial t} = f(X) & \text{a.e. in Q} \\ u(x,0) = 0 & \text{in } \Omega \\ u \geq 0, \ \dfrac{\partial u}{\partial n} \geq 0, \ u \cdot \dfrac{\partial u}{\partial n} = 0 & \text{on } \partial\Omega \times]0,\mathrm{T}[\end{cases}$$

is uniquely solvable in $H^{2,1}(Q)$.

4 Proof of the theorems

We denote by $\mathcal{B}$ the closure in the space $H^{2,1}(Q)$ of the class

$$W = \{u \in C^{2,1}(\overline{Q}) \bigcap C^3(Q) : \ u(x,0) = 0 \text{ in } \Omega,$$

$$u \geq 0, \ \frac{\partial u}{\partial n} \geq 0, \ u \cdot \frac{\partial u}{\partial n} = 0 \ on \ \partial\Omega \times]0,T[\}$$

Now we prove the Lemma

Lemma 4.1 *Let $\Omega \subset \mathbb{R}^2$ be an open, bounded and convex set with C^2-smooth boundary $\partial\Omega$; let $Q = \Omega \times]0,T[$ $(T > 0)$. For each $u \in W$ it results*

$$\int_Q \sum_{i,j=1}^{2} \left(\frac{\partial^2 u}{\partial x_i \partial x_j} \right)^2 dx\, dt \leq \int_Q |\Delta u|^2 \, dx\, dt. \tag{4.1}$$

Proof. Let $u \in W$, then for each fixed t in $(0,T)$, the function u belongs to $W^{2,2}(\Omega)$ and satisfies the conditions $u \geq 0$ $\frac{\partial u}{\partial n} \geq 0$, $u \cdot \frac{\partial u}{\partial n} = 0$ on $\partial\Omega$. By virtue of Lemma 4.1 by S.Giuffrè [3] it results

$$\int_\Omega \sum_{i,j=1}^{2} \left(\frac{\partial^2 u}{\partial x_i \partial x_j} \right)^2 dx \leq \int_\Omega |\Delta u|^2 \, dx.$$

In order to obtain estimate (4.1) it is enough to integrate the last inequality in $(0,T)$.

We observe that estimate (4.1) is still true for all $u \in \mathcal{B}$.

Now we prove, using estimate (4.1) and condition (A), that the mapping $\mathcal{A}(X, H(u))$ and Δu, both considered as operators from $\mathcal{B}$ in $L^2(Q)$, are near.

In fact we get

$$\left(\int_Q |\Delta(u-v) - \alpha \left[\mathcal{A}(X, H(u)) - \mathcal{A}(X, H(v)) \right]|^2 dx\, dt \right)^{\frac{1}{2}}$$

$$\leq \left(\int_Q |\gamma \, \|H(u-v)\| + \delta \, |\Delta(u-v)||^2 \, dx\, dt \right)^{\frac{1}{2}}$$

$$\leq \gamma \left(\int_Q \|H(u-v)\|^2 \, dx\, dt \right)^{\frac{1}{2}} + \delta \left(\int_Q |\Delta(u-v)|^2 \, dx\, dt \right)^{\frac{1}{2}}$$

$$\leq (\gamma+\delta)\left(\int_Q |\Delta(u-v)|^2\,dx\,dt\right)^{\frac{1}{2}}$$

with $\gamma+\delta<1$.

In order to prove that $\mathcal{A}(X,H(u))-\dfrac{\partial u}{\partial t}$ is near to $\Delta u-\alpha\dfrac{\partial u}{\partial t}$, using Theorem 3.2, it is enough to prove that $-\dfrac{\partial u}{\partial t}$ is a monotone operator with respect to Δu, that is

$$-\int u_t\,\Delta u\,dx\,dt\geq 0 \quad \forall u\in\mathcal{B}. \tag{4.2}$$

In order to obtain estimate (4.2), let us set

$$p_i=\frac{\partial u}{\partial x_i},\quad \mathrm{i}=1,2;\quad p_{ij}=\frac{\partial^2 u}{\partial x_i\partial x_j},\quad \mathrm{i,j}=1,2.$$

Taking into account that it results

$$\frac{\partial u}{\partial t}p_{ii}=\frac{\partial}{\partial x_i}(u_t\,p_i)-p_i\frac{\partial}{\partial x_i}u_t \quad i=1,2$$

and the Green-Gauss formula, we obtain

$$-\int_0^T\int_\Omega u_t\,\Delta u\,dx\,dt$$

$$=-\left(\int_0^T dt\int_0^L (u_tp_1x_2'-u_tp_2x_1')\,d\varphi-\frac{1}{2}\int_\Omega dx\int_0^T\frac{\partial}{\partial t}(p_1^2+p_2^2)\,dt\right)$$

$$=-\int_0^T\int_0^L u_t\frac{\partial u}{\partial n}\,dx\,dt+\frac{1}{2}\int_\Omega[p_1^2(x,T)+p_2^2(x,T)]dx.$$

Therefore it is enough to prove that

$$-\int_0^T\int_0^L u_t\,\frac{\partial u}{\partial n}\,dx\,dt\geq 0.$$

It results

$$-\int_0^T\int_0^L u_t\,\frac{\partial u}{\partial n}\,dx\,dt$$

$$=-\int_{\{(\varphi,T)\in[0,L]\times[0,T]:\frac{\partial u}{\partial n}(x_1(\varphi),x_2(\varphi),t)>0\}}u_t\,\frac{\partial u}{\partial n}\,dx\,dt$$

$$+\int_{\{(\varphi,T)\in[0,L]\times[0,T]:\frac{\partial u}{\partial n}(x_1(\varphi),x_2(\varphi),t)=0\}}u_t\,\frac{\partial u}{\partial n}\,dx\,dt.$$

The second integral is obviously equal to zero.
As it concerns the first integral, since $u \in \mathcal{B}$, in each point $(\varphi^*, t^*) \in [0, L] \times [0, T]$ where it results $\frac{\partial u}{\partial n}(x_1(\varphi^*), x_2(\varphi^*), t^*) > 0$, we obtain that $u(x_1(\varphi^*), x_2(\varphi^*), t^*) = 0$; on the other hand, since $u(x_1(\varphi), x_2(\varphi), t) \geq 0$ $\forall(\varphi, t) \in [0, L] \times [0, T]$, these points are minimum points for the function u, then $\frac{\partial u}{\partial t}(x_1(\varphi^*), x_2(\varphi^*), t^*) = 0$. Therefore we obtain estimate (4.2).
In virtue of Theorem 3.3 the operator $\Delta u - \alpha u_t$ is a bijection between $\mathcal{B}$ and $L^2(Q)$, therefore also the operator $\mathcal{A}(X, H(u)) - \frac{\partial u}{\partial t}$ is a bijection between the same spaces.

References

[1] H. Brezis, *"Problèmes Unilatéraux"*, J. Math. Pures et Appl. **51**, Fasc 1 (1972), pp. 1-168.

[2] S. Campanato, *"On the condition of nearness between operators"*, Ann. Mat. Pura Appl., **167** (1994), pp. 243-256.

[3] S. Giuffrè, *"On the strong solvability of a unilateral boundary value problem for nonlinear discontinuous operators in the plane"*, Equilibrium Problems: Nonsmooth Optimization and Variational Inequality Models, Kluwer Academic Publishers, F. Giannessi A. Maugeri P. Pardalos Eds., 2001.

SOLVING A SPECIAL CLASS OF DISCRETE OPTIMAL CONTROL PROBLEMS VIA A PARALLEL INTERIOR–POINT METHOD

Carla Durazzi
Department of Mathematics, University of Ferrara
via Machiavelli 35
44100 Ferrara, Italy
e-mail: c.durazzi@unife.it

Valeria Ruggiero
Department of Mathematics, University of Ferrara
via Machiavelli 35
44100 Ferrara, Italy
e-mail: v.ruggiero@unife.it

Gaetano Zanghirati
Department of Mathematics, University of Ferrara
via Machiavelli 35
44100 Ferrara, Italy
e-mail: g.zanghirati@unife.it

Abstract This paper is concerned with a parallel primal–dual Interior–Point method combined with the preconditioned conjugate gradient algorithm. Recently, the global convergence of this scheme has been proved and this approach has been used for solving linear stochastic programming and robust optimization problems on parallel computers.

In this paper, we analyse the method for a more general formulation of a constrained quadratic programming problem and we provide the hypotheses for the global convergence of the obtained scheme. A large-scale application related to a class of discrete quadratic optimal control problems is presented: the numerical results on parallel architectures confirm the effectiveness of the IPPCG method on programs with special structure.

Keywords: Interior–Point methods, inexact Newton methods, preconditioned conjugate gradient method, discrete optimal control problems, parallel computing.

1 Introduction

A crucial issue for an efficient implementation of Interior-Point (IP) methods is the solution of the large sparse linear system arising at each iteration for finding the search direction. The present state-of-the-art codes on IP methods [15] for linear and quadratic programming (LP, QP) problems use direct inner solvers, while iterative inner solvers are considered not competitive. Iterative methods can be used, however, with success on special classes of large-scale problems, having a particular structure in the matrix of the linear constraints. In [10], [11], [12], [13], efficient implementations of IP methods for network linear programs have been proposed: they are based on solving approximately the normal equations, using the Preconditioned Conjugate Gradient (PCG) method; a number of inexpensive and efficient preconditioners suggested by the underlying network structure of the problem have been presented.

A similar approach is used in an infeasible path-following IP method combined with the PCG algorithm, named IPPCG method [6], for solving linear stochastic programming and robust optimization problems. The numerical results obtained by a parallel code on Cray T3E/256 and SGI Origin 3800/32 show the effectiveness of the IPPCG method.

Furthermore, the paper [6] points out that the IPPCG method can be viewed as a very general formulation of an infeasible path-following IP method for an LP or QP problem. Indeed, using the well-known results on the global convergence of inexact Newton methods [7], a sufficient condition for an approximate solution of the normal equations is devised so that the global convergence of the obtained scheme is assured. In the case of a direct inner solver, the condition is obviously met, since the "exact" solution is determined; for an iterative inner solver, the condition can be used as an "adaptive" stopping rule that enables to avoid unnecessary inner iterations when we are far from the solution. In this work, we consider a more general formulation of an LP or QP problem with respect to that in [6] and we describe the consequent modifications in the IPPCG scheme, in the hypotheses and in the proof of its global convergence. Furthermore, we present a large–scale application concerning another class of programs with special structure. By a parallel implementation of the IPPCG method, we compute the numerical solution of two optimal control problems: the first is related to improving water quality by means of in–stream aeration process [9] and the second is related to a "diffusion-convection" process with a linear state equation and a source term independent on the solution. For the transcription of these control problems as QP problems by means of a finite difference discretization, see [5]. The numerical experiments carried out on Cray T3E/256, confirm the effectiveness of the method and the good performance of the parallel code as shown in [6] for the class of stochastic linear and robust optimization problems.

In the sequel of the work, we use the following notations. Given the vectors x, Δx and the scalar α, we denote with $x(\alpha)$ the vector $x + \alpha\Delta x$ and with $x_i(\alpha)$ an element of $x(\alpha)$. We indicate with $\mathbf{1}$ a size compatible vector with all entries equal to 1; when x is a generic n-vector, X denotes the $n \times n$ diagonal matrix $diag(x_1, \ldots, x_n)$. Finally, we denote the iteration count by a superscript in vectors and matrices, while a subscript is used for scalars.

2 Framework of the Method

Consider the linearly constrained quadratic programming (QP) problem

$$\begin{aligned} \min \quad & f(x) = \frac{1}{2}x^T Q x + c^T x, && \text{(1a)} \\ \text{s.t.} \quad & A_{\mathrm{I}} x \geq b_{\mathrm{I}}, && \text{(1b)} \\ & A_{\mathrm{E}} x = b_{\mathrm{E}}, && \text{(1c)} \\ & \Pi_{\mathrm{L}} x \geq \ell, && \text{(1d)} \\ & \Pi_{\mathrm{U}} x \leq u && \text{(1e)} \end{aligned}$$

where Q is a symmetric positive semidefinite matrix of order n, $A = (A_{\mathrm{I}}^T, A_{\mathrm{E}}^T)^T$ is an $m \times n$ matrix with $A_{\mathrm{I}} \in \mathbb{R}^{m_{\mathrm{I}} \times n}$, $A_{\mathrm{E}} \in \mathbb{R}^{m_{\mathrm{E}} \times n}$, $m_{\mathrm{I}} + m_{\mathrm{E}} = m$, c and x are n-vectors, $b = (b_{\mathrm{I}}^T \; b_{\mathrm{E}}^T)^T$ is an m-vector, $\ell \in \mathbb{R}^{n_{\mathrm{L}}}$, $u \in \mathbb{R}^{n_{\mathrm{U}}}$, $0 \leq n_{\mathrm{L}}, n_{\mathrm{U}} \leq n$.

All the entries of the matrices $\Pi_{\mathrm{L}} \in \mathbb{R}^{n_{\mathrm{L}} \times n}$ and $\Pi_{\mathrm{U}} \in \mathbb{R}^{n_{\mathrm{U}} \times n}$ are equal to zero except for $p_{ji}^L = 1$ if $x_i \geq \ell_j$ and $p_{ji}^U = 1$ if $x_i \leq u_j$. We define L as the index set of the variables bounded below and U as the index set of the variables bounded above and we assume that each variable is bounded below or/and above, i.e. $L \cup U = \{1, \ldots, n\}$, and that, when $\ell_j \leq x_i \leq u_k$, $i \in L \cap U$, then $\ell_j \neq u_k$.

The equality constraint (1c) can be formulated as

$$b_{\mathrm{E}} \geq A_{\mathrm{E}} x \geq b_{\mathrm{E}} \tag{2}$$

When $Q = 0$, (1) is a linear programming (LP) problem.

In order to derive an infeasible primal–dual path–following Interior–Point (IP) method, we rewrite the QP problem (1a), (1b), (2), (1d), (1e) introducing slack variables as appropriate to replace all the inequality constraints with simple nonnegativity constraints:

$$
\begin{aligned}
\min \quad & f(x), \\
\text{s.t.} \quad & Ax - s = b, \\
& p_{\text{E}} + s_{\text{E}} = 0, \\
& \Pi_{\text{L}} x - z_{\text{L}} = \ell, \\
& \Pi_{\text{U}} x + z_{\text{U}} = u \\
& s \geq 0, p_{\text{E}} \geq 0, z_{\text{L}} \geq 0, z_{\text{U}} \geq 0
\end{aligned} \tag{3}
$$

where $s = (s_{\text{I}}^T, s_{\text{E}}^T)^T \in \mathbb{R}^m$, $s_{\text{I}} \in \mathbb{R}^{m_{\text{I}}}$, $s_{\text{E}}, p_{\text{E}} \in \mathbb{R}^{m_{\text{E}}}$, $z_{\text{L}} \in \mathbb{R}^{n_{\text{L}}}$, $z_{\text{U}} \in \mathbb{R}^{n_{\text{U}}}$. The Lagrangian function of the problem (3) is given by:

$$
\begin{aligned}
L(x, s, p_{\text{E}}, z_{\text{L}}, z_{\text{U}}, y, q_{\text{E}}, v_{\text{L}}, v_{\text{U}}, \lambda_s, \lambda_{p_{\text{E}}}, \lambda_{z_{\text{L}}}, \lambda_{z_{\text{U}}}) = \\
= \; & (1/2) x^T Q x + c^T x \\
& - y^T (Ax - s - b) + q_{\text{E}}^T (p_{\text{E}} + s_{\text{E}}) - v_{\text{L}}^T (\Pi_{\text{L}} x - z_{\text{L}} - \ell) \\
& + v_{\text{U}}^T (\Pi_{\text{U}} x + z_{\text{U}} - u) - \lambda_s^T s - \lambda_{p_{\text{E}}}^T p_{\text{E}} - \lambda_{z_{\text{L}}}^T z_{\text{L}} - \lambda_{z_{\text{U}}}^T z_{\text{U}}
\end{aligned}
$$

where $y, \lambda_s \in \mathbb{R}^m$, $q_{\text{E}}, \lambda_{p_{\text{E}}} \in \mathbb{R}^{m_{\text{E}}}$, $v_{\text{L}}, \lambda_{z_{\text{L}}} \in \mathbb{R}^{n_{\text{L}}}$, $v_{\text{U}}, \lambda_{z_{\text{U}}} \in \mathbb{R}^{n_{\text{U}}}$ and they are the vectors of multipliers corresponding to the constraints of (3). From the Karush–Kuhn–Tucker (KKT) conditions for (3), we obtain the following constrained nonlinear system:

$$
\begin{aligned}
Qx + c - A^T y - \Pi_{\text{L}}^T v_{\text{L}} + \Pi_{\text{U}}^T v_{\text{U}} &= 0 && \text{(4a)} \\
y - \lambda_s + \Pi_{\text{E}}^T q_{\text{E}} &= 0 && \text{(4b)} \\
Ax - s &= b && \text{(4c)} \\
p_{\text{E}} + s_{\text{E}} &= 0 && \text{(4d)} \\
\Pi_{\text{L}} x - z_{\text{L}} &= \ell && \text{(4e)} \\
\Pi_{\text{U}} x + z_{\text{U}} &= u && \text{(4f)} \\
\Lambda_s S \mathbf{1} &= 0 && \text{(4g)} \\
Q_{\text{E}} P_{\text{E}} \mathbf{1} &= 0 && \text{(4h)} \\
V_{\text{L}} Z_{\text{L}} \mathbf{1} &= 0 && \text{(4i)} \\
V_{\text{U}} Z_{\text{U}} \mathbf{1} &= 0 && \text{(4j)} \\
\lambda_s \geq 0, q_{\text{E}} \geq 0, v_{\text{L}} \geq 0, v_{\text{U}} &\geq 0 && \text{(4k)} \\
s \geq 0, p_{\text{E}} \geq 0, z_{\text{L}} \geq 0, z_{\text{U}} &\geq 0 && \text{(4l)}
\end{aligned}
$$

where $\Pi_{\text{E}} = (0_{m_{\text{E}} \times m_{\text{I}}} \; I_{m_{\text{E}}}) \in \mathbb{R}^{m_{\text{E}} \times m}$ and $\lambda_{p_{\text{E}}} = q_{\text{E}}$, $\lambda_{z_{\text{L}}} = v_{\text{L}}$, $\lambda_{z_{\text{U}}} = v_{\text{U}}$ are used.

The system (4) can be written in compact form as

$$H(w) = 0 \quad w_{\mathrm{P}} \geq 0 \quad w_{\mathrm{D}} \geq 0 \tag{5}$$

where $w = (x^T, y^T, w_{\mathrm{D}}^T, w_{\mathrm{P}}^T)^T \in \mathbb{R}^{n+m+2n_{\mathrm{C}}}$, $w_{\mathrm{D}} = (\lambda_s^T, q_{\mathrm{E}}^T, v_{\mathrm{L}}^T, v_{\mathrm{U}}^T)^T \in \mathbb{R}^{n_{\mathrm{C}}}$, $w_{\mathrm{P}} = (s^T, p_{\mathrm{E}}^T, z_{\mathrm{L}}^T, z_{\mathrm{U}}^T)^T \in \mathbb{R}^{n_{\mathrm{C}}}$, $n_{\mathrm{C}} = m + m_{\mathrm{E}} + n_{\mathrm{L}} + n_{\mathrm{U}}$,

$$H(w) = \begin{bmatrix} G_{\mathrm{D}}(w) \\ G_{\mathrm{P}}(w) \\ W_{\mathrm{D}} W_{\mathrm{P}} \mathbf{1} \end{bmatrix}, \quad G_{\mathrm{D}}(w) = \begin{bmatrix} Qx + c - A^T y - \Pi_{\mathrm{L}}^T v_{\mathrm{L}} + \Pi_{\mathrm{U}}^T v_{\mathrm{U}} \\ y - \lambda_s + \Pi_{\mathrm{E}}^T q_{\mathrm{E}} \end{bmatrix}$$

$$\text{and} \quad G_{\mathrm{P}}(w) = \begin{bmatrix} Ax - s - b \\ p_{\mathrm{E}} + s_{\mathrm{E}} \\ \Pi_{\mathrm{L}} x - z_{\mathrm{L}} - \ell \\ \Pi_{\mathrm{U}} x + z_{\mathrm{U}} - u \end{bmatrix}.$$

The constrained nonlinear system (5) also gives the KKT conditions for the following problem that is the dual problem of (3):

$$\begin{aligned} \max \quad & \varphi(x, y, w_{\mathrm{D}}) = -\frac{1}{2} x^T Q x + b^T y + \ell^T v_{\mathrm{L}} - u^T v_{\mathrm{U}}, && \text{(6a)} \\ \text{s.t.} \quad & Qx + c - A^T y - \Pi_{\mathrm{L}}^T v_{\mathrm{L}} + \Pi_{\mathrm{U}}^T v_{\mathrm{U}} = 0, && \text{(6b)} \\ & y - \lambda_s + \Pi_{\mathrm{E}}^T q_{\mathrm{E}} = 0, && \text{(6c)} \\ & \lambda_s \geq 0, q_{\mathrm{E}} \geq 0, v_{\mathrm{L}} \geq 0, v_{\mathrm{U}} \geq 0 && \text{(6d)} \end{aligned}$$

In particular, the formulas (4a), (4b), (4k) represent the conditions of the dual feasibility, while (4c), (4d), (4e), (4f), (4l) are the conditions of the primal feasibility and (4g), (4h), (4i), (4j) are the complementary conditions.

The main task of each iteration of a primal–dual path–following IP method consists in solving, by a step of the Newton method, the perturbed KKT conditions for (3), given by

$$H(w) = \sigma \mu \boldsymbol{e}, \quad w_{\mathrm{P}} \geq 0, \; w_{\mathrm{D}} \geq 0, \quad \sigma, \mu > 0 \tag{7}$$

with $\boldsymbol{e} = (\mathbf{0}^T \mathbf{1}^T)^T, \mathbf{1} \in \mathbb{R}^{n_{\mathrm{C}}}, \mathbf{0} \in \mathbb{R}^{n+m+n_{\mathrm{C}}}$.

On the other hand, following [1] and [4], we can view the task of solving the nonlinear system (7) as one step of an inexact Newton method applied to solve the exact KKT conditions (5) with a residual term equal to $\sigma\mu\boldsymbol{e}$. In practice, starting from a point w^0, such that $w_{\mathrm{P}}^0 > 0$, $w_{\mathrm{D}}^0 > 0$, at each iteration k of an IP method, we have to find the solution Δw^* of the linear system

$$H'(w^k) \Delta w = -H(w^k) + \sigma_k \mu_k \boldsymbol{e}, \tag{8}$$

where

$$H'(w^k) = \begin{pmatrix} (\nabla_{(x^T,\, y^T)^T} G_{\mathrm{D}}(w^k))^T & (\nabla_{w_{\mathrm{D}}} G_{\mathrm{D}}(w^k))^T & (\nabla_{w_{\mathrm{P}}} G_{\mathrm{D}}(w^k))^T \\ (\nabla_{(x^T,\, y^T)^T} G_{\mathrm{P}}(w^k))^T & (\nabla_{w_{\mathrm{D}}} G_{\mathrm{P}}(w^k))^T & (\nabla_{w_{\mathrm{P}}} G_{\mathrm{P}}(w^k))^T \\ 0 & W_{\mathrm{P}}^k & W_{\mathrm{D}}^k \end{pmatrix}$$

with

$$(\nabla_{(x^T,\, y^T)^T} G_{\mathrm{D}}(w^k))^T = \begin{pmatrix} Q & -A^T \\ 0 & I \end{pmatrix}$$

and

$$(\nabla_{(x^T,\, y^T)^T} G_{\mathrm{P}}(w^k))^T = \begin{pmatrix} A & 0 \\ 0 & 0 \\ \Pi_{\mathrm{L}} & 0 \\ \Pi_{\mathrm{U}} & 0 \end{pmatrix}.$$

Here $(\nabla_{x_1} F(\bar{x}))^T$ denotes the Jacobian of $F(x)$ with respect to x_1, where $x = (x_1^T, x_2^T)^T$, computed in $\bar{x}$.
In particular, dropping the iteration index k, the linear system (8) has the form

$$\begin{aligned} Q\Delta x - A^T \Delta y - \Pi_{\mathrm{L}}^T \Delta v_{\mathrm{L}} + \Pi_{\mathrm{U}}^T \Delta v_{\mathrm{U}} &= \alpha \\ \Delta y - \Delta\lambda_s + \Pi_{\mathrm{E}}^T \Delta q_{\mathrm{E}} &= \beta \\ A\Delta x - \Delta s &= \gamma \\ \Pi_{\mathrm{E}} \Delta s + \Delta p_{\mathrm{E}} &= \delta \\ \Pi_{\mathrm{L}} \Delta x - \Delta z_{\mathrm{L}} &= \varepsilon \\ \Pi_{\mathrm{U}} \Delta x + \Delta z_{\mathrm{U}} &= \zeta \\ \Lambda_s \Delta s + S\Delta\lambda_s &= \eta \\ Q_{\mathrm{E}} \Delta p_{\mathrm{E}} + P_{\mathrm{E}} \Delta q_{\mathrm{E}} &= \theta \\ V_{\mathrm{L}} \Delta z_{\mathrm{L}} + Z_{\mathrm{L}} \Delta v_{\mathrm{L}} &= \nu \\ V_{\mathrm{U}} \Delta z_{\mathrm{U}} + Z_{\mathrm{U}} \Delta v_{\mathrm{U}} &= \xi \end{aligned} \tag{9}$$

where

$$\begin{aligned} (\alpha^T, \beta^T)^T &= -G_{\mathrm{D}}(w), \\ (\gamma^T, \delta^T, \varepsilon^T, \zeta^T)^T &= -G_{\mathrm{P}}(w) \\ (\eta^T, \theta^T, \nu^T, \xi^T)^T &= -W_{\mathrm{D}} W_{\mathrm{P}} \mathbf{1} + \mu\sigma\mathbf{1}. \end{aligned}$$

Using an elimination technique on the system (9), we can write

$$
\begin{aligned}
\Delta z_{\mathrm{U}} &= V_{\mathrm{U}}^{-1}\xi - V_{\mathrm{U}}^{-1} Z_{\mathrm{U}} \Delta v_{\mathrm{U}} \\
\Delta z_{\mathrm{L}} &= V_{\mathrm{L}}^{-1}\nu - V_{\mathrm{L}}^{-1} Z_{\mathrm{L}} \Delta v_{\mathrm{L}} \\
\Delta p_{\mathrm{E}} &= Q_{\mathrm{E}}^{-1}\theta - Q_{\mathrm{E}}^{-1} P_{\mathrm{E}} \Delta q_{\mathrm{E}} \\
\Delta \lambda_s &= S^{-1}\eta - S^{-1}\Lambda_s \Delta s \\
\Delta q_{\mathrm{E}} &= P_{\mathrm{E}}^{-1}(Q_{\mathrm{E}}(\Pi_{\mathrm{E}}\Delta s - \delta) + \theta) \\
\Delta v_{\mathrm{L}} &= Z_{\mathrm{L}}^{-1}(V_{\mathrm{L}}(\varepsilon - \Pi_{\mathrm{L}}\Delta x) + \nu) \\
\Delta v_{\mathrm{U}} &= Z_{\mathrm{U}}^{-1}(V_{\mathrm{U}}(\Pi_{\mathrm{U}}\Delta x - \zeta) + \xi) \\
\Delta s &= E(S^{-1}\eta - \Delta y + \beta + \Pi_{\mathrm{E}}^T P_{\mathrm{E}}^{-1}(Q_{\mathrm{E}}\delta - \theta))
\end{aligned}
\tag{10}
$$

where $E = (S^{-1}\Lambda_s + \Pi_{\mathrm{E}}^T P_{\mathrm{E}}^{-1} Q_{\mathrm{E}} \Pi_{\mathrm{E}})^{-1}$ is a diagonal positive definite matrix when $w^k > 0$.
In this way, the system (8) is reduced as follows

$$D\Delta x - A^T \Delta y = \varphi, \tag{11a}$$
$$A\Delta x + E\Delta y = \psi \tag{11b}$$

where

$$
\begin{aligned}
D &= Q + \Pi_{\mathrm{L}}^T Z_{\mathrm{L}}^{-1} V_{\mathrm{L}} \Pi_{\mathrm{L}} + \Pi_{\mathrm{U}}^T Z_{\mathrm{U}}^{-1} V_{\mathrm{U}} \Pi_{\mathrm{U}}, \\
\varphi &= \Pi_{\mathrm{L}}^T Z_{\mathrm{L}}^{-1}(V_{\mathrm{L}}\varepsilon + \nu) - \Pi_{\mathrm{U}}^T Z_{\mathrm{U}}^{-1}(\xi - V_{\mathrm{L}}\zeta) + \alpha, \\
\psi &= E(S^{-1}\eta + \beta + \Pi_{\mathrm{E}}^T P_{\mathrm{E}}^{-1}(Q_{\mathrm{E}}\delta - \theta)) + \gamma.
\end{aligned}
$$

Since $L \cup U = \{1, 2, \dots, n\}$, $\Pi_{\mathrm{L}}^T Z_{\mathrm{L}}^{-1} V_{\mathrm{L}} \Pi_{\mathrm{L}} + \Pi_{\mathrm{U}}^T Z_{\mathrm{U}}^{-1} V_{\mathrm{U}} \Pi_{\mathrm{U}}$ is a diagonal matrix with positive diagonal entries when $w^k > 0$. Then, D is a positive definite matrix and we can solve (11a) with respect to Δx, obtaining

$$\Delta x = D^{-1}(\varphi + A^T \Delta y) \tag{12}$$

Hence, the direction Δw is completely determined once Δy is computed by the linear system

$$(AD^{-1}A^T + E)\Delta y = \psi - AD^{-1}\varphi \tag{13}$$

We observe that, at each iteration k, $A(D^k)^{-1}A^T + E^k$ is a symmetric positive definite matrix of order m when $w^k > 0$.
In general, the solution of (13) (or (11)) and, then, the direction Δw^k are computed "exactly" within the working precision, using direct methods. This approach is used by the majority of the codes for QP and LP problems (see [15] or [14] for a survey).
In [6], we suggest to solve (13) by an iterative method, in particular by the Preconditioned Conjugate Gradient (PCG) algorithm that is well suited

to solve positive definite systems on parallel computers. To avoid unnecessary iterations when w^k is far from the solution, the inner iterative scheme must be stopped by an adaptive termination rule that determines an "approximate" vector Δy^k with an accuracy depending on the "quality" of the outer iterate w^k. In this case the computed solution Δw^k is an "approximate" direction for the IP method.
After the determination of an exact or approximate direction Δw^k, the next step of the kth iteration of the IP method consists in computing a positive scalar $\tilde{\alpha}_k \in (0,1]$, assuring that the vectors $w_{\text{D}}^k + \tilde{\alpha}_k \Delta w_{\text{D}}^k$ and $w_{\text{P}}^k + \tilde{\alpha}_k \Delta w_{\text{P}}^k$ are strictly positive. This aim is obtained if $\tilde{\alpha}_k$ is computed as the largest value in $(0,1]$ such that the following centrality conditions [8] are simultaneously met for any $\alpha \in (0, \tilde{\alpha}_k]$

$$\min_{i=1,\ldots,n_{\text{C}}} (w_{\text{D}}^k(\alpha))_i (w_{\text{P}}^k(\alpha))_i \geq \frac{\tau_1}{n_{\text{C}}} (w_{\text{D}}^k(\alpha))^T w_{\text{P}}^k(\alpha) \tag{14a}$$

$$(w_{\text{D}}^k(\alpha))^T w_{\text{P}}^k(\alpha) \geq \tau_2 \left\| \begin{matrix} G_{\text{D}}(w^k(\alpha)) \\ G_{\text{P}}(w^k(\alpha)) \end{matrix} \right\|_2 \tag{14b}$$

where $0 < \tau_1 < 1$, $\tau_1 \leq n_{\text{C}} \min_{i=1,\ldots,n_{\text{C}}} (w_{\text{D}}^0)_i (w_{\text{P}}^0)_i / (w_{\text{D}}^0)^T w_{\text{P}}^0$ and $0 < \tau_2 \leq (w_{\text{D}}^0)^T w_{\text{P}}^0 / \|((G_{\text{D}}(w^k(\alpha)))^T, (G_{\text{P}}(w^k(\alpha)))^T)^T\|_2$.
If $((w_{\text{D}}^k)^T, (w_{\text{P}}^k)^T)^T > 0$, we observe that, for any $\alpha \in (0, \tilde{\alpha}_k]$, $w^k(\alpha) > 0$ if and only if $\|H(w^k(\alpha))\|_2 > 0$, i.e. the IP method does not break down. See [8] for additional details.
An approach to the proof of the global convergence of the described iterative IP scheme is to show that this method belongs to the class of the inexact Newton methods [7, p.394]: so, the proof is reduced to find conditions such that the following inequality, referred as "inexact Newton condition", holds:

$$\|H'(w^k)\tilde{\alpha}_k \Delta w^k + H(w^k)\|_2 \leq \eta_k(\tilde{\alpha}_k) \|H(w^k)\|_2, \tag{15}$$

where $\eta_k(\tilde{\alpha}_k)$ is uniformly less than 1.
The well-known backtracking technique, which can be stated as

1. set $\alpha_k = \tilde{\alpha}_k$; $\eta_k = \eta_k(\tilde{\alpha}_k)$; $\vartheta, \beta \in (0,1)$;

2. while

$$\|H(w^k + \alpha_k \Delta w^k)\|_2 > (1 - \beta(1-\eta_k)) \|H(w^k)\|_2, \tag{16}$$

   update $\alpha_k = \vartheta \alpha_k$ and $\eta_k = 1 - \vartheta(1 - \eta_k)$;

3. $w^{k+1} = w^k + \alpha_k \Delta w^k$;

can be used within the IP algorithm: in this way, the method can be viewed as the INB algorithm in [7, p.410] and its global convergence follows from the Theorem 6.1 [7].
In the next section, we report the conditions assuring that (15) holds with $\eta_k(\tilde{\alpha}_k) \leq \bar{\eta} < 1$ for any $k \geq 0$.
In the following, the IP method described in this section with the PCG algorithm as inner solver is named IPPCG method. We observe that the general form (1) of a linearly constrained QP problem includes a large number of particular cases arising in real applications.
A crucial point for the above formulation of the IP method is that the matrices W_{D}^k and W_{P}^k must have strictly positive diagonal entries so that D^k and E^k are positive definite matrices. Then, the equality constraints must be written as inequalities and all the entries of x must be bounded above or below ($L \cup U = \{1, 2, \ldots, n\}$). Standard techniques always enable us to formulate a QP problem with prefixed or free variables as in (1) so that the matrix D^k is positive definite (see p. 3 or 16 of [15]).

3 Global convergence

The global convergence of the IPPCG method is proved in [6] for a simplified QP problem ($m_{\mathrm{E}} = 0$, $n_{\mathrm{L}} = n$, $n_{\mathrm{U}} = 0$, $\ell = 0$). In this section, we report the conditions required for the global convergence of the method when the QP problem has the general form in (1). The proofs of the propositions given below run following standard arguments as in [6] and, then, are omitted.
We introduce the closed set $\Omega(\epsilon)$, $\epsilon \geq 0$

$$\Omega(\epsilon) = \Big\{ w \in \mathbb{R}^{m+n+2n_{\mathrm{C}}} : \epsilon \leq \|H(w)\|_2 \leq \|H(w^0)\|_2, \quad \text{such that } w \text{ satisfies the conditions (14)} \Big\} \tag{17}$$

We assume that the following basic condition holds:

A1. The sequences $\{x^k\}$, $\{y^k\}$ and $\{q_{\mathrm{E}}^k\}$ are bounded above in $\Omega(\epsilon)$, $\epsilon \geq 0$.

To control A1, the following termination criterium is used to stop the IPPCG algorithm: stop if $\|H(w^k)\|_2 < \epsilon^*$ or $\|(x^k, y^k)\|_1 + \|q_{\mathrm{E}}^k\|_1 > \omega^*$, where ϵ^* is a prefixed tolerance and ω^* is a large positive scalar.
When we use the PCG method to solve (13) at the kth iteration of the IP method, the "approximate" solution Δy^k is the "exact" solution of the system

$$(A(D^k)^{-1}A^T + E^k)\Delta y = \psi^k - A(D^k)^{-1}\varphi^k + \hat{r}^k, \tag{18}$$

where $\hat{r}^k$ is the opposite of the residual vector at the last iteration of the PCG method. The direction Δw^k, computed by backward substitution in the formulas (12) and (10), is the exact solution of the system

$$H'(w^k)\Delta w = -H(w^k) + r^k, \tag{19}$$

where

$$r^k = \begin{bmatrix} 0_{m+n} \\ \hat{r}^k \\ 0_{m_\mathrm{E}+n_\mathrm{L}+n_\mathrm{U}} \\ \sigma_k\mu_k\mathbf{1} \end{bmatrix}. \tag{20}$$

We assume to choose μ_k in the following range (see also [4, 5])

$$\mu_k^{(1)} = \frac{(w_\mathrm{D}^k)^T w_\mathrm{P}^k}{n_\mathrm{C}} \le \mu_k \le \frac{\|H(w^k)\|_2}{\sqrt{n_\mathrm{C}}} = \mu_k^{(2)}. \tag{21}$$

Under the assumptions

A2. $\delta_k \in [0, \delta_{max}]$, $\sigma_k \in [\sigma_{min}, \sigma_{max}]$, $0 < \bar{\delta} \le \sigma_{min}$, $\delta_{max} + \sigma_{max} < 1$ for any $k \ge 0$;

A3. $\|\hat{r}^k\|_2 \le \delta_k \|H(w^k)\|_2$ for any $k \ge 0$;

from (19), (20) and (21), we have that

$$\begin{aligned} \|H'(w^k)\tilde{\alpha}_k\Delta w^k + H(w^k)\|_2 = \\ &= \|\tilde{\alpha}_k(-H(w^k) + r^k) + H(w^k)\|_2 \\ &\le (1-\tilde{\alpha}_k)\|H(w^k)\|_2 + \tilde{\alpha}_k\|\hat{r}^k\|_2 + \tilde{\alpha}_k\sigma_k\mu_k\sqrt{n_C} \\ &\le (1-\tilde{\alpha}_k(1-(\delta_k+\sigma_k)))\|H(w^k)\|_2, \end{aligned}$$

where $\eta_k(\tilde{\alpha}_k) \le 1$ for any $k \ge 0$.

The assumptions A1, A2 and A3 imply the following Lemma.

Lemma 3.1 *Let $\{w^k\}$ be generated by the IPPCG algorithm. Then*

(a) $\{w^k\} \subset \Omega(0)$;

(b) $\{w_{\text{D}}^k\}$ *and* $\{w_{\text{P}}^k\}$ *are componentwise bounded above and below away from zero in* $\Omega(\epsilon)$, $\epsilon > 0$; *

(c) *if* $\{w^k\} \subset \Omega(\epsilon)$, *with* $\epsilon > 0$, *the matrix* $A(D^k)^{-1}A^T + E^k$ *is well defined and positive definite,* $\{(H'(w^k))^{-1}\}$ *is bounded and, then,* $\{\Delta w^k\}$ *is bounded.*

Now, we have to find the condition which garantees that $\tilde{\alpha}_k$ is bounded below away from zero, i.e., $\tilde{\alpha}_k \geq \alpha^* > 0$, so that $\eta_k(\tilde{\alpha}_k) \leq (1 - \alpha^*(1 - (\delta_{max} + \sigma_{max}))) < 1$ for any $k \geq 0$.
In particular we assume that

A4.

$$\sigma_k \geq \frac{\sqrt{2}\tau_2\delta_k}{\min(1, \tau_2)} + \bar{\delta} \tag{22}$$

for any $k \geq 0$.†
Consequently, the following Lemma holds.

Lemma 3.2 *If* $\{w^k\} \subset \Omega(\epsilon)$, $\epsilon > 0$, *under the above assumptions, the sequence* $\{\tilde{\alpha}_k\}$ *is bounded away from zero and* $\eta_k(\tilde{\alpha}_k)$ *is uniformly less than one.*

Then, the assumptions A1, A2, A3 and A4 imply that the condition (15) holds for any $k \geq 0$. As observed in the previous section, the global convergence of the IPPCG method combined with the backtracking technique (16) follows immediately from the Theorem 6.1 in [7].

Theorem 3.1 *The IPPCG algorithm, with* $\epsilon^* = 0$, *generates a sequence* $\{w^k\}$ *such that* $\{\|H(w^k)\|_2\}$ *converges to zero and each limit point of* $\{w^k\}$ *satisfies the KKT conditions for (3) and (6).*

When $\lim_{k\to\infty} w^k = w^*$ and $H'(w^*)$ is a nonsingular matrix, from the Corollary 3.5 in [3], it follows that the IPPCG algorithm has a superlinear local convergence if $\sigma_k = \mathcal{O}(\|H(w^k)\|_2^t)$, $0 < t < 1$, and $\delta_k = \mathcal{O}(\|H(w^k)\|_2)$.
Finally, we observe that, when $\delta_k = 0$ for any $k \geq 0$, the IPPCG method is a standard IP method. Its global convergence holds under the above assumptions with $\delta_k = 0$.

*In order to prove that v_{L}^k and v_{U}^k are bounded below in $\Omega(\epsilon)$, $\epsilon > 0$, we observe from (4a), that the boundedness of $(v_{\text{L}}^k)_i$, $i \in L \setminus (L \cap U)$ and $(v_{\text{U}}^k)_j$, $j \in U \setminus (L \cap U)$, follows from the boundedness of $\{x^k\}$, $\{y^k\}$ and $\{H(w^k)\}$. Now we consider $i \in L \cap U$, i.e. $\ell_j \leq x_i \leq u_s$; $(v_{\text{L}}^k)_j$ and $(v_{\text{U}}^k)_s$ are both bounded or both unbounded. If they were unbounded, $\limsup_{k\to\infty}(v_{\text{L}}^k)_j = +\infty$ and $\limsup_{k\to\infty}(v_{\text{U}}^k)_s = +\infty$. In this last case, $\limsup_{k\to\infty}(z_{\text{L}}^k)_j = 0$ and $\limsup_{k\to\infty}(z_{\text{U}}^k)_s = 0$. From (4e) and (4f), it follows $x_i^k \to \ell_j$ and $x_i^k \to u_s$. This is a contradiction, since $\ell_j \neq u_s$ for the initial assumption.

†We observe that, if $\delta_{max} < (1 - \bar{\delta})/(1 + \sqrt{2}\tau_2/\min(1, \tau_2))$ and $\sigma_{max} \geq \bar{\delta} + \sqrt{2}\tau_2\delta_{max}/\min(1, \tau_2)$, the condition $\sigma_{max} + \delta_{max} < 1$ is satisfied.

4 A special class of discrete optimal control problems

We consider a special class of optimal control problems which are particularly suited to be solved with the IPPCG method on parallel architectures because of their special structure. We examine two optimal control problems which have been already studied in [5]: the first is the well-known problem of improving water quality by means of in-stream aeration process [9], while the other one concerns the study of diffusion-convection processes. Both these classical problems have a continuous formulation where the objective function is given by a double integral, the dynamic system is described by a system of partial differential equations and the unknown state and control depend on time and space.

In order to apply the IPPCG method, we consider a discretization of these problems, dividing the space interval into N subintervals of width Δh and the time interval into M subintervals of width Δt. The discretization is carried out in two phases: in the first one, the system of partial differential equations is discretized only respect to the space mesh points so that we obtain a continuous dynamic system of first order differential equations; then using a finite difference scheme, we obtain a system of difference equations.

For a more detailed description of the problems and of their discretization see [5].

So the water quality problem becomes:

$$\begin{aligned} \min \quad & \frac{1}{2}x^T Q x + c^T x, && \text{(23a)} \\ \text{s.t.} \quad & Ax - b = 0, && \text{(23b)} \\ & 0 \le Gx \le C_{max}\mathbf{1} && \text{(23c)} \end{aligned}$$

where $x = (y_1^T, y_2^T, \dots, y_M^T, u_0^T, \dots, u_{M-1}^T)^T \in \mathbb{R}^{2MN}$, $y_i \in \mathbb{R}^N$ are the state vectors and $u_i \in \mathbb{R}^N$ are the control vectors, $G = \begin{pmatrix} I & 0 \end{pmatrix} \in \mathbb{R}^{MN \times 2MN}$, $Q = \begin{pmatrix} Q_y & 0 \\ 0 & \rho Q_u \end{pmatrix}$ where Q_y and Q_u are $MN \times MN$ diagonal matrices and ρ is a positive parameter,

$$A = \left(\begin{array}{cccc|cccc} I & & & & -\Delta t I & & & \\ -\Omega & I & & & & \ddots & & \\ & \ddots & \ddots & & & & \ddots & \\ & & -\Omega & I & & & & -\Delta t I \end{array} \right) \in \mathbb{R}^{MN \times 2MN}$$

with $\Omega = I + \Delta t R$ and

$$R = \begin{pmatrix} -\left(\frac{v_1}{\Delta h} + k_3\right) & & & \\ \frac{v_2}{\Delta h} & -\left(\frac{v_2}{\Delta h} + k_3\right) & & \\ & \ddots & \ddots & \\ & & \frac{v_N}{\Delta h} & -\left(\frac{v_N}{\Delta h} + k_3\right) \end{pmatrix},$$

while $v \in \mathbb{R}^N$, $k_3, C_{max} \in \mathbb{R}$ are parameters related to the problem.
To obtain the asymptotical stability of the discrete dynamic system, we have to state the condition $\Delta t < 2/(\max_i v_i/\Delta h + k_3)$.
Analogously, the diffusion-convection process has the following form:

$$\min \quad \frac{1}{2} x^T Q x, \tag{24a}$$

$$\text{s.t.} \quad Ax - b = 0, \tag{24b}$$

$$0 \le Gx \le C_{max}\mathbf{1} \tag{24c}$$

where $x = (y_1^T, y_2^T, \dots, y_M^T, u_0^T, \dots, u_{M-1}^T)^T \in \mathbb{R}^{2M(N-1)}$, $y_i \in \mathbb{R}^{N-1}$ are the state vectors and $u_i \in \mathbb{R}^{N-1}$ are the control vectors, Q has the same diagonal form as that of the previous problem, $G = \begin{pmatrix} I & 0 \end{pmatrix} \in \mathbb{R}^{MN \times 2M(N-1)}$,

$$A = \left(\begin{array}{cccc|ccc} \Omega & & & & -\Delta t\, B & & \\ -I & \Omega & & & & \ddots & \\ & \ddots & \ddots & & & & \ddots \\ & & -I & \Omega & & & -\Delta t\, B \end{array}\right)$$

is an $M(N-1) \times 2M(N-1)$ matrix, where $\Omega = I - \Delta t R$ and

$$R = \begin{pmatrix} -\left(\frac{2a}{\Delta h^2} + c\right) & \left(\frac{a}{\Delta h^2} + \frac{b}{2\Delta h}\right) & & & \\ \left(\frac{a}{\Delta h^2} - \frac{b}{2\Delta h}\right) & -\left(\frac{2a}{\Delta h^2} + c\right) & \left(\frac{a}{\Delta h^2} + \frac{b}{2\Delta h}\right) & & \\ & \ddots & \ddots & \ddots & \\ & & \ddots & \ddots & \left(\frac{a}{\Delta h^2} + \frac{b}{2\Delta h}\right) \\ & & & \left(\frac{a}{\Delta h^2} - \frac{b}{2\Delta h}\right) & -\left(\frac{2a}{\Delta h^2} + c\right) \end{pmatrix}$$

are matrices of dimension $N-1$ and a, b, c, C_{max} are parameters related to the problem ($a > 0$ and $c \ge 0$).
In this case, if $\Delta h < 2a/|b|$ the asymptotical stability of the continuous dynamic system is guaranteed.
The effectiveness of the IPPCG method depends on the choice of a preconditioner M^k for the PCG algorithm. This preconditioner must be

easily solvable and, at the same time, a "good" approximation of the matrix $A(D^k)^{-1}A^T + E^k$. The structure of A in (23) or in (24) suggests to use a block–diagonal preconditioner $M^k = diag(M_i^k)$, $i = 1, \ldots, M$, with $M^k \in \mathbb{R}^{\bar{N}\times\bar{N}}$ (where $\bar{N} = N$ for (23) and $\bar{N} = N-1$ for (24)), given by

$$\begin{aligned} M_1^k &= (D_1^k)^{-1} + \Delta t^2 (D_{M+1}^k)^{-1} + E_1^k \\ M_i^k &= \Omega (D_{i-1}^k)^{-1}\Omega^T + (D_i^k)^{-1} + \Delta t^2 (D_{M+i}^k)^{-1} + E_i^k,\ i = 2, \ldots, M \end{aligned} \tag{25}$$

for the problem (23) and

$$\begin{aligned} M_1^k &= \Omega (D_1^k)^{-1}\Omega^T + \Delta t^2 B (D_{M+1}^k)^{-1} B^T + E_1^k \\ M_i^k &= (D_{i-1}^k)^{-1} + \Omega (D_i^k)^{-1}\Omega^T + \Delta t^2 B (D_{M+i}^k)^{-1} B^T + E_i^k,\ i = 2, \ldots, M \end{aligned} \tag{26}$$

for the problem (24). Here, $D^k = diag(D_1^k, \ldots, D_{2M}^k)$, with $D_i^k \in \mathbb{R}^{\bar{N}\times\bar{N}}$, $i = 1, \ldots, 2M$, $E^k = diag(E_1^k, \ldots, E_M^k)$, where $E_i^k \in \mathbb{R}^{\bar{N}\times\bar{N}}$, $i = 1, \ldots, M$ and M_i^k, $i = 1, \ldots, M$ are sparse, structured matrices (tridiagonal or pentadiagonal matrices).

In the parallel implementation of the IPPCG algorithm, we assume that any n-vector w and m-vector v is partitioned as follows

$$\begin{aligned} w &= (w_1^T, \ldots, w_M^T, w_{M+1}^T, \ldots, w_{2M}^T)^T, \quad w_i \in \mathbb{R}^{\bar{N}} \\ v &= (v_1^T, \ldots, v_M^T)^T, \quad v_i \in \mathbb{R}^{\bar{N}} \end{aligned}$$

and that the number p of the available processors is such that $M = s * p$. We allocate the m-vectors and the n-vectors (included the nonzero diagonal elements of E^k, D^k and Q) so that the local memory of the processor i, holds the blocks

$$(v_{(i-1)s+1}^T, \ldots, v_{is}^T)^T \text{ and } (w_{(i-1)s+1}^T, \ldots, w_{is}^T, w_{M+(i-1)s+1}^T, \ldots, w_{M+is}^T)^T,$$

$i = 1, \ldots, p$. Each processor also holds a copy of Ω, Δt, C_{max} and B. This static allocation of data among the p available processors gives a well balanced workload.

The most part of the calculations are trivially carried out using parallel vector "saxpy" operations. The factorization of $M_{(i-1)s+1}^k$, …, M_{is}^k and the solution of the related systems are computed locally by the ith processor and do not require communications.

Some global reduction functions‡ (synchronous multinode accumulation [2]) are required at each iterations of IPPCG. In particular, one global

‡A global reduction function takes as input ℓ vectors $(x_1^i, \ldots, x_\ell^i)$ $i = 1, \ldots, p$ distributed in every processor, executes a specific function on them ($y = \sum_{i=1}^p x_j^i$, $y = \prod_{i=1}^p x_j^i$, $y = \max_i(x_j^i)$, $y = \min_i(x_j^i)$, $j = 1, \ldots, \ell$) and leaves a single ℓ vector $(y_1 \cdots y_\ell)$ at every node.

minimization is executed to obtain α_k and one global sum has to be performed to complete the computation of $\|H(w^k)\|_2$, $(w_{\mathrm{D}}^k)^T w_{\mathrm{P}}^k$, $f(x^k)$, $f(x^k)-\varphi(x^k, y^k, w_{\mathrm{D}}^k)$, $\|G_{\mathrm{D}}(w^k)\|_2/(\|c\|_2+1)$, $\|G_{\mathrm{P}}(w^k)\|_2/(\|b\|_2+1)$. Moreover, at each iteration of the PCG algorithm, three global reductions are required (to form two scalar-products and to compute the euclidean norm of the residual). Although the number of global reductions is small, we observe that the time spent for them increases as p increases.

The parallel implementation of the IPPCG scheme requires three procedures that depend on the structure of the problem. The first and the second are the parallel computation of the matrix-vector products $v = Aw$ and $w = A^T v$ and the third is the parallel computation of the preconditioner.

In the following, we report the details of the corresponding procedures for the problem (23). The routine "send(*temp*,*temp*1,*np*)" sends the elements of the local vector *temp* from the current processor to the processor *np* that stores the received elements in the local vector *temp*1.

`mxv` procedure

The processor i holds the $2s\bar{N}$ vector

$$(w_{(i-1)s+1}^T, \dots, w_{is}^T, w_{M+(i-1)s+1}^T, \dots, w_{M+is}^T)^T$$

and computes the $s\bar{N}$ vector $(v_{(i-1)s+1}^T, \dots, v_{is}^T)^T$.

On processor $i = 1, \dots, p$ in parallel

if $i \neq p$

  $temp = -\Omega\ w_{is}$;

  send(*temp*,*temp*1,$i+1$);

end;

$(v_{(i-1)s+1}^T, \dots, v_{is}^T)^T = (w_{(i-1)s+1}^T, \dots, w_{is}^T)^T - \Delta t\ (w_{M+(i-1)s+1}^T, \dots, w_{M+is}^T)^T$;

for $j = 2, s$

  $v_{(i-1)s+j} = v_{(i-1)s+j} - \Omega\ w_{(i-1)s+j-1}$;

end;

synchronization;

if $i \neq 1$

  $v_{(i-1)s+1} = v_{(i-1)s+1} + temp1$;

end;

`mtxv` procedure

The processor i holds the $s\bar{N}$ vector $(v_{(i-1)s+1}^T, \dots, v_{is}^T)^T$ and computes the $2s\bar{N}$ vector $(w_{(i-1)s+1}^T, \dots, w_{is}^T, w_{M+(i-1)s+1}^T, \dots, w_{M+is}^T)^T$.

On processor $i = 1, \dots, p$ in parallel

if $i \neq 1$

  $temp = -\Omega\ v_{(i-1)s+1}$;

send($temp$,$temp1$,$i-1$);
end;
$(w_{(i-1)s+1}^T,\dots,w_{is}^T)^T = (v_{(i-1)s+1}^T,\dots,v_{is}^T)^T$;
$(w_{M+(i-1)s+1}^T,\dots,w_{M+is}^T)^T = -\Delta t\ (v_{(i-1)s+1}^T,\dots,v_{is}^T)^T$;
for $j=1,s-1$
$w_{(i-1)s+j} = w_{(i-1)s+j} - \Omega\ v_{(i-1)s+j+1}$;
end;
synchronization;
if $i \neq p$
$w_{is} = w_{is} + temp1$;
end;

<u>**precond** procedure</u>
The processor i holds the $2s\bar{N}$ diagonal elements of the blocks

$$D_{(i-1)s+1}^k,\dots,D_{is}^k, D_{M+(i-1)s+1}^k,\dots,D_{M+is}^k$$

and the $s\bar{N}$ diagonal elements of the blocks $E_{(i-1)s+1}^k,\dots,E_{is}^k$ and computes the s blocks $M_{(i-1)s+1}^k,\dots,M_{is}^k$.
On processor $i=1,\dots,p$ in parallel
if $i \neq p$
send(D_{is}^k,$temp$,$i+1$);
end;
for $j=1,s$
$M_{(i-1)s+j}^k = D_{(i-1)s+j}^k + \Delta t^2\ D_{M+(i-1)s+j}^k + E_{(i-1)s+j}^k$;
end;
for $j=2,s$
$M_{(i-1)s+j}^k = M_{(i-1)s+j}^k + \Omega\ D_{(i-1)s+j-1}^k\ \Omega^T$;
end;
synchronization;
if $i \neq 1$
$M_{(i-1)s+1}^k = M_{(i-1)s+1}^k + \Omega\ diag(temp)\ \Omega^T$;
end;

The above procedures require that each processor (unless one) sends $\bar{N}$ elements to the previous or the next processor. These communications are started by the sending processor before the data are required from the receiving processor. This send–ahead strategy enables to partially overlap the communication with the computation. A global synchronization assures the correct sequence of the operations.
Nevertheless the number $\bar{N}$ of the sent elements does not depend on the available processors. On the contrary, the computational complexity of each parallel task decreases as p increases. Then, it is not always advantageous to use a number of processors equal to M; indeed we can

observe a degradation of the performance of the parallel code as $p \to M$ (the maximum degree of parallelism of the algorithm). It is preferable to choose $p < M$ so that the granularity of the parallel tasks prevents the communication and synchronization penalty.

5 Numerical experiments

The IPPCG algorithm has been implemented in a parallel code to test its behaviour on the problems described in the previous section.
The interprocessor communications are based on the SHared MEMory access library (SHMEM) [16] of Cray T3E/256 and Origin 3800/32 at the InterUniversity Computing Center CINECA (Bologna, Italy).
The choice for the parameters involved in the algorithm is almost the same as in [6]: τ_1 and τ_2 ($\tau_1 = 10^{-5} n_{\mathrm{C}} \min_i ((w^0_{\mathrm{D}})_i (w^0_{\mathrm{P}})_i / ((w^0_{\mathrm{P}})^T w^0_{\mathrm{D}}))$, $\tau_2 = 10^{-3} (w^0_{\mathrm{P}})^T w^0_{\mathrm{D}} / \|((G_{\mathrm{D}}(w^0))^T, (G_{\mathrm{P}}(w^0))^T)^T\|_2$) are small enough so that we do not have to check the centrality conditions; the steplength is chosen as $\alpha_k = 0.995 / \max_i (-(\Delta w^k_{\mathrm{D}})_i / (w^k_{\mathrm{D}})_i, -(\Delta w^k_{\mathrm{P}})_i / (w^k_{\mathrm{P}})_i)$.
The value for μ_k is established by the rule:

$$\mu_k = \begin{cases} (w^k_{\mathrm{D}})^T w^k_{\mathrm{P}} / n_{\mathrm{C}} & \text{if } \left\| \begin{matrix} G_{\mathrm{D}}(w^k) \\ G_{\mathrm{P}}(w^k) \end{matrix} \right\|_2 < 10^{-1} \|H(w^k)\|_2 \\ & \text{or } \|H(w^k)\|_2 < 10^{-4}, \\ \|H(w^k)\|_2 / \sqrt{n_{\mathrm{C}}} & \text{otherwise,} \end{cases}$$

while

$$\begin{aligned} \delta_{\max} &= 0.8 / \big(1 + \sqrt{2}\tau_2 / \min(1, \tau_2)\big), \\ \sigma_{\max} &= 1.1\sqrt{2}\delta_{\max}\tau_2 / \min(1, \tau_2), \\ \delta_k &= \begin{cases} \min(\delta_{\max}, 0.8\|H(w^k)\|_2) & \text{if } k = 1, \\ \min\big(\delta_{\max}, \max(5 \cdot 10^{-5}, \|H(w^k)\|_2, \\ \qquad 0.5\|H(w^k)\|_2^2 / \|H(w^{k-1})\|_2^2)\big) & \text{otherwise,} \end{cases} \\ \sigma_k &= \max\Big(\min\big(\sigma_{\max}, \max(1.1\tau_2\delta_k\sqrt{2}/\min(1, \tau_2), 0.1\|H(w^k)\|_2^{1/2})\big), 10^{-7}\Big), \end{aligned}$$

The algorithm is stopped when one of the following criteria is true:

$$\begin{aligned} & \|H(w^k)\|_2 < 10^{-8}, \\ \text{or } & |f(x^k) - \varphi(x^k, y^k, w^k_{\mathrm{D}})| / (|f(x^k)| + 1) < 9 \cdot 10^{-8}, \\ \text{or } & k > 200 \text{ iterations.} \end{aligned}$$

Table 1. Relative speedup for problem (23) on Cray T3E.

Procs	Relative speedup $M = 150$ $N = 120$	Relative speedup $M = 300$ $N = 240$
2	1.8	1.7
5	4.4	4.2
10	8.8	8.1
15	13.2	12.0
30	24.2	24.4
50	31.9	40.2

Table 2. Relative speedup for problem (24) on Cray T3E.

Procs	Relative speedup $M = 120$ $N = 121$	Relative speedup $M = 180$ $N = 181$	Relative speedup $M = 240$ $N = 241$
4	3.8	3.8	3.9
6	5.8	5.6	5.8
12	11.9	11.7	11.2
20	19.9	19.6	19.2
30	28.2	29.1	28.6
60	40.6	51.9	54.6

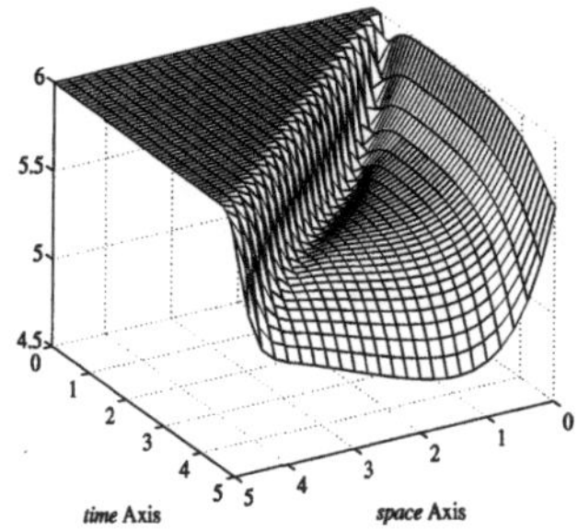

Figure 1: Problem (23): optimal oxygen concentration.

The value for the parameters in the definition of the discrete optimal control problems is chosen as in [5, sections 5.1 and 5.2], while ρ is set as 0.5 for both the "water quality problem" (23) and the "diffusion–convection problem" (24). The size of the considered test–problems are reported in the tables.

The Tables 1. and 2. show the relative speedup for problems (23)–(24) on Cray T3E, that is the ratio of the elapsed time for the execution of the IPPCG method on one processor to the time on a prefixed number of processors (the prefixed number of processors is denoted by "Procs" in the tables).

As shown by the tables, the algorithm exhibits a good scalability: in particular, when the number p of processors increases, we note that the performance of the algorithm improves if large size problems are considered and p is chosen less than M, while the speedup degrades with many processors and a small problem. This is due to the different granularity of the computational work for each processor, while the overhead due to the communication and the synchronization time increases (even if slowly) as p increases.

We have also tested the IPPCG algorithm for the problem (23) on Origin 3800/32 so that we can compare it with LOQO [15], a well-known serial code for quadratic programming problems. We have observed that when $M = 150$ and $N = 120$ on six processors the IPPCG method takes half a time with respect to LOQO, while with $M = 300$ and $N = 240$ the time required by the IPPCG method on six processors is eight times less than that needed by LOQO.

In the end, in Figure 1. we report the diagram of the solution for problem (23), that is the optimal dissolved oxygen concentration, obtained by the IPPCG algorithm: the obtained concentration agrees with that reported in [9].

6 Conclusions

This paper describes a generalization of the parallel Interior–Point method combined with the Preconditioned Conjugate Gradient algorithm, introduced in [6]. The hypotheses for the global convergence of the obtained scheme have been provided and the related theory has been developed. Furthermore, an application, concerning discrete quadratic optimal control problems, is presented: since these large-scale problems have a special structure in the linear constraints, they are particularly suited to be solved by the IPPCG algorithm on parallel computers, as shown by the numerical results reported in the previous section. These considerations point out that the IPPCG algorithm is an effective method for large-scale quadratic programs with special structure.

References

[1] S. Bellavia, *Inexact Interior–Point Method*, Journal of Optimization Theory and Applications, Vol. 96, pp. 109–121, 1998.

[2] D.P. Bertsekas and J.N. Tsitsiklis, *Parallel and Distributed Computation – Numerical Methods*, Prentice–Hall, Englewood Cliffs, New Jersey, 1989.

[3] R.S. Dembo, EISENSTAT, S. C., and STEIHAUG, T., *Inexact Newton Methods*, SIAM Journal on Numerical Analysis, Vol. 19, pp. 400–408, 1982.

[4] C. Durazzi, *On the Newton Interior–Point Method for Nonlinear Programming Problems*, Journal of Optimization Theory and Applications, Vol. 104, No 1, pp. 73–90, 2000.

[5] C. Durazzi and E. Galligani, *Nonlinear Programming Methods for Solving Optimal Control Problems*, Equilibrium Problems and Variational Models, Edited by F. Giannessi, A. Maugeri, and P. M. Pardalos, Kluwer Academic Publishers, Dordrecht, Netherlands, 2001.

[6] C. Durazzi, V. Ruggiero and G. Zanghirati, *Parallel Interior–Point Method for Linear and Quadratic Programs with Special Structure*, Journal of Optimization Theory and Applications, Vol. 110, No 2, 2001.

[7] S.C. Eisenstat and H.F. Walker, *Globally Convergent Inexact Newton Methods*, SIAM Journal on Optimization, Vol. 4, pp. 393–422, 1994.

[8] A.S. El–Bakry, R.A. Tapia, T. Tsuchiya and Y. Zhang, *On the Formulation and Theory of Newton Interior–Point Method for Nonlinear Programming*, Journal of Optimization Theory and Applications, Vol. 89, pp. 507–541, 1996.

[9] W. Hullet, *Optimal Estuary Aeration: An Application of Distributed Parameter Control Theory*, Applied Mathematics & Optimization, Vol. 1, No 1, pp. 20–63, 1974.

[10] S. Mehrotra, J.–S. Wang, *Conjugate Gradient Based Implementation of Interior Point Methods for Network Flow Problems*, Technical Report 95–70.1, Department of Industrial Engineering and Management Sciences, Northwestern University, Evanston, Ill., 1995.

[11] L.F. Portugal, M.G.C. Resende, J. Veiga and J.J. Júdice, *A truncated primal–feasible dual–infeasible interior–point network flow method*, Technical Report, Mathematical Science Research Center, AT&T Bell Laboratories, Murray Hill, N.J., 1994.

[12] M.G.C. Resende and C. Veiga , *An efficient implementation of a network interior point method*, AT&T Bell Laboratories, Murray Hill, USA, 1992.

[13] M.G.C. Resende and C. Veiga, *An implementation of the dual affine scaling algorithm for minimum cost flow on bipartite uncapacited networks*, SIAM Journal on Optimization, 3, pp. 516–537, 1993.

[14] S. Wright, *Primal–Dual Interior–Point Methods*, SIAM, Philadelphia, Pennsylvania, 1997.

[15] F. Potra, C. Roos and T. Terlaky, Editors, *Special Issue on Interior Point Methods*, Optimization Method & Software, Vols. 11–12, pp. 451–484, 1999.

[16] *SHMEM Technical Note for Fortran*, Cray Research, Seattle, Washington, 1994.

SOLVING LARGE SCALE FIXED CHARGE NETWORK FLOW PROBLEMS

Burak Ekşioğlu
Center of Applied Optimization,
University of Florida,
Gainesville, FL 32611
e-mail: duni@cao.ise.ufl.edu

Sandra Duni Ekşioğlu
Center of Applied Optimization,
University of Florida,
Gainesville, FL 32611
e-mail: eksioglu@cao.ise.ufl.edu

Panos M. Pardalos
Center of Applied Optimization,
University of Florida,
Gainesville, FL 32611
e-mail: pardalos@cao.ise.ufl.edu

Abstract This paper presents an approximation algorithm to solve large scale Fixed Charge Network Flow Problems (FCNFP). Fixed charge type of cost functions are used to model several problems of Supply Chain such as production planning and inventory management, transportation, plant location, and lot sizing. Our algorithm uses the Dynamic Slope Scaling Procedure (DSSP) to generate an initial solution to the problem, and then three different Local Search approaches contribute to finding a suboptimal solution. Computational experience with bipartite and layered networks are reported. The results indicate that our approach is suitable for large scale problems.

Keywords: Fixed Charge, Nonconvex Network Flow Problem, Computational Experiments, Branch and Bound, Mixed Integer Programming, Supply Chain.

1 Introduction

Many of the Supply Chain (*SC*) and logistics problems can be best modeled using fixed charge cost functions. Often on these problems the involved costs are assumed to be linear and/or convex. However, this assumption fails to hold in many of the applications encountered in practice. Important activities such as production planning and inventory management, transportation, plant location, etc. incur a fixed cost (setup cost) whenever the activity is performed, plus some variable unit cost. The fixed charge cost function is used to model these situations.

An example is the transportation problem. The transportation problem with linear cost functions has been studied in detail [1]. This problem aims to minimize the total cost of transporting products from a warehouse to a given set of retailers satisfying their demand. Since the transportation cost includes the fixed cost of setting-up a shipment and the variable cost (fuel consumption, driver's wage, etc.), this can be best represented by a fixed charge objective cost function.

The Fixed Charge Network Flow Problem (FCNFP) was first formulated by Hirsch and Dantzig in 1954 [17]. This problem falls into the category of Minimum Concave Cost Network Flow Problems (MCCNFP), therefore it achieves its optimal solution at the extreme points of the feasible region [18, 33]. A detailed survey on MCCNFP can be found in [15].

Most of the exact solution approaches to FCNFP transform the problem into an equivalent $0-1$ Mixed Integer Linear Program (MILP) [17] and use branch and bound techniques to solve optimaly. Branch and bound type of algorithms for fixed charge transportation problems are presented by Barr et al. [5], Cabot and Erenguc [7], Gray [14], Kennington and Unger [21], Lamar and Wallace [27], and Palekar et al. [30]. Bell et al. [6] present a branch and bound algorithm for capacitated fixed charge transportation problems. Most of these branch and bound algorithms are focused on conditional penalties called *up* and *down* penalties, that contribute in finding good lower bounds. Other exact solution algorithms used to solve FCNFP are vertex enumeration techniques and dynamic programming. Erickson et al. [11] use a dynamic programming approach called the send-and-split method, and Murty [29] uses an extreme point ranking technique, based on the simplex method, to solve general fixed charge problems.

Since the FCNFP is known to be an *NP*-hard problem [20], increasing the size of the problem exponentially increases the CPU requirements to solve it. Heuristic procedures have been developed by Belinski [3], Kuhn and Baumol [26], Cooper and Drebes [8], Denzler [9], Steinberg [34],

Walker [35], Diaby [10], and Khang and Fujuwara [22]. These heuristics use linear approximations of the objective function, or extend the simplex algorithm to adjust the processing to count for the fixed charge. Recently Kim and Pardalos [23], provide a dynamic slope scaling procedure (DSSP) to solve FCNFP. The DSSP consists of solving a sequence of linear programs that efficiently represent the combination of the fixed charge and the variable costs. This approach generates close-to-optimal solutions.

The approach we propose can also be used to solve Piecewise Linear Minimum Concave Cost Network Flow Problems (PLMCCNFP) [24]. The PLMCCNFP can be transformed to a FCNFP in an extended network. The size of the extended network depends on the number of the linear pieces in each arc cost function. Given a connected and directed graph $G(N, A)$ consisting of a set N of m nodes and a set A of n arcs, the number of arcs in the extended network $G_e(N, A_e)$ is equal to:

$$n_e = |A_e| = \sum_{(i,j)\in A} r_{ij}$$

where r_{ij} is the number of linear pieces in arc (i, j) of the original graph. Therefore the original concave piecewise linear objective function can be expressed as the sum of fixed charge arc cost functions:

$$\begin{aligned} f(x) &= \sum_{(i,j)\in A} f_{ij}(x_{ij}) \\ &= \sum_{(i,j)\in A} \sum_{k=1}^{r_{ij}} f_{ij}^k(x_{ij}^k), \end{aligned}$$

where $f_{ij}(x_{ij})$ presents the piecewise linear cost function on arc (i, j), and $f_{ij}^k(x_{ij}^k)$ presents the fixed charge cost function on arc $(i, j)^k$ on the extended network. Kim and Pardalos [25] extend the DSSP to solve Piecewise Linear Nonconvex Network Flow Problems.

The purpose of this paper is to develop fast approximation algorithms for large scale FCNFP. We analyze the effect of the s/c ratio (the fixed over the variable cost) on the CPU times and the quality of the solutions. We show how the network size and structure affect the performance of the algorithm. The paper is divided into 5 sections. The problem definition and formulation are given in section 2. A detailed description of the algorithm is presented in section 3. We describe the DSSP and the local search techniques. Section 4 presents the results of an extensive computational study, and finally section 5 concludes the paper.

2 Problem Definition and Formulation

Given a connected and directed graph $G(N, A)$ as before, we consider a transportation problem with a single warehouse, a set of trans-shipment nodes, a set of of retailers, a vector $C = (c_{ij}) \in R^n$, where c_{ij} represents the variable shipping cost for each unit transported from warehouse i to retailer j, and a vector $S = (s_{ij}) \in R^n$, where s_{ij} represents the fixed cost paid whenever a shipment is initiated. The objective is to find the optimal shipment schedule such that the retailers' demand would be satisfied at minimum cost. The FCNFP is formulated as follows:

$$\begin{array}{ll} \min & f(x) = \sum_{(i,j)\in A} f_{ij}(x_{ij}) \\ \text{s.t.} & \\ & \sum_{(k,i)\in A} x_{ki} - \sum_{(i,l)\in A} x_{il} = b_i \quad \forall i \in N \\ & 0 \leq x_{ij} \leq u_{ij} \qquad\qquad \forall (i,j) \in A. \end{array}$$

The objective function $f(x)$ is separable, and for each arc (i, j):

$$f_{ij}(x_{ij}) = \begin{cases} 0 & \text{if } x_{ij} = 0 \\ s_{ij} + c_{ij}x_{ij} & \text{if } 0 < x_{ij} \leq u_{ij} \end{cases}$$

where x_{ij} is the decision variable and u_{ij} is the upper bound for the flow on arc (i, j). The demand (supply) at the ith demand (supply) node is given by b_i.

Due to the discontinuity of the cost function $f(x)$, the FCNFP can be transformed into a $0-1$ MILP problem by introducing n binary variables. Assuming $s_{ij} > 0$, the cost function $f_{ij}(x_{ij})$ for each arc, can be replaced with:

$$f_{ij}(x_{ij}) = c_{ij}x_{ij} + s_{ij}y_{ij}$$

where

$$y_{ij} = \begin{cases} 0 & \text{if } x_{ij} = 0 \\ 1 & \text{if } x_{ij} > 0. \end{cases}$$

The MILP formulation of the FCNFP is as follows:

$$\begin{array}{ll} \min & f(x) = \sum_{(i,j)\in A}(c_{ij}x_{ij} + s_{ij}y_{ij}) \\ s.t. & \\ & \sum_{(k,i)\in A} x_{ki} - \sum_{(i,l)\in A} x_{il} = b_i, \quad \forall i \in N \\ & 0 \leq x_{ij} \leq u_{ij}y_{ij}, \qquad \forall (i,j) \in A \\ & y_{ij} \in \{0, 1\} \qquad\qquad \forall (i,j) \in A. \end{array}$$

The MILP formulation is used to solve the FCNFP optimally. Branch and bound algorithms can be employed to solve the MILP formulation. This formulation is useful in practice since the relaxed problems are Linear Cost NFPs.

3 Solution Procedure

This section presents the approach used to find approximate solutions to large scale FCNFP. The initial step of the algorithm is to find a feasible solution to the problem, and then use local search procedures to improve this solution. We use the DSSP to find a good starting point for the local search. For small size networks, the heuristic solution is compared with the exact MILP solution to check the quality of the algorithm. For large size networks, since it is not possible to get an exact solution, the LP relaxation of the MILP formulation is used as a lower bound. Applying Lagrangean relaxation to problems such as FCNFP would generate lower bounds that are no better than those found using the LP relaxation [12].

3.1 The DSSP

Pardalos and Kim [23] develop a heuristic approach (DSSP) that exploits the special structure of fixed-charge costs. The DSSP solves a sequence of LPs keeping the constraint set but changing the objective function cost coefficients each time. At each iteration, the linear objective function is updated until some stopping criteria are satisfied. Although no complexity results are given, and there is no proof of convergence, the computational results show that the heuristic solutions have extremely small gaps with the optimal. The idea behind the DSSP relies on the fact that there is an LP with appropriate objective coefficients that has the same solution as the FCNFP. We modify the DSSP and solve a series of Linear Cost Network Flow Problems (LNFPs) instead of LPs. This improves the CPU time significantly.

The linear cost coefficients used to generate the LNFPs are calculated in such a way that they would reflect the variable costs (c_{ij}) and the fixed costs (s_{ij}) simultaneously. Given a feasible solution $\hat{x}$, the objective function of FCNFP is expressed in linear form as follows:

$$\begin{aligned} f(\hat{x}) = \overline{c}^T(\hat{x})\hat{x} &= \textstyle\sum_{(i,j)\in A} f_{ij}(\hat{x}_{ij}) \\ &= \textstyle\sum_{(i,j)\in A} \overline{c}_{ij}(\hat{x}_{ij})\hat{x}_{ij} \end{aligned}$$

where the linear factor $\overline{c}$ is presented as:

$$\overline{c}_{ij}(\hat{x}_{ij}) = \begin{cases} c_{ij} + s_{ij}/\hat{x}_{ij} & \text{if } \hat{x}_{ij} > 0 \\ M & \text{if } \hat{x}_{ij} = 0. \end{cases}$$

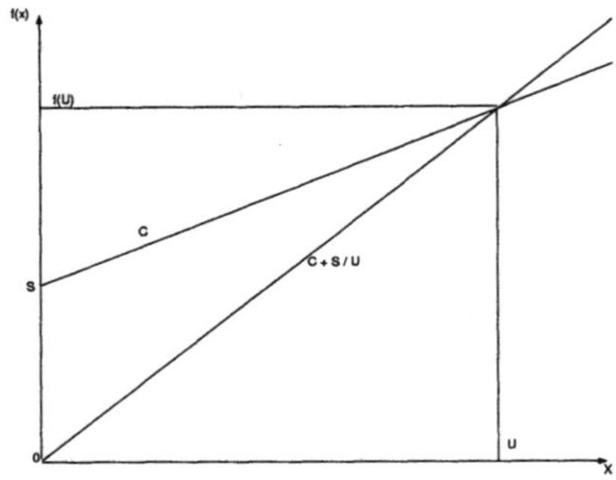

Figure 1: Linearization of FC Cost Function

M is some large number, $\overline{c}(\hat{x})$ is a modified objective function coefficient that distributes the fixed cost s evenly among all units of flow $\hat{x}$. Geometrically, $\overline{c}_{ij}(\hat{x}_{ij})$ represents the slope of the line connecting the origin with $(\hat{x}_{ij}, f(\hat{x}_{ij}))$ on arc (i, j) (see Figure 1).

The DSSP heuristic is based on three main decisions:

1. *Initial Solution.* Kim and Pardalos [23] propose two different ways to initiate the algorithm. The initiation scheme we use is shown to provide better results when used for large scale problems. The initial value of the linear factor is:

$$\overline{c}_{ij}^{0} = c_{ij} + s_{ij}/u_{ij} \text{ for j} = 1, 2, ..., n.$$

In this case the initial LNFP is nothing but an under-estimate of the FC cost.

2. *Updating Scheme.* The updating scheme used is given below:
If iteration $k = 0$, then:

$$\overline{c}_{ij}^{k+1} = \begin{cases} c_{ij} + s_{ij}/\overline{x}_{ij}^{k} & \text{if } \overline{x}_{ij}^{k} > 0 \\ c_{ij} & \text{if } \overline{x}_{ij}^{k} = 0 \end{cases}$$

if iteration $k \geq 1$, then:

$$\overline{c}_{ij}^{k+1} = \begin{cases} c_{ij} + s_{ij}/\overline{x}_{ij}^{k} & \text{if } \overline{x}_{ij}^{k} > 0 \\ max_{1 \leq l \leq k}\{\overline{c}_{ij}^{l} | \overline{x}_{ij}^{l} > 0\} & \text{if } \overline{x}_{ij}^{k} = 0. \end{cases}$$

3. *Stopping Criteria.* Two stopping criteria are used. The first stopping criterion is:

$$\overline{x}_{ij}^{k-1} = \overline{x}_{ij}^{k}, \text{ for all } (ij) \in A.$$

If consecutive solutions of FCNFP are equal, then the calculated linear cost coefficients and objective function values of consecutive iterations will be equal as well. As a result, once the above condition is satisfied, there can be no more improvement. The second stopping criterion is to terminate the algorithm after a certain number of iterations.

3.2 Local Search

As noted earlier, due to the complexity of the FCNFP, finding an exact global optimum becomes infeasible as the problem gets larger. Therefore, finding a good local optimum solution is aimed for large size instances.

A solution x to a FCNFP is locally optimal if no better solution exists within a specified neighborhood of x. Different neighborhood definitions result in different criteria for a local optimum. The standard definition of neighborhood for a solution x which is called the ϵ-neighborhood can be formulated as:

$$N_\epsilon(x) = \{x' | x' \text{ is feasible and } ||x - x'|| \leq \epsilon\}$$

where, $||.||$ is the Euclidean norm and ϵ is a positive constant. Local search based on N_ϵ for concave optimization is studied by Minoux [28] and Yaged [36]. In the case of single commodity FCNFP all extreme point solutions are local optimal in N_ϵ. This led to the following neighborhood definitions:

$$N_{AEF} = \{x' | x' \text{ is feasible and } x' \text{ is an adjacent extreme flow to } x\}.$$

Under this definition of neighborhood by Gallo and Sodini [13], x' is an adjacent extreme flow to an extreme solution x if x' is an extreme flow, and $x \cup x'$ contains a single undirected cycle.

Guiswite and Pardalos [16] give a more relaxed definition of neighborhood:

$$N_{AF} = \{x' | x' \text{ is feasible and } x' \text{ is an adjacent flow to } x\}$$

x' is an adjacent flow (not necessarily an adjacent extreme flow) to an extreme solution x if x' results from re-routing a single sub-path of flow within x.

The local search algorithms we implement, search for a local optimum over N_{AF}. Guiswite and Pardalos [16] give four algorithms that find locally optimum solutions over neighborhoods N_{AEF} and N_{AF}. The results indicate that algorithms over N_{AF} are faster in general. We implement two of the local search algorithms presented by Guiswite and Pardalos,

Layout of Local Search Algorithms

Best Local Search:
Find an initial extreme feasible solution X
WHILE (X is not a local optimum over N_{AF})
  Move to the best neighboring flow X' in N_{AF}
  Adjust X' to be extreme $\Rightarrow X''$
  $X \Leftarrow X''$

Better Local Search:
Find an initial extreme feasible solution X
WHILE (X is not a local optimum over N_{AF})
  Move to the first better neighboring flow X' in N_{AF}
  Adjust X' to be extreme $\Rightarrow X''$
  $X \Leftarrow X''$

Extended Better Local Search:
Find an initial extreme feasible solution X
WHILE (X is not a local optimum over N_{AF})
  Avoid revisiting previous neighbors and
  Move to the first better neighboring flow X' in N_{AF}
  Adjust X' to be extreme $\Rightarrow X''$
  $X \Leftarrow X''$

Figure 2: Layout of Local Search Algorithms

called *Best* and *Better*. We propose a new local search approach which takes advantage of the neighboring structure and basically takes a different path than the previous searches. This local search is called *Extended Better*, and it is based on the N_{AF} neighborhood. Although *Better* and *Extended Better* are the same in principle (see Figure 2), the *Extended Better* improves the solution time. This is achieved by trying to avoid revisiting the common neighbors when we move from one solution to another.

Best exhaustively searches the N_{AF} neighborhood of x and moves to the best neighboring flow x'; *Better* on the other hand moves to the first better neighboring flow x'; *Extended Better* is a variation of *Better* and offers a different option in terms of the order in which it evaluates adjacent solutions. The next section presents the findings of the implementation of these three algorithms.

4 Computational Results

The purpose of our computational experiments is to compare the DSSP and the three types of local search approaches empirically, and investigate the parameters that impact the behavior of the algorithms. The experimental plan and the results are described below.

We test our algorithm on randomly generated problems. We generate two types of networks:

- *Single source uncapacitated bipartite networks* which consist of a single source node, m_1 trans-shipment nodes, and m_2 demand nodes.
- *Single source uncapacitated layered networks* which consist of a single source node, m_2 layers, and m_1 demand nodes per layer (see Figure 3).

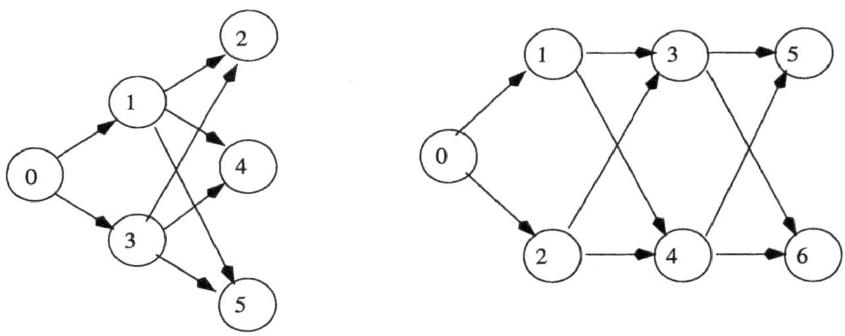

Figure 3: Bipartite and Layered Networks

All the test problems are fully dense. All demands, fixed charges, and variable costs for each test problem were generated randomly from different uniform distributions with pre-specified lower and upper limits as shown in Table 1. The characteristics of these test problems are the same with those developed by Kennington and Unger [21] and Palekar et al. [30]. The algorithm is programmed in the C language, and CPLEX 7.0 [19] callable libraries are used to solve the linear NFPs and the MILPs. The MILP solutions are within 0.01% to the optimal. The algorithms are compiled and executed on an IBM computer with 4 Power PC processors, 332 Mhz each.

Table 1. Characteristics of Test Problems

Set	Range of c_{ij}	Range of s_{ij}	Range of b_i
1	2-12	100-200	5-60
2	2-12	200-400	5-60
3	2-12	400-800	5-60
4	2-12	2000-4000	5-60
5	0-0	100-200	5-60
6	100-200	100-200	5-60
7	100-200	2-12	5-60
8	2-12	20-40	10-60
9	2-12	40-80	10-60
10	2-12	200-400	10-60
11	20-40	20-40	10-60
12	20-40	2-12	10-60
13	1-5	100-200	10-25
14	1-5	40-60	10-25
15	100-200	100-200	10-25
16	100-200	10-20	10-25

For a given type of network and for a given set of problem characteristics, we generate problems of different size. The problems are classified into small and large size problems (see Tables 2 & 3). For the small size problems an exact solution from the MILP formulation is obtained using CPLEX callable libraries, while for the large size problems we use the LP gap to check the quality of the solution. The LP gap is calculated by using the LP relaxation of the MILP as a lower bound.

For each particular problem type, we randomly generate 10 problem instances. We report on the average CPU times it takes to find an initial solution using the DSSP, and to perform the local search using each of the techniques. Because of the limitations of CPLEX, the largest problem we can solve optimally using the MILP formulation has around 30000 integer variables. This number changes depending on the problem characteristics. We show that the ratio of the fixed to variable cost, the size and structure of the network determine the numerical difficulty of the problem. We also present results on the improvement of local search approaches to the DSSP solution.

Table 2. Test Problems for Large Size Networks

	Bipartite			Layered		
	m_1	m_2	Nr. of arcs	m_1	m_2	Nr. of arcs
1	200	300	60200	50	10	22550
2	200	400	80200	60	10	32460
3	300	800	240300	175	10	275800
4	250	350	87750	40	15	22440
5	250	500	125250	50	15	35050
6	350	1000	350350	150	15	202650
7				30	20	11430
8				40	20	30440
9				125	20	297000

A sample of our results is presented in Tables 4-11. The results show that the CPU time it takes for a problem of a given size and structure changes as the s/c ratio changes. The problem becomes more difficult as the s/c ratio increases. However, problem set 5 is easy to solve although $s/c = \infty$. This result is consistent with what Palker et al. [30] observed. This suggests that the problem becomes easy when all variable costs are zero. Figure 4 plots the average CPU time (in sec.) versus s/c ratio for a small layered network with 915 arcs. Figure 5, presents the same results as Figure 4, excluding the CPU times for the MILP solution.

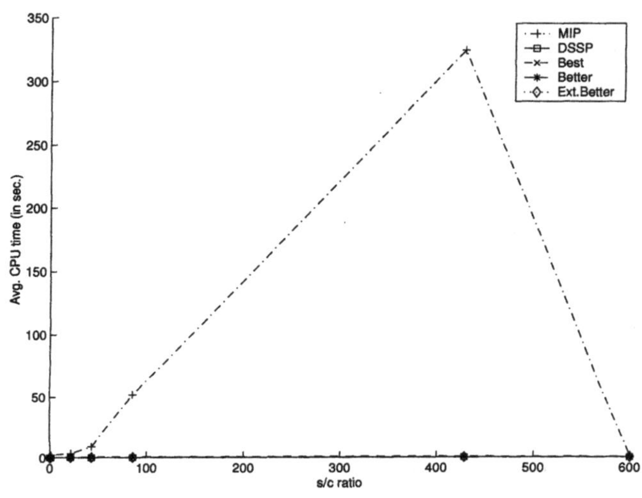

Figure 4: CPU Times vs. s/c Ratio for a Layered Network

Tables 8 and 9 present the average time it takes to solve each of the small size problems in bipartite and layered networks, using the Extended Better local search. The results show that the CPU times increase as the network grows, but for a given network size, the CPU times are almost constant. The average CPU times to solve the problems in small layered networks using the Extended Better local search varies from 0.01 to 1.21 CPU seconds. To solve the same problems exactly using the MILP formulation, it takes on average from 0.1 to 594.84 CPU seconds.

Tables 10 and 11 present the quality of the solution for small size problems, when the DSSP and the Extended Better local search are used. The quality of the solution for the small size problems is measured by:

$$Opt.Gap = \frac{Heuristic\_Solution}{Optimal\_Solution}.$$

The maximum Opt.Gap for the Better and the Extended Better local searches is 22.97%. The maximum Opt.Gap for the Best local search, however, is 18.28% which corresponds to approximately 26% improvement over the DSSP solution. For layered networks the largest Opt.Gap encountered is 17.56% for the DSSP and 8.25% for all of the three local searches. For easy problems with s/c ratio less than 20, the optimal solution is found in most of the cases. However, as the s/c ratio increases we are not able to obtain the optimal solution from CPLEX, hence we do not report optimality gaps for these cases. The most difficult set of problems were the ones where s/c = 428. For these problems we could find the optimal solution with up to 915 arcs in layered networks and 5000 arcs in bipartite networks.

The largest bipartite network we consider has 350350 arcs, and the largest layered network has 297000 arcs. However, we report the Best local search for bipartite networks with 87750 arcs and for layered networks with 22550 arcs. Table 4 presents a summary of these results for the bipartite network, and Table 5 gives the results for the layered network. The quality of the solution is measured in terms of the average CPU time and the average Lp.Gap:

$$Lp.Gap = \frac{Heuristic\_Solution}{LP\_Relaxation\_Solution}.$$

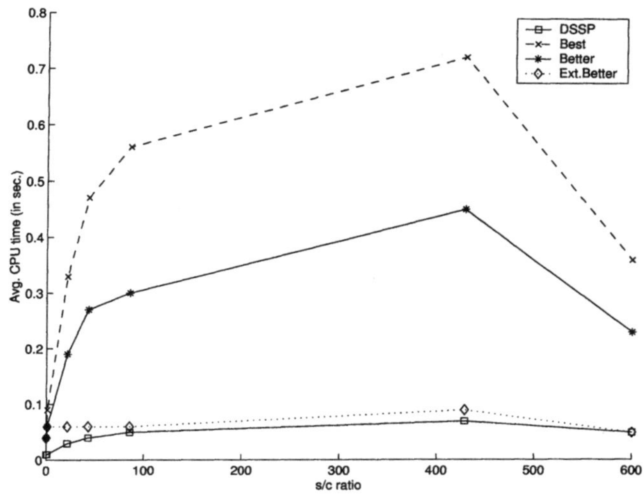

Figure 5: CPU Times vs. s/c Ratio for a Layered Network

Table 3. Test Problems for Small Size Networks

	Bipartite			Layered		
	m_1	m_2	Nr. of arcs	m_1	m_2	Nr. of arcs
1	10	40	410	5	10	230
2	20	30	620	5	15	355
3	25	25	650	5	20	480
4	30	20	630	5	25	605
5	40	10	440	5	30	730
6	20	80	1620	10	5	410
7	35	65	2310	10	10	910
8	50	50	2550	10	15	1410
9	65	35	2340	10	20	1910
10	80	20	1680	15	5	915
11	30	120	3630	15	10	2040
12	50	100	5050	15	15	3165
13	75	75	5700	20	5	1620
14	100	50	5100	20	10	3620
15	120	30	3720			
16	50	150	7550			
17	75	125	9450			
18	100	100	10100			
19	125	75	9500			
20	100	200	20100			
21	150	150	22650			
22	200	100	20200			
23	150	200	30150			
24	200	150	30200			

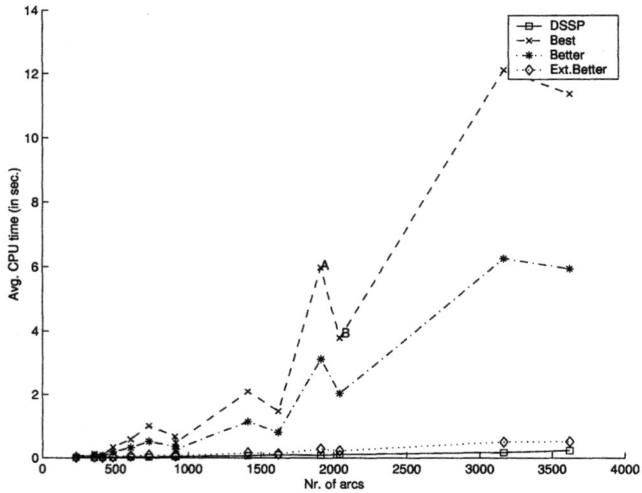

Figure 6: Average CPU Time vs. Number of Arcs for Layered Networks

Table 4. Summary of computational results for large size bipartite graph

	LB	**DSSP**		**Best**		**Better**		**Extended**	
s/c	**CPU**	**CPU**	**Lp.Gap**	**CPU**	**Lp.Gap**	**CPU**	**Lp.Gap**	**CPU**	**Lp.Gap**
0.05	0.13	0.30	1.0001	17.68	1.0001	17.48	1.0001	17.16	1.0001
1.00	0.13	0.56	1.0020	29.50	1.0020	26.18	1.0020	24.40	1.0020
21.43	0.14	0.74	1.0580	67.65	1.0573	57.82	1.0574	21.55	1.0574
42.86	0.14	1.29	1.0789	131.0	1.0777	99.24	1.0777	22.11	1.0777
85.71	0.15	1.98	1.0923	215.9	1.0900	163.2	1.0900	21.91	1.0900
428.57	0.15	2.99	1.1439	226.9	1.1421	172.1	1.1421	21.71	1.1421
Inf.	0.16	0.88	1.2250	727.3	1.2095	520.7	1.2101	22.52	1.2101
0.23	0.13	0.31	1.0004	30.04	1.0004	24.44	1.0004	20.51	1.0004
1.00	0.14	0.35	1.0019	34.49	1.0018	27.82	1.0018	21.44	1.0018
4.29	0.14	0.43	1.0171	70.11	1.0142	47.35	1.0144	22.64	1.0144
8.57	0.14	0.46	1.0289	46.90	1.0263	37.43	1.0265	20.44	1.0265
42.86	0.14	0.93	1.0778	127.1	1.0766	89.13	1.0766	21.75	1.0766
0.10	0.15	0.30	1.0004	22.97	1.0004	18.41	1.0004	18.34	1.0004
1.00	0.15	0.67	1.0034	31.86	1.0034	25.79	1.0034	20.33	1.0034
16.67	0.14	0.48	1.1094	26.02	1.1092	21.58	1.1092	18.27	1.1092
50.00	0.14	0.57	1.1485	206.2	1.1455	138.2	1.1455	21.74	1.1455

Table 5. Summary of computational results for large size layered graph

	LP	DSSP		Best		Better		Extended	
s/c	**CPU**	**CPU**	**Lp.Gap**	**CPU**	**Lp.Gap**	**CPU**	**Lp.Gap**	**CPU**	**Lp.Gap**
0.05	0.04	0.09	1.0003	47.04	1.0003	24.10	1.0003	7.79	1.0003
1.00	0.04	0.65	1.0072	174.57	1.0070	87.29	1.0070	8.04	1.0070
21.43	0.04	2.32	1.3027	371.71	1.2900	196.95	1.2901	7.52	1.2901
42.86	0.04	2.96	1.5868	460.98	1.5533	244.55	1.5537	7.71	1.5537
85.71	0.04	3.19	2.1026	517.69	2.0227	277.85	2.0234	7.85	2.0234
428.57	0.04	3.87	4.7034	721.25	4.3402	386.82	4.3492	11.66	4.3483
Inf.	0.04	2.91	8.5877	335.60	8.5226	198.91	8.5225	10.46	8.5225
0.23	0.04	0.08	1.0014	180.90	1.0013	79.67	1.0013	7.67	1.0013
1.00	0.04	0.36	1.0064	196.92	1.0063	82.17	1.0063	7.12	1.0063
4.29	0.04	0.63	1.0571	242.01	1.0555	124.34	1.0555	7.70	1.0555
8.57	0.03	1.04	1.1143	271.69	1.1109	140.37	1.1109	7.43	1.1109
42.86	0.04	2.81	1.5473	424.38	1.5178	235.97	1.5180	7.72	1.5180
0.10	0.04	0.12	1.0013	45.52	1.0013	24.31	1.0013	7.79	1.0013
1.00	0.04	0.60	1.0135	234.03	1.0132	119.22	1.0132	8.02	1.0132
16.67	0.04	2.45	1.3739	257.36	1.3698	146.67	1.3698	9.13	1.3698
50.00	0.04	2.98	2.9488	358.72	1.9232	212.52	1.9233	7.37	1.9233

*Table 6.*Lp.Gap for a small size bipartite graph (1620 arcs).

	MILP	DSSP	Best	Better	Extended
s/c	**Lp.Gap**	**Lp.Gap**	**Lp.Gap**	**Lp.Gap**	**Lp.Gap**
0.05	1.0001	1.0001	1.0001	1.0001	1.0001
1.00	1.0019	1.002	1.0019	1.0019	1.0019
21.43	1.0236	1.0432	1.0428	1.0428	1.0428
42.86	1.0544	1.0574	1.0568	1.0568	1.0568
85.71	1.0727	1.0774	1.075	1.075	1.075
428.57	1.1106	1.1269	1.1198	1.1198	1.1198
Inf.	1.1305	1.2203	1.2163	1.2163	1.2163
0.23	1.0004	1.0004	1.0004	1.0004	1.0004
1.00	1.0018	1.0019	1.0018	1.0018	1.0018
4.29	1.0118	1.0132	1.012	1.012	1.012
8.57	1.0205	1.0231	1.0215	1.0215	1.0215
42.86	1.052	1.0552	1.0546	1.0546	1.0546
0.10	1.0004	1.0004	1.0004	1.0004	1.0004
1.00	1.0034	1.0034	1.0034	1.0034	1.0034
16.67	1.0388	1.0496	1.048	1.048	1.048
50.00	1.0748	1.0867	1.0856	1.0856	1.0856

Table 7. Lp.Gap for a small size layered graph (910 arcs).

	MILP	**DSSP**	**Best**	**Better**	**Extended**
s/c	**Lp.Gap**	**Lp.Gap**	**Lp.Gap**	**Lp.Gap**	**Lp.Gap**
0.05	1.0003	1.0003	1.0003	1.0003	1.0003
1.00	1.0061	1.0062	1.0061	1.0061	1.0061
21.43	1.2418	1.2522	1.2439	1.244	1.244
42.86	1.4448	1.4777	1.452	1.4528	1.4528
85.71	1.7557	1.8613	1.7974	1.7991	1.7991
428.57	2.7957	3.4152	3.1411	3.1411	3.1411
Inf.	4.8705	5.0334	4.9437	4.9437	4.9437
0.23	1.0012	1.0012	1.0012	1.0012	1.0012
1.00	1.0056	1.0057	1.0057	1.0057	1.0057
4.29	1.0473	1.0482	1.0476	1.0476	1.0476
8.57	1.094	1.0962	1.0946	1.0946	1.0946
42.86	1.4178	1.4451	1.4232	1.4235	1.4235
0.10	1.0011	1.0011	1.0011	1.0011	1.0011
1.00	1.0114	1.0115	1.0114	1.0114	1.0114
16.67	1.342	1.351	1.3462	1.3462	1.3462
50.00	1.839	1.903	1.8493	1.8495	1.8495

The Lp.Gaps for the three local searches are almost the same, but the CPU times differ. The Lp.Gaps are really large, in particular on layered networks with high s/c ratio. Tables 6 and 7 present the Lp.Gaps for small size bipartite and layered networks. The gap between MILP solution and lower bounds, in small size bipartite networks goes up to 13% and in layered networks it goes up to 487%, this suggests that a large portion of the Lp.Gap for large size problems is caused by the poor LP lower bound. Note that the Lp.Gaps for set 5 are extremely high. The variable cost for problems in this set are all zero, thus the LP relaxation gives poor lower bounds. Another interesting observation is that the CPU times for the Extended Better are almost constant and insensitive to the s/c ratio, while the quality of the solutions for the three local searches is almost the same, which supports our observations for the small size problems.

Figure 6 plots the CPU time versus the problem size for layered networks. In general the problems become harder to solve as the size of the network increases. Point A in the figure corresponds to a network with 20 layers and 10 nodes per layer, whereas point B corresponds to a network with 10 layers and 15 nodes per layer. Problem instance B has more arcs, but the problem is solved in less amount of time, because of the difference in the network structure of the two problems. The problem becomes more difficult as the number of layers increases.

Table 8. Average CPU time for Small Size Bipartite Networks

	Demand-Unif.(5-60)							Demand-Unif.(10-60)						Demand-Unif.(10-25)		
Prob. \ s/c	**0.05**	**1.00**	**21.43**	**42.86**	**85.71**	**428.57**	**Inf.**	**0.23**	**1.00**	**4.29**	**8.57**	**42.86**	**0.10**	**1.00**	**16.67**	**50.00**
1	0.01	0.01	0.02	0.02	0.02	0.02	0.02	0.01	0.01	0.02	0.02	0.02	0.01	0.01	0.02	0.01
2	0.01	0.01	0.01	0.02	0.02	0.02	0.02	0.01	0.01	0.01	0.02	0.02	0.02	0.01	0.02	0.02
3	0.02	0.01	0.02	0.01	0.01	0.01	0.02	0.01	0.01	0.02	0.01	0.02	0.01	0.01	0.02	0.02
4	0.01	0.01	0.01	0.01	0.01	0.01	0.02	0.01	0.02	0.02	0.01	0.01	0.01	0.01	0.01	0.01
5	0.01	0.00	0.01	0.01	0.01	0.01	0.01	0.00	0.01	0.01	0.01	0.00	0.01	0.01	0.01	0.01
6	0.09	0.07	0.10	0.11	0.13	0.13	0.12	0.10	0.11	0.13	0.11	0.12	0.08	0.10	0.11	0.12
7	0.09	0.07	0.12	0.14	0.16	0.15	0.16	0.09	0.10	0.12	0.13	0.15	0.08	0.09	0.10	0.15
8	0.10	0.10	0.12	0.13	0.12	0.11	0.13	0.10	0.11	0.13	0.12	0.13	0.10	0.12	0.12	0.13
9	0.08	0.05	0.08	0.09	0.08	0.08	0.09	0.08	0.08	0.08	0.09	0.09	0.06	0.08	0.09	0.09
10	0.03	0.03	0.04	0.03	0.03	0.03	0.03	0.03	0.03	0.04	0.04	0.04	0.03	0.04	0.03	0.03
11	0.29	0.27	0.31	0.44	0.45	0.44	0.44	0.36	0.36	0.38	0.37	0.42	0.28	0.31	0.37	0.40
12	0.36	0.29	0.50	0.50	0.52	0.50	0.52	0.39	0.47	0.48	0.41	0.51	0.29	0.41	0.44	0.51
13	0.41	0.24	0.45	0.45	0.45	0.42	0.45	0.40	0.40	0.43	0.46	0.41	0.29	0.40	0.34	0.46
14	0.24	0.14	0.24	0.26	0.26	0.21	0.28	0.24	0.26	0.26	0.22	0.27	0.19	0.25	0.27	0.27
15	0.09	0.09	0.13	0.12	0.10	0.09	0.12	0.11	0.12	0.12	0.12	0.12	0.10	0.12	0.12	0.11
16	1.03	0.82	1.18	1.26	1.30	1.26	1.29	1.13	1.21	1.22	1.08	1.28	0.87	1.14	1.07	1.28
17	1.30	1.09	1.42	1.42	1.49	1.45	1.47	1.08	1.31	1.32	1.23	1.47	1.16	1.28	0.99	1.44
18	1.20	0.83	1.24	1.25	1.29	1.27	1.28	1.14	1.25	1.22	1.23	1.35	0.94	1.19	1.18	1.27
19	0.87	0.57	0.86	0.86	0.90	0.88	0.89	0.84	0.91	0.88	0.89	0.95	0.61	0.87	0.86	0.87
20	4.63	4.05	5.60	5.60	6.05	5.80	5.87	5.48	5.48	5.31	5.69	6.01	4.84	5.08	3.65	5.78
21	4.17	3.94	4.53	4.79	5.16	4.93	5.04	4.66	4.92	5.02	4.84	5.09	3.89	4.58	4.05	4.93
22	2.90	2.03	2.83	2.84	3.04	2.77	2.98	2.90	2.91	2.97	3.02	3.01	2.58	3.01	2.53	9.22
28	8.21	6.05	8.41	8.43	9.02	8.73	8.82	7.79	5.63	8.83	8.94	8.06	6.83	7.88	7.94	8.62
24	5.91	5.56	6.37	6.39	6.88	6.56	6.70	8.55	6.54	6.71	6.80	6.82	5.83	6.13	6.37	6.93

Table 9. Average CPU time for Small Size Layered Networks

	Demand-Unif.(5-60)							Demand-Unif.(10-60)						Demand-Unif.(10-25)		
Prob. \ s/c	**0.05**	**1.41**	**21.43**	**22.86**	**85.79**	**428.57**	**Inf.**	**0.23**	**4.07**	**4.29**	**8.57**	**42.86**	**0.40**	**1.00**	**16.67**	**50.02**
1	0.01	0.01	0.01	0.01	0.01	0.02	0.01	0.01	0.01	0.01	0.01	0.01	0.01	0.04	0.01	0.02
2	0.01	0.01	0.05	7.03	0.03	6.03	0.03	8.03	0.02	0.04	0.03	0.03	0.01	0.02	0.02	0.02
0	0.03	9.04	0.54	0.04	6.05	0.06	0.04	0.04	0.04	0.04	0.02	0.04	0.03	0.04	0.04	0.04
4	0.04	0.09	0.07	0.06	0.06	0.07	4.06	0.76	0.06	0.07	0.77	0.07	0.04	0.06	0.06	0.07
3	0.06	0.09	0.10	0.49	0.10	0.15	0.09	0.09	0.99	0.10	0.10	0.10	0.36	0.99	0.09	0.14
6	0.01	0.02	0.02	0.82	0.22	0.08	0.02	0.02	0.02	0.85	0.02	0.02	0.01	0.02	0.02	0.02
7	0.04	0.07	0.07	0.07	0.07	0.07	0.07	0.07	0.07	0.07	0.07	0.07	0.05	0.07	0.07	0.07
8	0.11	0.15	0.16	0.16	0.16	0.26	0.15	0.16	0.16	0.16	0.15	0.17	0.11	0.16	0.15	0.16
9	0.22	0.28	0.29	0.29	0.30	0.58	0.27	0.28	0.28	0.28	0.29	0.29	0.22	0.28	0.26	0.30
10	0.04	0.06	0.06	0.06	0.06	0.09	0.05	0.05	0.05	0.06	0.05	0.06	0.04	0.06	0.06	0.05
11	0.16	0.22	0.23	0.23	0.23	0.40	0.22	0.23	0.22	0.24	0.22	0.24	0.15	0.21	0.21	0.22
12	0.46	0.49	0.51	0.51	0.54	1.21	0.52	0.50	0.49	0.49	0.49	0.52	0.41	0.49	0.48	0.54
13	0.09	0.12	0.13	0.13	0.13	0.25	0.13	0.12	0.13	0.13	0.12	0.13	0.08	0.12	0.12	0.13
14	0.47	0.49	0.52	0.52	0.52	0.65	0.51	0.51	0.51	0.51	0.51	0.52	0.47	0.50	0.51	0.51

Table 10. Average Optimality Gap for Small Size Bipartite Networks

	Demand-Unif.(5-60)							Demand-Unif.(10-60)						Demand-Unif.(10-25)		
Prob. \ s/c	**0.05**	**1.00**	**21.43**	**42.86**	**85.71**	**428.57**	**Inf.**	**0.23**	**1.00**	**4.29**	**8.57**	**42.86**	**0.10**	**1.00**	**16.67**	**50.00**
1	1.0000	1.0000	1.0028	1.0031	1.0017	1.0108	1.0820	1.0000	1.0000	1.0008	1.0022	1.0032	1.0000	1.0000	1.0141	1.0091
2	1.0000	1.0000	1.0083	1.0088	1.0081	1.0496	1.0906	1.0000	1.0000	1.0003	1.0018	1.0089	1.0000	1.0002	1.0167	1.0221
3	1.0000	1.0000	1.0083	1.0145	1.0175	1.0713	1.0867	1.0000	1.0000	1.0010	1.0030	1.0176	1.0000	1.0000	1.0308	1.0203
4	1.0000	1.0000	1.0144	1.0172	1.0245	1.0950	1.1215	1.0000	1.0001	1.0018	1.0053	1.0133	1.0000	1.0001	1.0320	1.0126
5	1.0000	1.0000	1.0167	1.0298	1.0539	1.1498	1.1060	1.0000	1.0001	1.0000	1.0014	1.0377	1.0000	1.0002	1.0329	1.0505
6	1.0000	1.0000	1.0039	1.0023	1.0021	1.0083	1.1132	1.0000	1.0000	1.0002	1.0010	1.0025	1.0000	1.0000	1.0089	1.0100
7	1.0000	1.0000	1.0055	1.0068	1.0051	1.0426	1.1004	1.0000	1.0000	1.0005	1.0018	1.0080	1.0000	1.0000	1.0117	1.0256
8	1.0000	1.0000	1.0108	1.0119	1.0131	1.0777	1.0777	1.0000	1.0001	1.0012	1.0039	1.0118	1.0000	1.0001	1.0446	1.0306
9	1.0000	1.0001	1.0106	1.0117	1.0352	1.0920	1.0972	1.0000	1.0000	1.0027	1.0047	1.0095	1.0000	1.0003	1.0612	1.0261
10	1.0000	1.0001	1.0180	1.0308	1.0672	1.1342	1.0888	1.0000	1.0000	1.0037	1.0079	1.0308	1.0000	1.0004	1.0560	1.0331
11	1.0000	1.0000	1.0040	1.0032	1.0037	1.0078	1.1320	1.0000	1.0000	1.0005	1.0013	1.0035	1.0000	1.0000	1.0057	1.0096
12	1.0000	1.0000	1.0079	1.0064	1.0050	1.0496	N/A	1.0000	1.0000	1.0010	1.0028	1.0053	1.0000	1.0000	1.0135	1.0209
13	1.0000	1.0001	1.0118	1.0138	1.0093	N/A	1.0950	1.0000	1.0001	1.0015	1.0047	1.0126	1.0000	1.0001	1.0463	1.0315
14	1.0000	1.0000	1.0164	1.0166	1.0288	1.0927	1.0934	1.0000	1.0002	1.0020	1.0051	1.0136	1.0000	1.0002	1.0723	1.0271
15	1.0000	1.0001	1.0296	1.0331	1.0585	1.0876	1.1030	1.0000	1.0001	1.0052	1.0105	1.0311	1.0000	1.0002	1.0618	1.0512
16	1.0000	1.0000	1.0034	1.0043	1.0029	1.0252	N/A	1.0000	1.0000	1.0007	1.0019	1.0044	1.0000	1.0000	1.0062	1.0139
17	1.0000	1.0000	1.0065	1.0071	1.0051	N/A	N/A	1.0000	1.0001	1.0009	1.0031	1.0085	1.0000	1.0001	1.0216	1.0234
18	1.0000	1.0000	1.0129	1.0112	1.0087	N/A	N/A	1.0000	1.0000	1.0021	1.0062	1.0133	1.0000	1.0001	1.0552	1.0329
19	1.0000	1.0001	1.0183	1.0093	1.0236	N/A	N/A	1.0000	1.0001	1.0033	1.0085	1.0108	1.0000	1.0002	1.0754	1.0239
20	1.0000	1.0000	1.0067	1.0077	1.0051	N/A	N/A	1.0000	1.0000	1.0010	1.0026	1.0076	1.0000	1.0000	N/A	N/A
21	1.0000	1.0000	1.0116	1.0142	1.0101	N/A	N/A	1.0000	1.0001	1.0017	1.0043	1.0165	1.0000	1.0001	N/A	N/A
22	1.0000	1.0001	1.0239	1.0175	N/A	N/A	N/A	1.0000	1.0002	1.0035	1.0085	1.0181	1.0000	1.0002	N/A	N/A
23	1.0000	1.0000	1.0089	1.0110	N/A	N/A	N/A	1.0000	1.0001	1.0017	1.0046	1.0111	1.0000	1.0001	N/A	N/A
24	1.0000	1.0001	1.0158	1.0140	N/A	N/A	N/A	1.0000	1.0002	1.0023	1.0056	1.0144	1.0000	1.0001	N/A	N/A

Table 11. Average Optimality Gap for Small Size Layered Networks

	Demand-Unif.(5-60)							Demand-Unif.(10-60)						Demand-Unif.(10-25)		
Prob. \ s/c	**0.05**	**1.00**	**21.43**	**42.86**	**85.71**	**428.57**	**Inf.**	**0.23**	**1.00**	**4.29**	**8.57**	**42.86**	**0.10**	**1.00**	**16.67**	**50.00**
1	1.0000	1.0000	1.0010	1.0028	1.0101	1.0394	1.0170	1.0000	1.0000	1.0002	1.0003	1.0023	1.0000	1.0000	1.0017	1.0108
2	1.0000	1.0000	1.0008	1.0026	1.0075	1.0487	1.0177	1.0000	1.0000	1.0001	1.0002	1.0021	1.0000	1.0000	1.0011	1.0095
3	1.0000	1.0000	1.0010	1.0025	1.0070	N/A	1.0179	1.0000	1.0000	1.0001	1.0002	1.0022	1.0000	1.0000	1.0011	1.0075
4	1.0000	1.0000	1.0005	1.0018	N/A	N/A	1.0182	1.0000	1.0000	1.0001	1.0002	1.0015	1.0000	1.0000	1.0011	N/A
5	1.0000	1.0000	1.0005	1.0017	N/A	N/A	1.0186	1.0000	1.0000	1.0001	1.0001	1.0015	1.0000	1.0000	1.0009	N/A
6	1.0000	1.0000	1.0023	1.0077	1.0169	1.0564	1.0159	1.0000	1.0000	1.0004	1.0007	1.0062	1.0000	1.0000	1.0017	1.0140
7	1.0000	1.0000	1.0018	1.0056	N/A	N/A	1.0150	1.0000	1.0000	1.0003	1.0005	1.0040	1.0000	1.0000	1.0031	N/A
8	1.0000	1.0000	1.0021	N/A	N/A	N/A	1.0150	1.0000	1.0000	1.0003	1.0007	N/A	1.0000	1.0000	1.0039	N/A
9	1.0000	1.0000	1.0019	N/A	N/A	N/A	1.0135	1.0000	1.0000	1.0003	1.0006	N/A	1.0000	1.0000	1.0032	N/A
10	1.0000	1.0000	1.0022	1.0063	1.0140	1.0584	1.0144	1.0000	1.0000	1.0003	1.0006	1.0057	1.0000	1.0000	1.0040	1.0159
11	1.0000	1.0000	1.0040	N/A	N/A	N/A	1.0132	1.0000	1.0000	1.0007	1.0014	N/A	1.0000	1.0000	1.0043	N/A
12	1.0000	1.0000	N/A	N/A	N/A	N/A	1.0118	1.0000	1.0000	1.0007	1.0012	N/A	1.0000	1.0000	1.0034	N/A
13	1.0000	1.0000	1.0043	1.0116	N/A	N/A	1.0130	1.0000	1.0000	1.0005	1.0011	1.0102	1.0000	1.0000	1.0047	N/A
14	1.0000	1.0000	N/A	N/A	N/A	N/A	1.0107	1.0000	1.0000	1.0010	1.0021	N/A	1.0000	1.0000	1.0046	N/A

5 Concluding Remarks

In this paper we use a Dynamic Slope Scaling and a local search procedure to solve large scale fixed charge network flow problems. The problems can be formulated as a MILP, but it is difficult to solve large instances optimaly. We find approximate solutions for the FCNFP in bipartite and layered networks. We show that the local search improves the DSSP solution. This improvement goes up to 26% in bipartite networks. We show that the s/c ratio affects the CPU times and the performance of the algorithm. The *Extended Better* local search is shown to give almost the same quality of solution as the two other local search approaches, but in terms of CPU time, *Extended better* is faster. We are able to solve large size bipartite networks with up to 350350 arcs and layered networks with up to 297000 arcs in a reasonable amount of time.

In addition it is observed that the structure and size of the networks influence the performance of the algorithms. Large bipartite networks are easier to solve, than large layered networks.

References

[1] K.R. Ahuja, T.L. Magnanti, J.B. Orlin, *Network Flows: Theory, Algorithms and Applications.* Prentice-Hall, Upper Saddle River, New Jersey, 1993.

[2] A. Balakrishnan, S. Graves, A composite algorithm for a concave-cost network flow problem. *NETWORKS*, Vol. 19, pp. 175-202, 1989.

[3] M.L. Balinski, Fixed cost transportation problems. *Naval Research Logisitics Quarterly.* Vol. 8, pp. 41-54, 1961.

[4] R.H. Ballou, *Business Logistics Management.* Prentice-Hall, Upper Saddle River, New Jersey, 1993.

[5] R.S. Barr, F. Glover, D. Klingman, A new optimization method for large scale fixed charge transportation problems. *Operations Research*, Vol. 29, pp. 448-463, 1981.

[6] G.J. Bell, B.W. Lamar, C.A. Wallace, Capacity improvement, penalties, and the fixed charge transportation problem. *Naval Research Logistics*, Vol. 46, pp. 341-355, 1999.

[7] A.V. Cabot, S.S. Erenguc, Some branch-and-bound procedures for fixed-cost transportation problems. *Naval Research Logistics Quarterly*, Vol. 31, pp. 145-154, 1984.

[8] L. Cooper, C. Drebes, An approximate solution method for the fixed charge problem. *Naval Research Logistics Quarterly*, Vol. 14, pp. 101-113, 1967.

[9] D.R. Denzler, An approximative algorithm for the fixed charge problem. *Naval Research Logistics Quarterly*, Vol. 16, pp. 411-416, 1964.

[10] M. Diaby, Successive linear approximation procedure for generalized fixed-charge transportation problem. *Journal of the Operational Research Society*, Vol. 42, pp. 991-1001, 1991.

[11] R.E. Erickson, C.L. Monma, A.F. Veinott Jr., Send-and-Split method for Minimum-Concave-Cost Network Flows. *Mathematics of Operations Research*, Vol. 12, pp. 634-664, 1987.

[12] M. Fisher, An Applications Oriented Guide to Lagrangian Relaxation. *INTERFACES*, Vol. 15, pp. 10-21, 1985.

[13] G. Gallo, C. Sodini, Adjacent extreme flows and application to min concave cost flow problems. *NETWORKS*, Vol. 9, pp. 95-121, 1979.

[14] P. Gray, Exact solution of the fixed-charge transportation problem. *Operations Research*, Vol. 19, pp. 1529-1538, 1971.

[15] G.M. Guisewite, P.M. Pardalos, Minimum concave-cost network flow problems: applications, complexity, and algorithms. *Annals of Operations Research*, Vol. 25, pp. 75-100, 1990.

[16] G.M. Guisewite, P.M. Pardalos, Algorithms for the single-source uncapacitated minimum concave cost network flow problem. *Journal of Global Optimization*, Vol. 1, pp. 245-265, 1991.

[17] W.M. Hirsch, G.B. Dantzig, The Fixed Charge Problem. *Naval Research Logistics Quarterly*, Vol. 15, pp. 413-424, 1968.

[18] R. Horst, P.M. Pardalos (eds.), *Handbook of Global Optimization.* Kluwer Academic Publishers, Dordrecht, 1995.

[19] *ILOG CPLEX 7.0*, ILOG, France, 1999.

[20] D.S. Johnson, J.K. Lenstra, A.H.G. Rinnooy Kan, The complexity of the network design problem. *NETWORKS* Vol. 8, pp. 279-285, 1978.

[21] J. Kennington, E. Unger, A new branch-and-bound algorithm for the fixed charge transportation problem. *Management Science*, Vol. 22, pp. 1116-1126, 1976.

[22] D.B. Khang, O. Fujiwara, Approximate solutions of capacitated fixed-charge minimum cost network flow problems. *NETWORKS*, Vol. 21, pp. 689-704, 1991.

[23] D. Kim, P.M. Pardalos, A solution approach for the fixed charge network flow problem using a dynamic scaling procedure. *Operations Research Letters*, Vol. 24, pp. 195-203, 1999.

[24] D. Kim, P.M. Pardalos, Dynamic slope scaling and trust interval techniques for solving concave piecewise linear network flow problems. *NETWORKS* Vol. 35, pp. 216-222, 2000.

[25] D. Kim, P.M. Pardalos, A dynamic domain contraction algorithm for nonconvex piecewise linear network flow problems. *Journal of Global Optimization*, Vol. 17, pp. 225-234, 2000.

[26] H. Kuhn, W.J. Baumol, An approximate algorithm for the fixed charges transportation problem. *Naval Research Logistics Quarterly*, Vol. 9, pp. 1-15, 1962.

[27] B.W. Lamar, C.A. Wallace, Revised-modified penalties for fixed charge transportation problems. *Management Science*, Vol. 43, pp. 1431-1436, 1997.

[28] M. Minoux, Multiflots De Coût Minimal Avec Fonctions De Coût Concaves. *Annals of Telecommunication*, Vol. 31, pp. 1183-1191, 1976.

[29] K.G. Murty, Solving the Fixed Charge Problem by Ranking the Extreme Points. *Operations Research*, Vol. 16, pp. 268-279, 1968.

[30] U.S. Palekar, M.H. Karwan, S. Zionts, A branch-and-bound method for fixed charge transportation problem. *Management Science*, Vol. 36, pp. 1092-1105, 1990.

[31] P.M. Pardalos, Enumerative techniques for solving some nonconvex global optimization problems. *OR Spectrum*, Vol. 10, pp. 29-35, 1988.

[32] P.M. Pardalos, J.B. Rosen, Global minimization of large-scale constrained concave quadratic problems by separable programming. *Mathematical Programming*, Vol. 34, pp. 163-174, 1986.

[33] P.M. Pardalos, J.B. Rosen, *Constrained Global Optimization: Algorithms and Applications*, Lecture Notes in Computer Science 268, Springer-Verlag, Berlin, 1987.

[34] D.I. Steinberg, The fixed charge problem. *Naval Research Logistics Quarterly*, Vol. 17, pp. 217-236, 1970.

[35] W.E. Walker, A heuristic adjacent extreme point algorithm for the fixed charge problem. *Management Science*, Vol. 22, pp. 587-596, 1976.

[36] B. Yaged, Minimum cost routing for static network models. *NETWORKS*, Vol. 1, pp. 139-172, 1971.

VARIABLE PROJECTION METHODS FOR LARGE–SCALE QUADRATIC OPTIMIZATION IN DATA ANALYSIS APPLICATIONS

Emanuele Galligani
Department of Mathematics
University of Modena and Reggio Emilia
Via Campi 213/B, 41100 Modena, Italy
e-mail: galligani@unimo.it

Valeria Ruggiero
Department of Mathematics
University of Ferrara
Via Machiavelli 35, 44100 Ferrara, Italy
e-mail: rgv@dns.unife.it

Luca Zanni
Department of Mathematics
University of Modena and Reggio Emilia
Via Campi 213/B, 41100 Modena, Italy
e-mail: zanniluca@unimo.it

Abstract This paper concerns with the numerical evaluation of the variable projection method for quadratic programming problems in three data analysis applications. The three applications give rise to large–scale quadratic programs with remarkable differences in the Hessian definition and/or in the structure of the constraints.

Keywords: Large–scale quadratic programming, variable projection method, support vector machines, image restoration problem, bivariate interpolation problem.

1 Introduction

Consider the linearly constrained convex quadratic programming (QP) problem:

$$\begin{array}{ll} \textit{minimize} & f(\boldsymbol{x}) = \frac{1}{2}\boldsymbol{x}^T G\boldsymbol{x} + \boldsymbol{q}^T\boldsymbol{x} \\ \textit{subject to} & C\boldsymbol{x} = \boldsymbol{d}, \qquad A\boldsymbol{x} \geq \boldsymbol{b}, \end{array} \tag{1}$$

where G is a symmetric positive semidefinite matrix of order n, C is an $m_e \times n$ matrix of full row rank ($m_e \leq n$) and A is an $m_i \times n$ matrix. We assume that the feasible region $K = \{\boldsymbol{x} \in R^n \mid C\boldsymbol{x} = \boldsymbol{d};\ A\boldsymbol{x} \geq \boldsymbol{b}\}$ is a nonempty polyhedral set and that $f(\boldsymbol{x})$ is bounded from below on K.
In some recent papers ([42], [43]), for the solution of large–scale QP problems of the form (1) we propose the Variable Projection Method (VPM), that includes as special cases the classical projection and splitting methods and the scaled gradient projection methods [6]. The VPM has the following iterative scheme (see also [44], [45] for theoretical and numerical features about some variants of the method):

1. let D be a symmetric positive definite matrix, $\boldsymbol{x}^{(0)}$ be an arbitrary vector and ρ_1 be an arbitrary positive constant; $k \leftarrow 1$;

2. compute the unique solution $\boldsymbol{y}^{(k)}$ of the subproblem

$$\begin{array}{ll} \textit{minimize} & \frac{1}{2}\boldsymbol{x}^T \frac{D}{\rho_k}\boldsymbol{x} + (\boldsymbol{q} + (G - \frac{D}{\rho_k})\boldsymbol{x}^{(k-1)})^T \boldsymbol{x} \\ \textit{subject to} & C\boldsymbol{x} = \boldsymbol{d}, \qquad A\boldsymbol{x} \geq \boldsymbol{b}; \end{array} \tag{2}$$

3. set $\boldsymbol{d}^{(k)} = \boldsymbol{y}^{(k)} - \boldsymbol{x}^{(k-1)}$;

4. if $(G\boldsymbol{x}^{(k-1)} + \boldsymbol{q})^T \boldsymbol{d}^{(k)} < 0$ and $k \neq 1$, compute the solution θ_k of the problem

$$\min\{f(\boldsymbol{x}^{(k-1)} + \theta \boldsymbol{d}^{(k)});\quad \theta \in (0, 1]\}; \tag{3}$$

   else

$$\theta_k = 1;$$

5. compute

$$\boldsymbol{x}^{(k)} = \boldsymbol{x}^{(k-1)} + \theta_k \boldsymbol{d}^{(k)}; \tag{4}$$

6. terminate if $\boldsymbol{x}^{(k)}$ satisfies a stopping rule, otherwise update ρ_{k+1} by an appropriate rule; then $k \leftarrow k+1$ and go to step 2.

The R–linear convergence of the sequence $\{\boldsymbol{x}^{(k)}\}$ generated by the VPM to a solution of (1) is proved in [45] under the very general hypotheses that D is a symmetric positive definite matrix and the sequence $\{\rho_k\}$ is bounded from below and above by positive constants.
The iterative scheme of the VPM requires essentially the matrix–vector product $G\boldsymbol{y}^{(k)}$ and the solution of the QP subproblem (2) at each iteration. Of course, from the practical point of view, the matrix D must be an easily solvable matrix (diagonal or block diagonal matrix) in order to make each QP subproblem easier than the original problem. As

consequence, the VPM turns the complexity of the original problem to the choice of a solver for separable (or nearly separable) QP problems which is specialized for the structure of the feasible region (see [32], [38] as examples of solvers for separable quadratic programs with special feasible region defined by box constraints and a single equality constraint). When the constraints do not present a particular structure, we can formulate each inner subproblem as a mixed linear complementarity problem (LCP), that can be solved by sequential or parallel splitting methods [10], [20], [21], [28], [29]. Another important consideration is concerned with the choices of D and of the sequence of projection parameters $\{\rho_k\}$. Special choices of D and of the scalar parameters * give rise to the classical splitting and projection–type methods that are particular cases of the general VPM scheme.

The possibility to use a variable projection parameter at each step can be exploited to decrease the number of iterations and, consequently, to overcome the slow convergence rate exhibited by other splitting and projection–type methods (see [43]). To this aim, the following nonexpensive and efficient updating rules for ρ_k, $k = 2, 3, \ldots$ are devised on the basis of heuristic considerations in [44]:

$$\rho_k = \begin{cases} \rho_{k-1} & \text{for } \|G\boldsymbol{d}^{(k-1)}\|^2 \le \psi \|\boldsymbol{d}^{(k-1)}\|^2 \\ \dfrac{\boldsymbol{d}^{(k-1)^T} G \boldsymbol{d}^{(k-1)}}{\boldsymbol{d}^{(k-1)^T} G D^{-1} G \boldsymbol{d}^{(k-1)}} & \text{otherwise} \end{cases} \tag{5}$$

$$\rho_k = \begin{cases} \rho_{k-1} & \text{for } \|G\boldsymbol{d}^{(k-1)}\|^2 \le \psi \|\boldsymbol{d}^{(k-1)}\|^2 \\ \dfrac{\boldsymbol{d}^{(k-1)^T} D \boldsymbol{d}^{(k-1)}}{\boldsymbol{d}^{(k-1)^T} G \boldsymbol{d}^{(k-1)}} & \text{otherwise} \end{cases} \tag{6}$$

where ψ is a prefixed small tolerance.

An extensive numerical experimentation on large and sparse test problems has been carried out to show the effectiveness of the variable projection methods combined with the rules (5) and (6) (for a survey, see [41] and references therein). Nevertheless, the considered test problems are randomly generated with assigned features [11] or are chosen from the CUTE library [8].

The aim of this work is to show the numerical behaviour of the variable projection methods on a set of applications arising in the framework of data analysis.

---

*We recall that for $\rho_k = 1, \theta_k = 1$, for any k, and $2D - G$ positive definite, we have the splitting method [26] (see also [18], [19]); for $\rho_k = \rho < \frac{2}{\lambda_{max}(D^{-1/2}GD^{-1/2})}$, $\theta_k = 1$ for any k, we have the projection method [30]; for $\rho_k = \rho$, we have the scaled gradient projection method with a "limited minimization rule" [6].

In particular, in the next section, we describe the application of the VPM to the quadratic program arising in training the learning machines named Support Vector Machines (SVMs) [9]. In this QP problem the matrix G is dense and the feasible region is defined by box constraints and a single equality constraint. In the special instance of SVMs with Gaussian kernels, by introducing an appropriate updating rule for the projection parameters ρ_k, a good convergence rate of the VPM is obtained even on this dense problem.
The second application concerns with the quadratic program arising in the numerical solution of an image restoration problem with a point–spread–function without a specific form [1]. The feasible region is defined as in the previous problem but the matrix G is sparse with a particular structure that is well exploited by the VPM.
The application described in the final section is related to a constrained bivariate interpolation problem [16]. A well known approach to this problem requires to solve a quadratic program with sparse block matrix G and sparse constraint matrices A and C. In this case, the special definition of the entries of G suggests a very efficient splitting method for solving the QP problem. The effectiveness of the VPM is evaluated by comparison with this suitable splitting method.

2 Large QP Problems in Training Support Vector Machines

The Support Vector Machine is a technique for solving pattern recognition problems [7],[9],[36],[50]. Given a training set of labeled examples,

$$D = \{(\boldsymbol{p}_i, y_i),\ i = 1, \dots, n, \quad \boldsymbol{p}_i \in R^m,\ y_i \in \{-1, 1\}\},$$

the SVM learning technique performs pattern recognition by finding a decision surface, $F : R^m \to \{-1, 1\}$, obtained by solving a quadratic program of the form:

$$\begin{array}{ll} \textit{minimize} & \frac{1}{2}\boldsymbol{x}^T G \boldsymbol{x} - \sum_{i=1}^{n} x_i \\ \textit{subject to} & \sum_{i=1}^{n} y_i x_i = 0, \qquad 0 \le x_j \le C, \qquad j = 1, \dots, n, \end{array} \tag{7}$$

where G is a symmetric positive semidefinite matrix related to the chosen decision surface. Since the matrix G is dense with size equal to the number of training examples, the problem is challenging ($n \gg 10000$) in many interesting applications of the SVMs.

Among the various approaches proposed to overcome the difficulties involved in this problem, we may recognize two main classes. The first class collects the techniques based on different formulations of the optimization problem that gives rise to the separating surface (see [15] and references therein). These reformulations lead to more treatable optimization problems, but use criteria for determining the decision surface which, in some cases, are considerably different respect to the one of the standard SVM. The second class includes algorithms that exploit the special structure of problem (7) [14], [23], [37], [39]. In particular, the approaches proposed in [23], [37], [39] are based on special decomposition techniques that avoid explicit storage of G by splitting the original problem into a sequence of smaller QP subproblems of the form (7) with Hessian matrices equal to principal submatrices of G. These techniques differ in the strategy employed for identifying the variables to update at each iteration and in the size chosen for the subproblems. In [39] the subproblems have size 2 and can be solved analytically, while in [23], [37] the subproblem size is a parameter of the procedure and a numerical QP solver is required.

In order to explain how the VPM can be a suitable solver for these QP subproblems we briefly sketch the SVM technique. Starting from the case of linear decision surface and linearly separable examples, the goal is to determine the hyperplane that leaves all the examples with the same label on the same side and maximizes the margin between the two classes, where the margin is defined as the sum of the distances of the hyperplane from the closest point of the two classes. Thus [9], the linear classifier

$$F(\boldsymbol{p}) = sign\left(\boldsymbol{w}^{*T}\boldsymbol{p} + b^*\right), \qquad \boldsymbol{w}^* \in R^m, \quad b^* \in R,$$

is obtained from the solution $(\boldsymbol{w}^*, b^*)$ of the following problem

$$\begin{array}{ll} minimize & \frac{1}{2}\boldsymbol{w}^T\boldsymbol{w} \\ subject\ to & y_i(\boldsymbol{w}^T\boldsymbol{p}_i + b) \geq 1 \quad i = 1, \ldots, n \end{array} \tag{8}$$

When the two classes are nonseparable we can determine the hyperplane that maximizes the margin and minimizes a quantity proportional to the number of misclassification errors. The trade-off between the largest margin and the lowest number of errors is controlled by a positive constant C that has to be chosen beforehand [40]. In this case, the linear classifier comes from the solution $(\boldsymbol{w}^*, b^*, \boldsymbol{\xi}^*)$ of

$$\begin{array}{lll} minimize & \frac{1}{2}\boldsymbol{w}^T\boldsymbol{w} + C\sum \xi_i & \\ subject\ to & y_i(\boldsymbol{w}^T\boldsymbol{p}_i + b) \geq 1 - \xi_i & i = 1, \ldots, n \\ & \boldsymbol{\xi} \geq 0 & \end{array} \tag{9}$$

The pair $(\boldsymbol{w}^*, b^*)$ can be obtained by applying a QP solver to (9) (or to (8) in the separable case); nevertheless, in order to generalize the procedure to nonlinear decision surfaces, it is useful to look at the dual problem. The dual problem of (9) is a convex QP problem of the form (7) with $G_{ij} = y_i y_j \boldsymbol{p}_i^T \boldsymbol{p}_j$ (the dual of (8) differs from (7) for the absence of the upper bound on the variables). If $\boldsymbol{x}^*$ denotes the solution of the dual problem, by using the KKT conditions for the primal problem, we may express $(\boldsymbol{w}^*, b^*)$ as

$$\begin{aligned} \boldsymbol{w}^* &= \textstyle\sum_{i=1}^n x_i^* y_i \boldsymbol{p}_i, \\ b^* &= y_j - \boldsymbol{w}^{*T} \boldsymbol{p}_j \qquad \text{for any } j \text{ such that} \qquad 0 < x_j < C. \end{aligned}$$

Thus, the classifier can be written as a linear combination of the training vectors associated to the nonzero multipliers. These training vectors, named Support Vectors (SVs), include all the information contained in the training set which is needed to classify new data points.

The previous technique can be extended to the general case of nonlinear separating surfaces. This is easily done by mapping the input points into a space Z, called feature space, and by formulating the linear classification problem in the feature space. Typically Z is a Hilbert space of finite or infinite dimension. If $\boldsymbol{p} \in R^m$ is an input point, let $\varphi(\boldsymbol{p})$ be the corresponding feature point with φ a mapping from R^m to Z. The solution to the classification problem in the feature space will have the following form

$$F(\boldsymbol{p}) = sign\left(\sum_{i=1}^n x_i^* y_i \varphi(\boldsymbol{p})^T \varphi(\boldsymbol{p}_i) + b^*\right)$$

and therefore it will be nonlinear in the original input variables. In this case, the coefficients x_i^* are the solution of a QP problem of the form (7) where

$$G_{ij} = y_i y_j \varphi(\boldsymbol{p}_i)^T \varphi(\boldsymbol{p}_j).$$

At first sight it might seem that the nonlinear separating surface can not be determined unless the mapping φ is completely known. Nevertheless, since φ appears only in scalar products between two feature points, if we find an expression for the scalar product in feature space which uses the points in input space only, that is

$$\varphi(\boldsymbol{p}_i)^T \varphi(\boldsymbol{p}_j) = K(\boldsymbol{p}_i, \boldsymbol{p}_j), \tag{10}$$

then, it is not necessary a full knowledge of the function φ. The symmetric function K in (10) is called kernel. We may conclude that the extension of the theory to the nonlinear case is reduced to finding kernels

Figure 1: Gaussian SVMs: 3–D pattern of G

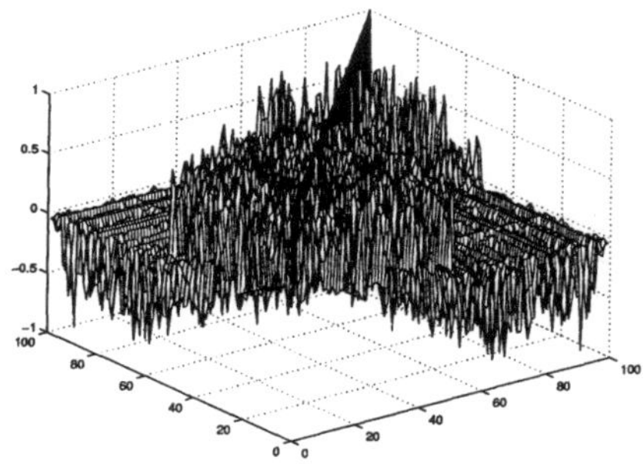

which identify certain families of decision surfaces and satisfy equation (10). Two frequently used kernels are the polynomial kernel, $K(\boldsymbol{p}_i, \boldsymbol{p}_j) = (1+\boldsymbol{p}_i^T \boldsymbol{p}_j)^d$, $d \in N\backslash\{0\}$, and the Gaussian kernel, $K(\boldsymbol{p}_i, \boldsymbol{p}_j) = exp(-\|\boldsymbol{p}_i - \boldsymbol{p}_j\|^2/(2\sigma^2)$, $\sigma \in R$. The separating surface in input space is a polynomial surface of degree d for the polynomial kernel and a weighted sum of Gaussians centered on the support vectors for the Gaussian kernel. Thus, from the computational point of view, the training of a SVM in the nonlinear case requires to solve a problem of form (7) where the entries of G are defined by special kernel functions.

In case of Gaussian SVMs, an example of the pattern of G is shown in Figure 1. We may observe that the matrix is dense but with some particular features: the main diagonal has entries equal to 1, the other entries are in $]-1, 1[$ and many of them are much less than 1 in absolute value. These considerations suggest that the VPM, with D diagonal matrix, may be an iterative solver suited for exploiting both the structure of the constraints and the particular nature that the Hessian matrix presents in the case of Gaussian SVMs. In fact, if D is a diagonal matrix, each VPM subproblem is a separable QP problem subject to a single linear equality constraint and box constraints, that is, a nonexpensive task that may be faced by efficient method [32], [38]. Furthermore, in [51] it is shown that, for the QP problem arising in training Gaussian SVMs, a good convergence rate of the VPM is obtained by using the following updating rule for the projection parameter ρ_{k+1}:

$$\rho_{k+1} = \begin{cases} \rho_k & \text{if} \quad \|G\boldsymbol{d}^{(k)}\|^2 \leq \epsilon \|\boldsymbol{d}^{(k)}\|^2 \\ \begin{cases} \dfrac{\boldsymbol{d}^{(k)T} G \boldsymbol{d}^{(k)}}{\boldsymbol{d}^{(k)T} G S^{-1} G \boldsymbol{d}^{(k)}} & \text{if} \quad \text{mod } (k, \tilde{k}) < \dfrac{\tilde{k}}{2} \\ \dfrac{\boldsymbol{d}^{(k)T} S \boldsymbol{d}^{(k)}}{\boldsymbol{d}^{(k)T} G \boldsymbol{d}^{(k)}} & \text{otherwise} \end{cases} \end{cases} \tag{11}$$

It is interesting to observe that, on this dense problem, an effective updating rule for ρ_{k+1} is obtained by combining the two rule (5) and (6) that, on sparse problems, exhibit good performance even if singly used. Unfortunately, in case of polynomial kernels, since the different definition of the dense Hessian matrix, the proposed updating rule is not so effective.

In the following, we report the results of a comparison between the VPM and the solver pr_LOQO actually used as inner QP solver in the decomposition scheme implemented in the package SVM^{light} [23]. The pr_LOQO is a version of the primal-dual infeasible interior point method of Vanderbei [49], appropriately implemented by Smola [46] for the dense problems of form (7). We consider small to medium test problems; of course, for large scale problems the two methods may be used as inner solvers of the decomposition techniques. The test problems are obtained by training Gaussian SVMs on the MNIST database of handwritten digits from AT&T Research Labs [27] and on the UCI Adult data set [31]. We construct test problems of size $N = 200, 400, 800, 1600$, by considering the first $N/2$ inputs of the digits 8 and 9. The UCI Adult data set allows to train a SVM to predict whether a household has an income greater than \$50000. We consider the versions of the data set with size $N = 1605, 2265, 3185$. We use $C = 10$, $\sigma = 1800$ for the MNIST data set and $C = 1$, $\sigma^2 = 10$ for the UCI Adult data set.

The results in Tables 1 and 2 are obtained on a workstation Compaq XP1000 at 667MHz with 1GB of RAM, with code written in standard C. In the VPM, the matrix D is chosen equal to the identity matrix, the stopping rule consists in the fulfillment of the KKT conditions within a tolerance of 0.001 (in general, a higher accuracy does not improve significantly the SVM performance [23],[39]) and for the parameter ρ_k we use $\rho_1 = 1$, $\epsilon = 10^{-14}$ and $\tilde{k} = 6$ in the updating rule (11). The pr_LOQO solver was run with the parameter *sigfig_max* = 8 (a lower value implies an accuracy not comparable with that of the VPM). In the tables we report the number of iterations, the computational time in seconds required by the solvers, the number of support vectors and bounded support vectors (BSV), i.e. support vectors with $x_i^* = C$.

Table 1: Test problems from the MNIST data set

	n	iterations	time	SV	BSV
pr_LOQO	200	13	0.15	76	0
	400	14	1.2	120	0
	800	15	10.0	176	0
	1600	16	111.5	239	0
VPM	200	57	0.02	76	0
	400	82	0.1	120	0
	800	124	0.3	176	0
	1600	232	1.7	238	0

Table 2: Test problems from the UCI Adult data set

	n	iterations	time	SV	BSV
pr_LOQO	1605	15	131.6	691	584
	2265	15	383.9	1011	847
	3185	15	1081.4	1300	1109
VPM	1605	136	2.3	691	584
	2265	171	5.8	1011	847
	3185	237	14.6	1299	1113

From Tables 1 and 2 we may observe the higher effectiveness of the VPM especially when the size of the problem increases. In particular, the VPM allows the solution of medium size problems ($N > 1000$) in few seconds. Thus, this method may be an useful inner QP solver in the decomposition techniques for very large-scale SVMs [51]. Finally, we emphasize that the method is easily parallelizable, since, in this application, each iteration consists essentially in a matrix-vector product and in the solution of a nonexpensive separable QP subproblem (eventually solvable in parallel [32]): hence, new parallel decomposition schemes can be based on the VPM [52].

3 Numerical Solution of Image Restoration Problem

The image degradation process can be described as a system which operates on an input image (object) $\phi(\xi, \eta)$ to produce an output image (degraded image) $\gamma(u, v)$. An additive noise term $\epsilon(u, v)$ is also included. Thus, the problem can be formulated as a first kind Fredholm integral equation:

$$\gamma(u,v) = \int_{-\infty}^{+\infty}\int_{-\infty}^{+\infty} h(u,\xi,v,\eta)\phi(\xi,\eta)d\xi d\eta + \epsilon(u,v) \tag{12}$$

Additional constraints must hold; first the radiant energy distributions for input and output images must be positive or zero, i.e.

$$\phi(\xi,\eta) \geq 0 \qquad \text{for any } (\xi,\eta) \tag{13}$$

$$\gamma(u,v) \geq 0 \qquad \text{for any } (u,v) \tag{14}$$

Second, according to an energy conservation concept, the total energy in the object is preserved in the image, that is

$$\int_{-\infty}^{+\infty}\int_{-\infty}^{+\infty} \phi(\xi,\eta)d\xi d\eta = \int_{-\infty}^{+\infty}\int_{-\infty}^{+\infty} \gamma(u,v)dudv \tag{15}$$

The problem of image restoration is the determination of the original object distribution ϕ given the image γ, the point–spread degradation function h and the noise ϵ. If we suppose that the function $\phi, \gamma, h, \epsilon$ are sampled uniformly over a finite rectangular grid to form arrays of size $n = \mu \times \nu$ (μ and ν are the numbers of inner grid points in u and v directions respectively) and use the rectangular rule for the approximation of the integral in (12), we obtain the following system of linear equations:

$$\boldsymbol{\gamma} = H\boldsymbol{\phi} + \boldsymbol{\epsilon} \tag{16}$$

Here, $\boldsymbol{\phi}$ is an n–vector such that

$$0 \leq \phi_i \leq \Phi \qquad i = 1,\dots,n \tag{17}$$

where Φ is the maximum gray level that we consider, and, from (15),

$$\sum_{i=1}^{n} \phi_i = \sum_{i=1}^{n} \gamma_i \tag{18}$$

The image restoration problem is a ill-conditioned problem (see [1]). A well known method for dealing with such problem is the method of regularization by Phillips and Tikhonov [48], that consists in solving the following minimization problem [1]

$$minimize\ \|H\boldsymbol{\phi} - \boldsymbol{\gamma}\|^2 + \lambda\|\Lambda\boldsymbol{\phi}\|^2 \tag{19}$$

where the matrix Λ is the regularization operator (usually chosen equal to the discretization of the Laplace or biharmonic operator) and λ is

Figure 2: Pattern of the nonzero entries of G

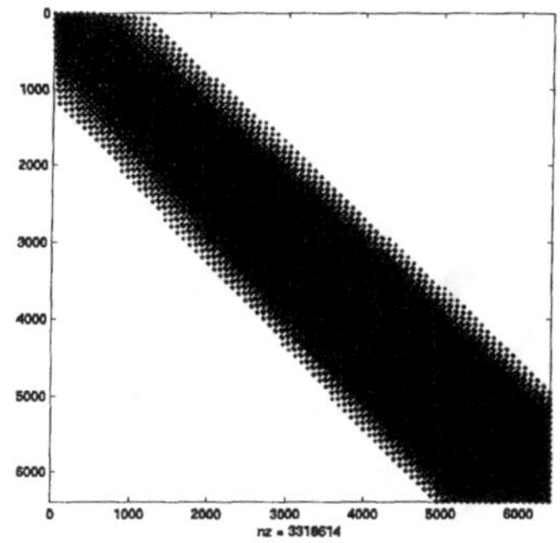

the regularization parameter ($\lambda > 0$; in the subsequent experiments $\lambda = 10^{-4}$).
In many image restoration problem the point–spread function h has the following general form, i.e., h is "space–variant" and "non–separable":

$$h(u, \xi, v, \eta) = \frac{N}{2\pi\sigma(u,v)} \exp\left(-\frac{(\xi - u)^2 + (\eta - v)^2}{2\sigma^2(u,v)}\right) \tag{20}$$

where N is a normalization constant for h and

$$\sigma^2(u,v) = \begin{cases} l|\sin(u+v)| & \text{if } \sin(u+v) \neq 0 \quad l \in [1,2] \\ l & \text{otherwise} \end{cases} \tag{21}$$

or

$$\sigma^2(u,v) = l|u+v| \qquad l \in [0.01, 0.02] \tag{22}$$

Here, l is a positive constant that determines the extent of blur in the image; indeed for increasing values of l, each image point is obtained with the contribution of a larger number of object points (superposition principle).
As for the application in the previous section, the VPM appears convenient for solving the QP problem (19) with the single equality constraint (18) and the simple bounds (17). In this case, the objective matrix $G = H^T H + \lambda \Lambda^T \Lambda$ is sparse with a special structure; for example, when $n = 6400$ and $\sigma^2(u,v)$ as in (22), the pattern of the nonzero entries of G is showed in Figure 2 (here nz denotes the number of nonzero entries of G). Figure 3 is a detail of the previous pattern for the leading principal submatrix of order 320 of G while Figure 4 displays in 3–D the entries of this submatrix.

Figure 3: Detail of the pattern of the nonzero entries of G

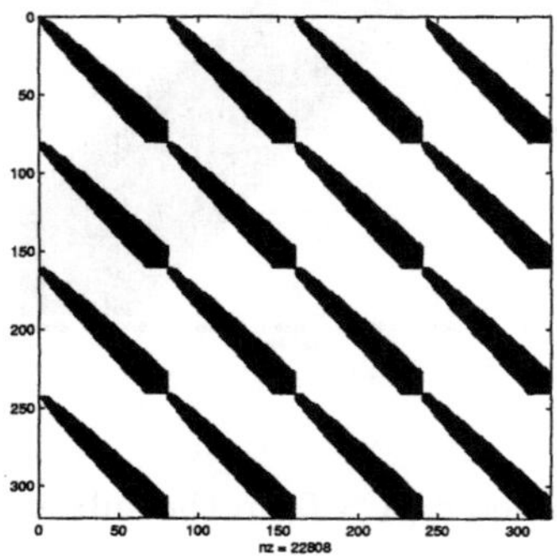

Figure 4: 3–D pattern of the detail of G

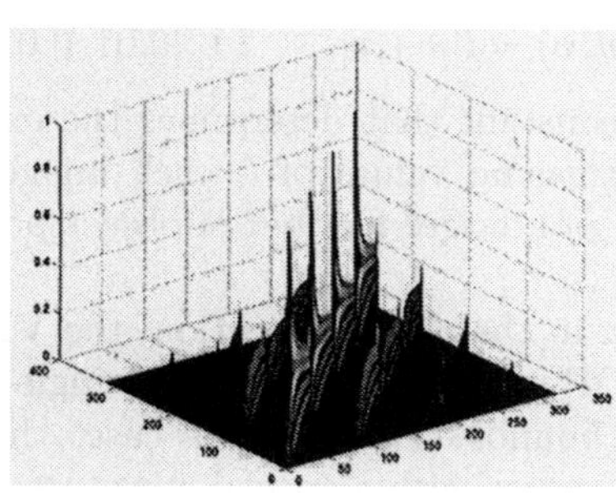

Figure 5: Original images for the test problems

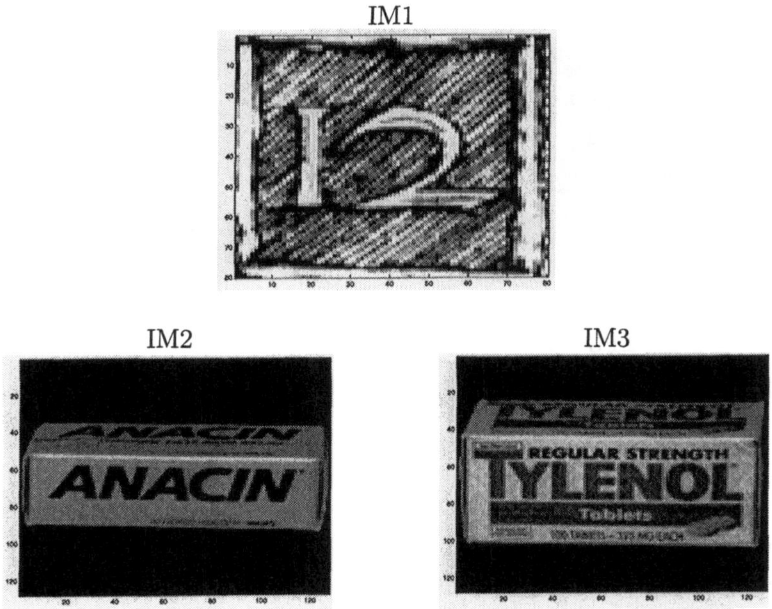

From practical point of view, it is convenient to store in compressed form only the nonzero elements of H and to avoid the explicit computation of $H^T H$ in G; the matrix–vector multiplication $G\boldsymbol{y}^{(k)}$ required by the VPM, may be computed by routines that exploit the sparsity of H.
In this implementation of the VPM we use the identity matrix as D, the updating rule (5) for $\{\rho_k\}$ and the solver in [38] for the inner subproblems. The three images in Figure 5 (see also Table 3) have been used to generate the test problems for evaluating the effectiveness of this approach.

Figure 6 shows four blurred images and Table 4 reports the parameters

Table 3: Features of the original images

image	size	gray level
IM1	80×80	64
IM2	128×128	256
IM3	128×128	256

Figure 6: Blurred images for the test problems

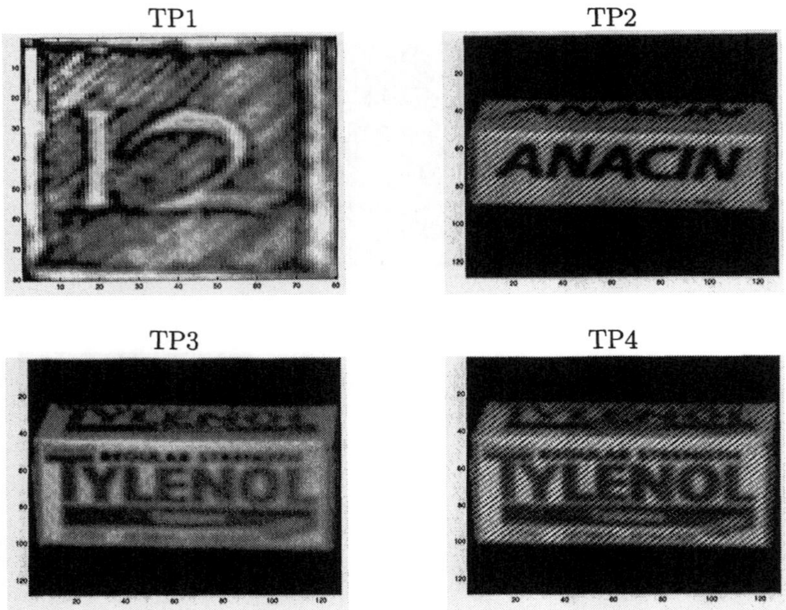

used in the related point–spread function and the three picture distance measures D, R, E, defined by Herman in [22]:

$$\begin{aligned} D &= \frac{\sum_{\xi=1}^{\mu}\sum_{\eta=1}^{\nu} |\phi(\xi,\eta)-\gamma(\xi,\eta)|}{\sum_{\xi=1}^{\mu}\sum_{\eta=1}^{\nu} |\phi(\xi,\eta)-\tau|} \\ R &= \frac{\sum_{\xi=1}^{\mu}\sum_{\eta=1}^{\nu} |\phi(\xi,\eta)-\gamma(\xi,\eta)|}{\sum_{\xi=1}^{\mu}\sum_{\eta=1}^{\nu} \phi(\xi,\eta)} \\ E &= \max_{\xi,\eta} |\phi(\xi,\eta)-\gamma(\xi,\eta)| \end{aligned}$$

where $\gamma(\xi,\eta)$ is the blurred image and τ denotes the average density of

Table 4: Features of the blurred images

test problem	original image	formula for σ^2	D	R	E
TP1	IM1	(22) $l=0.01$	0.46	0.19	33
TP2	IM2	(21) $l=1.3$	0.37	0.28	143
TP3	IM3	(22) $l=0.01$	0.23	0.14	110
TP4	IM3	(21) $l=1.5$	0.43	0.28	155

Figure 7: Restored images for the test problems

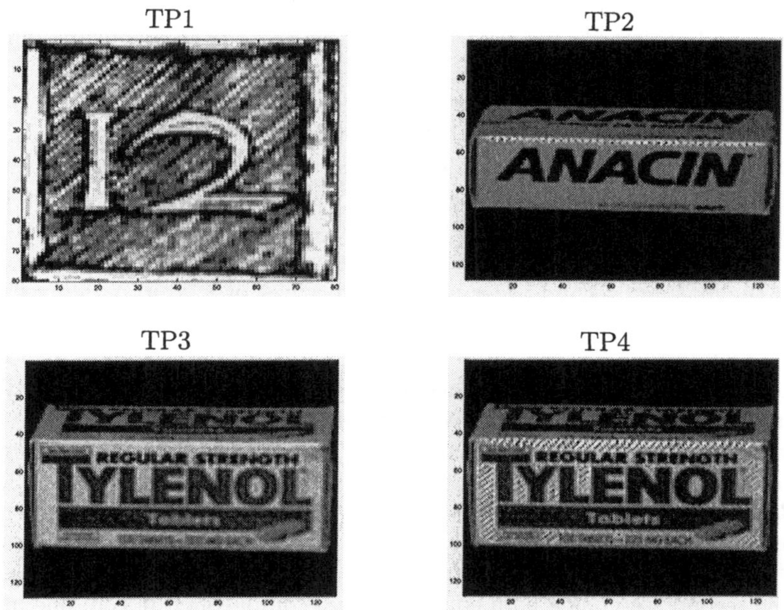

the digitized test image.

Table 5: Features of the restored images

test problem	*it*	*time*	*D*	*R*	*E*
TP1	101	35.0	0.21	$7.44 \cdot 10^{-2}$	18
TP2	138	135.1	$7.15 \cdot 10^{-2}$	$2.98 \cdot 10^{-2}$	124
TP3	131	245.8	0.11	$5.68 \cdot 10^{-2}$	94
TP4	169	193.9	0.13	$5.79 \cdot 10^{-2}$	115

Figure 7 shows the restored images and Table 5 summarize the results obtained by the VPM method. In particular *it* denotes the number of required iterations to satisfy the Karush–Kuhn–Tucker conditions within a tolerance 10^{-3}, while *time* is the elapsed time expressed in seconds required for the restoration on a Compaq XP1000 workstation. D, R and E are the picture distance measures between the original image $\phi(\xi, \eta)$ and the restored image $\gamma(\xi, \eta)$.

4 A Bivariate Interpolation Problem

A bivariate interpolation problem can be stated as follows: given N distinct and arbitrarily spaced points V_i, $i = 1, \ldots, N$ in a convex domain Ω in the x–y plane and N real numbers z_i, consider the problem of computing a C^1 bivariate function $F(x, y)$ whose values V_i are exactly z_i. This problem can easily have thousands of interpolating points V_i.

When some specific information is available for computing C^1 interpolants which are shape–preserving or when step gradients are implied by the data or values of $F(x, y)$ are rapidly varying in some subset of the domain Ω, appropriate equality and inequality constraints are imposed on the first and second order partial derivatives of $F(x, y)$ on a given set of points. In this case we have a bivariate interpolation problem with constraints.

A general method for interpolation of scattered data can be described by three steps. *Triangulation*: divide the domain into subdomains by connecting the points V_i appropriately ([25], [2]). *Curve network*: compute a smooth function on the boundary of each subdomain (e.g. see [33], [34]). *Blending*: construct the patch that interpolates the functions and their derivatives over each subdomain and fit the patches together to form a C^1 surface [4], [5], [24]).

The second step has a high computational complexity: indeed, one must determine the first order partial derivatives of $F(x, y)$ in the interpolation points V_i, $i = 1, ..., N$, and define the constraints on the C^1 surface $z = F(x, y)$. That is, one must solve a convex quadratic programming problem, which, generally, has more than a hundred and perhaps thousands of variables and constraints.

If $F_x(V_i)$ and $F_y(V_i)$ denote the first order partial derivatives of $F(x, y)$ in V_i and $\boldsymbol{u}$ is the $2N$–vector

$$\boldsymbol{u} = (F_x(V_1), F_y(V_1), F_x(V_2), F_y(V_2), ..., F_x(V_N), F_y(V_N))^T$$

we can express the value of $F(x, y)$ in a point Q in the triangle of vertices V_i, V_j and V_k as a linear combination of $F(V_i)$, $F(V_j)$, $F(V_k)$, $F_x(V_i)$, $F_y(V_i)$, $F_x(V_j)$, $F_y(V_j)$, $F_x(V_k)$ and $F_y(V_k)$.

The first and the second order partial derivatives of $F(x, y)$ in Q are expressed analogously. Thus the constraints on the surface $F(x, y)$ can be expressed as a set of linear equalities and inequalities of the form

$$\begin{aligned} \boldsymbol{c}_i^T \boldsymbol{u} &= d_i, \qquad && i = 1, ..., m_e; \\ \boldsymbol{a}_i^T \boldsymbol{u} &\geq b_i, \qquad && i = 1, ..., m_i; \end{aligned} \tag{23}$$

where m_e and m_i are the number of equality and inequality constraints respectively.

Nielson in [33] suggests a method to compute partial derivatives based upon a minimum norm network. This method consists in computing a set of cubic Hermite polynomial on the domain E, union of all the edges of the triangulation of Ω, which interpolate z_i in V_i, $i = 1, ..., N$ and in computing first order partial derivatives by minimizing an appropriate functional on E.
Analogous to the minimum pseudonorm property of univariate splines, the proposed quadratic functional is

$$J(F) = \sum_{ij \in N_e} \int_{e_{ij}} \left(\frac{\partial^2 F}{\partial e_{ij}^2} \right)^2 d\xi_{ij} \tag{24}$$

where $d\xi_{ij}$ represents the arc length along the edge e_{ij} in the triangulation with endpoints V_i and V_j and N_e is a list of indices representing the edges of the triangulation. Let $C[E]$ be the set of all C^1 functions F defined on Ω restricted to E, such that the univariate function f, obtained as restriction of F on e_{ij}, is an element of $H[e_{ij}]$, with $H[e_{ij}] = \{f : f \in C[e_{ij}],\ f' \text{ absolutely continuous},\ f'' \in L^2[e_{ij}]\}$.
It is convenient to view each $F \in C[E]$ as a collection of univariate functions

$$f_{ij}(\tau) = F\left((1-\tau)V_i + \tau V_j\right) \qquad i, j \in N_e \qquad 0 \leq \tau \leq 1$$

and, thus, the problem of minimizing the functional (24) can be written as the following quadratic programming problem:

$$\min \frac{1}{2} \boldsymbol{u}^T G \boldsymbol{u}^T + \boldsymbol{u}^T \boldsymbol{q} \tag{25}$$

where $G = (G_{ij})$, $\boldsymbol{q} = (\boldsymbol{q}_i)$, $i = 1, ..., N$, and

$$G_{ii} = \begin{pmatrix} \tilde{a}_i & \tilde{c}_i \\ \tilde{c}_i & \tilde{b}_i \end{pmatrix} \qquad G_{ij} = \begin{cases} \begin{pmatrix} \tilde{a}_{ij} & \tilde{c}_{ij} \\ \tilde{c}_{ij} & \tilde{b}_{ij} \end{pmatrix} & \text{if} \quad i, j \in N_i \\ 0 & \text{otherwise} \end{cases} \tag{26}$$

with

$$\begin{aligned} \tilde{a}_{ij} &= \tfrac{(x_j - x_i)^2}{2\|e_{ij}\|^3} & \tilde{a}_i &= 2\textstyle\sum_{ij \in N_i} \tilde{a}_{ij} \\ \tilde{c}_{ij} &= \tfrac{(x_j - x_i)(y_j - y_i)}{2\|e_{ij}\|^3} & \tilde{c}_i &= 2\textstyle\sum_{ij \in N_i} \tilde{c}_{ij} \\ \tilde{b}_{ij} &= \tfrac{(y_j - y_i)^2}{2\|e_{ij}\|^3} & \tilde{b}_i &= 2\textstyle\sum_{ij \in N_i} \tilde{b}_{ij}. \end{aligned} \tag{27}$$

and

$$\boldsymbol{q}_i = \left(-\frac{3}{2} \sum_{ij \in N_i} \frac{(x_j - x_i)}{\|e_{ij}\|^3}(z_j - z_i), \quad -\frac{3}{2} \sum_{ij \in N_i} \frac{(y_j - y_i)}{\|e_{ij}\|^3}(z_j - z_i) \right)$$

Figure 8: Pattern of the nonzero entries of G

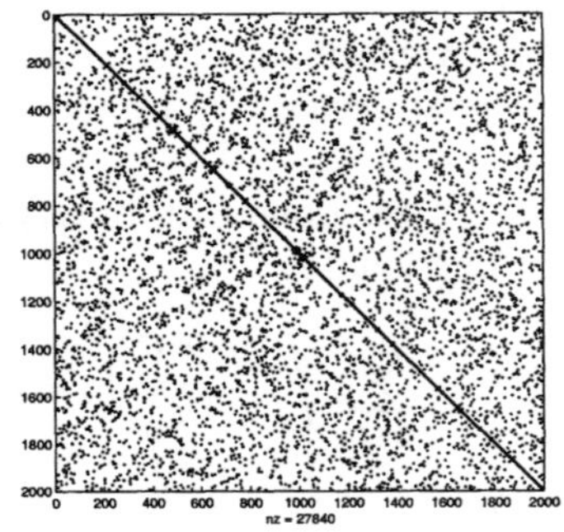

Figure 9: Detail of the pattern of the nonzero entries of G

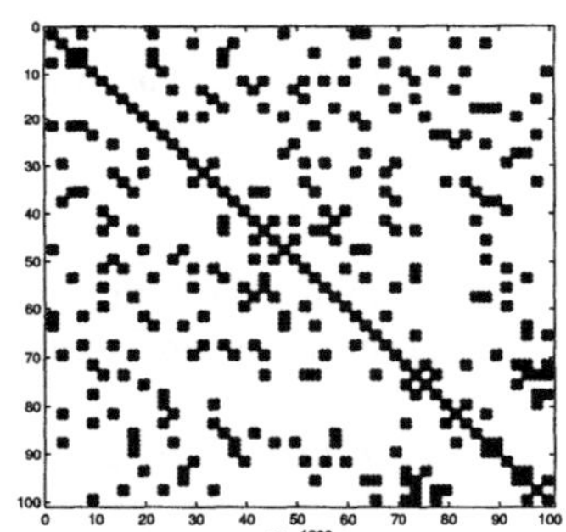

In Figure 8 it is showed the pattern of the nonzero entries (denoted as nz) of the matrix G with $N = 1000$; in Figure 9 a pattern of the nonzero entries of a principal submatrix of G of order 100 is displayed; the entries of this submatrix are also displayed in 3–D in Figure 10
It is possible to prove ([33]) that, among all functions $F \in C[E]$, $F(V_i) = z_i,\ \ i = 1, ..., N$, the function $S \in C[E]$ that uniquely minimizes $J(F)$ in (24) is a set of univariate cubic Hermite interpolants ($i, j \in N_e$, $0 \leq \tau \leq 1$)

$$\begin{aligned} s_{ij}(\tau) &= S\left((1-\tau)V_i + \tau V_j\right) \\ &= \tau^2(3-2\tau)z_j + (1-\tau)^2(2\tau+1)z_i + \\ &\quad +\tau(1-\tau)^2[(x_j - x_i)S_x(V_i) + (y_j - y_i)S_y(V_i)] + \\ &\quad +\tau^2(\tau-1)[(x_j - x_i)S_x(V_j) + (y_j - y_i)S_y(V_j)] \end{aligned}$$

Figure 10: 3–D pattern of the detail of G

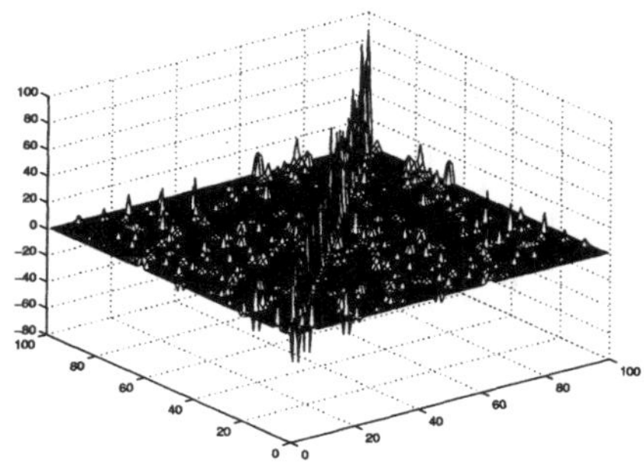

Thus, the vector $\boldsymbol{u}^* = (\boldsymbol{u}_1^{*T}, ..., \boldsymbol{u}_N^{*T})^T$, with $u_i^* = (S_x(V_i), S_y(V_i))^T$, $i = 1, ..., N$, is the solution of the quadratic problem (25)–(23).
In [16] it is proved that the symmetric matrix G is positive definite.
If we set $D = \text{diag}(G_{11}, ..., G_{NN})$ and $H = G - D$, in [18] it is proved that the symmetric matrix $D - H$ is positive definite (i.e. (D, H) is a P–regular splitting [35, p. 122]), thus splitting method for the strictly convex quadratic problem (25)–(23) can be applied (see [18], [19]).
Sometimes, it is convenient to compute the estimates of the first partial order derivatives by minimizing, in the class of exponential Hermite polynomials ([34]) or in the class of cubic Hermite polynomials ([3]) the tensioned quadratic functional

$$J_\alpha(F) = \sum_{ij \in N_e} \left[\int_{e_{ij}} \left(\frac{\partial^2 F}{\partial e_{ij}^2} \right)^2 d\xi_{ij} + (\alpha_{ij})^2 \int_{e_{ij}} \left(\frac{\partial F}{\partial e_{ij}} \right)^2 d\xi_{ij} \right]$$

where α_{ij} are nonnegative tension parameters, that can assume different values at each edge of the triangulation e_{ij}. In general it is used a normalized tension parameter $\alpha = \alpha_{ij} \| e_{ij} \|$.
In both the cases (cubic or exponential Hermite polynomials) the Hessian matrices have the same block structure of (26). In [17] it is proved that, in the case of exponential Hermite polynomials, the Hessian matrix is symmetric positive definite and a P–regular splitting can be derived for the functional (24) by the 2×2 block diagonal part of the Hessian matrix. In [17], it is also proved that, in the case of cubic Hermite polynomial, the Hessian matrix is block strictly (or block irreducibly) diagonally dominant (see [13]).

In order to solve problem (25)–(23) by splitting method, we have to solve at each iteration a convex quadratic subproblem

$$\begin{array}{ll} \textit{minimize} & \frac{1}{2}\boldsymbol{u}^T D\boldsymbol{u} + (\boldsymbol{q} + H\boldsymbol{u}^{(k-1)})^T\boldsymbol{u} \\ \textit{subject to} & C\boldsymbol{u} = \boldsymbol{d}, \qquad A\boldsymbol{u} \geq \boldsymbol{b}; \end{array} \tag{28}$$

where

$$C = \begin{bmatrix} \boldsymbol{c}_1^T \\ \vdots \\ \boldsymbol{c}_{m_e}^T \end{bmatrix} \qquad A = \begin{bmatrix} \boldsymbol{a}_1^T \\ \vdots \\ \boldsymbol{a}_{m_i}^T \end{bmatrix}$$

This subproblem can be formulate as mixed linear complementarity problem

$$\begin{array}{c} \begin{pmatrix} \boldsymbol{s} \\ 0 \end{pmatrix} = \begin{pmatrix} \boldsymbol{z}_1^{(k)} \\ \boldsymbol{z}_2^{(k)} \end{pmatrix} + M \begin{pmatrix} \boldsymbol{\lambda} \\ \boldsymbol{\mu} \end{pmatrix} \\ \boldsymbol{s} \geq 0; \qquad \boldsymbol{\lambda} \geq 0; \qquad \boldsymbol{s}^T\boldsymbol{\lambda} = 0 \end{array} \tag{29}$$

where

$$\boldsymbol{z}_1^{(k)} = -\boldsymbol{b} - AD^{-1}(\boldsymbol{q} + H\boldsymbol{u}^{(k-1)}); \qquad \boldsymbol{z}_2^{(k)} = -\boldsymbol{d} - CD^{-1}(\boldsymbol{q} + H\boldsymbol{u}^{(k-1)})$$

and the symmetric positive semidefinite matrix M of order $m_i + m_e$ is given by

$$M = \begin{pmatrix} AD^{-1}A^T & AD^{-1}C^T \\ CD^{-1}A^T & CD^{-1}C^T \end{pmatrix}$$

If

$$\begin{pmatrix} \boldsymbol{\lambda}^{(k)} \\ \boldsymbol{\mu}^{(k)} \end{pmatrix}$$

denotes the solution of (29), the solution of (28) can be computed by

$$\boldsymbol{u}^{(k)} = -D^{-1}(\boldsymbol{q} + H\boldsymbol{u}^{(k-1)} - A^T\boldsymbol{\lambda}^{(k)} - C^T\boldsymbol{\mu}^{(k)})$$

We observe that the matrix $CD^{-1}C^T$ is a symmetric positive definite matrix.
In the case of equality constraints only, splitting method is examined in [12]; the mixed LCP (29) is reduced to a symmetric positive definite linear system that can be solved by conjugate gradient method.
In order to solve the large–scale mixed LCP (29) (such as the one which arises from the bivariate interpolation problem), iterative schemes such as Cryer's SOR ([10]), are very efficient.
We can also consider parallel algorithms for LCP such as Parallel–SOR ([28]), gradient projection–SOR ([29]) and overlapping parallel-SOR ([20]).
A deep analysis on the evaluation of the performances of the parallel algorithms above on a distributed memory system (such as Cray MPP T3E at the CINECA Supercomputing Center in Bologna), has been developed

in [21] for the mixed LCP which arises from the bivariate interpolation problem.

In the numerical experiments in Table 6, we report the behaviour of the splitting method applied to the bivariate interpolation problem by solving the inner mixed LCP by Cryer's SOR.

The stopping rule for the iterations of the splitting method (outer iterations) is that the Karush–Kuhn–Tucker optimality conditions of the QP problem (25)–(23) hold within the tolerance 10^{-9}

$$\|G\boldsymbol{u}^{(k)} + \boldsymbol{q} - A^T\boldsymbol{\lambda}^{(k)} - C^T\boldsymbol{\mu}^{(k)}\|_\infty \leq 10^{-9}$$

The stopping rule for the iterations of the inner solver (inner iterations) is an adaptive criteria related to how the current approximation of the solution and the associate multipliers satisfy the KKT optimality conditions.

The computational experiments have been carried out on a Cray C90 (machine precision $\approx 7 \cdot 10^{-15}$).

Table 6: Behaviour of Splitting Method

	Splitting Method	
	outer iterations	time (seconds)
$n = 2000$ spars.$(G) = 99.3\%$; spars.$(C^T A^T) = 99.1\%$		
$m_e = 200$; $m_i = 300$	162	0.743
$m_e = 500$; $m_i = 500$	133	1.68
$m_e = 1000$; $m_i = 1000$	113	5.94
$n = 5000$ spars.$(G) = 99.7\%$; spars.$(C^T A^T) = 99.6\%$		
$m_e = 500$; $m_i = 500$	167	20.3
$m_e = 2500$; $m_i = 2500$	125	24.7
$n = 10000$ spars.$(G) = 99.8\%$; spars.$(C^T A^T) = 99.8\%$		
$m_e = 3000$; $m_i = 4000$	135	41.0

In Table 7, it is reported a comparison between the splitting method and the variable projection method (both with Cryer's SOR as inner solver) for the solution of a bivariate interpolation problem of size $n = 3000$. The comparison between the two methods has been done in terms of outer iterates, total amount of inner iterates (reported in brackets) and elapsed time expressed in seconds.

The sparsity of the matrices G and $(C^T A^T)$ are 99.6% and 99.7% respectively; the condition number of the matrix G is $K(G) = 146.5$.

Here the stopping rule for both the methods is:

$$\frac{\|\boldsymbol{u}^{(k+1)} - \boldsymbol{u}^{(k)}\|}{\|\boldsymbol{u}^{(k+1)}\|} \leq 10^{-6}$$

The experiments have been carried out on a Digital–Compaq workstation 500au.

Table 7: Comparison Splitting Method – Variable Projection Method

		Splitting Method		Variable Projection Method			
		$D = \text{diag}\{G_{ii}\}$		$D = \text{diag}\{G_{ii}\}$		$D = I$	
m_e	m_i	*iterations*	*time*	*iterations*	*time*	*iterations*	*time*
200	100	216(435)	1.3	116(190)	1.0	207(1226)	1.7
500	500	234(434)	5.5	116(228)	3.3	197(1232)	6.9
1000	1000	106(312)	12.6	95(261)	10.8	156(1231)	28.5

5 Conclusions

The following conclusions can be drawn:

- The effectiveness of the variable projection method has been evaluated by numerical experiments on test problems from CUTE library and randomly generated (by varying size, conditioning and sparsity of Hessian and constraints matrices) in [41].

  The comparison with other similar iterative schemes (classical splitting and projection methods, modified projection method ([47]), scaled gradient projection method) emphasized the definite advantage of VPM which uses a variable projection parameter ρ_k at each iteration.

- The efficiency of all the above iterative schemes is strictly related with an efficient solver for the inner subproblem of each iteration (inner solver). This is strictly connected with the structure of the constraints or how to reformulate the inner subproblem.

- The applications in sections 2 and 3 lead to quadratic programs with special structure of constraints but different sparsity of Hessian matrices. Efficient direct solvers for the inner subproblem (quadratic knapsack problem) are available in literature.

  The VPM solves efficiently the QP problem of Section 3 (Hessian matrix is very sparse) and it seems to be attractive also for the solution of the problem of Section 2, despite the density of the Hessian matrix.

- In the application of Section 4, the constraints are general and the Hessian matrix has a particular block structure. The inner subproblem is reformulated as a mixed linear complementarity problem; the inner iterative solver is very efficient because of the sort of block diagonally dominance of the Hessian matrix which "positively influences" the complementarity matrix.

  Because of this sort of block diagonally dominance of the Hessian matrix, there is no considerable advantage to use the variable projection method respect to the classical splitting method suggested by the special structure of the Hessian (see Table 7).

References

[1] HC. Andrews, BR. Hunt, Digital Image Restoration. Prentice Hall, Englewood Press, NJ, 1977.

[2] H. Akima, *A method of bivariate interpolation and smooth surface fitting for irregularly distributed data points.* ACM Transactions on Mathematical Software 1978; 4:148–159.

[3] L. Bacchelli Montefusco, G. Casciola, C^1 *surface interpolation.* ACM Transanctions on Mathematical Software 1989; 15:365–374.

[4] RE. Barnhill, G. Birkhoff, WJ. Gordon, *Smooth interpolation in triangles.* Journal of Approximation Theory 1973; 8:114–128.

[5] RE. Barnhill, JA. Gregory, *Compatible smooth interpolation in triangles.* Journal of Approximation Theory 1975; 15:214–225.

[6] DP. Bertsekas, Nonlinear Programming. Athena Scientific, Belmont, MA, 1995.

[7] CJC. Burges, *A Tutorial on Support Vector Machines for Pattern Recognition.* Data Mining and Knowledge Discovery 1998; 2(2):121–167.

[8] I. Bongartz, AR. Conn, N. Gould, PhL. Toint, *CUTE: constrained and unconstrained testing environment.* ACM Transactions on Mathematical Software 1995; 21:123–160.

[9] C. Cortes, VN. Vapnik, *Support vector network.* Machine Learning 1995; 20:1–25.

[10] CW. Cryer, *The solution of a quadratic programming problem using systematic overrelaxation.* SIAM Journal on Control 1971; 9:385–392.

[11] C. Durazzi, V. Ruggiero, L. Zanni, *A random generator for large-scale linearly constrained quadratic programming test problems.* MURST Project 1998–2000: "Analisi Numerica: Metodi e Software Matematico", Monograph 6, Ferrara, 2000.

[12] N. Dyn, J. Ferguson, *The numerical solution of equality–constrained quadratic programming problems.* Mathematics of Computation 1983; 41:165–170.

[13] DG. Feingold, RS. Varga, *Block diagonally dominant matrices and generalization of the Gerschgorin circle theorem.* Pacific Journal of Mathematics 1962; 12:1241–1250.

[14] MC. Ferris, TS. Munson, *Interior point methods for massive support vector machines*, Data Mining Institute Technical Report 00-05, Computer Sciences Department, University of Wisconsin, Madison, Wisconsin, 2000.

[15] G. Fung, OL. Mangasarian, *Proximal support vector machines classifiers*, Technical Report 01-02, Data Mining Institute, Computer Sciences Department, University of Wisconsin, Madison, Wisconsin, 2001.

[16] E. Galligani, *C^1 surface interpolation with constraints.* Numerical Algorithms 1993; 5:549–555.

[17] E. Galligani, *On a quadratic functional occurring in a bivariate scattered data interpolation problem.* Annali Università di Ferrara, Sez. VII, Scienze Matematiche (to appear).

[18] E. Galligani, V. Ruggiero, L. Zanni, *Splitting methods for quadratic optimization in data analysis.* International Journal of Computer Mathematics 1997; 63:289–307.

[19] E. Galligani, V. Ruggiero, L. Zanni, *Splitting methods for constrained quadratic programs in data analysis.* Computers and Mathematics with Applications 1996; 32:1–9.

[20] E. Galligani, V. Ruggiero, L. Zanni, *Splitting methods and parallel solution of constrained quadratic programs.* Rend. Circ. Matem. Palermo, (Equilibrium Problems with Side Constraints: Lagrangean Theory and Duality II, Giannessi F, Maugeri A, De Luca M eds.), Ser. II, 1997; 48:121–136.

[21] E. Galligani, V. Ruggiero, L. Zanni, *Parallel solution of large scale quadratic programs.* High Performance Algoritms and Software in Nonlinear Optimization (De Leone R, Murli A, Pardalos PM, Toraldo G eds.), Applied Optimization 24, Kluwer Academic Publishers, Dordrecht, 1998; 189–205.

[22] TG. Herman, Image Reconstruction from Projections. Academic Press, New York, 1980.

[23] T. Joachims, *Making large–scale SVM learning practical.* Advances in Kernel Methods–Support Vector Learning (Schölkopf B, et al. eds.), MIT Press, Cambridge, MA, 1998.

[24] IM. Klucewicz, *A Piecewise* C^1 *interpolant to arbitrarily spaced data.* Computer Graphics and Image Processing 1978; 8:92–112.

[25] CL. Lawson, *Software for* C^1 *surface interpolation.* Mathematical Software III (Rice JR ed.), Academic Press, New York 1977; 161–194.

[26] YY. Lin, JS. Pang, *Iterative methods for large convex quadratic programs: a survey.* SIAM Journal on Control and Optimization 1987; 25: 383–411.

[27] Y. LeCun, *MNIST Handwritten Digit Database.* Available at www.research.att.com/~yann/ocr/mnist.

[28] OL. Mangasarian, R. De Leone, *Parallel successive overrelaxation methods for symmetric linear complementarity problems and linear programs.* Journal of Optimization Theory and Applications 1987; 54:437–446.

[29] OL. Mangasarian, R. De Leone, *Parallel gradient projection successive overrelaxation for symmetric linear complementarity problems and linear programs.* Annals of Operations Research 1988; 14:41–59.

[30] P. Marcotte, JH. Wu, *On the convergence of projection methods: application to the decomposition of affine variational inequalities.* Journal of Optimization Theory and Applications 1995; 85:347–362.

[31] PM. Murphy, DW. Aha, *UCI Repository of Machine Learning Databases,* 1992. Availabe at www.ics.uci.edu/~mlearn/MLRepository.html.

[32] SN. Nielsen, SA. Zenios, *Massively parallel algorithms for single constrained convex programs.* ORSA Journal on Computing 1992; 4:166–181.

[33] GM. Nielson, *A method for interpolating scattered data based upon a minimum norm network.* Mathematics of Computation 1983; 40:253–271.

[34] GM. Nielson, R. Franke, *A method for construction of surfaces under tension.* Rocky Mountain Journal of Mathematics 1984; 14:203–221.

[35] JM. Ortega, Numerical Analysis: A Second Course. Academic Press, New York, 1972.

[36] E. Osuna, R. Freund, F. Girosi, *Support vector machines: training and applications.* A.I. Memo 1602, MIT Artificial Intelligence Laboratory, 1997.

[37] E. Osuna, R. Freund, F. Girosi, *Training support vector machines: an application to face detection.* Proceedings of the IEEE Conference on Computer Vision and Pattern Recognition CVPR'97, 1997; 130–136.

[38] P.M. Pardalos, N. Kovoor, *An algorithm for a single constrained class of quadratic programs subject to upper and lower bounds.* Mathematical Programming 1990; 46:321–328.

[39] JC. Platt, *Fast training of support vector machines using sequential minimal optimization*, Advances in Kernel Methods – Support Vector Learning (Schölkopf B, Burges C, Smola A eds.), MIT Press, Cambridge, Mass., 1998.

[40] M. Pontil, A. Verri, *Properties of Support Vector Machines.* Neural Computation 1998; 10:977–996.

[41] V. Ruggiero, E. Galligani, L. Zanni, *Projection–type methods for large convex quadratic programs: theory and computational experience.* MURST Project 1998–2000: "Analisi Numerica: Metodi e Software Matematico", Monograph 5, Ferrara, 2000.

[42] Ruggiero V, Zanni L. *A modified projection algorithm for large strictly convex quadratic programs.* Journal of Optimization Theory and Applications 2000; 104:281–299.

[43] V. Ruggiero, L. Zanni, *On the efficiency of splitting and projection methods for large strictly convex quadratic programs.* Nonlinear Optimization and Related Topics (Di Pillo G, Giannessi F eds.), Applied Optimization 36, Kluwer Academic Publishers, Dordrecht, 2000; 401–413.

[44] V. Ruggiero, L. Zanni, *An overview on projection–type methods for convex large–scale quadratic programs.* Equilibrium Problems: Nonsmooth Optimization and Variational Inequality Models (Giannessi F, Maugeri A, Pardalos PM eds.), Nonconvex Optimization and Its Applications 58, Kluwer Academic Publishers, Dordrecht, 2001; 269–300

[45] V. Ruggiero, L. Zanni, *Variable projection methods for large convex quadratic programs*, Recent Trends in Numerical Analysis (Trigiante D ed.), Advances in the Theory of Computational Mathematics 3, Nova Science Publishers Inc., Huntington, New York 2000; 299-313.

[46] AJ. Smola, *pr_LOQO optimizer.* Available at www.kernel-machines.org/code/prloqo.tar.gz.

[47] MV. Solodov, P. Tseng, *Modified Projection–type methods for monotone variational inequalities.* SIAM Journal on Control and Optimization 1996; 34:1814–1830.

[48] AN. Tikhonov, VY. Arsenin, Solution of Ill–Posed Problems. John Wiley & Sons, New York, 1977.

[49] R. Vanderbei, *LOQO: An Interior Point Code for Quadratic Programming*, Technical Report SOR 94-15, Princeton University, 1994.

[50] VN. Vapnik, Statistical Learning Theory. John Wiley & Sons, New York, 1998.

[51] G. Zanghirati, L. Zanni, *Large quadratic programs in training Gaussian support vector machines*, Technical Report Nr. 320, Department of Mathematics, University of Ferrara, 2002.

[52] G. Zanghirati, L. Zanni, *A parallel solver for large quadratic programs in training support vector machines*, Technical Report Nr. 47, Department of Mathematics, University of Modena, 2002.

STRONG SOLVABILITY OF BOUNDARY VALUE PROBLEMS IN ELASTICITY WITH UNILATERAL CONSTRAINTS

Sofia Giuffrè
D.I.M.E.T. – Università di Reggio Calabria,
Via Graziella, Loc. Feo di Vito - 89100 REGGIO CALABRIA (ITALIA)
e-mail: giuffre@ing.unirc.it

Abstract Strong solvability in Sobolev spaces is proved for a unilateral boundary value problem for a class of nonlinear discontinuous operators. The operator is assumed to be of Carathéodory type and to satisfy a suitable ellipticity condition. Only measurability with respect to the independent variable x is required.
The main tool of the proof is an estimate for the second derivatives of the functions which satisfy unilateral boundary conditions, that allows us to obtain an uniqueness and existence theorem, extending the result in [6], which holds only for $n = 2$.

Keywords: Signorini problem, unilateral problems, nonlinear elliptic equations.

2000 Mathematics Subject Classification. Primary 35J85; Secondary 35R05.

1 Introduction

The paper deals with the strong solvability of the following boundary value problem with unilateral boundary conditions

$$\begin{cases} \mathcal{A}(x, D^2u) - \lambda u = f(x) & \text{a.e. in } \Omega \subset \mathbb{R}^n \\ u \geq 0, \ \dfrac{\partial u}{\partial \chi} \geq 0, \ u \cdot \dfrac{\partial u}{\partial \chi} = 0 & \text{on } \partial\Omega \end{cases} \tag{1.1}$$

where the number λ is greater than zero and χ is the unit outward normal to the boundary $\partial\Omega$.

Let us observe that when $\mathcal{A}(x, D^2u) = \Delta u$ the problem

$$\begin{cases} \Delta u - \lambda u = f(x) & \text{a.e. in } \Omega \\ u \geq 0, \; \dfrac{\partial u}{\partial \chi} \geq 0, \; u \cdot \dfrac{\partial u}{\partial \chi} = 0 & \text{on } \partial\Omega \end{cases} \tag{1.2}$$

is the celebrated problem of Signorini, and, for each $f \in L^2(\Omega)$, it admits a unique solution $u \in W^{2,2}(\Omega)$ (see [1], [11]).

This problem represents the conceptual model of an elastic body Ω with boundary $\partial\Omega$ which is in contact with a rigid support body and is subject to volume force f. These forces produce a deformation of Ω and a displacement on $\partial\Omega$ with the normal component non negative. If we denote by u the displacement field on $\partial\Omega$, one of the following two condition systems holds

$$\begin{cases} u \cdot \dfrac{\partial u}{\partial \chi} = 0, \\ u = 0, \; \dfrac{\partial u}{\partial \chi} > 0 \end{cases} \tag{1.3}$$

or

$$\begin{cases} u \cdot \dfrac{\partial u}{\partial \chi} = 0, \\ u > 0, \; \dfrac{\partial u}{\partial \chi} = 0. \end{cases} \tag{1.4}$$

These conditions have been called by A.Signorini ambiguous boundary conditions, since, in a fixed point of $\partial\Omega$, it is not known a priori which system holds.

Conditions (1.3) mean that, in the considered point, the elastic body, in its equilibrium position, is in contact with the rigid support body, and, then, the reaction of the constraint has a positive component along the normal.

On the other hand conditions (1.4) mean that the body, keeping its equilibrium, goes away from the support body so that it does not react anymore on the elastic body.

Obviously conditions $u \geq 0$, $\dfrac{\partial u}{\partial \chi} \geq 0$, $u \cdot \dfrac{\partial u}{\partial \chi} = 0$ on $\partial\Omega$ represent unilateral constraints, since the body Ω may be in contact or go away from the support body, but it can not move down in any admissible movement.

Our aim is to prove $W^{2,2}(\Omega)$- solvability also for problem (1.1).

We assume the operator $\mathcal{A}(x, \xi)$, $(x, \xi) \in \Omega \times \mathbb{R}^{n^2}$, to be a Carathéodory function and to satisfy ellipticity condition (A), that has been introduced by S.Campanato ([3]) (see section n.2).

Let us note that condition (A) does not imply the continuity of the mapping $\mathcal{A}(x,\xi)$ with respect to x, and, in view of Carathéodory's condition, only measurability of $\mathcal{A}(x,\xi)$ with respect to x is required.
The main tool of the proof is the following estimate (see Theorem 2.1)

$$\int_\Omega \sum_{i,j=1}^{2} (D_{ij}u)^2 \, dx \le \int_\Omega |\Delta u|^2 \, dx,$$

$$\forall u \in W^{2,2}(\Omega) \ : \ u \ge 0, \ \frac{\partial u}{\partial \chi} \ge 0, \ u \cdot \frac{\partial u}{\partial \chi} = 0 \text{ on } \partial\Omega$$

that has interest in itself and, together with condition (A), ensures the solvability of problem (1.1).
The novel fact is that the constant in the above estimate is equal to 1 (see section n.4).
Finally it is well known that Signorini problem (1.2), in a weak formulation, can be expressed by a variational inequality (see [1], [11]):
find $u \in K$ such that

$$a(u, v-u) \ge (f, v-u) \quad for\ v \in K$$

where

$$K = \left\{u \in H^1(\Omega) \ : \ u \ge 0 \ on\ \partial\Omega\right\},$$

$$a(u,v) = \int_\Omega \sum_{i=1}^{2} D_i u D_i v dx - \lambda \int_\Omega uv dx$$

and

$$(f, v) = \int_\Omega f \, v \, dx.$$

2 Basic assumptions and main results

Let $\Omega \subset \mathbb{R}^n$, $n > 2$, be a bounded, convex and open set with C^2-smooth boundary $\partial\Omega$.
Let us set $\chi = (X_1, \ldots, X_n)$ for the unit outward normal to $\partial\Omega$ and $H(x)$ for the mean curvature of $\partial\Omega$.
For a real valued function $u : \Omega \to \mathbb{R}$, we denote by $Du = \{D_i u\}_{i=1,\ldots,n}$ and $D^2 u = \{D_{ij} u\}_{i,j=1,\ldots,n}$ its gradient and Hessian matrix, respectively.
Let $\mathcal{A}(x,\xi) : \Omega \times \mathbb{R}^{n\times n} \to \mathbb{R}$, a real valued function satisfying Carathéodory's condition, i.e. it is measurable with respect to x for all $\xi \in \mathbb{R}^{n\times n}$ and continuous in the other variables for a.a. $x \in \Omega$.
We are aimed at the investigation of the strong solvability of problem (1.1).

As usual by strong solution $u \in W^{2,2}(\Omega)$, we mean a twice weakly differentiable function with L^2 summable derivatives.
We define W to be the closure in $W^{2,2}(\Omega)$ of the class $V = \{u \in C^2(\overline{\Omega}) \bigcap C^3(\Omega) \; : u \geq 0, \; \frac{\partial u}{\partial \chi} \geq 0, \; u \cdot \frac{\partial u}{\partial \chi} = 0 \text{ on } \partial\Omega\}$.
The first result we may prove is the validity, also in the case $n > 2$, of the following estimate for the second derivatives of $u \in W$.

Theorem 2.1 *Let $\Omega \subset \mathbb{R}^n$ be an open, bounded and convex set with C^2-smooth boundary $\partial\Omega$. Then for each $u \in W$, it results*

$$\int_\Omega \sum_{i,j=1}^{n} (D_{ij}u)^2 \, dx \leq \int_\Omega |\Delta u|^2 \, dx. \tag{2.5}$$

As it concerns the operator $\mathcal{A}(x, \xi)$, in order to prove the solvability of (1.1), we assume the following ellipticity condition, introduced by S.Campanato (see [3]):
there exist three constants α, γ, $\delta > 0$, with $\gamma + \delta < 1$, such that, for almost all $x \in \Omega$, for all ξ, $\tau \in \mathbb{R}^{n \times n}$, one has

$$\left| \sum_{i=1}^{n} \xi_{ii} - \alpha \left[\mathcal{A}(x, \xi + \tau) - \mathcal{A}(x, \tau) \right] \right|^2 \leq \gamma \, \|\xi\|_{n^2}^2 + \delta \left| \sum_{i=1}^{n} \xi_{ii} \right|^2 \tag{A}$$

where $\|\xi\|_{n^2} = \left(\sum_{i,j=1}^{n} \xi_{ij}^2 \right)^{\frac{1}{2}}$ and

$$\mathcal{A}(x, 0) = 0.$$

Using estimate (2.5) and condition (A) we may prove the following uniqueness and existence result.

Theorem 2.2 *Let condition (A) be fulfilled. Then for each $f \in L^2(\Omega)$ and for each constant $\lambda > 0$, the problem*

$$\begin{cases} \mathcal{A}(x, D^2u) - \lambda u = f(x) & \text{a.e. in } \Omega \\ u \in W \end{cases} \tag{2.6}$$

is uniquely solvable.

Let us note that condition (A) is equivalent to the following strong monotonicity conditions (see [4] and [8]):

$$|\mathcal{A}(x, \xi + \tau) - \mathcal{A}(x, \tau)| \leq \frac{\sqrt{\gamma} + \sqrt{n}(1 + \sqrt{\delta})}{\alpha} \, \|\xi\|_{n^2} \tag{2.7}$$

and

$$(\mathcal{A}(x,\xi+\tau)-\mathcal{A}(x,\tau))\sum_{i=1}^{n}\xi_{ii}\geq\frac{1-\delta}{2\alpha}\left|\sum_{i=1}^{n}\xi_{ii}\right|^{2}-\frac{\gamma}{2\alpha}\|\xi\|_{n^2}^{2} \tag{2.8}$$

for all $\xi,\tau\in\mathbb{R}^{n^2}$ and a.a. $x\in\Omega$.

This characterization is important since (2.7) and (2.8) are the appropriate elliptic conditions in the theory of nonlinear divergence form equations.

Moreover from conditions (2.7) and (2.8) it follows that the function $\xi\to\mathcal{A}(x,\xi)$ is differentiable almost everywhere in $\mathbb{R}^{n^2}$, the derivatives $\dfrac{\partial\mathcal{A}(x,\xi)}{\partial\xi_{ij}}$ belong to $L^\infty(\Omega\times\mathbb{R}^{n\times n})$, and the strong ellipticity condition holds:

$$\sum_{i,j=1}^{n}\frac{\partial\mathcal{A}(x,\xi)}{\partial\xi_{ij}}\zeta_i\zeta_j\geq\frac{1-(\gamma+\delta)}{2\alpha}\sum_{i=1}^{n}\zeta_i^2 \tag{2.9}$$

for a.a. $x\in\Omega$, $\forall\xi\in\mathbb{R}^{n\times n}$, $\forall\zeta\in\mathbb{R}^n$.

Moreover let us observe that, concerning linear second order differential operators $\mathcal{A}(x,D^2u)=\sum_{i,j=1}^{2}a_{ij}(x)D_{ij}u$, Cordes condition:

$$\frac{\sum_{i,j=1}^{n}a_{ij}^2(x)}{\left(\sum_{i=1}^{n}a_{ii}(x)\right)^2}<\frac{1}{n-1+\varepsilon},\quad a.e.\Omega$$

implies condition (A).

3 Preliminary results

In the proof of Theorem 2.2 we will use the nearness theory. Then we start recalling the definition of near operators.

Definition 3.1 *Let $\mathcal{B}$ be a set and $\mathcal{B}_1$ a real Banach space. Consider two mappings A and B defined on $\mathcal{B}$ with values in $\mathcal{B}_1$. The mapping A is said to be near to B if there exist two positive constants α and k, with $k\in(0,1)$, such that $\forall u,v\in\mathcal{B}$ it implies*

$$\|B(u)-B(v)-\alpha[A(u)-A(v)]\|_{\mathcal{B}_1}\leq\|B(u)-B(v)\|_{\mathcal{B}_1}.$$

The following theorem holds.

Theorem 3.1 *The mapping $A:\mathcal{B}\to\mathcal{B}_1$ is injective, or surjective, or bijective if and only if it is near to a mapping $B:\mathcal{B}\to\mathcal{B}_1$, which is injective, or surjective, or bijective.*

We also recall the definition of monotone operators.

Definition 3.2 *Let $\mathcal{B}$ be a set, $\mathcal{B}_1$ a real Hilbert space, A, B two operators from $\mathcal{B}$ into $\mathcal{B}_1$. A is said to be monotone with respect to B if, $\forall u, v \in \mathcal{B}$, we have*

$$(A(u) - A(v), B(u) - B(v))_{\mathcal{B}_1} \geq 0.$$

We also need the following result.

Theorem 3.2 *Let A, B, C be operators from $\mathcal{B}$ into $\mathcal{B}_1$, with $\mathcal{B}_1$ being a real Hilbert space. If:*
A is near to B with constants α e k,
C is monotone with respect to B,
then the mapping $A + C$ is near to $B + \alpha C$ with the same constants α e k.

In order to obtain Theorem 2.2 we will use the following existence and uniqueness theorem.

Theorem 3.3 *For all $f \in L^2(\Omega)$, for all $\lambda > 0$, Signorini problem*

$$\begin{cases} \Delta u - \lambda u = f(x) & \text{a.e. in } \Omega \\ u \geq 0, \ \dfrac{\partial u}{\partial \chi} \geq 0, \ u \cdot \dfrac{\partial u}{\partial \chi} = 0 & \text{on } \partial\Omega \end{cases}$$

admits a unique solution $u \in W^{2,2}(\Omega)$.

4 Proof of the theorems

Proof of Theorem 2.1.
Let $u \in V$. Bearing in mind the identity

$$\sum_{i,j=1}^{n} (D_{ij}u)^2 + \sum_{i,j=1}^{n} (D_{ii}uD_{jj}u - (D_{ij}u)^2) = (\Delta u)^2,$$

in order to obtain estimate (2.5), we have to prove that

$$\int_\Omega \sum_{i,j=1}^{n} (D_{ii}uD_{jj}u - (D_{ij}u)^2)dx \geq 0. \tag{4.10}$$

Taking into account the following identity, where δ^{ij} are the Kronecker delta,

$$\sum_{i,j=1}^{n} (D_{ii}uD_{jj}u - (D_{ij}u)^2) = \sum_{i,j,r,s=1}^{n} (\delta^{ij}\delta^{rs} - \delta^{ir}\delta^{js})\, D_j(D_iuD_{rs}u)$$

by means of Gauss theorem, we get

$$\int_\Omega \sum_{i,j=1}^{n} (D_{ii}uD_{jj}u - (D_{ij}u)^2)dx = \int_{\partial\Omega} \sum_{i,j,r,s=1}^{n} (\delta^{ij}\delta^{rs} - \delta^{ir}\delta^{js})\, D_i u D_{rs} u X_j d\sigma. \tag{4.11}$$

Fixing a local coordinate system $\{t_1, \ldots, t_{n-1}\}$ at an arbitrary point of $\partial\Omega$, since $\partial\Omega$ is of class C^2, it is represented by the equations

$$x_i = x_i(t_1, \ldots, t_{n-1}) \quad i = 1, \ldots, n, \quad x_i \in C^2$$

with $(t_1, \ldots, t_{n-1})$ taken on a suitable $(n-1)$-dimensional domain T. Hence

$$\frac{\partial x}{\partial t_i}\frac{\partial x}{\partial t_j} = \begin{cases} 0 & i \neq j, \quad i, j = 1, \ldots, n-1 \\ E_i = \left|\dfrac{\partial x}{\partial t_i}\right|^2 & i = j. \end{cases} \tag{4.12}$$

To calculate the surface integral (4.11), we consider the operators:

$$\delta_i : C^1(\overline{\Omega}) \to C^0(\partial\Omega)$$

given by

$$\delta_i(u) = u_i = D_i u - u_0 X_i, \quad u_0 = \frac{\partial u}{\partial \chi}$$

and $\delta u = (u_1, \ldots, u_n)$.
It results

$$\delta u = Du - u_0 \chi \tag{4.13}$$

and

$$\delta u \cdot \chi = 0 \quad \text{on } \partial\Omega. \tag{4.14}$$

Hence, $\delta u(x)$ is the projection of the gradient $Du(x)$ onto the tangent hyperplane to $\partial\Omega$ at the point x and δu is called the tangential gradient of u.
From (4.12), (4.13) it results

$$\delta u = \sum_{j=1}^{n-1} \frac{1}{E_j}(\delta u \frac{\partial x}{\partial t_j})\frac{\partial x}{\partial t_j} = \sum_{j=1}^{n-1} \frac{1}{E_j}(Du \frac{\partial x}{\partial t_j})\frac{\partial x}{\partial t_j} = \sum_{j=1}^{n-1} \frac{1}{E_j}\frac{\partial u}{\partial t_j}\frac{\partial x}{\partial t_j} \tag{4.15}$$

whence

$$u_i = \sum_{j=1}^{n-1} \frac{1}{E_j}\frac{\partial x_i}{\partial t_j}\frac{\partial u}{\partial t_j}, \quad i = 1, \ldots, n. \tag{4.16}$$

As it is shown in [8], [9], [10], for the surface integral (4.11) the following identity holds:

$$\int_\Omega \sum_{i,j=1}^n (D_{ii}uD_{jj}u - (D_{ij}u)^2)dx = \int_{\partial\Omega} u_0 \sum_{r=1}^n u_{rr}d\sigma - \int_{\partial\Omega} \delta u \delta u_0 d\sigma$$

$$- \int_{\partial\Omega} |\delta u|^2 K_\chi(\delta u)d\sigma - (n-1)\int_{\partial\Omega} u_0^2 H d\sigma \tag{4.17}$$

where $K_\chi(\delta u)$ is the normal curvature of $\partial\Omega$ in the direction δu and

$$u_{rr} = \sum_{k,j=1}^{n-1} \frac{1}{E_j}\frac{\partial E_k^{-1}}{\partial t_j}\frac{\partial x_r}{\partial t_j}\frac{\partial x_r}{\partial t_k}\frac{\partial u}{\partial t_k}$$

$$+ \sum_{k,j=1}^{n-1} \frac{1}{E_j}\frac{1}{E_k}\frac{\partial x_r}{\partial t_j}\frac{\partial x_r}{\partial t_k}\frac{\partial^2 u}{\partial t_k \partial t_j} \tag{4.18}$$

$$+ \sum_{k,j=1}^{n-1} \frac{1}{E_j}\frac{1}{E_k}\frac{\partial u}{\partial t_k}\frac{\partial x_r}{\partial t_j}\frac{\partial^2 x_r}{\partial t_k \partial t_j}.$$

Since Ω is a convex set, $K_\chi(\delta u) \leq 0$ and $H \leq 0$, then the last two terms in (4.17) are positive.
As it concerns the integral

$$\int_{\partial\Omega} u_0 \sum_{r=1}^n u_{rr}d\sigma \tag{4.19}$$

setting

$$\partial\Omega^+ = \{(t_1, \ldots, t_{n-1}) \in T \ : \ u(x_1(t_1, \ldots, t_{n-1}), \ldots, x_n(t_1, \ldots, t_{n-1})) > 0\}$$

$$\partial\Omega^0 = \{(t_1, \ldots, t_{n-1}) \in T \ : \ u(x_1(t_1, \ldots, t_{n-1}), \ldots, x_n(t_1, \ldots, t_{n-1})) = 0\},$$

since $u \geq 0$ on $\partial\Omega$, we have

$$\int_{\partial\Omega} u_0 \sum_{r=1}^n u_{rr}d\sigma = \int_{\partial\Omega^+} u_0 \sum_{r=1}^n u_{rr}d\sigma$$

$$+ \int_{\partial\Omega^0} u_0 \sum_{r=1}^n u_{rr}d\sigma. \tag{4.20}$$

The first integral in the right hand side of (4.20) is equal to zero, because $u \in V$, then $u \neq 0$ implies $u_0 = 0$.

As it concerns the second integral, since $u \geq 0$ on $\partial\Omega$, each point $x^* = (x_1(t_1^*, \dots, t_{n-1}^*), \dots, x_n(t_1^*, \dots, t_{n-1}^*))$, with $(t_1^*, \dots, t_{n-1}^*) \in int(T)$, such that $u(x^*) = 0$, is a global minimum point.
Then

$$\frac{\partial u}{\partial t_k}(x^*) = 0, \tag{4.21}$$

$$\sum_{k=1}^{n-1} \frac{\partial^2 u}{\partial t_k^2}(x^*) \geq 0 \tag{4.22}$$

and from (4.18), (4.21) on $\partial\Omega^0$ it results

$$u_{rr} = \sum_{k,j=1}^{n-1} \frac{1}{E_j}\frac{1}{E_k}\frac{\partial x_r}{\partial t_j}\frac{\partial x_r}{\partial t_k}\frac{\partial^2 u}{\partial t_k \partial t_j}. \tag{4.23}$$

From (4.12), (4.20), (4.22) and (4.23) it follows

$$\begin{aligned} &\int_{\partial\Omega} u_0 \sum_{r=1}^{n} u_{rr} d\sigma \\ &= \int_{\partial\Omega^0} u_0 \sum_{r=1}^{n}\sum_{k,j=1}^{n-1} \frac{1}{E_j E_k}\frac{\partial x_r}{\partial t_j}\frac{\partial x_r}{\partial t_k}\frac{\partial^2 u}{\partial t_k \partial t_j} d\sigma \\ &= \int_{\partial\Omega^0} u_0 \sum_{k=1}^{n-1} \frac{1}{E_k}\frac{\partial^2 u}{\partial t_k^2} d\sigma \geq 0, \end{aligned} \tag{4.24}$$

and hence the first integral at the right hand side of (4.17) is non negative.
It remains to prove that the integral

$$\int_{\partial\Omega} \delta u \delta u_0 d\sigma \tag{4.25}$$

is non positive.
Taking into account (4.12) and (4.16), it results

$$\begin{aligned} \delta u \delta u_0 &= \sum_{i=1}^{n} u_i u_{0i} \\ &= \sum_{i=1}^{n}\sum_{k,j=1}^{n-1} \frac{1}{E_j E_k}\frac{\partial x_i}{\partial t_j}\frac{\partial x_i}{\partial t_k}\frac{\partial u}{\partial t_j}\frac{\partial u_0}{\partial t_k} = \sum_{k=1}^{n-1} \frac{1}{E_k}\frac{\partial u}{\partial t_k}\frac{\partial u_0}{\partial t_k}. \end{aligned} \tag{4.26}$$

Then, since $u \in V$

$$\int_{\partial\Omega} \delta u \delta u_0 d\sigma$$

$$= \int_{\partial\Omega^0} \sum_{k=1}^{n-1} \frac{1}{E_k} \frac{\partial u}{\partial t_k} \frac{\partial u_0}{\partial t_k} d\sigma \tag{4.27}$$

$$+ \int_{\partial\Omega^+} \sum_{k=1}^{n-1} \frac{1}{E_k} \frac{\partial u}{\partial t_k} \frac{\partial u_0}{\partial t_k} d\sigma.$$

From (4.21) we get that the first integral at the right hand side of (4.27) is equal to zero.
We may prove that also the second integral is equal to zero.
In fact, since $u \in V$, $u_0 \geq 0$ on $\partial\Omega$. Then each point $\tilde{x} = (x_1(\tilde{t}_1, \ldots, \tilde{t}_{n-1}), \ldots, x_n(\tilde{t}_1, \ldots, \tilde{t}_{n-1}))$, with $(\tilde{t}_1, \ldots, \tilde{t}_{n-1}) \in int(T)$, such that $u_0(\tilde{x}) = 0$, is a global minimum point.
Then

$$\frac{\partial u_0}{\partial t_k}(\tilde{x}) = 0. \tag{4.28}$$

Since $u > 0$ implies $u_0 = 0$, from (4.28) we obtain that the integral

$$\int_{\partial\Omega^+} \sum_{k=1}^{n-1} \frac{1}{E_k} \frac{\partial u}{\partial t_k} \frac{\partial u_0}{\partial t_k} d\sigma = 0. \tag{4.29}$$

It follows (4.10) and then estimate (2.5) for $u \in V$.

It remains to prove that estimate (2.5) holds true also for $u \in W$.
Let $u_k \in V$, such that $u_k \to u$ in $W^{2,2}(\Omega)$. We have

$$\left\|D^2 u_k\right\|_{L^2(\Omega)} \leq \|\Delta u_k\|_{L^2(\Omega)},$$

and hence, when $k \to \infty$, we get

$$\left\|D^2 u\right\|_{L^2(\Omega)} \leq \|\Delta u\|_{L^2(\Omega)}.$$

Proof of Theorem 2.2.
Using estimate (2.5) and condition (A), it is possible to prove that the mapping $\mathcal{A}(x, D^2u)$ is near to the Laplacian, both considered as operators from W in $L^2(\Omega)$. In fact:

$$\left(\int_\Omega \left|\Delta(u-v) - \alpha\left[\mathcal{A}\left(x, D^2u\right) - \mathcal{A}\left(x, D^2v\right)\right]\right|^2 dx\right)^{\frac{1}{2}}$$

$$\leq \left(\int_\Omega \left(\gamma \left\| D^2(u-v) \right\|_{n^2}^2 + \delta \left| \Delta(u-v) \right|^2 \right) dx \right)^{\frac{1}{2}}$$

$$\leq \gamma \left(\int_\Omega \left\| D^2(u-v) \right\|_{n^2}^2 dx \right)^{\frac{1}{2}} + \delta \left(\int_\Omega \left| \Delta(u-v) \right|^2 dx \right)^{\frac{1}{2}}$$

$$\leq (\gamma + \delta) \left(\int_\Omega \left| \Delta(u-v) \right|^2 dx \right)^{\frac{1}{2}}$$

with $\gamma + \delta < 1$.
Moreover, by means of Theorem 3.2, to achieve that $\mathcal{A}(x, D^2u) - \lambda u$ is near to $\Delta u - \alpha\lambda u$, it is enough to prove that $-\lambda u$ is a monotone operator with respect to Δu.
Thus, since from Theorem 3.3 it follows that the operator $\Delta u - \alpha\lambda u$ is a bijective one from W into $L^2(\Omega)$, we obtain that $\mathcal{A}(x, D^2u) - \lambda u$ is bijective and then Theorem 2.2.
Then we have just to prove that $-\lambda u$ is monotone with respect to Δu, that is

$$-\lambda \int_\Omega u \, \Delta u \, dx \geq 0.$$

By the identity

$$u\Delta u = [\frac{\partial}{\partial x_1}(uD_1u) + \ldots + \frac{\partial}{\partial x_n}(uD_nu)] - ((D_1u)^2 + \ldots + (D_nu)^2)$$

and the Green-Gauss formula, we get

$$-\lambda \int_\Omega u \, \Delta u \, dx$$

$$= \lambda \int_\Omega ((D_1u)^2 + \ldots + (D_nu)^2) dx - \lambda \int_{\partial\Omega} (uD_1uX_1 + \ldots + uD_nuX_n) d\sigma.$$

Since $uD_1uX_1 + \ldots + uD_nuX_n = u \cdot \dfrac{\partial u}{\partial \chi} = 0$, the integral becomes

$$-\lambda \int_\Omega u \, \Delta u \, dx = \lambda \int_\Omega ((D_1u)^2 + \ldots + (D_nu)^2) dx \geq 0,$$

and then Theorem 2.2 follows.

References

[1] C. Baiocchi - A. Capelo, *Variational and Quasivariational Inequalities: applications to free boundary problems*, J. Wiley and Sons, Chichester, 1984.

[2] H. Brezis, *Problèmes Unilatéraux*, J. Math. Pures et Appl. 51 (1972), 1-168.

[3] S. Campanato, *On the condition of nearness between operators*, Ann. Mat. Pura Appl. 167 (1994), 243-256.

[4] S. Campanato, *Non variational differential systems. A condition for local existence and uniqueness*, Ric. Mat. 40 (1991), 129-140.

[5] G. Fichera, *Boundary value problems in elasticity with unilateral constraints* Handbuch der Physik VI a/2 Springer-Verlag, Berlin, Heidelberg New York (1972), 347-389.

[6] S. Giuffrè, *On the strong solvability of a unilateral boundary value problem for nonlinear discontinuous operators in the plane*, Equilibrium Problems: Non smooth Optimization and Variational Inequality Models, Kluwer Academic Publishers, F.Giannessi- A.Maugeri- P.Pardalos Eds. (2001), 119-127.

[7] J. L. Lions - G. Stampacchia, *Variational Inequalities*, Comm. Pure Appl. Math. 20 (1967), 493-519.

[8] A. Maugeri - D.K. Palagachev - L. Softova, *Elliptic and parabolic equations with discontinuous coefficients*, Wiley, VCH Publishers, 2000.

[9] C. Miranda, *Sulle equazioni ellittiche del secondo ordine di tipo non variazionale, a coefficienti discontinui*, Ann. Mat. Pura Appl. (4) 63 (1963), 353-386.

[10] C. Miranda, *Su di una particolare equazione ellittica del secondo ordine a coefficienti discontinui*, An. Sti. Univ. Al. I. Cuza Iasi, N. Ser., Sect. 1a 11B (1965), 209-215.

[11] A. Signorini, *Questioni di elasticità non lineare e semilinearizzata*, Rend. Mat. 18 (1959), 95-139.

[12] G. Talenti, *Problemi di derivata obliqua per equazioni ellittiche in due variabili*, Boll. Un. Mat. Ital. 22 (1967), 505-526.

TIME DEPENDENT VARIATIONAL INEQUALITIES – SOME RECENT TRENDS

Joachim Gwinner
Institut für Mathematik, Fakultät für Luft- und Raumfahrttechnik,
Universität der Bundeswehr München, Werner–Heisenberg–Weg 39,
D-85577 Neubiberg, Germany.
e-mail: joachim.gwinner@unibw-muenchen.de

Dedicated to the memory of Professor W. Oettli

Abstract In this paper we attempt to survey several classes of time dependent variational inequalities that model various constrained evolution problems, in particular with unilateral constraints encountered in applied sciences. Here we are mainly concerned with the challenging evolution problems where the state of the system lives in an *infinite* dimensional space; the wide field of differential inclusions in finite dimensional space is outside of the scope of the present paper. Also we refrain here from covering the literature on the existence and regularity theory of parabolic and hyperbolic evolution inequalities.

At first we address time dependent variational inequalities where time enters as an additional parameter in the variational inequality. This class of time dependent variational inequalities has been recently introduced to study certain time dependent traffic flow problems. In addition we draw the attention to work on constrained evolution problems that include the time history via memory terms.

Then we turn to time dependent variational inequalities that generalize classical ordinary differential equations. In particular, we report on recent extensions of the sweeping process introduced by Moreau. Moreover, we deal with projected dynamical systems in a Hilbert space framework. Quite recently, this class of time dependent variational inequalities has been introduced and studied in finite dimensions to treat various time dependent network problems in operations research, particularly in traffic science. It is shown that projected dynamical systems are equivalent to a class of differential inclusions that were already analysed twenty years ago.

Finally we deal with a central issue of evolution problems and equilibrium theory, namely the asymptotics of the time processes and their convergence to steady-state solutions.

1991 Mathematics Subject Classification. 34A60, 34G20, 49J40, 65M12.

1 Introduction

Equilibrium problems are encountered in many different fields: mechanics, physics, chemistry, biology (in particular population dynamics), economics, traffic science, management science, and operations research. For rational economics, respectively for traffic science the history of the concept of equilibrium can be traced back to the work of Walras (1874), respectively to the work of Wardrop (1952). Equilibrium problems have been analyzed with mathematical rigour first by vector calculus and differential equations, then by optimization methods and more generally by methods of game theory (Nash equilibrium).

Then in the seventies of the 20th century, the theory of variational inequalities arose. This theory has originally been invented for the mathematical analysis of nonlinear boundary value problems with partial differential equations of elliptic type, in particular with nonlinear boundary conditions of unilateral type (Signorini conditions). By the pioneering work of S. Dafermos and M.J. Smith it became apparent that the methodology of variational inequalities is well-suited for the mathematical modelling of traffic equilibrium problems and related equilibrium problems on networks. Here one has to face the requirement of a conservation law (conservation of mass and the like) and possibly capacity constraints, in addition. One distinguishes between a system-optimization approach to traffic equilibria and user-oriented traffic equilibria. In the system-optimization approch one considers an objective, like total cost, set by society or by some authority. Traffic equilibria following this approach can also be cast as more traditional optimization problems. On the contrary, user-oriented traffic equilibria, when abandoning artificial assumptions of symmetry and thus abandoning the existence of a potential, cannot be formulated as simple optimization problems. Instead when every agent in traffic (or economics) strives for the minimization of his individual cost function, the overall flow pattern is ruled by equilibrium conditions that fit very well in the framework of the theory of variational inequalities.

Equilibrium per definitionem excludes *time*. On the other hand, time is central in our reality, in the physical-technological world as well as in the socio-economical world. Thus we are led to the study of time-dependent variational inequalities or evolutionary variational inequalities that model various constrained evolution problems, in particular with unilateral constraints encountered in applied sciences.

It is the purpose of the present paper to survey *some recent* trends in this research area. Here we are mainly interested in the challenging evolution problems where the state of the system itself in its time dependence lives in an *infinite* dimensional space; the wide field of differential inclusions in finite dimensional space (see [16], also [3], [4], [22]) is outside of the scope of the present paper. Also we refrain here from covering the literature on the existence and regularity theory of parabolic and hyperbolic evolution inequalities.

At first we address time dependent variational inequalities where time enters as an additional parameter in the variational inequality. This class of time dependent variational inequalities has been recently introduced to study certain time dependent traffic flow problems. Here we are not only concerned with existence theory, but also with numerical procedures (subgradient method) and with differentiability properties of the solution path with respect to time. Moreover we draw the reader's attention to the issues of existence and uniqueness of solutions to variational inequalities of *Volterra type*, in particular to recent work on constrained evolution problems that include the time history via memory terms.

Then we turn to time dependent variational inequalities that generalize classical ordinary differential equations. In particular, we report on recent extensions of the sweeping process, which has been introduced by Moreau [37]. Here we restrict our attention to infinite dimensional existence theory for problems of first order in time; for a more comprehensive survey we refer to [29].

Moreover in our paper, we deal with projected dynamical systems that have been recently introduced and studied in finite dimensions [39] to treat various time dependent network problems in operations research, particularly in traffic science. Here we discuss projected dynamical systems in a Hilbert space framework and thus lift this class of time dependent variational inequalities to a more abstract level. It turns out that projected dynamical systems are equivalent to a class of differential inclusions known as differential variational inequalities [4] that were already analysed over twenty years ago.

Finally we deal with a central issue of evolution problems and equilibrium theory, namely the asymptotics of the time processes and their convergence to steady-state solutions. Here we treat exponential decay to the steady-state solution in the presence of a strongly monotone operator, what extends to the discretization process with respect to space and time. In addition, we report on finite convergence and (weaker) monotone convergence for continuous-time subgradient procedures under specific assumptions for the solution of convex optimization problems.

Throughout the paper we use the following notation and conventions. $\mathbb{N}$, respectively $\mathbf{R}$ denote the natural numbers, respectively the field of real numbers. $\mathbf{R}_+$, respectively $\mathbf{R}_{++}$ is the closed half axis $[0, +\infty)$, respectively the open half axis $(0, +\infty)$. With $T \in \mathbf{R}_{++}$ being given, $S = [0, T]$ denotes a fixed time interval.

All vector spaces are implicitly understood to be real. Linear topological spaces, linear normed spaces, and Banach spaces are denoted by $V, W, \ldots$; Hilbert spaces by $H, G, \ldots$; V' is the continuous dual space of V, where $\langle \cdot, \cdot \rangle$ is the duality (form) on $V \times V'$. When no confusion arises, a Banach space norm, respectively a Hilbert space norm is denoted by $\| \cdot \|$, respectively by $| \cdot |$ belonging to the salar product $\langle \cdot, \cdot \rangle$. If not otherwise specified, int, respectively cl denotes the topological interior, respectively the topological closure with respect to the norm topology. When a Hilbert space H is identified with its dual H', then a Gelfand triple $\{V, H, V'\}$ means that the triple V, H, V' of linear topological spaces satisfies

$$V \subset H \subset V'$$

with continuous and dense embeddings.

Examples are provided by the function spaces $H = L^2(\Omega)$ and Sobolev spaces $V = W^{1,p}(\Omega)$ (see [1]) with parameter $p \in [1, \infty)$, where $\Omega \subset \mathbf{R}^d$ ($d = 2$ or $d = 3$ in applications) is a bounded open set with Lipschitz boundary $\partial\Omega$.

If not otherwise specified, $A, B, \ldots$ denote linear or nonlinear operators. K, Z, respectively $C, D \ldots$ stand for nonempty closed convex subsets, respectively closed convex cones with vertex at 0; K is sometimes assumed to be compact. For any fixed vector x, $[x]$ is the vector subspace generated by x.

For $X, Y \subset V$, the Hausdorff-distance d_H between X and Y is defined as

$$d_H(X, Y) = \max \{ \sup_{y \in Y} \operatorname{dist}(y, X), \sup_{x \in X} \operatorname{dist}(x, Y) \}.$$

For any subset $M \subset V$, the polar $M^0 \subset V'$ is

$$M^0 = \{ \xi \in V' : \langle \xi, x \rangle \leq 1 \text{ for all } x \in M \},$$

which simplifies for a cone C to

$$C^0 = \{ \xi \in V' : \langle \xi, x \rangle \leq 0 \text{ for all } x \in C \} =: C^-$$

and for a subspace M to

$$M^0 = \{ \xi \in V' : \langle \xi, x \rangle = 0 \text{ for all } x \in M \} =: M^\perp;$$

in particular

$$[x]^\perp = \{ \xi \in V' : \langle \xi, x \rangle = 0 \}.$$

The indicator function δ_Z is defined by

$$\delta_Z(x) = \begin{cases} 0 & \text{if } x \in Z\,; \\ +\infty & \text{otherwise.} \end{cases}$$

Note that the Fenchel conjugate (see e.g. [26] for terminology of convex analysis) of δ_Z is

$$\delta_Z^*(\xi) = \sup_{z \in Z} \langle \xi, z \rangle\,, \qquad \xi \in V'\,,$$

known as the barrier function of Z and the subdifferential of δ_Z at $x \in V$ is the normal cone to Z at x denoted by $N_Z(x)$. Thus $N_Z(x) = \emptyset$ for $x \notin Z$ and

$$\begin{aligned} N_Z(x) &= \{\xi \in V' : \langle \xi, z - x \rangle \leq 0 \text{ for all } z \in Z\} \\ &= \{\xi \in V' : \langle \xi, x \rangle = \delta_Z^*(\xi)\} \end{aligned}$$

for $x \in Z$. The formula above shows that for $x \in Z$

$$N_Z(x) = S_Z(x)^-\,,$$

where $S_Z(x)$ denotes the support cone (tangent cone, contingent cone) to Z at x, given by

$$S_Z(x) = \operatorname{cl}\left(\bigcup_{t>0} \frac{1}{t}(Z - x)\right)\,.$$

Further $\operatorname{proj}(x, Z)$ is the (nearest) projection of $x \in H$ onto a closed and convex $Z \subset H$, i.e., $y = \operatorname{proj}(x, Z)$ if and only if $y \in Z$ as well as $\langle x - y, z - y \rangle \leq 0$ for all $z \in Z$. A special case is $\operatorname{proj}(0, Z) =: Z^{\#}$, which is the element of Z with minimal norm. If no confusion arises, the nonexpansive operator $\operatorname{proj}(\cdot, Z)$ is abbreviated by P.

2 Time - an additional parameter in variational inequalities

In the following we concentrate on evolutionary variational inequalities, where time is only involved as an additional scalar parameter, but no time derivative occurs in the variational inequalities. This section is subdivided in three subsections. First we report on a recent field of time-dependent variational inequalities, where the contraints and the right hand side depend on time and the state lives in a finite dimensional space. This new arising theory is largely motivated by market and network equilibrium problems. Then we draw the attention of the reader to the older paper of Haraux [25] that treats a more specialized class of one-parameter dependent variational inequalities, where only the right hand side depends (but not the constraints) on the scalar parameter. There the differentiability of the solution path with respect to the parameter and formulas characterizing the derivative are the main object of the study, Finally in this section, we are concerned with classes of variational inequalities of *Volterra type.* Here we focus on the recent work of Steinbach [50, 51] on evolutionary variational inequalities with memory terms motivated by an application in electro-chemical machining.

2.1 Time-dependent variational inequalities and quasi-variational inequalities

In the paper [14] Daniele and Maugeri extend a disequilibrium market model for spatially distributed markets, which is due to Nagurney and Zhao and has been analyzed in a refined form in [24], to the time-dependent case. The terminology "dynamical equilibrium problem" in the title may be a little disleading, since there is no dynamics like in dynamical systems, in particular neither time derivatives nor integral terms as memory terms appear. Instead the driving forces of the markets, namely total supply, total demand as well as the unit costs are considered time-dependent on a fixed time interval $[0, T]$. Consequently, the response of the system, described by supply prices, demand prices, and transportation fluxes are time-dependent, too. Here the system is assumed to respond to changes of the driving forces so gradually in the considered time scale that, at each instant, equilibrium conditions prevail.

As the main result of this paper (Theorem 3.1), the dynamic equilibrium, which is the natural extension of the market equilibrium in the static case, is characterized by a variational inequality that involves the L^2-scalar product on $(0, T)$. One can ask whether a similar characterization also holds in a $C[0, T]$-setting.

The paper [15] by Daniele, Maugeri, and Oettli also contributes to this line of research. It starts from the following model of a traffic network on a finite directed graph. There is given a set $\mathcal{W}$ of origin-destination pairs and a set $\mathcal{R}$ of routes. Each route $r \in \mathcal{R}$ links some origin-destination pair $w \in \mathcal{W}$. This leads to the (nonvoid) set $\mathcal{R}(w)$ of all $r \in \mathcal{R}$ that link a given $w \in \mathcal{W}$. The topology of the network is described by the pair-route incidence matrix $\Phi = (\Phi_{w,r})$ with $w \in \mathcal{W}, r \in \mathcal{R}$, where

$$\Phi w, r = \begin{cases} 1 & \text{if route } r \text{ links the pair } w, \\ 0 & \text{else.} \end{cases}$$

While the topology remains fixed, one considers time dependent flow vectors $f(t) \in \mathbb{R}^{\mathcal{R}}$, where t varies in the fixed time interval $\mathcal{T} := (0, T)$ and the component $f_r(t)$ gives the flow in route $r \in \mathcal{R}$ at time t. Thus one arrives at the flow trajectory $f : \mathcal{T} \to \mathbb{R}^{\mathcal{R}}$. Now *feasible* flows have to satisfy time-dependent capacity constraints and demand requirements, namely in [15] that almost everywhere on $\mathcal{T}$, with respect to the nonnegative orthant in $\mathbb{R}^{\mathcal{R}}$ as order cone,

$$\lambda(t) \le f(t) \le \mu(t) \text{ and } \Phi f(t) = \rho(t),$$

where the bounds $\lambda \le \mu$ as well as the demand $\rho = (\rho_w)_{w \in \mathcal{W}} \ge 0$ are given. Considering a L^p setting with $p \in (1, \infty)$ one assumes that λ and μ belong to $E := L^p(\mathcal{T}, \mathbb{R}^{\mathcal{R}})$ and that ρ lies in $L^p(\mathcal{T}, \mathbb{R}^{\mathcal{W}})$. In addition, assume with respect to the nonnegative orthant in $\mathbb{R}^{\mathcal{W}}$ as order cone,

$$\Phi\lambda(t) \le \rho(t) \le \Phi\mu(t) \text{ a.e. on } \mathcal{T}. \tag{2.1}$$

The latter assumption guarantees the nonvoidness of

$$K := \left\{ f \in E \mid \lambda(t) \le f(t) \le \mu(t), \Phi f(t) = \rho(t) \text{ a.e. on } \mathcal{T} \right\},$$

the set of feasible flows (shortly for flow trajectories). This can be seen as follows. Let

$$F := \left\{ f \in E \mid \lambda(t) \le f(t) \le \mu(t) \text{ a.e. on } \mathcal{T} \right\},$$

a nonvoid convex, closed, bounded, hence weakly compact subset of E. Then $K = \emptyset$ means that ρ does not belong to the nonvoid convex, weakly compact set ΦF. The separation theorem implies the existence of $\xi = (\xi_w)_{w \in \mathcal{W}} \in L^q(\mathcal{T}, \mathbb{R}^{\mathcal{W}})$ $(1/p + 1/q = 1)$, $\xi \ne 0; \alpha \in \mathbb{R}$ such that

$$\langle \xi, \rho \rangle < \alpha \le \langle \xi, \Phi f \rangle, \ \forall \ f \in F. \tag{2.2}$$

Then choose $\tilde{f} \in F$ a.e. on $\mathcal{T}$ by

$$\tilde{f}_r(t) = \begin{cases} \lambda_r(t) & \text{if } r \in \mathcal{R}(w) \text{ and } \xi_w(t) \ge 0 \\ \mu_r(t) & \text{if } r \in \mathcal{R}(w) \text{ and } \xi_w(t) < 0 \end{cases}$$

and obtain by (2.1)

$$\langle \xi, \Phi\tilde{f} \rangle = \sum_{w \in \mathcal{W}} \int_{\xi_w > 0} \xi_w(t)(\Phi\lambda)_w(t)\, dt + \int_{\xi_w < 0} \xi_w(t)(\Phi\mu)_w(t)\, dt \le \langle \xi, \rho \rangle,$$

a contradiction to (2.2). This proves $K \ne \emptyset$.

Clearly, K is convex and weakly compact. Finally in this model of time-dependent equilibria there is a mapping $\gamma : K \to E^* =$

$L^q(\mathcal{T}, \mathbb{R}^{\mathcal{R}})$ $(1/p+1/q=1)$ which assigns to each flow (trajectory) $f \in K$ the cost trajectory $\gamma(f)$,

$$\langle \gamma(f), g \rangle = \int_{\mathcal{T}} \sum_{s \in \mathcal{R}} \gamma_s(f)(t)\, g_s(t)\, dt, \quad g \in E.$$

In this model, $h \in K$ satisfies *Wardrop's condition*, if and only if, for all $w \in \mathcal{W}$, all $r, s \in \mathcal{R}(w)$ and a.e. on $\mathcal{T}$ there holds

$$\gamma_r(h)(t) < \gamma_s(h)(t) \Rightarrow h_r(t) = \mu_r(t) \text{ or } h_s(t) = \lambda_s(t)\,. \tag{2.3}$$

This is a user-oriented equilibrium flow. It can be characterized (see [15, Theorem 3.1]) as a solution to the following variational inequality: Find $h \in K$ such that

$$\langle \gamma(h), f - h \rangle \geq 0, \ \forall\, f \in K. \tag{2.4}$$

The proof uses an argument from linear programming and runs as follows. Assume that (2.3) holds pointwise on a measurable subset $\hat{\mathcal{T}} \subset \mathcal{T}$, where $\mathcal{T} \setminus \hat{\mathcal{T}}$ is neglible. Fix $t \in \hat{\mathcal{T}}$. Let $w \in \mathcal{W}$ and

$$\begin{aligned} A &:= \Big\{ r \in \mathcal{R}(w) | h_r(t) < \mu_r(t) \Big\}, \\ B &:= \Big\{ s \in \mathcal{R}(w) | h_s(t) > \lambda_s(t) \Big\}. \end{aligned}$$

By (2.3), $\gamma_r(h)(t) \geq \gamma_s(h)(t)$ for all $r \in A, s \in B$. So there exists $\tau_w \in \mathbb{R}$ such that

$$\inf_{r \in A} \gamma_r(h)(t) \geq \tau_w \geq \sup_{s \in B} \gamma_s(h)(t)\,.$$

Let $f \in K$ be arbitrary. Then for every $s \in \mathcal{R}(w), \gamma_s(h)(t) < \tau_w$ implies $s \notin A$, $h_s(t) = \mu_s(t) \geq f_s(t)$, therefore $\Big(\gamma_s(h)(t) - \tau_w\Big)\Big(f_s(t) - h_s(t)\Big) \geq 0$. Likewise $\gamma_r(h)(t) > \tau_w$ implies $\Big(\gamma_r(h)(t) - \tau_w\Big)\Big(f_r(t) - h_r(t)\Big) \geq 0$. Hence for $t \in \hat{\mathcal{T}}$

$$\begin{aligned} \sum_{r \in \mathcal{R}(w)} \gamma_r(h)(t)\Big(f_r(t) - h_r(t)\Big) &\geq \sum_{r \in \mathcal{R}(w)} \tau_w \Big(f_r(t) - h_r(t)\Big) \\ = \tau_w \sum_{r \in \mathcal{R}(w)} \Big(f_r(t) - h_r(t)\Big) &= \tau_w \Big(\rho_w(t) - \rho_w(t)\Big) = 0\,. \end{aligned}$$

This gives (2.4) as follows:

$$\langle \gamma(h), f - h \rangle = \sum_{w \in \mathcal{W}} \int_{\mathcal{T}} \sum_{r \in \mathcal{R}(w)} \gamma_r(h)\,(f_r - h_r)\, dt \geq 0\,.$$

On the other hand, assume that (2.3) does not hold. Let $w \in \mathcal{W}$ and $r, s \in \mathcal{R}(w)$ such that on a nonneglible subset $\tilde{\mathcal{T}} \subset \mathcal{T}$ one has

$$\gamma_r(h) < \gamma_s(h),\ h_r < \mu_r,\ h_s > \lambda_s .$$

Define $\delta \in L^p(\mathcal{T})$ by

$$\delta(t) = \begin{cases} \min\{\mu_r - h_r, h_s - \lambda_s\}, & \text{if } t \in \tilde{\mathcal{T}} \\ 0 & \text{else} \end{cases}$$

and define $\tilde{f} \in E$ by

$$\tilde{f}_r := h_r + \delta, \tilde{f}_s := h_s - \delta, \tilde{f}_{\tilde{r}} = h_{\tilde{r}} \ \text{ if } \tilde{r} \neq r, s .$$

Then $\tilde{f} \in K$ and

$$\langle \gamma(h), \tilde{f} - h \rangle = \int_{\tilde{\mathcal{T}}} \delta\Big(\gamma_r(h) - \gamma_s(h)\Big) < 0,$$

completing the proof.

As another issue, the paper addresses the case of additional constraints on the (time-dependent) flows. Here duality methods known from convex optimization theory come into play. Under an appropriate regularity condition the existence of an abstract Lagrange multiplier is established such that the equilibrium solution in the restricted feasible set can be characterized as a solution of a variational inequality over the original feasible set, where the cost mapping has to be updated by this multiplier. This can be interpreted as a worsening of the costs due to the additional constraints.

Returning to the original time-dependent problem, an existence result (Corollary 5.1) is derived from an abstract existence result (Theorem 5.1) based on monotonicity methods under pseudomonotonicity and hemicontinuity assumptions. Finally the paper in the Hilbert space setting $L^2(0, T)$, proposes a solution method that extends the subgradient method of convex optimization. Here a gap function is introduced to incorporate the linear equality constraint (describing the demand requirements) into the equilibrium problem. Thus only the simple box constraints (describing the capacity constraints) remain as explicit contraints, which are easy to handle, since projections are readily available. Moreover using the weak compactness of such a relaxed feasible set, the boundedness of the subgradients of the gap function can be guaranteed, which by Theorem 6.1 results in the convergence of the proposed solution method.

In [17] DeLuca starts from Wardrop's equilibrium distribution for a network with asymmetric costs and elastic demands as the solution to the following quasi-variational inequality: Find x such that

$$\begin{cases} x \in C(x),\, \gamma(x)^T(y-x) \geq 0\,, & \forall y \in C(x) \\ C(x) := \{y \in \mathbf{R}^m_+ \,:\, \Phi y = \rho(x)\}\,, & \end{cases}$$

where

$\gamma : \mathbf{R}^m_+ \to \mathbf{R}^m_+$ is the cost function;
Φ is the $l \times m$ incidence matrix (l O/D pairs and m paths);

$C : K \Rightarrow \mathbf{R}^m_+$ is a multifunction with nonempty, closed, convex values and with K nonempty, compact, and convex $\subset \mathbf{R}^m_+$
$\rho : K \to \mathbf{R}^m_+$ is the elastic demand.

The aim of this paper is, as DeLuca put it, to extend some of the known existence results (in the static case) to a dynamic model for a network, that is to a model that takes into account changes in the time of the demand (due, e.g., to the income change, custom or activity changes, ...) in a fixed time interval. Again the network is supposed to change only gradually in the time scale such that acceleration or inertia effects do not appear.

In mathematical terms, the following formulation of an equilibrium distribution is considered:

$$(QVI)\begin{cases} \text{Find } u \in C(u) \quad \text{such that} & \\ \int_0^T \gamma(t,u(t))^T(v(t)-u(t))\,dt \geq 0\,, & \forall v \in C(u)\,, \end{cases}$$

This can be called a "weak formulation", since it relaxes the condition $\gamma(t,u(t))^T(v(t)-u(t)) \geq 0$ a.e., so we need not have a Wardrop equilibrium at each instant.

Here

$$C : K \Rightarrow L^2(0,T;\mathbf{R}^m_+)$$

is a multifunction with nonempty, closed, convex values, K is a nonempty, compact, and convex subset of $L^2(0,T;\mathbf{R}^m_+)$, $\Phi = (\phi)_{jk}$ is the $l \times m$ incidence matrix as above, and now $C(u)$ is the set of those $v \in L^2(0,T;\mathbf{R}^m_+)$ with $v_k(t) \geq 0$ a.e. in $[0,T]$, $k = 1,\cdots,m$ such that

$$\sum_{k=1}^m \phi_{jk} v_k(t) = \frac{1}{T}\int_0^T \rho_j(t,u(\tau))\,d\tau \text{ a.e. in } [0,T],\ j = 1,\cdots,l\,.$$

The question of existence of solutions for this quasi-variational inequality is dealt with the following result (see [17, Theorem 2.1]).

Theorem 2.1 *Assume that the functions*

$$\gamma : [0,T] \times \mathbb{R}^m_+ \to \mathbb{R}^m_+ \,, \qquad \rho : [0,T] \times \mathbb{R}^m_+ \to \mathbb{R}^l_+$$

satisfy the conditions:
i) $\gamma(t,z)$ *is measurable in* t *for all* $z \in \mathbb{R}^m_+$, *continuous in* z *for* t *a.e. in* $[0,T]$,

$$\exists a \in L^2(0,T), \alpha \geq 0 \,:\, |\gamma(t,z)| \leq a(t) + \alpha|z|;$$

ii) $\rho(t,z)$ *is measurable in* t *for all* $z \in \mathbb{R}^m_+$, *continuous in* z *for* t *a.e. in* $[0,T]$,

$$\exists b \in L^2(0,T]), \beta \geq 0 \,:\, |\rho(t,z)| \leq b(t) + \beta|z|;$$

iii) $C(v) \subseteq K, \forall v \in K$.

Then (QVI) *admits a solution.*

This result is proved in [17] from the following existence result for abstract quasi-variational inequalities due to Nguyen Xuan Tan [52].

Theorem 2.2 *Let* E *be a locally convex, Hausdorff topological vector space,* K *a non-empty , compact, convex subset of* $E, \gamma : K \to E$ *a continuous function, and* $C : K \Rightarrow K$ *a closed lower semicontinuous multifunction with* $C(x) \subset K$ *non-empty, compact, convex for all* $x \in K$. *Then there exists a solution to the following quasi-variational inequality*

$$x \in C(x) \,, \, \langle \gamma(x), y - x \rangle \geq 0 \,, \qquad \forall y \in C(x) \,.$$

Thus the crucial part of the existence proof of DeLuca is the proof of the lower semicontinuity of the multifunction C using the special structure of the incidence matrix Φ. More precisely, $C(u)$ as defined above is bounded, thus only weakly compact. Therefore the application of the abstract existence result of Tan gives only existence of a solution of the modified quasi-variational inequality where the multifunction C is replaced by $\tilde{C}(u) = C(u) \cap K$. However, compactness in the strong topology is very restrictive in infinite-dimensional spaces.

On the other hand, inspecting the constraints that define $C(u)$ above one may be led to work with the weak topology instead. But then the Nemitskii operators, associated to the functions γ and ρ, have to be weakly continuous. which - as is well-known in calculus of variations - restricts γ and ρ to be linear. A major difficulty lies then in the lower semicontinuity of C. Let us refer also to [35, Theorem 2.1] that covers the existence of a solution to (more standard) variational inequalities similarly in $L^2(0,T;\mathbb{R}^m)$. There the weak continuity of a (single-valued) Nemitskii operator C is required, which then necessarily is a linear operator (see [35] for more details).

The discussion above shows that the existence of solutions to time dependent quasi-variational inequalities in infinite-dimensional spaces is under realistic assumptions still an open challenge.

2.2 Some classic results on the differentiability of the projection on closed convex subsets in Hilbert space

A more specialized class of one-parameter dependent variational inequalities, where only the right hand side (but not the constraints) is differentiable with respect to the scalar parameter, has already earlier been treated by Haraux [25]. There the differentiability of the solution path with respect to the parameter and formulas characterizing the derivative are the main object of the study, To this end, the Gateaux differentiability of the projection on a closed convex subset in a Hilbert space is investigated.

Let us abbreviate P :=proj$(\cdot, Z)$ for a given closed convex subset Z of a Hilbert space H. For any fixed $x \in H$ let

$$\begin{aligned} C_Z(x) &= \bigcup_{\lambda>0} \lambda(Z - x), \\ S_Z(x) &= \overline{C_Z(x)}, \\ \Sigma_Z(x) &= S_Z(x) \cap [x - Px]^{\perp}. \end{aligned}$$

Then we can cite the following result about the Gateaux differentiability of the projection, which is of independent interest.

Proposition 2.1 *[25, Proposition 1] For any fixed $w \in H$, let γ be a weak limit of*

$$\gamma(t) = \frac{P(x + tw) - Px}{t}$$

as $t \to 0+$. Then

$$\begin{cases} \gamma \in \Sigma_Z(x), & \langle \gamma, w - \gamma \rangle \geq 0, \\ \langle w - \gamma, v \rangle \leq 0, & \forall v \in C_Z(Px) \cap [x - Px]^{\perp}. \end{cases}$$

The composition $P \circ f$ for a differentiable function f is treated in the following result.

Theorem 2.3 *[25, Theorem 2] Suppose that for all $x \in H$, $C_Z(Px) \cap [x - Px]^{\perp}$ is dense in $S_Z(Px) \cap [x - Px]^{\perp}$. Let $f : [0, T[\to H$ be right-differentiable at every point. Then $u(t) = P(f(t))$ is right-differentiable and*

$$\frac{d^+ u}{dt} = proj\,(\frac{d^+ f}{dt}, \Sigma_Z(f(t))\,.$$

The results above can be refined, when the convex set is replaced by a convex closed cone $K = \{x \in H, x \geq 0\}$ in a Hilbert lattice, where for any $x \in H$, $|x^+| \leq m|x|$ for some positive m and in the lattice as usual $x^+ = \sup\{x, 0\}$.

Consider for given $g : [0, T[\to H$ the variational inequality: Find $u(t) \in K$ such that for all $v \in K$

$$\langle g(t) - u(t), v - u(t) \rangle \leq 0 .$$

Theorem 2.4 *[25, Theorem 4] If g is right-differentiable , then the solution u of the above variational inequality is also right-differentiable. Its derivative $\gamma := \frac{d^+u}{dt}$ is given by the variational system*

$$\begin{cases} \gamma \in S_K(u(t)) \cap [g(t) - u(t)]^\perp , \\ \langle \gamma, w - u(t) \rangle \geq \langle \gamma, w - u(t) \rangle \qquad \forall w \in K \cap [g(t) - u(t)]^\perp , \\ |\gamma|^2 = \langle \frac{d^+g}{dt}, \gamma \rangle . \end{cases}$$

Based on this latter abstract result, Haraux [25] treats as applications the standard obstacle problem for the Laplace operator, respectively the boundary obstacle problem (Signorini problem) for the Helmholz equation, where in both cases the right hand side is assumed to be a right-differentiable curve.

2.3 Time-dependent variational inequalities with memory terms

In this subsection we draw the reader´s attention to the issues of existence and uniqueness of solutions to classes of variational inequalities of *Volterra type*, in particular to recent work on constrained evolution problems where time enters via memory terms.

First results on existence and uniqueness of solutions in $L^2(0, T; V)$ to the following class of variational inequalities of Volterra type

$$\int_0^T [A(t)u + \int_0^t B(t,s)u(s)\, ds, v - u]\, dt + \varphi(v) - \varphi(u) \geq \int_0^T [f, v - u]\, dt$$

with given linear bounded operators $A(t), B(t,s)\,; t, s \in [0, T]$ acting on a Hilbert space $\left(V, [\cdot,\cdot]\right)$ are due to Naumann [40]. A related class of first order evolution inequalities with the additional time derivative of the unknown u and with memory terms is treated in [41] and applied to a Volterra integrodifferential equation arising in heat conduction with memory.

Let us now focus on the more recent work of Steinbach [50, 51] on evolutionary variational inequalities with memory terms.

In [50] the existence and uniqueness of solutions of the following evolutionary variational inequality in the time interval $S = [0,T]$ are studied: Find $u \in \mathcal{K} = \{w \in C(S,V) : \mathrm{w(t)} \in K \text{ for all } t \in S\}$ such that

$$[A(t)\,u(t), v - u(t)] \geq [f(t), v - u(t)] + [\int_0^t (Bu)(\tau)\,d\tau, v - u(t)]$$

for all $v \in K$, $t \in S$, where K is a convex, closed and nonempty subset of a Hilbert space $\Big(V, [\cdot,\cdot]\Big)$. The Volterra operator $B \in (L_2(S,V) \to L_2(S,V))$ is assumed to satisfy a Lipschitz condition and $A(t) \in (V \to V)$ is a demicontinuous and strongly monotone operator for all $t \in S$. Moreover the above mentioned inequality, integrated over the compact time interval S, is discussed. In a second part the nonlinear operators A and B are replaced by continuous bilinear forms. Then, the penalization error between the solution u of the evolutionary variational inequality above and the corresponding penalty problem is estimated and regularity theorems are proved.

In [51] an obstacle problem with a memory term is studied in the framework of the variational inequality theory. This paper starts with a free boundary problem with space- and, in particular, time-dependent coefficients from an electro-chemical machining process. Then by a generalized Baiocchi-type transformation this problem is rewritten as an evolutionary variational inequality with time dependent convex sets $K(t)$ that are characterized by zero obstacle and an inhomogeneous Dirichlet boundary condition depending on the time t, but not on the space variable x. This gives rise to the following bilinear forms

$$\begin{aligned} a(t; v(t), w) &= \int_\Omega \sigma(x,t)\,\nabla v(x,t) \cdot \nabla w(x)\,dx\,, \\ b(\tau; v(\tau), w) &= \int_\Omega \partial_t\sigma(x,\tau)\,\nabla v(x,\tau) \cdot \nabla w(x)\,dx \end{aligned}$$

with the space- and time-dependent conductivity σ. After a shift in the boundary condition one obtains the time dependent convex set

$$\tilde{K}(t) = \big\{ w \in H_0^1(\Omega) : w \geq -g_D(t) \text{ a.e. on } \Omega \big\} \;, t \in [0,T]$$

with the obstacle $g_D \in C[0,T]$ given. One looks for $\tilde{u}(t) \in \tilde{K}(t)$, $t \in [0,T]$ that satisfies for all $v \in \tilde{K}(t)$

$$a\Big(t; \tilde{u}(t), v - \tilde{u}(t)\Big) \geq \Big[F(t), v - \tilde{u}(t)\Big] + \int_0^t b\Big(\tau; \tilde{u}(\tau), v - \tilde{u}(t)\Big)\,d\tau\,. \quad (2.5)$$

Here the following assumptions on σ are introduced:

$$\sigma \in W^1_\infty(0,T; L_\infty(\Omega)),\ \sigma(x,t) \geq \sigma_0 > 0 \qquad \forall x \in \Omega, t \in [0,T]\,,$$

which implies coercivity of $a(t;\cdot,\cdot)$ in $H_0^1(\Omega)$, i.e. there holds $a(t;v,v) \geq m\,\|v\|^2_{H^1(\Omega)}$ for all $v \in H_0^1(\Omega)$ and all $t \in [0,T]$ with some $m > 0$. Further nonnegativity of g_D and right hand regularity $F \in C([0,T], L^2(\Omega))$ are assumed.

Under the assumptions above the following result [51, Theorem 3.1] is established.

Theorem 2.5 *The evolutionary variational inequality (2.5) admits a unique solution $\tilde{u}$ with $\tilde{u}(t) \in \tilde{K}(t)$ and $\tilde{u} \in C\Big([0,T]; H_0^1(\Omega)\Big)$. Moreover, the estimate*

$$\|\tilde{u_1} - \tilde{u_2}\|_{C([0,T];H_0^1(\Omega))} \leq M\,\|F_1 - F_2\|_{C([0,T];H^{-1}(\Omega))} \tag{2.6}$$

holds, where M only depends on $m, T, \|\sigma\|_{W^1_\infty(0,T;L_\infty(\Omega))}$.

The *proof* consists of two parts. First in the case without memory term b (case $\partial_t\sigma = 0$) one shows as follows that the problem has a unique solution $\tilde{u} \in C\Big([0,T]; H_0^1(\Omega)\Big)$ for any right hand side $f \in C\Big([0,T]; H^{-1}(\Omega)\Big)$.

By the Lions-Stampacchia theorem in elliptic VI theory, the existence and uniqueness of $\tilde{u}(t)$ for each fixed t is guaranteed. To show the continuity of $\tilde{u} : [0,T] \ni t \to \tilde{u}(t) \in H_0^1(\Omega)$ an abstract stability result on convergence in Mosco's sense (see [49, Theorem 4.1, section 4.4]) is used in the following way. Let a sequence $\{t_n\} \subset [0,T]$ with $t_n \to t \in [0,T]$ be given. Owing to the continuity of g_D, $g_D(t_n) \to g_D(t)$ follows. Hence one concludes that $\tilde{K}(t_n) \to \tilde{K}(t)$ holds in Mosco's sense; in particular for any $v \in \tilde{K}(t)$ there exists a sequence $\{v_n\}$ with $v_n \in \tilde{K}(t_n)$ such that $v_n \to v$; indeed, take e.g. $v_n = \max\{v, -g_D(t_n)\}$. Furthermore one has $a(t_n, v_n, w) \to a(t, v, w)$ for any sequence $\{v_n\}$ with $\tilde{K}(t_n) \ni v_n \to v \in \tilde{K}(t)$. Together with the assumption $F \in C\Big([0,T]; L_2(\Omega)\Big)$ all conditions of the stability result are satisfied. Consequently $\tilde{u}(t_n) \to \tilde{u}(t)$ in $H^1(\Omega)$ for $\tilde{u}(t_n)$ and $\tilde{u}(t)$ as solutions for t_n and t respectively. In this case, the estimate (2.6) is with $M = m^{-1}$ a well-known consequence of coercivity.

The second part of the *proof* dealing with the full problem rests on an application of the Banach fixed point theorem as follows.

Consider the set

$$\mathcal{K} = \{w \in C\Big([0,T]; H_0^1(\Omega)\Big) \,|\, w(t) \in \tilde{K}(t),\ \forall t \in [0,1]\}\,.$$

Observe that for any $\alpha \in [0,1]$, the evolutionary variational inequality: Find $w \in \mathcal{K}$ that satisfies for all $t \in [0,T], v \in \tilde{K}(t)$

$$a\Big(t; w(t), v - w(t)\Big) \geq \Big[F(t), v - w(t)\Big] + \alpha \int_0^t b\Big(\tau; w(\tau), v - w(t)\Big)\, d\tau \tag{2.7}$$

admits an a priori estimate of the form (2.6). This can be obtained from coercivity using Gronwall's inequality. Further for $\alpha, \beta \in [0,1]$, define the mapping $w \in \mathcal{K} \mapsto \overline{w} = U_{\alpha,\beta} w \in \mathcal{K}$ given by

$$a(t; \overline{w}(t), v - \overline{w}(t)) - \alpha \int_0^t b\Big(\tau; \overline{w}(\tau), v - \overline{w}(t)\Big)\, d\tau \geq$$

$$\Big[F(t), v - \overline{w}(t)\Big] + \beta \int_0^t b\Big(\tau; w(\tau), v - \overline{w}(t)\Big)\, d\tau, \ \forall v \in \tilde{K}(t), \ t \in [0,T]\,.$$

Then by the a priori estimate of the form (2.6) applied to the right hand side above, the mapping $U_{\alpha,\beta}$ (if defined) is contractive for $0 < \beta \leq \beta_0$ with small enough $\beta_0 > 0$. Thus the following bootstrap argument works. Starting with $\alpha_0 = 0$ from the first part of the proof, with α_j being given, the mapping $U_{\alpha_j,\beta}$ is well-defined, contractive for $\beta = \beta_j = \min(\beta_0, 1-\alpha_j)$ and hence possesses a fixed point in $\mathcal{K}$. This gives a solution to (2.7) with $\alpha = \alpha_{j+1} = \alpha_j + \beta_j$, and another step runs. After a finite number of steps, $\alpha = 1$ is reached, and the theorem is proved.

Furthermore, the paper presents regularity results with respect to time and space by using a penalization method. For the quasi-stationary model of the electro-chemical machining problem one arrives at $u \in W^1_\infty(\Omega \times (0,T))$.

3 Ordinary Differential Inclusions with Convex Constraints: Sweeping Processes

In the following sections we do not cover the whole vast field of ordinary differential inclusions with convex constraints in infinite dimensional spaces. Instead we concentrate on two classes of evolutionary variational inequalities with (ordinary) time derivatives: sweeping processes and projected dynamical systems.

In this section we first describe moving convex sets and sweeping processes introduced by Moreau and sketch their relations to the modern theory of hysteresis operators useful in elastoplasticity. Then we turn to recent work on generalized sweeping processes, where we restrict our attention to infinite dimensional existence theory for problems of first order in time.

3.1 Moving convex sets and systems with hysteresis

Let $C : S = [0, T] \rightrightarrows H$ be a multifunction with nonempty, closed, convex values $C(t)$ in a (real) Hilbert space H. Then the *sweeping process* introduced by Moreau [37] consists in finding a function $v : S \to H$ that satisfies

$$v(t) \in C(t), \qquad t \in S \tag{3.1}$$

and moreover the differential inclusion

$$-\dot{v}(t) \in N_{C(t)}(v(t))\,. \tag{3.2}$$

This means that the single moving point described by v is not allowed to leave $C(t)$. If $v(t)$ lies in the topological interior (or in the algebraic interior) of $C(t)$, then the point rests; otherwise, $v(t)$ hits the boundary of $C(t)$ and then the moving point v is driven along an inward normal (= − outward normal) of $C(t)$ at $v(t)$ according to the definition of the normal cone N.

The special case

$$C(t) = w(t) + Z, \qquad t \in S, \tag{3.3}$$

where Z is a closed convex subset of a Hilbert space H containing 0 and $w : S \to H$ is a given function, describes a purely translational motion of Z from the initial position $C(0)$ to the final position $C(T)$. Thus the single moving point v is constrained by

$$v(t) = w(t) + z(t), \quad z(t) \in Z \qquad (t \in S)\,. \tag{3.4}$$

Here (3.2) reduces to the evolution variational inequality

$$\langle \dot{v}(t), \zeta - z(t) \rangle \geq 0, \quad \forall \zeta \in Z$$

to be solved together with (3.4). Treating z as the only unknown, one arrives at the evolution variational inequality

$$\langle \dot{z}(t), \zeta - z(t) \rangle \geq \langle g(t), \zeta - z(t) \rangle, \quad \forall \zeta \in Z\,,$$

where $g(t) = -\dot{w}(t)$ is given.

Instead of prescribing the motion of a convex set Z that drives the single point v, one can also couple the two motions the other way round. Namely, with v being prescribed, find the motion of the convex set compatible with (3.4), that is, find a function $w : S \to H$ that satisfies

$$\dot{w}(t) \in N_{C(t)}(v(t))\,. \tag{3.5}$$

By condition (3.5), the convex set $C(t)$ remains fixed as long as $v(t)$ lies in the topological interior (or in the algebraic interior) of $C(t)$. On the other hand, if $v(t) \in \partial C(t)$, hence $z(t) \in \partial Z$ by (3.3), (3.4) implies the decomposition $\dot{v}(t) = \dot{w}(t) + \dot{z}(t)$ with $\dot{z}(t)$ tangential to Z, and therefore $C(t)$ is forced to move along the nonzero outward normal $\dot{w}(t)$, provided $\dot{v}(t)$ points in a nontangential direction.

In the special case (3.3), similarly as above, (3.5) reduces to the evolution variational inequality

$$\langle \dot{w}(t), \zeta - z(t) \rangle \leq 0, \quad \forall \zeta \in Z$$

to be solved together with (3.4). Treating z as the only unknown, one arrives at the evolution variational inequality

$$\langle \dot{z}(t), \zeta - z(t) \rangle \geq \langle f(t), \zeta - z(t) \rangle, \quad \forall \zeta \in Z \,, \tag{3.6}$$

where $f(t) = \dot{v}(t)$ is given.

These evolution variational inequality problems have to be complemented by a suitable initial condition, here

$$z(0) = z_0 \tag{3.7}$$

with some given $z_0 \in Z$, to arrive at well-posed problems; see [37], [27, Theorem 3.1, Proposition 3.9], [9, Theorem 2.1] for existence and uniqueness results. Thus one is led to the solution operators associated to (3.5), respectively (3.6)

$$(v, z_0) \mapsto w = \mathcal{P}(v, z_0), \qquad (v, z_0) \mapsto z = \mathcal{S}(v, z_0) \,.$$

The operators $\mathcal{P}$, $\mathcal{S}$ are known in systems with hysteresis as the *play operator* and the *stop operator*, respectively. We refrain here from giving more details about the connections between hysteresis problems encountered in particular in elastoplasticity on one hand and convex analysis and variational inequality theory on the other; instead we refer to the comprehensive papers of Brokate [9] and of Krejci [28].

3.2 Sweeping processes and generalizations

In the monograph [36] Monteiro Marques - after collecting the necessary mathematical tools like functions with bounded variation, convergence of the filled-in graphs and some geometrical inequalities - investigates Moreau's sweeping process in the cases when either the moving convex set $t \to C(t)$ has right-continuous bounded variation or it has nonempty interior instead. Chapter 3 treats the problem of the dynamics of a mechanical system with a finite number of degrees of freedom, subject to a unique unilateral constraint and experiencing inelastic shocks (without friction or with Coulomb's isotropic friction). This leads to the study of related, but more involved differential inclusions, restricted to finite dimensions. Chapter 4 deals with related problems involving what may be called externally induced collisions. Some directions of research which can be linked to the above situation are discussed in chapter 5, in particular existence theory for some inclusions of second order.

More recent research on sweeping processes and extensions is in particular due to Castaing, Kunze, and again Monteiro Marques.

The paper of Castaing and Monteiro Marques [11] is concerned with topological properties of solution sets for functional-differential inclusions of sweeping process type:

$$\frac{du}{dt} \in -N_{K(t)}\big(u(t)\big) + F(t,u),$$

where K is a γ-Lipschitzean multifunction from $[0,T]$ to the set of nonempty compact convex subsets of H, N denotes as before the normal cone and $F : [0,T] \times \mathcal{C}_0 \rightrightarrows H$ is an upper semicontinuous convex weakly compact valued multifunction. Here $\mathcal{C}_0 = \mathcal{C}([-r,0],H)$ denotes the Banach space of continuous vector-valued functions defined on $[-r,0]$ taking values in a real separable Hilbert space H, where $r > 0$ is a finite delay. As an application, *periodic* solutions to such functional-differential inclusions are obtained, when K is T-periodic, i.e. when $K(0) = K(T)$ with $T \geq r$.

In the paper [5] several new existence results are given for differential inclusions in an infinite-dimensional space and for the sweeping process. These results are based on refined compactness and time discretization methods. One class of the differential inclusions treated is of the following form:

$$u'(t) \in F(u(t)) + G(u(t)) + g(t,u(t))$$

where F is an upper semicontinuous set-valued map defined on a separable Banach space which satisfies $F(x) \subset A(\partial\gamma(x))$ where A is a compact operator, $\partial\gamma$ is the subdifferential of a convex continuous function, G is a continuous multifunction not necessarily convex-valued, and g is a Caratheodory mapping. As another class, upper semicontinuous perturbations of the sweeping process in a Hilbert space of the form:

$$-u'(t) \in N_{C(t)}(u(t)) + F(t,u(t)) \tag{3.8}$$

are studied.

In [30] the multifunction $C : [0,T] \rightrightarrows H$ with closed convex nonvoid values $C(t)$ for every $t \in [0,T]$ is assumed to be continuous with respect to the Hausdorff metric d_H, shortly d_H-continuous. For $\xi_0 \in C(0)$ and every $\lambda > 0$ let u_λ denote the corresponding Yosida-Moreau-approximation of the sweeping process with moving set $C(\cdot)$, i.e., the absolutely continuous unique solution of

$$\begin{cases} \dfrac{du_\lambda}{dt}(t) + \dfrac{1}{\lambda}\Big[u_\lambda(t) - \text{ proj }\Big(u_\lambda(t), C(t)\Big)\Big] = 0 \quad \text{a.e. on } [0,T]\,, \\ u_\lambda(0) = \xi_0\,. \end{cases} \tag{3.9}$$

Then Kunze and Monteiro Marques prove in [30] the following result.

Theorem 3.1 *Let H be a Hilbert space and $T > 0$. Moreover, let $C : [0,T] \rightrightarrows H$ be a d_H-continuous multifunction such that $C(t)$ is closed convex and has nonempty interior for all $t \in [0,T]$. If $\xi_0 \in C(0)$ then, as $\lambda \to 0^+$, the Yosida-Moreau-approximations (3.9) are pointwise strongly convergent on $[0,T]$ to the unique solution u of the corresponding sweeping process*

$$-\frac{du}{|du|} \in N_{C(t)}\Big(u(t)\Big), \quad u(t) \in C(t), \quad u(0) = \xi_0. \tag{3.10}$$

Here, $u : [0,T] \to H$ is called a solution of (3.10) if u is continuous and of bounded variation (cbv) on $[0,T]$ such that $u(0) = \xi_0, u(t) \in C(t)$ for $t \in [0,T]$ and

$$-u'(t) \in N_{C(t)}\Big(u(t)\Big) \quad \text{for } |du| \text{ - almost all } t \in [0,T]\,.$$

Let us recall that the variation of a function $u : [0,T] \to H$ is defined as

$$\begin{aligned} \text{var}(u) = \quad & \sup\Big\{\sum_{\nu=1}^{N} |u(t_\nu) - u(t_{\nu-1})| : \\ & 0 = t_0 < t_1 < \cdots < t_{N-1} < t_N = T \text{ is a partition of } [0,T]\Big\}\,, \end{aligned}$$

and u is called of bounded variation, if $\text{var}(u) < \infty$. Further u' is understood as a Radon-Nikodỳm density of the Stieltjes measure (differential measure) du of the cbv function u with respect to its modulus $|du|$ of du, cf [36, section 0.1] for more details. From [36, Theorem 2.2.1, p. 46], it is known that under the given hypotheses such a unique solution to the sweeping process always exists.

The paper [6] presents an extension of the sweeping process to the nonconvex case. The existence of bounded variation and Lipschitzean solutions for the sweeping process associated to a nonconvex closed moving set and its application to a new class of evolution problems governed by a subdifferential of a Lipschitz function are discussed. In addition, convex and nonconvex perturbations of the preceding evolution problem are also studied.

The following papers by Kunze and Monteiro Marques are devoted to generalizations of the sweeping process and present some extensions of known existence results for the sweeping process.

In [32] evolutionary problems of the type

$$-\frac{du}{dr(t)} \in A(t)u(t) \quad dr - \text{a.e. in } [0,T] \tag{3.11}$$

are studied extending (3.10) as seen below, where

$$A(t) : D(A(t)) \subset H \rightrightarrows H$$

is a maximal monotone operator (mmop) in a Hilbert space H for every $t \in [0,T]$ and the dependence $t \to A(t)$ is - in some sense - of bounded variation.

More precisely it is assumed:

(H1) There exists a function $r : [0,T] \to [0,\infty)$ which is right-continuous and nondecreasing on $[0,T]$ with $r(T) < \infty$ such that

$$\text{dis } (A(t), A(s)) \le dr(]s,t]) = r(t) - r(s) \qquad \text{for } 0 \le s \le t \le T,$$

where dis (.,.) is the pseudo-distance between mmops introduced in [53] and given by

$$\text{dis } (A_1, A_2) =$$

$$= \sup \left\{ \frac{< y_1 - y_2, x_2 - x_1 >}{1 + |y_1| + |y_2|} : x_i \in D(A_i), y_i \in A_i x_i \, (i = 1,2) \right\}$$

for mmops A_1 and A_2 in H with domains $D(A_1)$, resp. $D(A_2)$.

It is noted that in the special case of a sweeeping process, the pseudo-distance reduces to the Hausdorff distance d_H. Namely, if $C(t)$ is closed and convex for $t \in [0,T]$, then with $A(t) = \partial\delta_{C(t)} = N_{C(t)}$ one arrives, see [53, Lemma 3.5] at

$$\text{dis } (A(t), A(s)) = d_H(C(t), C(s)) \qquad \forall t, s \in [0,T] .$$

Let $A(t,x)^{\#}$ denote the element of minimal norm of $A(t)x$ and suppose that

(H2) there exists a nonnegative $c \in L^1([0,T]; dr)$ such that

$$|A(t,x)^{\#}| \le c(t)(1 + |x|) \qquad \forall t \in [0,T], x \in D(A(t)) .$$

This is trivially satisfied for a sweeping process since in this case every $N_{C(t)}(x)$ is a cone and hence $0 \in N_{C(t)}(x)$ for $x \in C(t)$. Therefore the subsequent theorem encompasses known existence results for the sweeping process.

Theorem 3.2 *Suppose (H1) and (H2) with some functions r and c. Then for all $u_0 \in D(A(0))$ the evolution problem (3.11) with initial value $u(0) = u_0$ has a unique solution of bounded variation.*

As variants of this theorem a local existence result, where (H2) is replaced by a local condition, and an existence result for absolutely continuous (instead of bounded variation) solutions are proved.

In [33] multifunctions that depend on t and u are considered and the following evolution problem

$$-u'(t) \in N_{C(t,u(t))}(u(t)) \text{ a.e. in } [0,T], u(0) = u_0 \in C(0, u_0)$$

is studied in Hilbert space. Here the essential assumption is a Lipschitz condition with respect to the Hausdorff distance:

$$d_H(C(t,u), C(s,v)) \leq L_1|t-s| + L_2|u-v|$$

with $L_2 < 1$. (It is noted a counterexample for existence in the case $L_2 = 1$.)

This evolution problem generalizes Moreau's classical sweeping process $-u'(t) \in N_{C(t)}(u(t))$ as well as quasivariational inequalities of the following form: Find $v(t) \in C(v(t))$ such that

$$\begin{cases} \langle v'(t) + f(t), w - v(t) \rangle \geq 0 & \forall w \in C(v(t)); \\ v(0) = v_0 \in \Gamma(v(0)). \end{cases}$$

The study of the evolution problem above is in particular motivated by a variational model of sandpiles growth due to Prigozhin [44]. This variational model is related to the Bean critical-state model in superconductivity, also introduced by Prigozhin in [45]. Both models lead to evolutionary quasivariational inequalities. In special cases, where the quasivariational inequalities become variational ones, the existence and uniqueness of the solution have been already proved in [44], [45]. In the two-dimensional case of the Bean model in superconductivity, also Prigozhin [46] presents a numerical solution based on finite difference discretization in time and finite elements in space, solving the resulting constrained optimization problem by underrelaxation.

For a more comprehensive survey on the theory of first and in addition second order sweeping processes starting from an elementary level, respectively on numerical aspects of the sweeping process giving an introduction to solution procedures in contact dynamics we refer to the appearing paper [29], respectively to the recent paper of Moreau [38].

4 Projected dynamical systems

This subsection is devoted to a class of dynamical systems constrained to convex subsets in a Hilbert space. This class was introduced and investigated by Nagurney and Zhang [39], [56] in finite dimensions as a tool for the study of equilibrium problems. They coined the notion of a *projected dynamical system.* Here we treat projected dynamical systems in a Hilbert space setting and thus lift this class of problems to a more abstract level. To this end we need to refine the discussion of the differentiability of the projection onto closed convex subsets in Hilbert space. In particular, we introduce the notions of the *quasi interior* and the *quasi boundary* of a closed convex set. Thus without any restriction of dimension, we can characterize a stationary point of a projected dynamical system by a related variational inequality. To conclude this subsection we address the issue of well-posedness of projected dynamical systems. It turns out that projected dynamical systems are equivalent to a class of differential inclusions known as differential variational inequalities [4] that were already analysed over twenty years ago. Therefore earlier existence results [2] apply to projected dynamical systems in infinite dimensions.

4.1 Differentiability of the projection onto closed convex subsets revisited

Let us first recall some further well-known objects of convex analysis. For a closed convex subset Z of a Hilbert space H and for any $x \in Z$, the *support cone* (often termed *tangent cone* or *contingent cone*, see e.g. [4]) to Z at x, denoted by $S_Z(x)$, is the closure of the convex cone $\bigcup\left\{\lambda(Z-x) \, : \lambda > 0\right\}$. The set $S_Z(x)$ is clearly a closed convex cone with vertex 0 and is the smallest cone C whose translate $x+C$ has vertex x and contains Z. Note that taking polars with respect to the scalar product in H, $(S_Z(x))^0 = (S_Z(x))^- = N_Z(x)$, the normal cone to Z at x.

The utility of the support cone for projected dynamical systems stems from the fact that for the projection $P_Z = \text{proj}\,(Z, \cdot)$ onto Z there holds for any $x \in Z, h \in H$

$$P_Z(x + th) = x + tP_{S_Z(x)}\, h + o(t),\; t > 0$$

in a Hilbert space (see [55] Lemma 4.6, p. 300). This can even proved in a more general class of reflexive Banach spaces (see [43] Lemma 3). The related differentiability result (here Proposition 2.1) due to Haraux that covers the case $x \notin Z$ is not needed here.

On the other hand, in [39], also [19], the set of *unit inward normals* to Z at x is defined by

$$n_Z(x) = \{v : |v| = 1 \text{ and } \langle v, x-y\rangle \leq 0,\ \forall\, y \in Z\}$$

and the "projection" of the vector h at x with respect to Z is defined by the directional derivative

$$\Pi_Z(x,h) = \lim_{t\to 0} \frac{P_Z(x+th)-x}{t}\,.$$

Thus

$$\Pi_Z(x,h) = P_{S_Z(x)}\, h\,, \tag{4.1}$$

the (metric) projection onto the support cone $S_Z(x)$, and immediately

$$n_Z(x) = \partial B(0;1) \cap -\Big(S_Z(x)\Big)^0,$$

where $\partial B(0;1) = \Big\{z : |z| = 1\Big\}$ is the unit sphere.

Now we call the *quasi interior* of Z (denoted $qi\,Z$) the set of those $x \in Z$ for which $S_Z(x) = H$. Note that Zarantonello (p. 262) calls $x \in Z$ an "inner point" of Z, if $S_Z(x)$ is a subspace, only, and Borwein and Lewis ([7]) as an extension of the relative interior in finite dimensions, introduce the quasi relative interior of Z, denoted $qri\,Z$, as the set of those $x \in Z$ for which $S_Z(x)$ is a subspace, only. Observe that $qi\,Z = qri\,Z$ holds with an appropriate choice of the space H, which can always be arranged in the situation considered.

As is well-known, in infinite dimensions the (topological) interior as well as the relative algebraic interior of a convex set can be void. On the contrary, the quasi interior may be nonempty, as seen in the following

Example 4.1 *Let $H = L^2(T,\mu)$ (or more generally any $L^P(T,\mu)$ space with $1 \leq p < \infty$). Consider the closed convex cone $D = \Big\{x \in H : x(t) \geq 0 \text{ a.e.}\Big\}$. Then e.g. $1 \in qi\,C$. Indeed, by Lebesque's theorem of majorized convergence, any $z \in H$ can be approximated by the sequence $\{z_n\}$ of truncations,*

$$z_n(t) = \begin{cases} z(t) & \text{if } z(t) \geq -n \text{ a.e.;} \\ -n & \text{elsewhere,} \end{cases}$$

and clearly $z_n \in n(C-1)$.

In addition, we define the *quasi boundary* of a closed convex set Z (denoted by $qbdry\,Z$) as the set $Z \setminus qi\,Z$. Then, in virtue of the strong separation theorem, $x \in qbdry\,Z$, if and only if, $n_Z(x) \neq \emptyset$. Moreover, we have the following geometric interpretation that extends [39, Lemma 2.1], [18] to infinite dimensions.

Lemma 4.2 *i) If $x \in qi\, Z$, then for any $h \in H$,*

$$\Pi_Z(x, h) = h\,.$$

ii) If $x \in qbdry\, Z$, then for any $v \in H \setminus S_Z(x)$ there exists $n^*(x) \in n_Z(x)$ such that

$$\beta(x) = -\langle v, n^*(x)\rangle > 0 \text{ and } \Pi_Z(x, v) = v + \beta(x)\, n^*(x)\,.$$

Proof.

i) If $x \in qi\, Z$, then $S_Z(x) = H$ and $\Pi_Z(x, \cdot)$ is the identity.

ii) If $x \in qbdry\, Z$, then by (4.1) we have for

$$\hat{v} := \Pi_Z(x, v) = P_{S_Z(x)}\, v$$

that

$$\langle v - \hat{v}, w - \hat{v}\rangle \leq 0, \ \forall\, w \in S_Z(x)\,.$$

Since $S_Z(x)$ is a cone, it follows

$$\langle v - \hat{v}, \hat{v}\rangle = 0, \quad v - \hat{v} \in \left(S_Z(x)\right)^0. \tag{4.2}$$

Since $v - \hat{v} \neq 0$ by assumption, the latter relation implies the existence of some $n^* \in n(x), \beta > 0$ such that

$$\hat{v} - v = \beta n^*\,.$$

Moreover the orthogonality $\langle n^*, \hat{v}\rangle = 0$ implies

$$\beta = -\langle v, n^*\rangle\,.$$

proving the lemma.

From (4.2) we immediately obtain the following characterization.

Corollary 4.1 *Let $x \in Z$. Then for any $v \in H$*

$$\Pi_Z(x, v) = \left(v - N_Z(x)\right)^{\#}.$$

Proof. Let

$$\hat{v} = \Pi_Z(x, v)\,.$$

In the special case $x \in qi\, Z$, $N_Z(x)$ reduces to zero and hence the claim trivially follows. Thus suppose $x \in qbdry\, Z$. Then by (4.2) we obtain that

$$\hat{v} \in v - N_Z(x)\,.$$

Since by (4.1), $\hat{v} \in S_Z(x) = (N_Z(x))^0$, it follows from (4.2) moreover that

$$\langle \hat{v}, v - \hat{v} - z\rangle \geq 0, \ \forall\, z \in N_Z(x)\,.$$

This means that $\hat{v} = \text{proj}\ (0, v - N_Z(x))$ and proves the claim.

4.2 Projected dynamical systems and stationarity

Following [39] a projected dynamical system is an ordinary differential equation of the form

$$\dot{x} = \Pi_Z\Big(x, -F(x)\Big)\,, \tag{4.3}$$

where the convex closed set Z and the vector field $F : Z \to H$ are given. Here one looks for a function $x : [0, +\infty) \to Z$ that is absolutely continuous and satisfies $\dot{x}(t) = \Pi_Z\Big(x(t), -F\big(x(t)\big)\Big)$ except a set of Lebesgue measure zero. Like in the classical theory of ordinary differential equations, (4.3) has to be complemented by an initial condition

$$x(0) = x_0\,, \tag{4.4}$$

where $x_0 \in Z$ is given. Clearly, the sought solution depends on the initial value x_0. To stress this dependence one defines [39, Definition 2.6] a projected dynamical system $PDS(F, Z)$ as the map $\phi : Z \times [0, \infty) \to Z, (x_0, t) \mapsto x(t) =: \phi(x_0, t)$, where x is a solution of (4.3) and (4.4).

A vector $x^* \in Z$ is termed [39, Definition 2.7] an *equilibrium point* or *stationary point* of the projected dynamical system $PDS(F, Z)$, if x^* satisfies

$$\Pi_Z\Big(x^*, -F(x^*)\Big) = 0\,. \tag{4.5}$$

This means that once the projected dynamical system reaches x^* at some time $t^* \geq 0$, it will remain at x^* for all future times $t \geq t^*$.

Based on the Lemma 4.2 we can characterize the stationary points of a projected dynamical system $PDS(F, Z)$ as solutions of a related variational inequality $VI(F, Z)$. Thus we can extend [39, Theorem 2.4] by dispensing with its restriction to finite dimension and to convex polyhedra (= finite intersections of closed half spaces).

Proposition 4.1 *Let Z be closed convex and $x^* \in Z$. Then x^* is a stationary point of the projected dynamical system $PDS(F, Z)$, if and only if*

$$\langle F(x^*), z - x^* \rangle \geq 0,\ \forall\, z \in Z\,. \tag{4.6}$$

Proof. Let x^* solve (4.6) which is equivalent to

$$x^* = P_Z\Big(x^* - tF(x^*)\Big),\ \forall\, t > 0,$$

whence by definition, $\Pi_Z\Big(x^*, -F(x^*)\Big) = 0$ follows.

Conversely let x^* be a stationary point. Consider the case $x^* \in qbdry\, Z, -F(x^*) \notin S_Z(x^*)$. Then by the Lemma 4.2, for some $\beta^* > 0, n^* \in n_Z(x^*)$, there holds $F(x^*) = \beta^* n^*$ and hence x^* solves (4.6). In the remaining cases, we have $-F(x^*) \in S_Z(x^*)$, hence by (4.1), $F(x^*) = 0$, and x^* trivially solves (4.6).

4.3 Well-posedness for projected dynamical systems

The following well-posedness result for projected dynamical systems in finite dimensions is proved in [39].

Theorem 4.1 *[39, Theorem 2.5]*
Assume that there exists a $\beta > 0$ *such that the (continuous) vector field* $F : \mathbb{R}^n \to \mathbb{R}^n$ *satisfies the linear growth condition*

$$|F(x)| \leq \beta(1 + |x|), \quad \forall\, x \in K,$$

and also the monotonicity condition

$$\langle F(x) - F(y), x - y \rangle + \beta\, |x - y|^2 \geq 0, \quad \forall\, x, y \in K.$$

Then

i) for any $x_0 \in K$*, there exists a unique solution* x^0 *to the initial value problem (4.3), (4.4);*

ii) if $x_n \to x_0$ *as* $n \to \infty$*, then the associated solution* x^n *converges to* x^0 *uniformly on every compact set of* $[0, \infty)$.

Note that by (4.1) and corollary 4.1, (4.3) is equivalent to

$$\dot{x} = P_{S_Z(x)}\Big(-F(x)\Big) \equiv -\Big(N_Z(x) + F(x)\Big)^{\#}.$$

Thus the initial value problem (4.3), (4.4) consists in finding the "slow" solution (the solution of minimal norm) to the initial condition (4.4) and the *differential variational inequality* [4, chapter 6, section 6]

$$\dot{x}(t) \in -\Big(N_Z(x(t)) + F(x(t))\Big)$$

what is equivalent [13] to finding the "slow" solution to the initial condition (4.4) and the *projected variational inequality* [4, chapter 6, section 6]

$$\dot{x}(t) \in P_{S_Z(x(t))}\Big(-F(x(t))\Big),$$

where here the operator F is single-valued. Therefore earlier existence results (see [4] for more results and references) apply to projected dynamical systems in infinite dimensions. In particular, [2, Theorem 2] encompasses the above existence result.

5 Some Asymptotic Results

In this final section we deal with a central issue of evolution problems and equilibrium theory, namely the asymptotics of the time processes and their convergence to steady-state solutions. First, we recall some classical results due to Brézis on the exponential decay of the trajectory of a differential inclusion to the steady-state solution in case of a strongly monotone operator. Then we briefly describe a discretization procedure (with respect to space and time) of the differential inclusion that preserves the exponential decay to the steady-state solution. To conclude this section, we report on finite convergence and (weaker) monotonic convergence results for continuous-time subgradient procedures under specific assumptions for the solution of convex optimization problems. Here we concentrate on the work [20, 21]. Very recent asymptotic results on dynamical systems in optimization can be found in [12, 47].

5.1 Some classical results

Let $f \in L^1_{loc}([0,+\infty); H)$ and A be a (maximal) monotone operator identified with its graph. Further let u be a solution of

$$\frac{du}{dt} + Au \ni f\,. \tag{5.1}$$

Then the behaviour of $u(t)$ for $t \to +\infty$ is of interest.

First let us note that if $f_\infty = \lim_{t\to\infty} f(t)$ and $u_\infty = \lim_{t\to\infty} u(t)$ exist, then $[u_\infty, f_\infty] \in A$, provided A is maximal monotone. Indeed, a weak solution $u \in C\Big([0,+\infty); H\Big)$ of (5.1) can be characterized according to [8, Proposition 3.6] by the (relaxed) variational inequality

$$\frac{1}{2}|u(t)-x|^2 \le \frac{1}{2}|u(s)-x|^2 + \int_s^t \langle f(\sigma)-y, u(\sigma)-x\rangle\, d\sigma,$$
$$\forall [x,y] \in A,\ 0 \le s \le t.$$

which implies by Cauchy's inequality $\left(ab \le \frac{1}{2}a^2 + \frac{1}{2}b^2\right)$

$$\langle u(t)-u(s), u(s)-x\rangle \le \int_s^t \langle f(\sigma)-y, u(\sigma)-x\rangle\, d\sigma$$
$$\forall [x,y] \in A,\ 0 \le s \le t$$

and thus in particular

$$\langle u(t+1) - u(t), u(t) - x \rangle \le \int_t^{t+1} \langle f(\sigma) - y, u(\sigma) - x \rangle \, d\sigma$$
$$\forall [x,y] \in A, \; t \ge 0$$

Hence in the limit for $t \to \infty$, one arrives at

$$0 \le \langle f_\infty - y, u_\infty - x \rangle \qquad \forall [x,y] \in A \,,$$

proving the claim.

In general, even with $f \equiv 0, u(t)$ does not need to converge for $t \to \infty$. This is shown by the simple example of the rotation with angle $\frac{\pi}{2}$ as monotone operator in $\mathbb{R}^2$, that is

$$A = \begin{pmatrix} 0 & -1 \\ 1 & 0 \end{pmatrix},$$

what leads to a harmonic oscillation u.

However, under extra conditions, it is known by the work of Brézis ([8], chapter III.5) that $u(t)$ converges to a steady-state solution of a related variational inequality. One of these classical results [8, Theorem 3.9] asserts that in the case of a uniformly monotone operator (with constant α) one has exponential decay

$$|u(t) - u_\infty| \le e^{-\alpha t} \, |u(0) - u_\infty| + \int_0^t e^{\alpha(s-t)} \, |f(s) - f_\infty| \; ds, \tag{5.2}$$

where u_∞ is the unique solution of $Au_\infty \ni f_\infty$.

Let us point out that there is related work of Rodrigues [48] on the asymptotic convergence of the strong solution of some monotone parabolic variational inequalities to the solution of the corresponding stationary variational inequality. There the convergence is ensured by convergence properties of the inhomogeneous term. Also the convergence rate is estimated in terms of that one of the inhomogeneous term.

5.2 An asymptotic result for full discretization

In the following we report on the paper [23] which shows that the decay rate (5.2) to the steady-state solution can be preserved with a suitable discretization. Under asssumptions as close as possible to the continuous theory this asymptotic convergence behaviour can be proved, when the backward Euler's scheme for time discretization and a *conforming* (see below for more details) approximation of the convex set underlying the variational inequality are used. This conforming approximation, however, can only be realized in space discretization of obstacle problems, when the obstacle function is linear or piecewise linear and when piecewise linear finite elements are used for space approximation. For a more comprehensive discretization theory, however on a *fixed* time interval, that dispenses with these restrictions we refer to [10].

Let $(H, |\cdot|, \langle\cdot,\cdot\rangle)$ be a real Hilbert space. Let the right hand side f be in $L^1_{\text{loc}}(0,\infty;H)$, $K \subseteq H$ a nonvoid closed, convex set, and $\varphi : K \times K \to \mathbb{R}$ a given functional. Then under appropriate conditions on φ, in particular convexity and monotonicity conditions to be introduced later, with some given $u_0 \in K$, consider the evolution inequality: Find $u \in C([0,\infty);H)$ absolutely continuous on every compact time interval such that $u(0) = u_0$, $u(t) \in K$ for $t > 0$ and for all $v \in K$ a.e. on $(0,\infty)$ there holds

$$(E) \qquad \langle u'(t), v - u(t)\rangle + \varphi(u(t), v) \geq \langle f(t), v - u(t)\rangle .$$

It is assumed that, for any $u \in K$, $\varphi(u,.)$ is convex and continuous on K, and satisfies $\varphi(u,u) = 0$. Furthermore, the existence of constants $\mu, \langle > 0$ and $\kappa \geq 0$ is required such that for all $u, v, w \in K$ there holds

$$\text{(A1)} \qquad \mu\,|u - v|^2 \;\leq\; -\{\varphi(u,v) + \varphi(v,u)\} ,$$

$$\text{(A2)} \qquad |\varphi(u,v) - \varphi(u,w)| \;\leq\; (\kappa + \lambda|u|)|v - w| .$$

Note that in (A1) one can fix $u = \overline{u} \in K$ and obtain by a separation argument a subgradient of $\varphi(\overline{u},.)$ at $\overline{u}$ and thus with some nonnegative constant $\overline{c} = \overline{c}(\overline{u})$

$$-\varphi(v,\overline{u}) \geq \mu\,|v - \overline{u}|^2 - \overline{c}|v - \overline{u}| \quad \text{for all } v \in K .$$

Hence φ is *coercive*, that is

$$-\varphi(v,\overline{u})/|v| \to +\infty \quad \text{as } v \in K\,,\ |v|\rangle\infty .$$

Moreover (A1) implies that φ is *monotone*, that is

$$\varphi(u,v) + \varphi(v,u) \leq 0 \quad \text{for all } u, v \in K .$$

By (A2) one imposes Lipschitz continuity in the second variable. For the right hand side f one requires that $\lim_{t\to\infty} f(t) = \hat{f}$ exists in H.

Under the condition (A1) one can prove that for the unique solution u of the evolution inequality (E) above, we have $\lim_{t\to\infty} u(t) = \hat{u}$, where $\hat{u}$ solves the steady-state variational inequality: Find $\hat{u} \in H$ such that

$$\text{(S)} \qquad \hat{u} \in K \ , \ \varphi(\hat{u}, v) \geq \langle \hat{f}, v - \hat{u} \rangle \quad \text{for all } v \in K \,.$$

This is done as in the case of a strongly monotone operator A, see [8, Theorem 3.9], where using the weak compactness of the image sets $A(u)$ one defines

$$\varphi(u, v) = \max\{\langle \zeta, v - u \rangle : \zeta \in A(u)\} \,.$$

Instead of the continous evolution problem (E) one investigates its full space time discretization. For time discretization let $\{\tau_j\}_{j \in \mathbb{N}}$ be a sequence of positive step lengths. Then define $t_0 = 0$; $t_j = t_{j-1} + \tau_j$ ($j \in \mathbb{N}$). To approximate the right hand side f introduce

$$f_j = \frac{1}{\tau_j} \int_{t_{j-1}}^{t_j} f(t) dt \,.$$

For space discretization, let parameters h_j ($j \in \mathbb{N}$) converging to 0 and a sequence $\{H_{h_j}\}_{j \in \mathbb{N}}$ of closed subspaces of H be given. Assume that $H_{h_j} \subseteq H_{h_{j+1}}$ ($j \in \mathbb{N}$). Here a conforming approximation of the subset K is considered, that is, one has a sequence $\{K_j\}_{j \in \mathbb{N}} = \{K_{h_j}\}_{j \in \mathbb{N}}$ of closed convex nonempty subsets of $H_{h_j} \cap K$. For the steady-state solution $\hat{u}$ it is required that one has $u_j \in K_j$ such that $u_j \to \hat{u}$ ($j \to \infty$). In view of the hierarchical discretization scheme $\{H_{h_j}\}$ of H, one can assume that $\cap_{j \in \mathbb{N}} K_j \neq \emptyset$.

Altogether one replaces the evolution inequality (E) by the following discretization: For any $j \in \mathbb{N}$ find $q_j \in H_{h_j}$ such that $q_j \in K_j$ and for all $y \in K_j$ there holds

$$(P_j) \qquad \varphi(q_j, y) + \langle q_j', y - q_j \rangle \geq \langle f_j, y - q_j \rangle \,,$$

where one defines

$$q_j' = \frac{1}{\tau_j}(q_j - q_{j-1}) \ , \ q_0 = u_0 \,.$$

Note that because of the assumption $H_{h_j} \subseteq H_{h_{j+1}}$ ($j \in \mathbb{N}$) there is no need to project q_j into $H_{h_{j+1}}$. Thus $q_j \in K_j$ has to approximate $u(t_j) \in K$.

Then one constructs functions on $[0, \infty)$ by linear interpolation on $[t_{j-1}, t_j]$ ($j \in \mathbb{N}$) which preserves the ordering $\leq$. Thus the function $q(.)$ is given by $q(t_j) := q_j$ ($j = 0, 1, ...$); likewise the functions $u(.), f(.)$ from the sequences $\{u_j\}_{j \in \mathbb{N}}$, $\{f_j\}_{j \in \mathbb{N}}$, respectively, and $z(.) = q(.) - \hat{u}$ are defined. Instead of analysing $z(.)$ one equivalently estimates

$$\zeta(t) := |q(t) - u(t)|^2 = |z(t) - \{u(t) - \hat{u}\}|^2$$

from above using

$$\theta(t) := |u(t) - \hat{u}| + |u(t) - \hat{u}|^2 + |f(t) - \hat{f}|^2 + |u'(t)|^2 \,.$$

Note that $\lim_{t\to\infty} \theta(t) = 0$, since in particular $\lim_{t\to\infty} u'(t) = 0$ in view of $\lim_{j\to\infty} \frac{1}{\tau_j}(u_j - u_{j-1}) = 0$ because of $\tau_j \geq \overline{\tau} > 0$. Thus in [23] one obtains the following result.

Theorem 5.1 *Let φ satisfy the conditions (A1)–(A2). Let $u_j \in K_j$ ($j \in$ $\mathbb{N}$) be such that $\lim_{j\rangle\infty} u_j = \hat{u}$. Suppose that $\tau_j \geq \overline{\tau} > \frac{1}{\mu}$ ($j \in \mathbb{N}$). Then $\lim_{j\rangle\infty} |q_j - \hat{u}| = 0$ holds. More precisely with the functions ζ and θ given above, for any $\varepsilon > 0$ there exists a positive constant C only depending on $\varepsilon, \mu, \kappa, \langle, \overline{\tau}, |\hat{u}|, |\hat{f}|$ such that*

$$\zeta(t) \leq \zeta(0)\exp(-(\mu-\varepsilon)t) + C\int_0^t \theta(t)\exp[-(\mu-\varepsilon)(t-\tau)]d\tau\,.$$

Remarks. The proved asymptotic behaviour of the full discretization process is the same as in the continuous case, up to an arbitrary $\varepsilon > 0$ and a dependent constant $C > 0$. This asymptotic result applies to more concrete parabolic variational inequalities (see [23] for more details), where the given obstacle function is linear or piecewise linear and piecewise linear finite elements are employed for space discretization, what leads to a conforming discretization scheme. Thus the question arises, whether such an asymptotic convergence behaviour can also be obtained with a nonconforming discretization scheme, where the given convex subset K is approximated by sets $\{K_{h_j}\}_{j\in\mathbb{N}}$ that are no longer assumed to be contained in K.

5.3 Some convergence results for continuous-time subgradient procedures for convex optimization

When the general monotone operator A is replaced by a specific subgradient operator related to (classes of) convex optimization problems, more refined aymptotic results are available. In [20] Flam considers the standard convex program

$$(P) \qquad \text{minimize } f_0(x) \text{ subject to } F_i(x) \leq 0, i = 1, ..., m\,,$$

where $f_0, F_1, \ldots, F_m$ are finite-valued, convex, possibly nonsmooth functions defined on some real Hilbert space H. For the solution of (P), as a new primal method it is advocated to trace an absolutely continuous solution path of an ordinary differential inclusion of the following type

$$\dot{x}(t) \in -\sum_{i=0}^{m} \lambda_i(t, x(t))\,\partial f_i(x(t))\,,$$

where

$$f_i(x) := \max(0, F_i(x)) \text{ for all } i = 1, ..., m,$$

emanating from an arbitrary initial point $x(0) \in H$. Whenever this inclusion of subgradient type offers some freedom of choice, a steepest descent direction is selected.

Here implementing a closed-loop regime, the weights $\lambda_0(t, x(t)), \ldots, \lambda_m(t, x(t))$ are adapted in the time process to pursue a compromise between minimizing f_0 and obtaining feasibility $F_i \le 0$.

In the case of sharp constraints, i.e. if

$$\exists \alpha > 0 : \max_i F_i(x) \ge \alpha \ \operatorname{dist}(x, C), \forall x \notin C;$$

where $\operatorname{dist}(x, C)$ denotes the distance to the feasible domain $C = \{x \in H : F_i(x) \le 0\}$, it is shown that the solution path converges to an optimal solution in finite time.

It is noted that the constraints are sharp provided the Slater condition, $\inf_{x \in C} \max_i F_i(x) < 0$, holds and C is bounded. This can be seen as follows. Fix some $\hat{x} \in C$ such that $F(\hat{x}) := \max_i F_i(\hat{x}) < 0$. Then for any $x \notin C$ there exists $\tilde{x} \in [\hat{x}, x]$ such that $F(\tilde{x}) = 0$. By convexity of F,

$$F(\tilde{x}) - F(\hat{x}) \le \frac{\|\tilde{x} - \hat{x}\|}{\|x - \tilde{x}\| + \|\tilde{x} - \hat{x}\|} [F(x) - F(\hat{x})],$$

hence

$$F(x) \ge -F(\hat{x}) \frac{\|x - \tilde{x}\|}{\|\tilde{x} - \hat{x}\|} \ge -F(\hat{x}) \frac{\operatorname{dist}(x, C)}{\operatorname{diam} C},$$

and α can be taken as $\dfrac{-F(\hat{x})}{\operatorname{diam} C}$.

In [21] Flam and Seeger consider the more general cone-constrained convex program

$$(P) \begin{cases} \text{minimize } f_0(x) \\ \text{subject to } x \in X \text{ and } -F(x) \in C. \end{cases}$$

Here X is a convex closed subset in a Hilbert space H, C a closed convex cone strictly contained in another Hilbert space G, and $f_0 : H \to \mathbb{R}, F : H \to G$ are given functions under continuity and convexity assumptions specified later. For the solution of (P) a continuous-time, primal-dual subgradient procedure employing penalization, Lagrangian relaxation, and projection is proposed.

First to treat the constraint $-F(x) \in C$ an exterior penalty function $f : H \to Y$ is introduced that satisfies

$$f(H) \subseteq C$$
$$f(x) = 0 \Leftrightarrow -F(x) \in C.$$

Examples of such functions are constructed in [21] for concrete classes of (P) using isotone projection.

Then to scalarize the vector penalty function f, the Lagrangian $L : H \times G \to \mathbf{R}$ given by

$$L(x,y) = f_0(x) + \langle y, f(x) \rangle$$

is brought into play. This leads to the dual problem

$$\begin{array}{ll} (D) & \text{maximize } \inf\limits_{x \in X} L(x,y) \\ & \text{subject to } y \in Y, \end{array}$$

where $Y = C^+ = -C^0$ is the cone of "nonnegative" multipliers. Assuming henceforth that the cone Y spans G, that is

$$G = Y - Y = C^+ - C^-,$$

the problems (P) and

$$(\text{minsup}) \begin{cases} \text{minimize } \sup\limits_{y \in Y} L(x,y) \\ \text{subject to } x \in X \end{cases}$$

have the same optimal solutions and values according to [21, Proposition 1.2]. In what follows, f_0 and $\langle y, f(\cdot) \rangle$ are assumed to be convex and continuous for all $y \in Y$. Then the restriction of the Lagrangian L to $X \times Y$ is convex-concave and separately continuous in each variable. Consequently, the setvalued mapping

$$(x,y) \in X \times Y \Rightarrow \partial \mathcal{L}(x,y) = \Big(\partial_x \mathcal{L}(x,y), -\partial_y \mathcal{L}(x,y) \Big)$$

is maximal monotone, where

$$\mathcal{L}(x,y) = \begin{cases} L(x,y) & \text{if } x \in X \text{ and } y \in Y \\ +\infty & \text{if } x \notin X \text{ and } y \in Y \\ -\infty & \text{if } y \notin Y \end{cases}$$

is a closed proper saddle function.

Now to solve (P) it is proposed to follow a trajectory $(x,y)(t), t > 0$ of the *differential inclusion*

$$(DI) \begin{cases} \dot{x}(t) \in -\partial_x L\Big(x(t), y(t)\Big) - N_X\Big(x(t)\Big) \\ \dot{y}(t) \in \partial_y L\Big(x(t), y(t)\Big) \equiv f\Big(x(t)\Big), \end{cases}$$

where again N_X denotes the normal cone to X.

In addition, assume that the cone C is *acute*, that is $C \subseteq Y \equiv C^+$ in the Hilbert space $G \cong G'$ holds. Then one can show viability, existence and uniqueness of trajectories: From each point in $X \times Y$, there emanates a unique trajectory of (DI) that satisfies $(x, y)(t) \in X \times Y$ for all $t \geq 0$ [21, Corollary 2.2, Theorem 2.3].

This is a reasonable solution method, since with Y spanning G it can be easiliy shown [21, Proposition 3.1] that the stationary points of (DI) coincide with the saddle points of L with respect to $X \times Y$, which clearly give solutions of (P).

The main justification of the procedure comes from the following convergence result [21, Theorem 3.5].

Theorem 5.2 *Assume that C is acute and that Y spans the Hilbert space G. Assume that F maps bounded sets into bounded sets. Moreover assume that the admissible set*

$$A := \{x \in X : -F(x) \in C\}$$

is weakly closed and the set S of saddle points of L with respect to $X \times Y$ is nonempty. Then, any trajectory $z = (x, y)$ of (DI) emanating from an arbitrary initial point in $X \times Y$ is bounded and converges monotonically in norm to S, that is, the norm distance $d\Big(z(t), S\Big) \downarrow 0$ as $t \to \infty$. Also $y(t)$ converges weakly and Y-monotonically upwards to some $\overline{y} \in Y$, that is, $\overline{y} - y(t) \in Y$ for all $t \geq 0$.

Acknowledgment. The author thanks Bruce Calvert for constructive criticism on a previous version of the paper.

References

[1] R.A. Adams, *Sobolev Spaces*, Academic Press, New York, 1975.

[2] H. Attouch and A. Damlamian, *Problèmes d' évolution dans les Hilberts et applications*, J. Math. pures et appl., IX. Ser., **54**, 1974, 53-74.

[3] J.-P. Aubin, *Viability Theory*, Birkhäuser, Basel, Boston, 1991.

[4] J.-P. Aubin and A. Cellina, *Differential Inclusions. Set-valued Maps and Viability Theory*, Grundlehren der Mathematischen Wissenschaften, **264**, Springer-Verlag, Berlin etc., 1984.

[5] H. Benabdellah, C. Castaing and A. Salvadori, *Compactness and discretization methods for differential inclusions and evolution problems*, Atti Semin. Mat. Fis. Univ. Modena, **45**, 1, 1997, 9-51.

[6] H. Benabdellah, C. Castaing, A. Salvadori and A. Syam, *Nonconvex sweeping processes*, J. Appl. Anal., **2**, 2, 1996, 217-240.

[7] J.M. Borwein and A.S. Lewis, *Partially finite convex programming, part I: quasi relative interiors and duality theory*, Math. Programming, **57**, 1992, 15-48.

[8] H. Brézis, *Operateurs Maximaux Monotones*, North-Holland, Amsterdam, 1973.

[9] M. Brokate, *Elastoplastic constitutive laws of nonlinear kinematic hardening type*, Functional analysis with current applications in science, technology and industry, Brokate, M. and Siddiqi, A. H., Longman, Pitman Research Notes in Mathematics Series, **377**, 1998, 238-272.

[10] C. Carstensen and J. Gwinner, *A theory of discretization for nonlinear evolution inequalities applied to parabolic Signorini problems*, Ann. Mat. Pura Appl., **177**, 1999, 363-394.

[11] C. Castaing and M.D.P. Monteiro Marques, *Topological properties of solution sets for sweeping processes with delay*, Port. Math., **54**, 4, 1997, 485-507.

[12] R. Cominetti and O. Alemany, *Steepest descent evolution equations: asymptotic behavior of solutions and rate of convergence*, Trans. Am. Math. Soc., **351**, 12, 1999, 4847-4860.

[13] B. Cornet, *Existence of slow solutions for a class of differential inclusions*, J. Math. Anal. Appl., **96**, 1983, 130-147.

[14] P. Daniele and A. Maugeri, *On dynamical equilibrium problems and variational inequalities*, Equilibrium Problems: Nonsmooth Optimization and Variational Inequality Models, Kluwer Academic Publishers, F. Giannessi - A. Maugeri - P. Pardalos Eds., 2001, 59-69.

[15] P. Daniele, A. Maugeri and W. Oettli, *Time-dependent variational inequalities*, J. Optim. Theory Appl., **103**, 3, 1999, 543-555.

[16] K. Deimling, *Multivalued Differential Equations*, Walter de Gruyter, Berlin. New York, 1992.

[17] M. De Luca, *Existence of solutions for a time-dependent quasi-variational inequality*, Proceedings of the workshop on equilibrium problems with side constraints. Lagrangean theory and duality II, F. Giannessi et al. eds., Suppl. Rend. Circ. Mat. Palermo, II Ser. 48, 1997, 101-106.

[18] P. Dupuis, *Large deviations analysis of reflected diffusions and constrained stochastic approximation algorithms in convex sets*, Stochastics, **21**, 1987, 63-96.

[19] P. Dupuis and A. Nagurney, *Dynamical systems and variational inequalities*, Ann. Oper. Res., **44**, 1993, 9-42.

[20] S.D. Flam, *Solving convex programs by means of ordinary differential equations*, Math. Oper. Res., **17**, 1992, 290-302.

[21] S.D. Flam and A. Seeger, *Solving cone-constrained convex programs by differential inclusions*, Math. Program., **65A**, 1, 1994, 107-121.

[22] L. Gorniewicz, *Topological Fixed Point Theory of Multivalued Mappings*, Kluwer, Dordrecht, 1999.

[23] J. Gwinner, *A remark on the asymptotic behaviour of parabolic variational inequalities and their finite element approximation by the Courant element*, Finite Element Methods – Fifty Years of the Courant Element, M. Křížek and P. Neittaanmäki eds., Marcel Dekker, 1994, 235-242.

[24] J. Gwinner, *Stability of monotone variational inequalities with various applications*, Variational Inequalities and Network Equilibrium Problems, F. Giannessi and A. Maugeri eds., Plenum Publishing, 1995, 123-142.

[25] A. Haraux, *How to differentiate the projection on a convex set in Hilbert space. Some applications to variational inequalities*, J. Math. Soc. Japan, **29**, 1977, 615-631.

[26] A.D. Ioffe and V.M. Tihomirov, *Theory of Extremal Problems*, Studies in Mathematics and its Applications, **6**, North-Holland, Amsterdam, 1979.

[27] P. Krejci, *Hysteresis, Convexity and Dissipation in Hyperbolic Equations*, Gakkōtosho, Tokyo, 1996.

[28] P. Krejci, *Evolution variational inequalities and multidimensional hysteresis operators*, Nonlinear differential equations, Pavel Drabek et al. eds., Chapman & Hall/CRC, Res. Notes Math., **404**, 1999, 47-110.

[29] M. Kunze and M.D.P. Monteiro Marques, *An introduction to Moreau's sweeping process*, Impacts in Mechanical Systems. Analysis and Modelling, Brogliato, B. ed., Springer, **551**, Lect. Notes Physics, 2000, to appear.

[30] M. Kunze and M.D.P.Monteiro Marques, *Yosida-Moreau regularization of sweeping processes with unbounded variation*, J. Differ. Equations, **130**, 2, 1996, 292-306.

[31] M. Kunze and M.D.P. Monteiro Marques, *Existence of solutions for degenerate sweeping processes*, J. Convex Anal., **4**, 1, 1997, 165-176.

[32] M. Kunze and M.D.P. Monteiro Marques, *BV solutions to evolution problems with time-dependent domains*, Set-Valued Anal., **5**, 1, 1997, 57-72.

[33] M. Kunze and M.D.P. Monteiro Marques, *On parabolic quasi-variational inequalities and state-dependent sweeping processes*, Topol. Methods Nonlinear Anal., **12**, 1, 1998, 179-191.

[34] M. Kunze and M.D.P. Monteiro Marques, *On the discretization of degenerate sweeping processes*, Port. Math., **55**, 2, 1998, 219-232.

[35] A. Maugeri, *Monotone and nonmonotone variational inequalities*, Proceedings of the workshop on equilibrium problems with side constraints. Lagrangean theory and duality II, F. Giannessi et al. eds., Suppl. Rend. Circ. Mat. Palermo, II. Ser. 48, 1997, 179-184.

[36] M.D.P. Monteiro Marques, *Differential Inclusions in Nonsmooth Mechanical Problems: Shocks and Dry Friction*, Birkhäuser, Basel, Boston, 1993.

[37] J.J. Moreau, *Evolution problem associated with a moving convex set in a Hilbert space*, J. Differ. Equations, **26**, 1977, 347-374.

[38] J.J. Moreau, *Numerical aspects of the sweeping process*, Comput. Methods Appl. Mech. Engrg., **177**, 3-4, 1999, 329-349.

[39] A. Nagurney and D. Zhang, *Projected Dynamical Systems and Variational Inequalities with Applications*, Kluwer, Boston, Dordrecht, 1996.

[40] J. Naumann, *On a class of variational inequalities of Volterra type*, Rend. Mat., VI. Ser., **11**, 2, 1978, 161-186.

[41] J. Naumann, *On a class of first-order evolution inequalities arising in heat conduction with memory*, SIAM J. Math. Anal., **10**, 6, 1979, 1144-1160.

[42] W. Oettli and D. Schläger, *Generalized vectorial equilibria and generalized monotonicity*, Functional analysis with current applications in science, technology and industry, Brokate, M. and Siddiqi, A.

H. eds., Longman, Pitman Research Notes in Mathematics Series, Harlow, Essex, UK, **377**, 1998, 145-154.

[43] R.R. Phelps, *Metric projection and the gradient projection method in Banach spaces*, SIAM J. Control Optim., **23**, 1985, 973-977.

[44] L. Prigozhin, *Variational model of sandpiles growth*, European J. Appl. Math., **7**, 1996, 225-235.

[45] L. Prigozhin, *On the Bean critical-state model in superconductivity*, European J. Appl. Math., **7**, 1996, 237-247.

[46] L. Prigozhin, *The Bean model in superconductivity: Variational formulation and numerical solution*, J. Comput. Phys., **129**, 1, 1996, 190-200.

[47] S. Reich and A.J. Zaslavski, *Asymptotic behavior of dynamical systems with a convex Lyapunov function*, J. Nonlinear Convex Anal., **1**, 1, 2000, 107-113.

[48] J.-F. Rodrigues, *Some remarks on the asymptotic behaviour of strong solutions to monotone parabolic variational inequalities*, Rend. Math. Appl., VII. Ser., **4**, 1984, 458-470.

[49] J.F. Rodrigues, *Obstacle Problems in Mathematical Physics*, North-Holland, Amsterdam, 1987.

[50] J. Steinbach, *Evolutionary variational inequalities with a Volterra term*, Numer. Funct. Anal. Optimization, **12**, 3 &4, 1991, 395-411.

[51] J. Steinbach, *On a variational inequality containing a memory term with an application in electro-chemical machining* , J. Convex Anal., **5**, 1, 1998, 63-80.

[52] N.X. Tan, *Quasi-variational inequality in topological linear locally convex Hausdorff spaces*, Math. Nachr., **122**, 1985, 231-245.

[53] A.A. Vladimirov, *Nonstationary dissipative evolution equations in a Hilbert space*, Nonlinear Anal. T. M. A., **17**, 1991, 499-518.

[54] A.A. Vladimirov, V.S. Kozyakin, N.A. Kuznetsov, and A. Mandelbaum, *An investigation of the dynamic complementarity problem by methods of the theory of desynchronized systems. (English. Russian original)*, Russ. Acad. Sci., Dokl., Math., **47, No. 2**, 1993, 169-173.

[55] E.H. Zarantonello, *Projections on convex sets in Hilbert space and spectral theory*, Contributions to Nonlinear Functional Analysis, Zarantonello, E. H. ed., Academic Press, 1971, 237-424.

[56] D. Zhang and A. Nagurney, *On the stability of projected dynamical systems*, J. Optim. Theory Appl., **85**, 1995, 97-124.

ON THE CONTRACTIBILITY OF THE EFFICIENT AND WEAKLY EFFICIENT SETS IN R^2

Nguyen Quang Huy
Department of Mathematics, Hanoi Pedagogical Institute II,
Xuan Hoa, Me Linh, Vinh Phuc, Vietnam

Ta Duy Phuong
Hanoi Institute of Mathematics,
P.O. Box 631, 10000 Bo Ho, Hanoi, Vietnam

Nguyen Dong Yen
Hanoi Institute of Mathematics,
P.O. Box 631, 10000 Bo Ho, Hanoi, Vietnam

Abstract In this paper we give a detailed proof for a theorem due to D. T. Luc saying that the efficient and weakly efficient sets of compact convex subsets of R^2 are homeomorphic to certain simplexes. An example is designed to analyze the proof scheme of Luc for this theorem.

Keywords: Convex subset of R^2, efficient set, weakly efficient set, contractibility, simplex.

2000 Mathematics Subject Classification. 90C29.

1 Introduction

Contractibility of the efficient set of a convex set with respect to a pointed closed convex cone has been studied by several authors. Peleg [4] proved that if $A \subset R^n$ is a compact convex set then the efficient set $E(A|R^n_+)$ of A w.r.t. the cone R^n_+ (the nonnegative orthant of the n-dimensional Euclidean space R^n) is contractible. Afterwards, Luc [3] has extended this result to the case where A is a convex subset of a uniformly convex space which has compact sections w.r.t. the ordering cone. Meanwhile, the question on the contractibility of the weakly efficient sets of convex sets is at present far from being solved. If $A \subset R^2$ is a nonempty compact convex subset then the efficient set $E(A|C)$ of A w.r.t. a convex cone $C \subset R^2$ is homeomophic to a simplex, provided the closure $\overline{C}$ of C is a pointed cone. This strong theorem was obtained by Luc [1] (see also [2, Theorem 3.8, p.144]). From the result it follows that the set $E(A|C)$ is contractible. Unfortunately, the proof in [1] and [2] falls short of providing nessesary explanations of several key arguments. This does not allow the readers to understand the proof and, consequently, the interesting theorem in full extent. The aim of this paper is to give a detailed proof for the above-mentioned theorem of Luc. We also design an example to analyze the proof scheme of of Luc for the theorem. Section 2 presents some preliminaries. Section 3 provides a detailed exposition of Theorem 3.8 from [2, p.144]. In Section 4 an example is considered.

2 Preliminaries

We now explain some terminology that will be used thoughout this paper. Let $A \subset R^n$ be a nonempty convex subset, $C \subset R^n$ a nonempty convex cone. We write $y \leq_C x$ if $x - y \in C$.

Definition 2.1 *An efficient point of A w.r.t. C is an element $x \in A$ such that there exists no $y \in A \setminus \{x\}$ satisfying $y \leq_C x$. The set of all the efficient points of A w.r.t. C is denoted by $E(A|C)$. If the interior $intC$ of C is nonempty, we say that $x \in A$ is a weakly efficient point of A w.r.t. C if there exists no $y \in A \setminus \{x\}$ satisfying $y \leq_{intC \cup \{0\}} x$. The set of all the weakly efficient points of A w.r.t. C is denoted by $E^w(A|C)$.*

Definition 2.2 *A set $M \subset R^n$ is said to be contractible if there exists a continuous mapping $H : M \times [0,1] \longrightarrow M$ and a point $x_0 \in M$ such that $H(x,0) = x$ and $H(x,1) = x_0$ for every $x \in M$.*

It is easily proved that any convex set $M \subset R^n$ is contractible. As a consequence, any subset of R^n which is homeomorphic to a convex set, is contractible.

Definition 2.3 *A set $\Delta \subset C \setminus \{0\}$ is called a base of the cone C if for every $x \in C$ there exists a unique number $t = t(x) > 0$ such that $tx \in \Delta$.*

Definition 2.4 *If $C \cap (-C) = \{0\}$ then C is said to be a pointed cone.*

If C is a pointed closed convex cone then it has a compact convex base. The converse is also true. (See [2, Remark 1.6, p.4]).

In the next section we shall need the following result.

Lemma 2.1 (cf. [2, Corollary 3.11, p.50]) *For any nonempty compact subset $A \subset R^n$ and for any nonempty convex cone $C \subset R^n$, the set $E(A|C)$ is nonempty.*

3 Topological structure of the efficient sets of compact convex sets

The strong result stated in Theorem 3.1 below is due to D. T. Luc (see [1], [2]). It would be desirable to have a detailed proof for the result. Our purpose is to present such a proof in this section.

Theorem 3.1 ([1], [2, Theorem 3.8, p.144]) *Let $A \subset R^2$ be a nonempty compact convex subset, $C \subset R^2$ a nonempty convex cone whose closure $\overline{C}$ is a pointed cone. Then the set $E(A|C)$ is homeomorphic to an 0-dimensional or an 1-dimensional simplex.*

We shall divide the proof into several steps. As A is a compact set, there exists some $a \in R^2$ such that $a + A \subset intR^2_+$. Since $E(A|C) = E(a + A|C) - a$, for proving the theorem it suffices to show that $E(a + A|C)$ is homeomorphic to an 0-dimensional or an 1-dimensional simplex. Without restriction of generality we can assume that $A \subset intR^2_+$.

Lemma 3.1 *For $C = R^2_+$, the set $E^w(A|C)$ is homeomorphic to an 0-dimensional or an 1-dimensional simplex.*

Proof. First we prove that $E^w(A|C)$ is closed. Taking any sequence $\{a_k\}$, $a_k \in E^w(A|C)$ for every k, and

$$\lim_{k \to \infty} a_k = a.$$

By the closedness of A, $a \in A$. If $a \notin E^w(A|C)$ then there exists $b \in A$ such that $a \in b + intR^2_+$. As $b + intR^2_+$ is an open set, for all k sufficiently large we have $a_k \in b + intR^2_+$. This implies that $a_k \notin E^w(A|C)$ provided k is large enough, a contradiction.

Our next claim is that $K := cone(E^w(A|C))$, the cone generated by the weakly efficient set, is pointed, closed and convex.

Since $E^w(A|C) \subset A \subset intR^2_+$, K is a subcone of R^2_+. In particular, K is a pointed cone.

Since $E^w(A|C)$ is a closed subset of the compact set A, $E^w(A|C)$ is a compact set. Since $0 \notin A$ then $0 \notin E^w(A|C)$. From this we conclude that K is a closed cone.

It remains to prove that K is a convex cone. Let a_1 and b_1 be two different points from K. Let $c_1 = \theta_1 a_1 + (1-\theta_1)b_1$, where $\theta_1 \in (0,1)$. The proof of the convexity of K is completed by showing that $c_1 \in K$.

Since $a_1,\ b_1 \in K = cone(E^w(A|C))$, there exist $a,\ b \in E^w(A|C)$ and $\lambda,\ \mu > 0$ such that $a_1 = \lambda a,\ b_1 = \mu b$. Therefore

$$\begin{aligned} c_1 &= \theta_1\lambda a + (1-\theta_1)\mu b \\ &= (\theta_1\lambda + (1-\theta_1)\mu)\left(\frac{\theta_1\lambda}{\theta_1\lambda+(1-\theta_1)\mu}a + \frac{(1-\theta_1)\mu}{\theta_1\lambda+(1-\theta_1)\mu}b\right). \end{aligned}$$

Define

$$\theta = \frac{\theta_1\lambda}{\theta_1\lambda+(1-\theta_1)\mu}, \qquad c = \theta a + (1-\theta)b.$$

We have $\theta \in (0,1),\ c_1 = (\theta_1\lambda + (1-\theta_1)\mu)c$. If we prove that $c \in K$ then $c_1 \in K$.

Consider the following optimization problem

$$(P_c) \qquad \begin{cases} \text{minimize } t \\ \text{s.t.} \quad t \geq 0,\ tc \in A + R^2_+. \end{cases}$$

Note that the constraint set of (P_c), denoted by X_c, is a closed subset of R. Moreover, X_c is bounded below and $1 \in X_c$. On account of these remarks, there exists a unique optimal solution of (P_c) which is denoted by t_0. Since

$$A + R^2_+ \subset intR^2_+ + R^2_+ \subset intR^2_+,$$

$t_0 > 0$. The task is now to show that

$$t_0 c \in A. \tag{1}$$

We have $0 < t_0 \leq 1$. If $t_0 = 1$ then $t_0 c = c \in A$, and (1) is valid. We are reduced to considering the case $t_0 \in (0,1)$. Let $a = (a^1, a^2),\ b = (b^1, b^2)$. Since $a,\ b \in E^w(A|R^2_+)$, there is no loss of generality in assuming $a^1 > b^1,\ a^2 \leq b^2$. From $c = \theta a + (1-\theta)b,\ \theta \in (0,1)$, we deduce that

$$c^1 < a^1,\ c^2 \leq b^2. \tag{2}$$

It follows from $t_0 c \in A + R^2_+$ that $(t_0 c - R^2_+) \cap A \neq \emptyset$. Fix any $d \in (t_0 c - R^2_+) \cap A$. We have $d \in E^w(A|R^2_+)$. Indeed, if $d \notin E^w(A|R^2_+)$ then there is $d' \in A$ satisfying

$$d' \in d - intR^2_+ \subset t_0 c - R^2_+ - intR^2_+ \subset t_0 c - intR^2_+.$$

Hence $t_0 c \in d' + intR^2_+ \subset int(A + R^2_+)$. From this it follows that t_0 cannot be a solution of (P_c), contrary to the choice of t_0. We thus get $d \in E^w(A|R^2_+)$. Suppose $d = (d^1, d^2)$, $c = (c^1, c^2)$. Since $d \in t_0 c - R^2_+$ and $t_0 \in (0,1)$, taking account of (2) we have

$$d^1 < c^1 < a^1,\ d^2 < c^2 \leq b^2.$$

As a, $b \in E^w(A|R^2_+)$, from $d^1 < a^1$ and $d^2 < b^2$ it follows that $a^2 \leq d^2$ and $b^1 \leq d^1$. Combining these we obtain

$$b^1 \leq d^1 < c^1 < a^1 \tag{3}$$

and

$$a^2 \leq d^2 < c^2 \leq b^2. \tag{4}$$

If we can prove that

$$t_0 c \in conv\{a,\ b,\ d\}, \tag{5}$$

where $convM$ denotes the convex closure of a set M, then it is clear that (1) holds true.

Consider the affine function $\varphi_1(x)$ defined by the following formula

$$\varphi_1(x) = (a^1 - b^1)x^2 + (b^2 - a^2)x^1 - (b^2 - a^2)a^1 - (a^1 - b^1)a^2,$$

where $x = (x^1, x^2) \in R^2$. We proceed to show that

$$\varphi_1(a) = 0, \quad \varphi_1(b) = 0, \quad \varphi_1(d) < 0, \quad \varphi_1(t_0 c) < 0. \tag{6}$$

The first two equalities in (6) are obvious. From (3) and (4) it follows that

$$a^1 - d^1 > a^1 - c^1 > 0$$

and

$$c^2 - a^2 > d^2 - a^2 \geq 0.$$

Consequently,

$$(a^1 - d^1)(c^2 - a^2) > (a^1 - c^1)(d^2 - a^2).$$

Therefore

$$\frac{c^2 - a^2}{a^1 - c^1} > \frac{d^2 - a^2}{a^1 - d^1}. \tag{7}$$

From the equality $c = \theta a + (1 - \theta)b$ we have

$$\begin{aligned} c^1 - a^1 &= (1 - \theta)(b^1 - a^1) \\ c^2 - a^2 &= (1 - \theta)(b^2 - a^2), \end{aligned}$$

which implies that
$$\frac{c^2 - a^2}{a^1 - c^1} = \frac{b^2 - a^2}{a^1 - b^1}.$$
Combining the last equality with (7) yields
$$\frac{d^2 - a^2}{a^1 - d^1} < \frac{b^2 - a^2}{a^1 - b^1}.$$

This gives $(a^1 - b^1)(d^2 - a^2) < (b^2 - a^2)(a^1 - d^1)$, which establishes the first inequality in (6). Using the equality $c = \theta a + (1 - \theta)b$ and the equalities in (6) it is easy to verify that $\varphi_1(c) = 0$. By (3) and (4), $a^1 - b^1 > 0$ and $b^2 - a^2 > 0$. Then $\varphi_1(y) < \varphi_1(x)$ for all $x, y \in R^2$ satisfying $y \neq x$, $y \leq_{R^2_+} x$. As $t_0 \in (0,1)$ and $c \in intR^2_+$ we have $t_0 c \neq c$, $t_0 c \leq_{R^2_+} c$. Hence
$$\varphi_1(t_0 c) < \varphi_1(c) = 0,$$

which completes the proof of (6).

Consider the affine function $\varphi_2(x)$ defined by setting
$$\varphi_2(x) = (d^1 - b^1)x^2 + (b^2 - d^2)x^1 - (b^2 - d^2)d^1 - (d^1 - b^1)d^2,$$

where $x = (x^1, x^2) \in R^2$. We want to show that
$$\varphi_2(b) = 0, \quad \varphi_2(d) = 0, \quad \varphi_2(a) > 0, \quad \varphi_2(t_0 c) \geq 0. \tag{8}$$

The first two equalities in (8) are obvious. From the equality $c = \theta a + (1 - \theta)b$ we have $(\theta - 1)(b - c) = \theta(a - c)$, which implies that
$$\frac{b^2 - c^2}{b^1 - c^1} = \frac{a^2 - c^2}{a^1 - c^1}.$$

From this we deduce that
$$(b^2 - a^2)c^1 + (a^1 - b^1)c^2 - a^1 b^2 + a^2 b^1 = 0. \tag{9}$$

By (3) and (4), $b^2 - a^2 > 0$, $a^1 - b^1 > 0$, $c^1 > d^1$, and $c^2 > d^2$. From (9) it follows that
$$(b^2 - a^2)d^1 + (a^1 - b^1)d^2 - a^1 b^2 + a^2 b^1 < 0. \tag{10}$$

Therefore
$$(d^1 - b^1)a^2 + (b^2 - d^2)a^1 - (b^2 - d^2)d^1 - (d^1 - b^1)d^2 > 0$$

or, equivalently, $\varphi_2(a) > 0$. The first inequality in (8) has been proved. By (3) and (4), $d^1 - b^1 \geq 0$ and $b^2 - d^2 > 0$. Then $\varphi_2(y) \leq \varphi_2(x)$ for all

$x, y \in R^2$ satisfying $y \leq_{R^2_+} x$. By the choice of d, $d \leq_{R^2_+} t_0c$. From what has already been said, we obtain

$$\varphi_2(t_0c) \geq \varphi_2(d) = 0,$$

which completes the proof of (8).

Finally, consider the affine function $\varphi_3(x)$ defined by setting

$$\varphi_3(x) = (a^1 - d^1)x^2 + (d^2 - a^2)x^1 - (d^2 - a^2)d^1 - (a^1 - d^1)d^2,$$

where $x = (x^1, x^2) \in R^2$. We want to show that

$$\varphi_3(a) = 0, \quad \varphi_3(d) = 0, \quad \varphi_3(b) > 0, \quad \varphi_3(t_0c) \geq 0. \tag{11}$$

The first two equalities in (11) are obvious. From (10) it follows that

$$(a^1 - d^1)b^2 + (d^2 - a^2)b^1 - (d^2 - a^2)d^1 - (a^1 - d^1)d^2 > 0$$

or, equivalently, $\varphi_3(b) > 0$. The first inequality in (11) has been proved. By (3) and (4), $a^1 - d^1 > 0$ and $d^2 - a^2 \geq 0$. Then $\varphi_3(y) \leq \varphi_3(x)$ for all $x, y \in R^2$ satisfying $y \leq_{R^2_+} x$. By the choice of d, $d \leq_{R^2_+} t_0c$. From what has already been said, we obtain

$$\varphi_3(t_0c) \geq \varphi_3(d) = 0,$$

which completes the proof of (11).

Let us define three half-planes P_1^-, P_2^+, P_3^+ by setting

$$P_1^- = \{x \in R^2 : \varphi_1(x) \leq 0\}, \quad P_2^+ = \{x \in R^2 : \varphi_2(x) \geq 0\},$$

$$P_3^+ = \{x \in R^2 : \varphi_3(x) \geq 0\}.$$

From (6), (8) and (11) it follows that the points a, b, d and t_0c belong to the intersection of the half-planes P_1^-, P_2^+, P_3^+. It is also clear from (6), (8) and (11) that this intersection is exactly the set $conv\{a, b, d\}$ (the triangle with the vertexes a, b, d). The inclusion (5) has been proved. In particular, $t_0c \in A$.

We have

$$t_0c \in E^w(A|C). \tag{12}$$

Indeed, if $t_0c \notin E^w(A|C)$ then there exists $z \in A$ such that

$$t_0c \in z + intR^2_+ \subset int(A + R^2_+).$$

Hence there is $t \in (0, t_0)$ such that $tc \in int(A + R^2_+)$. This shows that t_0 cannot be a solution of (P_c), which contradicts the choice of t_0.

On account of (12), we have

$$c \in \frac{1}{t_0} E^w(A|C) \subset cone(E^w(A|C)) = K.$$

This implies that $c_1 \in K$. The convexity of K has been proved.

We are now in a position to show that $E^w(A|C)$ is homeomorphic to an 0-dimensional or an 1-dimensional simplex. We shall use a construction called *the radial projection mapping.* For each $x \in intR^2_+$, we set $L(x) = \{tx : t \geq 0\}$. Obviously, $L(x) \cap E^w(A|C) = \{x\}$ for every $x \in E^w(A|C)$. Let

$$\Delta := \{x = (x^1, x^2) \in R^2_+ : x^1 + x^2 = 1\} \cap cone(E^w(A|C)).$$

As Δ is a closed convex subset of an 1-dimensional simplex, Δ is an 0-dimesional or an 1-dimensional simplex. It is clear that Δ is a compact base of $cone(E^w(A|C))$. Since $A \subset intR^2_+$, for every $x \in E^w(A|C)$ there exists a unique positive number $t(x) > 0$ such that $t(x)x \in \Delta$. In fact, we have $L(x) \cap \Delta = \{t(x)x\}$. Define a mapping $f : E^w(A|C) \to \Delta$ by setting $f(x) = t(x)x$. From what has already been said, it follows that f is an one-to-one mapping. We proceed to show that f is a homeomorphism. Given any closed subset $D \subset \Delta$ we observe that $f^{-1}(D) = cone(D) \cap E^w(A|C)$ is a closed set. So f^{-1} is continuous mapping. Now, let M be an arbitrary closed subset from $E^w(A|C)$. Since $0 \notin M$ and M is a compact set (recall that $E^w(A|C)$ is a compact set), $cone(M)$ is a closed set. Therefore $f(M) = cone(M) \cap \Delta$ is a closed set. Summarizing all the above, we conclude that f^{-1} is a homeomorphism. Since Δ is an 0-dimensional or an 1-dimensional simplex, the proof is complete. $\triangle$

Lemma 3.2 *For $C = R^2_+$, the set $E(A|C)$ is homeomorphic to an 0-dimensional or an 1-dimensional simplex.*

Proof. We first claim that $E(A|C)$ is closed. Indeed, suppose the claim were false. Then we could find a sequence of vectors $\{a_k\}$ from $E(A|C)$ such that $\lim_{k\to\infty} a_k = a$, but $a \notin E(A|C)$. By the closedness of A, $a \in A$. As $a \notin E(A|C)$, there exists $b \in A$ such that $b \in a - R^2_+ \setminus \{0\}$. Let $a = (a^1, a^2)$, $b = (b^1, b^2)$. There is no loss in assumming that $a^1 > b^1$ and $a^2 = b^2$. Indeed, if $a^2 > b^2$ then $a^1_k > b^1$ and $a^2_k > b^2$ for k sufficiently large. This contradicts the fact that $a_k \in E(A|C)$.

From $\lim_{k\to\infty} a_k = a$ and $a_k \in E(A|C)$ it follows that, for k sufficiently large,

$$b^1 < a^1 < a^1_k, \quad a^2_k < a^2 = b^2. \tag{13}$$

or

$$b^1 < a_k^1 \leq a^1, \quad a_k^2 < a^2 = b^2, \tag{14}$$

Fix any k_0 such that, for $k = k_0$, either (13) or (14) is valid. In both cases, we see at once that there exists $\lambda_{k_0} \in (0,1)$ satisfying

$$\lambda_{k_0} > -\frac{a^1 - a_{k_0}^1}{a_{k_0}^1 - b^1}. \tag{15}$$

Let $d_{k_0} = \lambda_{k_0} b + (1 - \lambda_{k_0}) a_{k_0} \in A$. By (15),

$$\begin{aligned} d_{k_0}^1 &= a_{k_0}^1 - \lambda_{k_0}(a_{k_0}^1 - b^1) \\ &< a_{k_0}^1 + \frac{a^1 - a_{k_0}^1}{a_{k_0}^1 - b^1}(a_{k_0}^1 - b^1) = a^1. \end{aligned}$$

By (13) and (14),

$$\begin{aligned} d_{k_0}^2 = \lambda_{k_0} b^2 + (1 - \lambda_{k_0}) a_{k_0}^2 &= \lambda_{k_0} a^2 + (1 - \lambda_{k_0}) a_{k_0}^2 \\ &< \lambda_{k_0} a^2 + (1 - \lambda_{k_0}) a^2 = a^2. \end{aligned}$$

Thus $a \in d_{k_0} + intR_+^2$. Since $\lim_{k \to \infty} a_k = a$, this implies that $a_k \in d_{k_0} + intR_+^2$ for all k sufficiently large, contrary to the fact that $a_k \in E(A|C)$. The closedness of $E(A|C)$ has been proved. From the compactness of A it follows that $E(A|C)$ is a compact set.

Let $K_1 = cone(E(A|C))$. Since $0 \notin E(A|C)$ and $E(A|C)$ is a compact set, K_1 is a closed cone. For obtaining the convexity of K_1, let us fix two different points a_1 and b_1 from K_1. Let $c_1 = \theta_1 a_1 + (1 - \theta_1) b_1$, where $\theta_1 \in (0,1)$. The proof of the convexity of K_1 is completed by showing that $c_1 \in K_1$. Since $a_1,\ b_1 \in K_1 = cone(E(A|C))$, there exist $a,\ b \in E(A|C)$ and $\lambda, \mu > 0$ such that $a_1 = \lambda a,\ b_1 = \mu b$. Therefore

$$\begin{aligned} c_1 &= \theta_1 \lambda a + (1 - \theta_1) \mu b \\ &= (\theta_1 \lambda + (1 - \theta_1)\mu) \left(\frac{\theta_1 \lambda}{\theta_1 \lambda + (1 - \theta_1)\mu} a + \frac{(1 - \theta_1)\mu}{\theta_1 \lambda + (1 - \theta_1)\mu} b \right). \end{aligned}$$

Define

$$\theta = \frac{\theta_1 \lambda}{\theta_1 \lambda + \mu(1 - \theta_1)}, \qquad c = \theta a + (1 - \theta) b.$$

We have $\theta \in (0,1)$, $c_1 = (\theta_1 \lambda (1 - \theta_1))c$. If we prove that $c \in K_1$ then $c_1 \in K_1$.

Consider the optimization problem (P_c) defined in the proof of Lemma 3.1. From $a,\ b \in E(A|C)$ it follows that $a,\ b \in E^w(A|C)$. Let

$t_0 > 0$ be the unique solusion of (P_c). Repeated application of the arguments for proving the inclusion (1) in the proof of Lemma 3.1 enables us to conclude that $t_0c \in A$.

If $t_0c \in E(A|C)$ then $c \in K_1$, as required.

If $t_0c \notin E(A|C)$ then there exists $d \in A$ such that

$$d \leq_{R^2_+} t_0c, \ d \neq t_0c. \tag{16}$$

Clearly, $d \in (t_0c - R^2_+) \cap A$. We claim that $t_0 < 1$. On the the contrary, suppose that $t_0 = 1$. Since $a, \ b \in E(A|C)$ and $a \neq b$, there is no loss in assuming that

$$a^1 > b^1, \quad b^2 > a^2. \tag{17}$$

Consider the affine function $\varphi_1(x)$ defined as in the proof of Lemma 3.1. Namely,

$$\varphi_1(x) = (a^1 - b^1)x^2 + (b^2 - a^2)x^1 - (b^2 - a^2)a^1 - (a^1 - b^1)a^2,$$

for every $x = (x^1, x^2) \in R^2$. We have $\varphi_1(a) = \varphi_1(b) = \varphi_1(c) = 0$. From (16) and (17) it follows that

$$\varphi_1(d) < \varphi_1(t_0c) = \varphi_1(c) = 0, \quad \varphi_1(0) < 0. \tag{18}$$

Let $d' \in R^2$ be such that d and d' are symmetric with respect to the line passing through a and b. From the construction it follows that c belongs to the interior of the convex set $D := conv\{a, \ d, \ b, \ d'\}$, while 0 is outside that set. Consequently, the intersection of the line segment $[0, c]$ with the boundary of D is nonempty. Let $\tilde{c} = \tilde{t}c$, where $\tilde{t} \in (0, 1)$, be a point from that intersection. By (18), $\varphi_1(\tilde{c}) < 0$. This implies that $\tilde{c}$ belongs to the union of the line segments $[a, d]$ and $[b, d]$. In particular, $\tilde{t}c = \tilde{c} \in A$. Hence $\tilde{t}$ satisfies the constraints of (P_c). From this we conclude that $t_0 = 1$ cannot be a solution of (P_c), a contradiction. Our claim has been proved.

Let $\varphi_1(\cdot)$ be as above. Let $\varphi_2(\cdot)$ and $\varphi_3(\cdot)$ be as in the proof of Lemma 3.1. We have (6), (8) and (11). We wish to prove that

$$\varphi_2(t_0c) > 0, \ \varphi_3(t_0c) > 0. \tag{19}$$

By (3) and (4), $b^2 - d^2 > 0, \ d^1 - b^1 \geq 0$. If $d^1 - b^1 = 0$ then $b = (b^1, b^2)$ cannot be an element from $E(A|C)$, which is impossible. Thus $b^2 - d^2 > 0$ and $d^1 - b^1 > 0$. Hence it follows from (16) that $\varphi_2(t_0c) > \varphi_2(d) = 0$. The first inequality in (19) has been proved. Similarly, (3) and (4) imply that $a^1 - d^1 > 0, \ d^2 - a^2 \geq 0$. If $d^2 - a^2 = 0$ then $a = (a^1, a^2)$ cannot be an element from $E(A|C)$, which is impossible. Thus $a^1 - d^1 > 0$ and $d^2 - a^2 > 0$. On account of these inequality and (16), we have $\varphi_3(t_0c) > \varphi_3(d) = 0$. The proof of (19) is complete.

We conclude from (6), (8), (11) and (19) that $t_0 c$ belongs to the interior of the set $conv\{a,\ b,\ d\}$. In particular, $t_0 c \in int(A + R^2_+)$. The last inclusion shows that t_0 cannot be the solusion of (P_c), which is imposible. We have thus proved that $t_0 c \in E(A|C)$. This yields $c \in K_1$, hence $c_1 \in K_1$.

Let

$$\Delta_1 := \{x = (x^1, x^2) \in R^2_+ : x^1 + x^2 = 1\} \cap cone(E(A|C)).$$

Note that Δ_1 is a closed convex subset of an 1-dimensional simplex, so Δ_1 is an 0-dimensional or an 1-dimensional simplex. Let f be the homeomorphism defined in the proof of Lemma 3.1. It is clear that the restriction of f on $E(A|C)$ is an homeomorphism between $E(A|C)$ and Δ_1. The proof is complete. □

Lemma 3.3 *For* $C = \{(x, y) \in R^2 : x > 0,\ y \geq 0\} \cup \{(0, 0)\}$, *the set* $E(A|C)$ *is homeomorphic to an 0-dimensional or an 1-dimensional simplex.*

Proof. Arguing similarly as in the preceding proofs, we can show that the set $E(A|C)$ is closed and the cone $cone(E(A|C))$ is convex and closed. Observe that $E(A|C) \subset E^w(A|R^2_+)$. The restriction of the function f constructed in the proof of Lemma 3.1 maps $E(A|C)$ homeomorphically to the simplex $\Delta_2 := \{x = (x^1, x^2) \in R^2_+ : x^1 + x^2 = 1\} \cap cone(E(A|C))$. The proof is complete. □

Lemma 3.4 *For* $C = \{(x^1, x^2) \in R^2 : x^1 \geq 0,\ x^2 = 0\}$, *the set* $E(A|C)$ *is homeomorphic to an 0-dimensional or an 1-dimensional simplex.*

Proof. We first observe that $E(A|C)$ is closed. The proof of this fact is elementary, so it is omited. Let $\mu_1 = \min\{x^2 : \exists x^1 \ s.t.\ (x^1, x^2) \in A\}$, $\mu_2 = \max\{x^2 : \exists x^1 \ s.t.\ (x^1, x^2) \in A\}$, and $M = \{0\} \times [\mu_1, \mu_2]$. We set

$$f(u) = (0, u^2) \quad \text{for every } u = (u^1, u^2) \in E(A|C).$$

It is easily seen that $f : E(A|C) \to M$ is a continuous 1-1 mapping. Using the above-mentioned closedness of $E(A|C)$, it is a simple matter to show that f maps every closed subset $Q \subset E(A|C)$ to a closed subset $f(Q) \subset M$. From this we conclude that f^{-1} is a continuous mapping. We have thus proved that $E(A|C)$ is a homeomorphic to M, which is 0-dimensional or an 1-dimensional simplex. □

Proof of Theorem 3.1. By assumption, $C \subset R^2$ a nonempty convex cone whose closure $\overline{C}$ is a pointed cone. Let D be a compact convex base of $\overline{C}$. There are two possibilities: D consists of a unique point a, or D is a segment $[a,b]$, $a \neq b$. The intersection $D \cap C$ is a base of C. By interchanging the roles of a and b, we can assume that one of the following situations occurs: (i) $D \cap C = (a,b)$, (ii) $D \cap C = [a,b]$, (iii) $D \cap C = [a,b)$, (iv) $D \cap C = \{a\}$. Let $T: R^2 \to R^2$ be the unique nondegenerate linear transformation satisfying $T(a) = (1,0)$, $T(b) = (0,1)$. (If $D = \{a\}$ then instead of the condition $T(b) = (0,1)$ we require that $T((0,1)) = (0,1)$). Clearly, $T(E(A|C)) = E(T(A)|T(C))$. Applying Lemmas 3.1–3.4 with $T(A)$ and $T(C)$, respectively, in the place of A and C, we conclude that $T(E(A|C))$ is homeomorphic to an 0-dimensional or an 1-dimensional simplex. Then $E(A|C)$ also has such a property. The proof is complete. □

4 Example

The conclusion of Theorem 3.1 is no longer true if instead of R^2 one considers an Euclidean space of higher dimension, say R^3.

Example 4.1. Let $C = R^3_+$ and $A \subset R^3$ be defined by the following formula

$$A = conv\Big(\Big\{ \begin{array}{l} x = (x^1, x^2, x^3) : \\ (x^1 - 1)^2 + (x^3 - 1)^2 \leq 1,\ 0 \leq x^1 \leq 1,\ x^2 = 0, \\ 0 \leq x^3 \leq 1\} \cup \{(1,1,0)\} \cup \{(0,1,1)\Big\}\Big). \end{array}$$

Clearly,

$$A = \Big\{ \begin{array}{l} z \in R^3 : z = tx + (1-t)y,\ t \in [0,1],\ x = (x^1, x^2, x^3), \\ (x^1 - 1)^2 + (x^3 - 1)^2 \leq 1,\ 0 \leq x^1 \leq 1,\ x^2 = 0,\ 0 \leq x^3 \leq 1, \\ y = (y^1, y^2, y^3),\ y^1 = \lambda,\ y^2 = 1,\ y^3 = 1 - \lambda,\ \lambda \in [0,1]\Big\}, \end{array}$$

and A is a compact convex subset of R^3_+.
Setting

$$\begin{array}{l} D_1 = \{x = (x^1, x^2, x^3) : (x^1 - 1)^2 + (x^3 - 1)^2 \leq 1, \\ \qquad 0 \leq x^1 \leq 1,\ x^2 = 0,\ 0 \leq x^3 \leq 1\}, \\ D_2 = \{x = x^1, x^2, x^3) : x^1 = 1,\ 0 \leq x^2 \leq 1,\ x^3 = 0\}, \\ D_3 = \{x = x^1, x^2, x^3) : x^1 = 0,\ 0 \leq x^2 \leq 1,\ x^3 = 1\}, \end{array}$$

we observe that

$$D_2 \cap D^3 = \emptyset,\ D_1 \cap D_2 = \{(1,0,0)\},\ D_1 \cap D_3 = \{(0,0,1)\}.$$

Claim 1. $E^w(A|C) = D$, *where* $D := D_1 \cup D_2 \cup D_3$. *(This implies that* $E^w(A|C)$ *is not homeomorphic to any simplex.)* For proving the claim we set

$$F = \{z \in R^3 : z = tx + (1-t)y,\ t \in [0,1),\ x = (x^1, x^2, x^3),\\ (x^1-1)^2 + (x^3-1)^2 \le 1,\ 0 \le x^1 \le 1,\ x^2 = 0,\ 0 \le x^3 \le 1,\\ y = (y^1, y^2, y^3),\ y^1 = \lambda,\ y^2 = 1,\ y^3 = 1-\lambda,\ \lambda \in (0,1)\}.$$

From the formulas defining A, F and D, it follows that

$$F = A \setminus D. \tag{20}$$

For any $z = (z^1, z^2, z^3) \in F$ there exist $t \in [0,1)$, $\lambda \in (0,1)$, $x = (x^1, x^2, x^3)$, and $y = (y^1, y^2, y^3)$ such that

$$(x^1-1)^2 + (x^3-1)^2 \le 1,\ 0 \le x^1 \le 1,\ x^2 = 0,\ 0 \le x^3 \le 1, \tag{21}$$

and

$$y^1 = \lambda,\ y^2 = 1,\ y^3 = 1-\lambda,\ z = tx + (1-t)y. \tag{22}$$

Define

$$\Omega = \{v = (v^1, v^2, v^3) \in R^3 : (v^1-1)^2 + (v^3-1)^2 \le 1,\ v^2 \ge 0\}.$$

Note that Ω is a closed convex set, and

$$int\Omega = \{v = (v^1, v^2, v^3) \in R^3 : (v^1-1)^2 + (v^3-1)^2 < 1,\ v^2 > 0\}.$$

By (21), $x = (x^1, x^2, x^3) \in \Omega$. By (22), $y^2 > 0$ and

$$(y^1-1)^2 + (y^3-1)^2 = (\lambda-1)^2 + \lambda^2 = 1 - 2\lambda(1-\lambda) < 1.$$

Hence $y = (y^1, y^2, y^3) \in int\Omega$. As $z = tx + (1-t)y,\ t \in [0,1)$, we have $z \in int\Omega$. That is

$$(z^1-1)^2 + (z^3-1)^2 < 1, \quad z^2 > 0. \tag{23}$$

It is immediate from (22) and the condition $t \in [0,1)$ that

$$0 < z^1 \le 1, \quad 0 < z^3 \le 1. \tag{24}$$

By (23) and (24), there exist $\tilde{z}^1 \in (0,1]$, $\tilde{z}^3 \in (0,1]$ such that $\tilde{z}^1 < z^1$, $\tilde{z}^3 < z^3$, and

$$(\tilde{z}^1-1)^2 + (\tilde{z}^3-1)^2 < 1.$$

Setting $\tilde{z} = (\tilde{z}^1, 0, \tilde{z}^3)$, we have $\tilde{z} \in A$ and $z = \tilde{z} + intR^3_+$. This shows that $z \notin E^w(A|C)$. As $z \in F$ is arbitrarily chosen, we conclude from (20) that $E^w(A|C) \subseteq D$.

For proving the inclusion $D \subseteq E^w(A|C)$, let us fix any $z \in D$. If $z \notin E^w(A|C)$ then there exists $\hat{z} = (\hat{z}^1, \hat{z}^2, \hat{z}^3) \in A$ such that

$$\hat{z}^1 < z^1, \quad \hat{z}^2 < z^2, \quad \hat{z}^3 < z^3. \tag{25}$$

If $z \in D_1$ then $z^2 = 0$. If $z \in D_2$ then $z^3 = 0$. If $z \in D_3$ then $z^1 = 0$. In these cases, it follows from (25) that $\hat{z}^2 < 0$, $\hat{z}^3 < 0$, $\hat{z}^1 < 0$, respectively. Each one of the last three inequalities yields $\hat{z} \notin A$, a contradiction. Claim 1 has been proved.

Claim 2. *For $a := (3,1,3)$, $A' := A + a$, one has $A' \subset intR^3_+$ and $cone(E^w(A'|C))$ is a nonconvex cone.* The first assertion of Claim 2 is obvious. To check the second assertion, choose $z_1 := (1,1,0) \in E^w(A|C)$ and $z_2 := (0,1,1) \in E^w(A|C)$. Since $E^w(A'|C) = E^w(A|C)+a$, $(4,2,3) = z_1 + a$ and $(3,2,4) = z_2 + a$ belong to $E^w(A'|C)$. Suppose, contrary to our claim, that $cone(E^w(A'|C))$ is a convex cone. Then we would have

$$w := \left(\frac{7}{2}, 2, \frac{7}{2}\right) = \frac{1}{2}(4,2,3) + \frac{1}{2}(3,2,4) \in cone(E^w(A'|C)).$$

This implies that there exists $t \geq 0$ such that

$$tw = t\left(\frac{7}{2}, 2, \frac{7}{2}\right) \in E^w(A'|C) = D'_1 \cup D'_2 \cup D'_3,$$

where $D'_i = D_i + a$ for $i = 1, 2, 3$. If $tw \in D'_1$ then $tw - a \in D_1$. This gives $t = \frac{1}{2}$ and $\frac{7}{4} - 3 \geq 0$, which is impossible. If $tw \in D'_2$ then $tw - a \in D_2$. This clearly forces $t = \frac{8}{7}$ and $t = \frac{6}{7}$, which is impossible. The same argument excludes the case $tw \in D'_3$. Claim 2 has been proved.

Analyzing once again the proofs of Lemmas 3.1–3.4 we see that the convexity of $cone(E^w(A|C))$ and $cone(E(A|C))$ is a vital property. When this property is violated, the proof scheme of the above-mentioned lemmas (which is originated to [1] and [2, pp.144-145]) fails to work. For the set A described in Example 4.1 we have shown that $cone(E^w(A+a|R^3_+))$, where $a = (3,1,3)$, is a nonconvex cone. This example shows that the radial projection technique used for proving Theorem 3.1 can hardly give us results on contractibility of the weakly efficient sets of compact convex subsets of R^n, $n \geq 3$. (As it has been said in the Introduction, the contractibility of the corresponding efficient sets has been established by Peleg [4] and Luc [3].)

Acknowledgements. Financial support of the National Basic Research Program in Natural Sciences (Vietnam) is gratefully acknowledged.

References

[1] D.T. Luc, *Structure of the efficient point sets*, Proc. Amer. Math. Soc. Vol.95, 1985, 433-440.

[2] D.T. Luc, *Theory of Vector Optimization*, Lecture Notes in Economics Mathematical Systems, Vol.319, Springer-Verlag, Berlin, 1989.

[3] D.T. Luc, *Contractibility of efficient sets in normed space*, Nonlinear Anal. Theory Methods & Appl. Vol.15, 1990, 527-535.

[4] B. Peleg, *Topological properties of the efficient point set*, Proc. Amer. Math. Soc. Vol.35, 1972, 531-536.

EXISTENCE THEOREMS FOR A CLASS OF VARIATIONAL INEQUALITIES AND APPLICATIONS TO A CONTINUOUS MODEL OF TRANSPORTATION

Francesco Marino
Dipartimento di Matematica - Università di Catania,
Viale A. Doria, 6 - 95125 CATANIA (ITALIA)
e-mail: fmarino@dmi.unict.it

Abstract In this paper we study a variational inequality concerning with the traffic equilibrium problem in the continuous case and we prove an existence theorem.

Keywords: Variational inequalities, Continuous transport problems.

2000 Mathematics Subject Classification. 49J40, 90C90.

1 Introduction

In the paper [6] the Author has considered the transportation equilibrium problem in the continuous case and has shown that, when the "user optimizing" approach is used, the equilibrium conditions are equivalent to the following problem "Find $u \in \mathbb{K}$ such that

$$\int_\Omega c(x, u(x)) \cdot (v(x) - u(x))\, dx \geq 0 \qquad \forall\, v \in \mathbb{K}\text{"}, \tag{1}$$

where

$$\mathbb{K} = \{u = (u_1, u_2) \in \mathcal{E}(\Omega, \mathbb{R}^2),\ u_1(x),\ u_2(x) \geq 0,\ \mathrm{div} u + t(x) = 0, \\ u_1|_{\partial\Omega} = \varphi_1(x),\ u_2|_{\partial\Omega} = \varphi_2(x)\} \tag{2}$$

with

$$\mathcal{E}(\Omega, \mathbb{R}^2) = \{v \in L^2(\Omega, \mathbb{R}^2), \quad \mathrm{div} v \in L^2(\Omega)\},$$

equipped with the norm

$$\|v\|_{\mathcal{E}(\Omega,\mathbb{R}^2)} = \|v\|_{L^2(\Omega,\mathbb{R}^2)} + \|\mathrm{div} v\|_{L^2(\Omega)}\,.$$

The aim of this paper is to show the following existence theorem for Problem (1).

Theorem 1.1 *Let us assume that the following conditions hold:*

(*i*) Ω *is a simply connected bounded domain in* $\mathbb{R}^2$ *of generic point* $x = (x_1, x_2)$, *with Lipschitz boundary* $\partial\Omega$;

(*ii*) $t(x) \in L^2(\Omega)$, $\varphi_1(x)$, $\varphi_2(x) \in L^1(\partial\Omega)$;

(*iii*) $c(x,u) : \Omega \times \mathbb{R}^2 \to \mathbb{R}^2$ *is a Carathéodory function such that*

$$\|c(x,u)\|_{\mathbb{R}^2} \leq \alpha(x) + \|u\|_{\mathbb{R}^2}, \qquad \alpha(x) \in L^2(\Omega)\,; \tag{3}$$

(*iv*) *there exists* $\nu > 0$ *such that*

$$c(x,u) \cdot u \geq \nu \|u\|^2_{\mathbb{R}^2} \qquad \forall\, u \in \mathbb{R}^2, \text{ a.e. in } \Omega\,; \tag{4}$$

(*v*) $(c(x,u) - c(x,v)) \cdot (u - v) \geq 0 \qquad \forall\, u, v \in \mathbb{R}^2$, *a.e. in* Ω.

Then the variational inequality (1) admits at least a solution.

The above existence theorem deals with the "user optimizing" approach; it is different from the "system optimization" approach which leads to a infinite dimensional constrained optimization problem (see [3], [4], and [5]).
In Section 2 we briefly present the continuous transportation model and the equilibrium condition and in Section 3 we give the proof of Theorem 1.1.

An interesting problem is to weak the coercitivity condition (1.4) and also to consider the function t depending on x and u. These problems will be object of a next paper.

2 Continuous transportation model.

We consider a dense urban street network that occupies a region Ω of $\mathbb{R}^2$ satisfying condition (*i*) in Section 1. We are interested in the case when the traffic flows only in the direction of increasing axes x_1 and x_2. The traffic condition at each point $x \in \Omega$ is described by a vector field $u(x)$, whose components $u_1(x)$ and $u_2(x)$ represent the traffic density through a neighbourhood of x in the directions x_1 and x_2 respectively. It will be $u_1(x) \geq 0$, $u_2(x) \geq 0$, and $u_1(x)$, $u_2(x)$ have non negative fixed traces $\varphi_1(x)$ and $\varphi_2(x)$ on $\partial\Omega$ (or on a part of $\partial\Omega$) respectively. We suppose $u(x) \in \mathcal{E}(\Omega, \mathbb{R}^2)$ and $\varphi(x) = (\varphi_1(x), \varphi_2(x)) \in L^1(\partial\Omega, \mathbb{R}^2)$. We associate with each point x of Ω a scalar field $t(x) \in L^2(\Omega)$ (vehicles unit area unit time) representing the density of the flow originating or terminating at a neighbourhood of x. The fields u and t are connected by the following flow conservation law

$$\iint_{\tilde{\Omega}} t(x)\, dx + \oint_{\partial\tilde{\Omega}} u \cdot n\, ds = 0, \tag{5}$$

where n is the normal at the point $x \in \partial\tilde{\Omega}$ and $\tilde{\Omega}$ is any subdomain of Ω with smooth boundary. Applying Green's theorem to (5), one obtain the flow conservation law

$$\mathrm{div} u + t(x) = \frac{\partial u_1}{\partial x_1} + \frac{\partial u_2}{\partial x_2} + t(x) = 0. \tag{6}$$

Then the set of feasible flows is given by (2). In order to introduce the "personal cost", we assume that a vehicle located at a point x and travelling in the direction of the x_i axis ($i = 1, 2$) incurs a travel cost c_i which depends on the traffic flow, i.e.

$$c_i = c_i(x, u(x))\,.$$

We say "personal cost" the vector valued function $c(x, u(x)) = (c_1(x, u(x)), c_2(x, u(x)))$. In our continuous transportation model users have predetermined origins but they are free to choose the destination as well as the travel paths in order to minimize their "travel cost". This choice mechanism leads to the following definition of equilibrium usually called "user optimizing" equilibrium from the user's point of view.

Definition 2.1 *$u \in \mathbb{K}$ is an equilibrium distribution flow if there exists a potential $\mu \in H^1(\Omega)$ with*

$$\frac{\partial \mu}{\partial x_i} \leq c_i(x, u(x)) \qquad i = 1, 2\,, \quad a.e.\ in\ \Omega$$

such that

$$\left(c_i(x, u(x)) - \frac{\partial \mu}{\partial x_i}\right) u_i(x) = 0 \qquad i = 1, 2\,, \quad a.e.\ in\ \Omega. \tag{7}$$

If we are interested in the "sistem optimizing" approach we can consider the "total cost" which can be expressed in the form

$$c_T(x, u(x)) = c_1(x, u(x))u_1(x) + c_2(x, u(x))u_2(x)\,,$$

and then we have to minimize the "global total cost" whose expression is

$$F(u) = \int_\Omega [c_1(x, u(x))u_1(x) + c_2(x, u(x))u_2(x)]\, dx\,.$$

As said in Section 1, in [6] is proved the following result

Theorem 2.1 *Let (i), (ii), and (iii) hold true, then $u \in \mathbb{K}$ is an equilibrium distribution if and only if*

$$\int_{\Omega} c(x, u(x)) \cdot (v(x) - u(x))\, dx \geq 0 \qquad \forall v \in \mathbb{K}.$$

Remark 1. μ measures the cost occurred when a network user travels from the point x to the boundary $\partial\Omega$ using the cheapest possible path.

Remark 2. Let's assume that $\frac{\partial c_i}{\partial x_j}$, and $\frac{\partial^2 \mu}{\partial x_i \partial x_j}$ exist, then from condition (7) we can deduce the following new version of the equilibrium condition.

Definition 2.2 *(see [1]). $u \in \mathbb{K}$ is an equilibrium distribution flow if*

$$\begin{cases} \frac{\partial c_1}{\partial x_2} - \frac{\partial c_2}{\partial x_1} = 0 & \text{if } u_1 > 0,\ u_2 > 0 \\ \frac{\partial c_1}{\partial x_2} - \frac{\partial c_2}{\partial x_1} > 0 & \text{if } u_1 = 0,\ u_2 > 0 \\ \frac{\partial c_1}{\partial x_2} - \frac{\partial c_2}{\partial x_1} < 0 & \text{if } u_1 > 0,\ u_2 = 0. \end{cases} \tag{8}$$

The meaning of (8) is the following: each user chooses whatever he is free to choose, such as his origin, his destination or his travel path, so as to minimize his "travel cost".

3 Existence Theorem.

We shall recall some definitions.
Let E be a real topological vector space, $K \subseteq E$ convex nonempty, and $C : K \to E^*$.

Definition 3.1 *We say C monotone iff, for all x, $y \in K$, one has*

$$\langle C(x) - C(y), x - y \rangle \geq 0\,.$$

Definition 3.2 *We say C pseudomonotone iff, for all x, $y \in K$, one has*

$$\langle C(x), y - x \rangle \geq 0 \Longrightarrow \langle C(y), x - y \rangle \leq 0.$$

Definition 3.3 *We say C hemicontinuous along line segments iff, for all x, $y \in K$, the function $t \to \langle C((1-t)x + ty), y - x \rangle$ is upper semicontinuous on $[0, 1]$.*

In order to prove Theorem 1 we will use the following result.

Theorem 3.1 *(see [2]). Let E be a real topological vector space, and let $K \subseteq E$ be convex, bounded and nonempty. Let $C : K \to E^*$ be pseudomonotone and hemicontinuous along line segments. Then, there exists $\bar{x} \in K$ such that*

$$\langle C(\bar{x}), y - \bar{x} \rangle \geq 0 \qquad \forall\, y \in K.$$

It is possible to prove, adopting a standard argument, and thanks to Theorem 3, the following result:

Theorem 3.2 *Let E be a real topological vector space, and let $K \subseteq E$ be convex, unbounded and nonempty. Let $C : K \to E^*$ be pseudomonotone, hemicontinuous along line segments, and let suppose that there exists $u_0 \in K$ such that*

$$\lim_{\substack{\|v\|_E \to +\infty \\ v \in K}} \langle C(v), v - u_0 \rangle = +\infty. \tag{9}$$

Then threre exists $\bar{x} \in K$ such that

$$\langle C(\bar{x}), y - \bar{x} \rangle \geq 0 \qquad \forall\, y \in K.$$

We can now prove Theorem 1. Let be $C : \mathbb{K} \to (\mathcal{E}(\Omega, \mathbb{R}^2))^*$ such that

$$\langle C(u), v \rangle = \int_\Omega c(x, u(x)) \cdot v(x)\, dx \qquad \forall\, u \in \mathbb{K},\ \forall\, v \in \mathcal{E}(\Omega, \mathbb{R}^2).$$

By hypothesis (v) of Theorem 1.1, for all u, $v \in \mathbb{K}$, we have

$$\int_\Omega (c(x, u(x)) - c(x, v(x))) \cdot (u(x) - v(x))\, dx \geq 0\,,$$

or equivalently

$$\langle C(u) - C(v), u - v \rangle \geq 0 \qquad \forall\, u, v \in \mathbb{K}\,,$$

therefore C is monotone and then pseudomonotone.
C is hemicontinuous along line segments, infact let $\{t_n\} \subseteq [0,1]$ be such that $t_n \to t_0 \in [0,1]$, thanks to (iii), for all $u, v \in \mathbb{K}$, it follows that

$$c(x, (1 - t_n)u(x) + t_n v(x)) \cdot (v(x) - u(x)) \to$$

$$\to c(x, (1 - t_0)u(x) + t_0 v(x)) \cdot (v(x) - u(x)) \text{ a.e. in } \Omega.$$

Moreover

$$\begin{aligned}&|c(x,(1-t_n)u(x)+t_nv(x))\cdot(v(x)-u(x))|\leq\\ &\leq \|c(x,(1-t_n)u(x)+t_nv(x))\|_{\mathbb{R}^2}\|v(x)-u(x)\|_{\mathbb{R}^2}\leq\\ &\leq \big(\alpha(x)+\|(1-t_n)u(x)+t_nv(x)\|_{\mathbb{R}^2}\big)\|v(x)-u(x)\|_{\mathbb{R}^2}\leq\\ &\leq \big(\alpha(x)+(1-t_n)\|u(x)\|_{\mathbb{R}^2}+t_n\|v(x)\|_{\mathbb{R}^2}\big)\|v(x)-u(x)\|_{\mathbb{R}^2}\\ &\forall n\in\mathbb{N}\,,\ \text{a.e. in } \Omega\,.\end{aligned}$$

Thanks to the $\{t_n\}$ and $\{1-t_n\}$ bounds, and thanks to Lebesgue Theorem we have

$$\langle C((1-t_n)\,u+t_n\,v),v-u\rangle \to \langle C((1-t_0)\,u+t_0\,v),v-u\rangle\,,$$

that is the function $t\to\langle C((1-t)\,u+t\,v),v-u\rangle$ is continuous on t_0 and then upper semicontinuous.
Finally C fulfils condition (9). Infact, let u_0 be an arbitrary element of $\mathbb{K}$, then, by hypothesis (iv), for all $v\in\mathbb{K}$ we have

$$\begin{aligned}\langle C(v),v-u_0\rangle &= \int_\Omega c(x,v(x))\cdot(v(x)-u_0(x))\,dx=\\ &=\int_\Omega c(x,v(x))\cdot v(x)\,dx-\int_\Omega c(x,v(x))\cdot u_0(x)\,dx\geq\\ &\geq\int_\Omega \nu\|v(x)\|^2_{\mathbb{R}^2}\,dx-\left|\int_\Omega c(x,v(x))\cdot u_0(x)\,dx\right|.\end{aligned}$$

Moreover

$$\begin{aligned}&\left|\int_\Omega c(x,v(x))\cdot u_0(x)\,dx\right|\leq\int_\Omega |c(x,v(x))\cdot u_0(x)|\,dx\leq\\ &\leq\int_\Omega \|c(x,v(x))\|_{\mathbb{R}^2}\|u_0(x)\|_{\mathbb{R}^2}\,dx\leq\int_\Omega\big(\alpha(x)+\|v(x)\|_{\mathbb{R}^2}\big)\|u_0(x)\|_{\mathbb{R}^2}\,dx=\\ &=\int_\Omega\alpha(x)\,\|u_0(x)\|_{\mathbb{R}^2}\,dx+\int_\Omega\|v(x)\|_{\mathbb{R}^2}\|u_0(x)\|_{\mathbb{R}^2}\,dx\leq\\ &\leq\|\alpha(x)\|_{L^2(\Omega)}\|u_0\|_{L^2(\Omega,\mathbb{R}^2)}+\|v\|_{L^2(\Omega,\mathbb{R}^2)}\|u_0\|_{L^2(\Omega,\mathbb{R}^2)}=\\ &=\gamma_1+\gamma_2\|v\|_{L^2(\Omega,\mathbb{R}^2)}.\end{aligned}$$

Therefore

$$\langle C(v),v-u_0\rangle\geq\nu\|v\|^2_{L^2(\Omega,\mathbb{R}^2)}-\gamma_1-\gamma_2\|v\|_{L^2(\Omega,\mathbb{R}^2)}\,.$$

Let $v\in\mathbb{K}$, we know that

$$\|v\|_{\mathcal{E}(\Omega,\mathbb{R}^2)}=\|v\|_{L^2(\Omega,\mathbb{R}^2)}+\|t\|_{L^2(\Omega)}\,,$$

then

$$\lim_{\substack{\|v\|_{\mathcal{E}(\Omega,\mathbb{R}^2)}\to+\infty\\ v\in\mathbb{K}}} < C(v),v-u_0 > = \lim_{\substack{\|v\|_{L^2(\Omega,\mathbb{R}^2)}\to+\infty\\ v\in\mathbb{K}}} < C(v),v-u_0 > = +\infty\,.$$

Theorem 1 is therefore a consequence of Theorem 2. □

References

[1] S.C. Dafermos, *Continuum Modeling of Transportation Networks,* Transp. Res., 14B (1980), 295–301.

[2] P. Daniele, A. Maugeri, W. Oettli, *Time-Dependent Traffic Equilibria,* J. Optimization Theory Appl., 103 (1999), 543–555.

[3] J. Gwinner, *A Hilbert Space Approach to some Flow Problems,* in "Recent Developments in Optimization, Seventh French German Conference on Optimization", R. Durier-C. Michelot (Eds.), Lecture Notes in Economics and Mathematical Systems, 429, Springer, Berlin, 1995.

[4] J. Gwinner, *On Continuum Modeling of Large Dense Networks in Urban Road Traffic,* in "Mathematics in Transport Planning and Control", J.D. Griffiths (Ed.), IMA Conference, Cardiff, 1998.

[5] A. Maugeri, *New Classes of Variational Inequalities and Applications to Equilibrium Problems,* Rend. Acc. Naz. dei XL, 11 (1987), 277–284.

[6] A. Maugeri, *Equilibrium Problems and Variational Inequalities,* in "Equilibrium Problems: Non smooth Optimization and Variational Inequality Models", F. Giannessi-A. Maugeri-P. Pardalos (Eds.), Kluwer Academic Publishers, 2000.

ON AUXILIARY PRINCIPLE FOR EQUILIBRIUM PROBLEMS

Giandomenico Mastroeni
Department of Mathematics, University of Pisa
Via Buonarroti, 2, Pisa - 56127 ITALY

Abstract The auxiliary problem principle introduced by Cohen is extended to a general equilibrium problem. In particular, applications to variational inequalities and to convex optimization problems are analysed.

1 Introduction

The analysis of a general equilibrium problem (for short, EP):

$$\text{find } y^* \in K \text{ s.t. } f(x, y^*) \geq 0, \quad \forall x \in K, \tag{EP}$$

where $f : K \times K \longrightarrow \mathbb{R}$, with $f(x,x) = 0$, for all $x \in K$ a closed convex subset of the reflexive Banach space X, has led to a unified approach in the development of the study of different topics in Optimization such as constrained extremum problems and variational inequalities (for short, VI). In particular VI have been extensively studied since they allow to generalize the classical optimality conditions for constrained extremum problems and to formalize equilibrium conditions for problems of different nature as network flow, economic and mechanical engineering equilibrium problems [7, 9].

If we define $f(x,y) := \langle F(y), x - y\rangle$ then EP collapses into the classic VI:

$$\text{find } y^* \in K \text{ s.t. } \langle F(y^*), x - y^*\rangle \geq 0, \quad \forall x \in K, \tag{VI}$$

where $F : K \longrightarrow X^*$, $K \subseteq X$ and $\langle \cdot, \cdot \rangle$ is the duality pairing between X and its topological dual X^*. If $f(x,y) := h(x) - h(y)$ then EP is equivalent to the optimization problem

$$\min_{x \in K} h(x) \quad \text{s.t. } x \in K,$$

where $h : K \longrightarrow \mathbb{R}$. We refer to [1] and references therein for a wide analysis concerning the existence results for the problem EP and for further examples of equivalent formulations obtained by means of an EP.

The aim of this paper is to show that, exploiting a suitable fixed point formulation of EP, it is possible to define a class of iterative methods which are a straightforward extension of those proposed by Cohen [3, 4] for solving VI and constrained extremum problems.

In Section 2 we will introduce a suitable regularization of EP which we will refer to as the auxiliary equilibrium problem (for short AEP). AEP is a further equilibrium problem which is equivalent to the original EP. By means of the fixed point formulation of AEP we will show that it is possible to define a sequence strongly convergent to a solution of AEP and therefore of EP (Section 3). The proposed method collapses into the one developed by Cohen [4] when EP represents a VI. Applications to constrained extremum problems are considered in Section 4.
We recall the main notations and definitions that will be used in the sequel.
A function $f : X \times X \longrightarrow \mathbb{R}$ is said strongly monotone on $K \in X$, with modulus $a > 0$, iff:

$$f(x, y) + f(y, x) \leq \ -a\|y - x\|^2, \quad \forall x, y \in K.$$

A functional $G : X \longrightarrow \mathbb{R}$ is said to be "differentiable" (in the sense of Gateaux) at the point $x^* \in X$ iff there exists finite:

$$\lim_{t \to 0} \frac{G(x^* + ty) - G(x^*)}{t} =: \langle G'(x^*), y \rangle,$$

with $G'(x^*) \in X^*$ (see [5]).

A function $h : X \longrightarrow \mathbb{R}$ is said to be "directionally differentiable" (in the sense of Dini) at the point $x^* \in X$ in the direction y, iff there exists finite:

$$\lim_{t \to 0^+} \frac{h(x^* + ty) - h(x^*)}{t} =: h'(x^*; y).$$

$h : K \longrightarrow \mathbb{R}$ is said strongly convex on K with modulus a ($a \geq 0$) iff $\forall x_1, x_2 \in K, \forall \lambda \in [0, 1]$,

$$h(\lambda x_1 + (1 - \lambda)x_2) \leq \lambda h(x_1) + (1 - \lambda)h(x_2) - a[\lambda(1 - \lambda)/2]\|x_1 - x_2\|^2.$$

If h is differentiable then h is strongly convex on K, with modulus $a > 0$, iff $\forall x_1, x_2 \in K$,

$$h(x_1) - h(x_2) \geq \langle \nabla h(x_2), x_1 - x_2 \rangle + a/2\|x_1 - x_2\|^2.$$

We will say that the mapping $F : X \longrightarrow X^*$ is monotone on $K \subseteq X$ iff:

$$\langle F(y) - F(x), y - x \rangle \geq 0, \quad \forall x, y \in K;$$

it is strictly monotone if strict inequality holds $\forall x \neq y$.

We will say that the mapping F is strongly monotone, with modulus $\mu > 0$, on K iff:

$$\langle F(y) - F(x), y - x\rangle \geq \mu\|y - x\|^2, \quad \forall x, y \in K;$$

F is Lipschitz continuous with modulus $L > 0$ over K iff

$$\|F(x) - F(y)\| \leq L\|x - y\|, \quad \forall x, y \in K.$$

2 The auxiliary equilibrium problem

Most of the algorithms developed for solving EP can be derived from equivalent formulations of the equilibrium problem. We will focus our attention on fixed-point formulations of EP: we will show that such formulations lead to a generalization of the methods developed by Cohen for variational inequalities and optimization problems.

Let us recall the following preliminary result which states the above mentioned equivalent formulation of EP.

Lemma 2.1 *Suppose that $f(x, x) = 0$, $\forall x \in K$. Then, the following statements are equivalent:*
i) there exists $y^* \in K$ s.t. $f(x, y^*) \geq 0, \quad \forall x \in K$.
ii) $y^* \in K$ is a solution of the problem

$$\min_{x \in K} f(x, y^*). \tag{1}$$

If we assume that, for any $y^* \in K$, (1) has a unique solution we can define the following general iterative method:

General Algorithm.

(i) Let $k = 0$, $y^0 \in K$;
(ii) let y^{k+1} be the solution of the problem:

$$\min_{x \in K} f(x, y^k). \tag{2}$$

(iii) if $\|y^{k+1} - y^k\| < \mu$, for some fixed $\mu > 0$, then STOP, otherwise put $k = k + 1$ and go to step (ii).

Unfortunately, in most of the cases, it is not possible (or not convenient) to apply the previous algorithm directly to the problem EP. It is necessary to introduce an auxiliary equilibrium problem, equivalent to the given one, for which the above procedure leads to a solution of EP.

Proposition 2.1 *Let $f(x,y)$ be a convex differentiable function with respect to x at $y = y^* \in K$ and $\epsilon > 0$. Let $H(x,y) : K \times K \longrightarrow \mathbb{R}$ be non negative, differentiable on the convex set K with respect to x and such that*
i) $H(y,y) = 0, \quad \forall y \in K;$
ii) $H'_x(y,y) = 0, \quad \forall y \in K.$
Then y^* is a solution of EP iff it is a solution of the auxiliary equilibrium problem (AEP):

$$\text{find } y^* \in K \quad s.t. \ \epsilon f(x,y^*) + H(x,y^*) \geq 0 \quad \forall x \in K.$$

Proof. It is obvious that if y^* is a solution of EP then it is also a solution of AEP.
Vice versa let y^* be a solution of AEP. Then y^* is a minimum point of the problem

$$\min_{x \in K} [\epsilon f(x,y^*) \ + \ H(x,y^*)]. \tag{3}$$

Since K is convex then y^* is an optimal solution for (3) iff

$$\langle \epsilon f'_x(y^*,y^*) + H'_x(y^*,y^*), x - y^* \rangle \geq 0, \ \forall x \in K,$$

so that

$$\langle \epsilon f'_x(y^*,y^*), x - y^* \rangle \geq 0, \ \forall x \in K. \tag{4}$$

Dividing by ϵ, we obtain that (4) implies, by the convexity of $f(\cdot, y^*)$, that

$$f(x,y^*) \geq f(y^*,y^*) = 0, \quad \forall x \in K.$$

□

Corollary 2.1 *y^* is a solution of EP iff y^* is an optimal solution of the extremum problem*

$$\min_{x \in K} [\epsilon f(x,y^*) \ + H(x,y^*)].$$

In the next section we will see that applying the General Algorithm to the auxiliary equilibrium problem AEP, for a suitable choice of the function H, it will be possible to define a sequence $\{y_k\}$ convergent to a solution of EP.

3 The auxiliary problem principle

Following the approach of Cohen [3, 4] we will extend the auxiliary problem principle to the equilibrium problem EP. Given $G : K \longrightarrow \mathbb{R}$, a strongly convex differentiable functional, $\epsilon > 0$, let us introduce the auxiliary problem AEP obtained putting $H(x,y) := G(x) - G(y) - \langle G'(y), x - y \rangle$:

find $y^* \in K$ such that

$$\epsilon f(x, y^*) \; - \; \langle G'(y^*), x - y^* \rangle + G(x) - G(y^*) \geq 0, \quad \forall x \in K.$$

Lemma 3.1 *Let $f(x,y)$ be a convex differentiable function with respect to x at $y = y^* \in K$. Then y^* is a solution of EP iff it is a solution of AEP.*

Proof. It follows from the Proposition 2.1 noticing that the function $H(x,y) := G(x) - G(y) - \langle G'(y), x - y \rangle$ fulfils the hypotheses i) and ii) of the Proposition. □

Taking into account Corollary 2.1 we have that y^* is a solution of EP iff y^* is an optimal solution of the extremum problem

$$\min_{x \in K} \, [\epsilon f(x, y^*) \; - \; \langle G'(y^*), x \rangle + G(x)].$$

Applying the General Algorithm to the problem AEP we obtain the following iterative method:

Algorithm 3.1.

(i) Let $k = 0$, $y^0 \in K$;
(ii) let y^{k+1} be the solution of the problem $P(k)$:

$$\min_{x \in K} \, [\epsilon f(x, y^k) \; - \; \langle G'(y^k), x \rangle + G(x)];$$

(iii) if $\|y^{k+1} - y^k\| < \mu$, for some fixed $\mu > 0$, then STOP, otherwise put $k = k + 1$ and go to step (ii).

Remark 3.1 We observe that $P(k)$ has a unique solution since G is a strongly convex function.

Theorem 3.1 *Suppose that the following conditions hold:*

(i) $f(x,y)$ is a convex l.s.c. function with respect to x, $\forall y \in K$;

(ii) $f(x,y)$ is continuous with respect to y, on every finite dimensional subspace of X, $\forall x \in K$;

(iii) f is strongly monotone on K with modulus a;

(iv) G is strongly convex on K with modulus b;
(v) there exist constants α, $\beta > 0$ *such that* $\forall x, y, z \in K$:

$$f(x,y) + f(y,z) \geq f(x,z) - \alpha\|x-y\|^2 - \beta\|y-z\|^2;$$

Then, if $\epsilon \leq \frac{b}{2\alpha}$ *and* $\beta < a$, *the sequence* $\{y^k\}$, *defined in the Algorithm 3.1,strongly converges towards the solution* y^* *of* EP.

Proof. In our hypotheses the solution y^* of EP exists and it is unique (see e.g. [1] Theorem 1), as well as the solution of $P(k)$. Consider the functional

$$\Lambda(x) := G(y^*) - G(x) - \langle G'(x), y^* - x\rangle \geq \frac{b}{2}\|x - y^*\|^2, \tag{5}$$

and the difference

$$\Lambda(y^k) - \Lambda(y^{k+1}) = G(y^{k+1}) - G(y^k) - \ \langle G'(y^k), y^* - y^k\rangle +$$

$$\langle G'(y^{k+1}), y^* - y^{k+1}\rangle \ - \ \langle G'(y^k), y^{k+1}\rangle \ + \ \langle G'(y^k), y^{k+1}\rangle \ =$$

$$G(y^{k+1}) - G(y^k) - \langle G'(y^k), y^{k+1} - y^k\rangle \ + \ \langle G'(y^{k+1}) - G'(y^k), y^* - y^{k+1}\rangle \ \geq$$

$$\frac{b}{2}\|y^{k+1} - y^k\|^2 + \epsilon(f(y^{k+1}, y^k) - f(y^*, y^k)).$$

The previous inequality is due to the strong convexity of G and from the fact that, since y^{k+1} solves $P(k)$, we have

$$\langle \epsilon f_x(y^{k+1}, y^k) - G'(y^k) + G'(y^{k+1}), x - y^{k+1}\rangle \geq 0 \quad \forall x \in K.$$

Computing the previous inequality for $x = y^*$, we obtain

$$\langle G'(y^{k+1}) - G'(y^k), y^* - y^{k+1}\rangle \ \geq \ \langle \epsilon f_x(y^{k+1}, y^k), y^{k+1} - y^*\rangle \geq$$

$\epsilon(f(y^{k+1}, y^k) - f(y^*, y^k))$, recalling that $f(\cdot, y^k)$ is convex.
Therefore

$$\Lambda(y^k) - \Lambda(y^{k+1}) \geq \frac{b}{2}\|y^{k+1} - y^k\|^2 +$$

$$\epsilon(f(y^{k+1}, y^k) + f(y^k, y^*)) - \epsilon(f(y^*, y^k) + f(y^k, y^*)).$$

Exploiting *(iv)* and *(v)*, we obtain

$$\Lambda(y^k) - \Lambda(y^{k+1}) \geq \frac{b}{2}\|y^{k+1} - y^k\|^2 +$$

$$\epsilon(f(y^{k+1}, y^*) - \alpha\|y^k - y^{k+1}\|^2 - \beta\|y^k - y^*\|^2) +$$

$$\epsilon a\|y^k - y^*\|^2 \geq (\frac{b}{2} - \epsilon\alpha)\|y^k - y^{k+1}\|^2 + \epsilon(a - \beta)\|y^k - y^*\|^2.$$

Therefore the sequence $\Lambda(y^k)$ is strictly decreasing and bounded from below,by (5), so that $\Lambda(y^k) \longrightarrow \bar{\Lambda}$ and $y^k \longrightarrow y^*$. □

In the next section we will see that the Algorithm 3.1, if applied to a variational inequality problem, collapses into the one proposed by Cohen in [4].

4 Applications to variational inequalities and optimization problems

Consider the generalized variational inequality (for short GVI):

$$\text{find } y^* \in K \text{ s.t. } \langle F(y^*), x - y^* \rangle \geq \phi(y^*) - \phi(x) \quad \forall x \in K, \qquad (GVI)$$

where $F : K \longrightarrow X^*$, $\phi : K \longrightarrow \mathbb{R}$ is a l.s.c. proper convex function. Put

$$f(x, y) := \langle F(y), x - y \rangle - \phi(y) + \phi(x),$$

then GVI is equivalent to EP so that we can apply the Algorithm 3.1 to the problem GVI.

Algorithm 4.1.

(i) Let $k = 0$, $y^0 \in K$;
(ii) let y^{k+1} be the solution of the problem :

$$\min_{x \in K} [\langle \epsilon F(y^k) - G'(y^k), x \rangle + G(x) + \epsilon \phi(x)].$$

(iii) if $\|y^{k+1} - y^k\| < \mu$, for some fixed $\mu > 0$, then STOP, otherwise put $k = k + 1$ and go to step (ii).

Algorithm 4.1 coincides with the one proposed by Cohen [4].

Theorem 4.1 *Under the hypotheses*
(i) F is a strongly monotone operator on K (with modulus a) which is weakly continuous over every finite-dimensional subspace of X;
(ii) G is differentiable and strongly convex on K with modulus b;
(iii) there exist constants α, $\beta > 0$ such that $\forall x, y, z \in K$:

$$\langle F(y) - F(z), x - y \rangle \geq -\alpha \|x - y\|^2 - \beta \|y - z\|^2.$$

Then, if $\epsilon \leq \frac{b}{2\alpha}$ and $\beta < a$, the sequence $\{y^k\}$, defined in the Algorithm 4.1, strongly converges towards the solution y^* of GVI.

Proof. It follows from Theorem 3.1, putting

$$f(x,y) := \langle F(y), x-y\rangle - \phi(y) + \phi(x).$$

□

Theorem 4.1 is a slight generalization of Theorem 2.2 of [4] since it is possible to show that, if F is Lipschitz continuous over K, then iii) holds for a suitable choice of the costants α and β.

Proposition 4.1 *Assume that F is Lipschitz continuous with modulus L over K, then condition iii) of Theorem 4.1 holds provided that $\sqrt{\alpha\beta} \geq L\backslash 2$.*

Proof. Since F is Lipschitz continuous, we have

$$\langle F(y)-F(z), x-y\rangle \geq -\|F(y)-F(z)\|\|x-y\| \geq -L\|y-z\|\|x-y\| \geq -2\sqrt{\alpha\beta}\|y-z\|\|x-y\| \geq -\alpha\|x-y\|^2 - \beta\|y-z\|^2, \quad \forall x,y,z \in K.$$

□

Remark 4.1 In order to apply Theorem 4.1 it is sufficient to choose α and β that fulfil the following system:

$$\sqrt{\alpha\beta} \geq L\backslash 2, \ \beta < a, \alpha > 0, \ \beta > 0.$$

The analysis of the variational inequality GVI allows to define a further algorithm for a constrained extremum problem of the form

$$\min_{x\in K} \, [\psi(x) + \phi(x)] \tag{P}$$

where $\psi : K \longrightarrow \mathbb{R}$ is a (Gateaux) differentiable strongly convex functional, $\phi : K \longrightarrow \mathbb{R}$ is convex with finite directional derivative $\phi'(x;z)$, $\forall x, z \in K$.
As already mentioned, (P) is equivalent to the equilibrium problem

$$f(x, y^*) \geq 0, \quad \forall x \in K,$$

provided that $f(x,y) = \psi(x) + \phi(x) - \psi(y) - \phi(y)$. We observe that condition iii) of Theorem 3.1 cannot be fulfilled whatever ψ and ϕ may be. An equivalent equilibrium problem can be stated as a first order optimality condition for (P).

Proposition 4.2 *y^* is a solution of P iff it a solution of the GVI*

$$\langle \psi'(y^*), x - y^*\rangle + \phi(x) - \phi(y^*) \geq 0, \quad \forall x \in K. \tag{6}$$

Proof. We observe that, since P is a convex problem, y^* is a solution of P if and only if

$$\langle \psi'(y^*), x - y^* \rangle + \phi'(y^*; x - y^*) \geq 0, \quad \forall x \in K. \tag{7}$$

Suppose that (7) holds. Since ϕ is convex we have (see [13])

$$\phi(x) - \phi(y^*) \geq \phi'(y^*; x - y^*), \quad \forall x \in K,$$

so that $\langle \psi'(y^*), x - y^* \rangle \geq -\phi'(y^*; x - y^*) \geq \phi(y^*) - \phi(x), \quad \forall x \in K.$ Vice versa if (6) holds then y^* is minimum point of the problem

$$\min_{x \in K} \left[\langle \psi'(y^*), x - y^* \rangle + \phi(x) - \phi(y^*) \right],$$

which first order optimality condition is given by (7) and the proof is complete. □

The variational inequality (6) can be solved by means of the Algorithm 4.1 which, in this setting, turns out to be a particular case of Algorithm 2.1 stated in [3] by Cohen.

5 Concluding remarks

We have considered a fixed point algorithm for solving a general equilibrium problem. We have shown that this method, that could be stated directly for the original problem, must be applied to an equivalent auxiliary equilibrium problem in order to achieve the convergence.

Many are the possible developments of the analysis: t he connections with the proximal methods for equilibrium problems [12]; the applications to decomposition algorithms [3, 6, 2]; the extensions to vector equilibrium problems (see e.g. [8]); the applications of the epiconvergence theory in the analysis of the auxiliary problem principle [14].

References

[1] E. Blum and W. Oettli, From Optimization and Variational Inequalities to Equilibrium Problems, The Mathematics Student, **Vol. 63**, 1-23, 1993;

[2] G. Chen and R.T. Rockafellar, Convergence Rates in Forward-Backward Splitting, *SIAM Journal of Optimization* , **Vol. 7**, 421-444, 1997;

[3] G. Cohen, Auxiliary Problem Principle and Decomposition of Optimization Problems, *Journal of Optimization Theory and Applications*, **Vol. 32**, 277-305, 1980;

[4] G. Cohen, Auxiliary Problem Principle Extended to Variational Inequalities, *Journal of Optimization Theory and Applications*, **Vol. 59**, 325-333, 1988;

[5] I. Ekeland and R. Temam, Convex Analysis and Variational Problems, North-Holland, Amsterdam, 1977;

[6] E.N. Farouq and G. Cohen, Progressive Regularization of Variational Inequalities and Decomposition Algorithms, *Journal of Optimization Theory and Applications*, **Vol. 97**, 407-433, 1998;

[7] F. Giannessi, Separation of sets and Gap Functions for Quasi-Variational Inequalities, in *"Variational Inequalities and Network Equilibrium Problems"*, F.Giannessi and A.Maugeri (eds), Plenum Publishing Co, 101-121, 1995;

[8] F. Giannessi, Vector Variational Inequalities and Vector Equilibria, Kluwer Acad. Publ., Dordrecht, Boston, London, 2000;

[9] P.T. Harker, J.S. Pang, Finite–Dimensional Variational Inequalities and Nonlinear Complementarity Problem: a Survey of Theory, Algorithms and Applications, *Mathematical Programming*, **Vol.48**, 161-220, 1990;

[10] S. Karamardian, An Existence Theorem for the Complementary Problem, *Journal of Optimization Theory and Applications*, **Vol.18**, 445-454, 1976;

[11] G. Mastroeni, Minimax and extremum problems associated to a Variational Inequality, *Supplemento ai Rendiconti del Circolo Matematico di Palermo*, **Suppl. 58** , 185-196, 1999;

[12] A. Moudafi, Proximal Point Algorithm Extended to Equilibrium Problems, *Journal of Natural Geometry*, **Vol. 15**, 91-100, 1999;

[13] R.T. Rockafellar, Convex Analysis, Princeton University Press, Princeton, 1970.

[14] G. Salmon, V.H. Nguyen and J.J. Strodiot, Coupling the Auxiliary Problem Principle and Epiconvergence Theory to Solve General Variational Inequalities, *Journal of Optimization Theory and Applications*, **Vol. 104**, 629-657, 2000.

MULTICRITERIA SPATIAL PRICE NETWORKS: STATICS AND DYNAMICS

Anna Nagurney
Department of Finance and Operations Management,
Isenberg School of Management, University of Massachusetts
Amherst, Massachusetts 01003

June Dong
Department of Marketing and Management, School of Business
State University of New York at Oswego
Oswego, New York 13126

Ding Zhang
Department of Marketing and Management, School of Business
State University of New York at Oswego
Oswego, New York 13126

Abstract In this paper, we develop a spatial price network equilibrium model in which consumers at the demand markets consider both the transportation cost and the transportation time associated with obtaining the particular commodity. We provide the governing equilibrium conditions for the multicriteria spatial price problem and derive the variational inequality formulation. We establish existence and uniqueness of the equilibrium commodity shipment and demand price pattern and then propose a dynamic tatonnement process whose set of stationary points coincides with the set of solutions of the variational inequality problem. An iterative scheme is described which provides a time discretization of the continuous time adjustment process and which converges to a stationary point. Numerical examples are given for illustrative purposes.

1 Introduction

Spatial price equilibrium problems have provided a basic formalism for the study of a wide variety of applications arising in agricultural markets, energy, and in interregional and international trade. The rigorous formulation of such problems dates to Samuelson (1952) and Takayama and Judge (1971) who considered problems in which the governing equilibrium conditions could be reformulated as an equivalent mathematical programming (optimization) problem.

In spatial price equilibrium problems, one assumes that the supply and demand markets are spatially separated, that the competition is perfect, and that, in equilibrium, a commodity produced at a supply market will be shipped to a demand market, where it is consumed, provided that the supply price plus the unit transportation cost is equal to the demand price. If the supply price at the supply market plus the unit cost of transportation exceeds the demand price that the consumers are willing to pay for the commodity, then the commodity will not be shipped between the pair of markets.

The basic models were subsequently extended to allow for the treatment of asymmetric price and transportation cost functions and multicommodity situations using a variational inequality framework (cf. Dafermos and Nagurney (1987), Florian and Los (1982), and Nagurney (1987), among others). Refer to Nagurney (1999) and the references therein for recent research on the formulation, theoretical analysis, and computation of solutions to spatial price equilibrium problems.

In this paper, we propose a multicriteria spatial price network equilibrium model. In the model, we assume that consumers in each distinct demand market may be faced with several criteria in selecting the commodity that is produced, specifically, not only the price of the commodity but also the time it takes to receive the commodity. Hence, the consumers are not only price-sensitive but also time-sensitive. We construct explicit demand functions which express these concerns and study the model both from a *static* perspective, from the point of view of the equilibrium pattern, using the theory of variational inequalities, as well as from a *dynamic* perspective through the use of a dynamic tatonnement process which reveals how the producers adjust their commodity shipments to the demand markets and how the generalized prices at the demand markets adjust. The theoretical analysis of the dynamics is conducted using projected dynamical systems theory (see Nagurney and Zhang (1996)).

We note that multicriteria network equilibrium models have been constructed for traffic networks and were introduced by Quandt (1967) and Schneider (1968) and explicitly consider that travelers may be faced with several criteria, notably, travel time and travel cost, in selecting their optimal routes of travel. The ideas were further developed by Dial (1979) who proposed an uncongested model and Dafermos (1981) who introduced congestion effects and derived an infinite-dimensional variational inequality formulation of her multiclass, multicriteria traffic network equilibrium problem, along with some qualitative properties.

Recently, there has been renewed interest in the formulation, analysis, and computation of multicriteria traffic network equilibrium problems.

Researchers who have considered an infinite-dimensional variational inequality formulation, motivated by Dafermos' (1981) multiclass model, have included Leurent (1993a) (see also Leurent (1993b)), who presented an elastic demand formulation but did not allow travel cost to be a function of flow. For an overview of multicriteria traffic network equilibrium problems and different formulations, see Leurent (1998) and Marcotte (1998).

In this paper, we build upon the recent work of Nagurney (2000) and Nagurney and Dong (2002) who developed, respectively, a multiclass, multicriteria traffic network equilibrium model with fixed travel demands and with elastic travel demands. However, due to the special structure of the spatial price network problem under consideration here we are able to obtain sharper results in the sense that we are able to establish, under quite reasonable conditions, strict monotonicity of the function that enters the variational inequality problem. Moreover, for the first time, we propose a dynamical system to describe the evolution of the trajectories for a multicriteria network equilibrium problem.

The paper is organized as follows. In Section 2, we present the multicriteria spatial price equilibrium model and derive the variational inequality formulation of the governing equilibrium conditions. In Section 3, we focus on the "statics" and obtain an existence result as well as a uniqueness result. In Section 4, we then describe a dynamic tatonnement process and relate the dynamic and static interpretations of the problem. In Section 5, we propose the Euler method, which is a discrete-time algorithm, and provide convergence results. Section 6 contains numerical examples which illustrate the model and the computational approach. Section 7 summarizes our results and presents the conclusions.

2 The Multicriteria Spatial Price Model

In this Section, we develop the multicriteria spatial price network equilibrium model. The model permits the consumers at each of the demand markets to weight the transportation cost and the transportation time associated with the shipment of the commodity from the supply markets in an individual manner. The equilibrium conditions are then shown to satisfy a finite-dimensional variational inequality problem (see, e.g., Kinderlehrer and Stampacchia (1980) and Nagurney (1999)).

We assume that a certain commodity is produced at m supply markets and is consumed at n demand markets. We denote a typical supply market by i and a typical demand market by j. Let s_i denote the supply of the commodity at supply market i and let Q_{ij} denote the nonnegative commodity shipment from supply market i to demand market j. We group the supplies into a column vector s in R^m and the commodity shipments into a column vector Q in R^{mn}.

We associate with each supply market i a supply price π_i and we group the supply prices into a row vector π in R^m. We assume that, in general, the supply price at a supply market i can depend on the supplies of the commodity at all the supply markets, that is,

$$\pi_i = \pi_i(s), \quad \forall i, \tag{1}$$

where π is a known smooth function.

The supply of the commodity at each supply market i must satisfy the following conservation of flow equation:

$$s_i = \sum_{j=1}^{n} Q_{ij}, \tag{2}$$

that is, the supply of the commodity at a supply market must be equal to the sum of the commodity shipments from the supply market to all the demand markets.

In view of (1) and (2), and, for simplicity of the subsequent derivations and notation, we define the supply price function $\hat{\pi}_i$, for each supply market i, which is a function of the commodity shipment pattern:

$$\hat{\pi}_i = \hat{\pi}_i(Q) \equiv \pi_i(s), \quad \forall i, \tag{3}$$

and we group these functions into the row vector $\hat{\pi} \in R^m$.

We introduce a unit transportation cost c_{ij} associated with shipping the commodity between supply market i and demand market j and the transportation time t_{ij} associated with the shipment. We group the transportation costs and times, respectively, into the row vectors c in R^{mn} and t in R^{mn}. We assume, in turn, that the unit cost of transportation depends on the quantity of the commodity shipped between the pair of markets, that is,

$$c_{ij} = c_{ij}(Q_{ij}), \quad \forall ij, \tag{4}$$

as does the transportation time, i.e.,

$$t_{ij} = t_{ij}(Q_{ij}), \quad \forall ij, \tag{5}$$

where the transportation costs and times are assumed to be known smooth functions.

We assume that each demand market represents a distinct class of consumer who perceives the transportation cost and time in an individual manner. Hence, consumers in one demand market may not be as concerned as to when the commodity is delivered provided that the transportation cost is low, whereas consumers at another demand market may be more time-sensitive and may be willing to pay a higher transportation cost provided that the commodity reaches them in a more timely manner. We let w_j^1 denote the weight associated with the transportation cost for demand market j and we let w_j^2 denote the weight associated with the transportation time to demand market j. We assume that the weights are positive for all demand markets.

We then construct the *generalized* cost associated with link (i, j) and denoted by $\hat{c}_{ij}$ as follows:

$$\hat{c}_{ij} = w_j^1 c_{ij} + w_j^2 t_{ij}. \tag{6}$$

Note that a possible weighting scheme may be one where the weights for each demand market sum to one, that is, $w_j^1 + w_j^2 = 1$, for all j. Dafermos (1981) utilized such a weighting scheme in the context of a traffic network equilibrium model.

We assume that the demand for the commodity at demand market j is determined according to:

$$d_j = d_j(\lambda), \quad \forall j, \tag{7}$$

where λ is the column vector of demand market *generalized* (since it reflects both time and cost) prices with demand market j's generalized price being denoted by λ_j. We group the demand fuctions into the row vector $d(\lambda) \in R^n$.

Multicriteria Spatial Price Network Equilibrium Conditions

The spatial price network equilibrium conditions in the case of known demand functions (see Takayama and Judge (1971), Nagurney and Zhao (1993), and Nagurney, Takayama, and Zhang (1995a, b)), in the generalized context of the multicriteria spatial price network equilibrium problem, take on the form: A pattern $(Q^*, \lambda^*) \in R_+^{mn+n}$ is an equilibrium pattern if for each pair of supply and demand markets (i, j) the following conditions hold:

$$\hat{\pi}_i(Q^*) + w_j^1 c_{ij}(Q_{ij}^*) + w_j^2 t_{ij}(Q_{ij}^*) \begin{cases} = \lambda_j^*, & \text{if} \quad Q_{ij}^* > 0 \\ \geq \lambda_j^*, & \text{if} \quad Q_{ij}^* = 0, \end{cases} \tag{8}$$

and

$$d_j(\lambda^*) \begin{cases} = \sum_{i=1}^{m} Q_{ij}^*, & \text{if} \quad \lambda_j^* > 0 \\ \leq \sum_{i=1}^{m} Q_{ij}^*, & \text{if} \quad \lambda_j^* = 0. \end{cases} \tag{9}$$

In other words, the commodity will be shipped between a pair of supply and demand markets if the supply price plus the generalized cost associated with shipping the commodity is equal to the generalized price at the demand market. If the supply price plus the generalized cost exceeds the generalized price at the demand market, then there will be no trade between the pair of markets. In addition, if the generalized price associated with a demand market is positive, then the market clears for that demand market; that is, the sum of the commodity shipments from the supply markets to that demand market is equal to the demand associated with that demand market; if the generalized price is zero, then the sum of the commodity shipments can exceed the demand for the commodity at the demand market. Henceforth, we refer to the generalized price simply as the price of the commodity at the particular demand market.

We define the feasible set $\mathcal{K}$ underlying the problem as $\mathcal{K} \equiv \{(Q,\lambda) \mid (Q,\lambda) \in R_+^{mn+n}\}$.

The equivalence between the multicriteria spatial price network equilibrium conditions and a variational inequality is now established.

Theorem 1: Variational Inequality Formulation
A multicriteria commodity shipment and demand price pattern $(Q^*,\lambda^*) \in \mathcal{K}$ is a spatial price network equilibrium, that is, satisfies equilibrium conditions (8) and (9) if and only if it satisfies the variational inequality problem:

$$\sum_{i=1}^{m}\sum_{j=1}^{n}(\hat{\pi}_i(Q^*) + w_j^1 c_{ij}(Q_{ij}^*) + w_j^2 t_{ij}(Q_{ij}^*) - \lambda_j^*) \times (Q_{ij} - Q_{ij}^*)$$

$$+\sum_{j=1}^{n}(\sum_{i=1}^{m} Q_{ij}^* - d_j(\lambda^*)) \times (\lambda_j - \lambda_j^*) \geq 0, \quad \forall (Q,\lambda) \in \mathcal{K}; \tag{10}$$

equivalently, in standard form:

$$\langle F(X^*), X - X^* \rangle \geq 0, \quad \forall X \in \mathcal{K}, \tag{11}$$

where $X \equiv (Q,\lambda)$, and $F(X) \equiv (F_Q(X), F_\lambda(X))$ with component ij of $F_Q(X)$, denoted by $F_Q(X)_{ij}$, given by:

$$F_Q(X)_{ij} = \hat{\pi}_i(Q) + w_j^1 c_{ij}(Q_{ij}) + w_j^2 t_{ij}(Q_{ij}) - \lambda_j, \quad \forall ij,$$

and component j of $F_\lambda(X)$, denoted by $F_\lambda(X)_j$, given by:

$$F_\lambda(X)_j = \sum_{i=1}^{m} Q_{ij} - d_j(\lambda), \quad \forall j.$$

The expression: $\langle \cdot,\cdot \rangle$ denotes the inner product in N-dimensional Euclidean space R^N where here $N = mn + n$.

Proof: Assume that (Q^*, λ^*) satisfies equilibrium conditions (8) and (9). Then we have from (8) that, for a fixed pair of supply and demand markets ij:

$$(\hat{\pi}_i(Q^*)+w_j^1c_{ij}(Q_{ij}^*)+w_j^2t_{ij}(Q_{ij}^*)-\lambda_j^*)\times(Q_{ij}-Q_{ij}^*) \geq 0, \quad \forall Q_{ij} \geq 0, \tag{12}$$

and from (9), that, for a fixed demand market j:

$$-(d_j(\lambda^*) - \sum_{i=1}^{m} Q_{ij}^*) \times (\lambda_j - \lambda_j^*) \geq 0, \quad \forall \lambda_j \geq 0. \tag{13}$$

Summing inequalities (12) over all pairs of markets ij, and summing (13) over all demand markets j, and adding the two resulting inequalities, yields

$$\sum_{i=1}^{m}\sum_{j=1}^{n}(\hat{\pi}_i(Q^*) + w_j^1c_{ij}(Q_{ij}^*) + w_j^2t_{ij}(Q_{ij}^*) - \lambda_j^*) \times (Q_{ij} - Q_{ij}^*)$$

$$-\sum_{j=1}^{n}(d_j(\lambda^*) - \sum_{i=1}^{m} Q_{ij}^*)) \times (\lambda_j - \lambda_j^*) \geq 0, \quad \forall (Q, \lambda) \in R_+^{mn+n}, \tag{14}$$

which is variational inequality (10).

Assume now that $(Q^*, \lambda^*) \in \mathcal{K}$ is a solution to variational inequality (10). Let $\lambda = \lambda^*$ and let $Q_{kl} = Q_{kl}^*$ for all $kl \neq ij$, and substitute these into variational inequality (10), yielding

$$(\hat{\pi}_i(Q^*)+w_j^1c_{ij}(Q_{ij}^*)+w_j^2t_{ij}(Q_{ij}^*)-\lambda_j^*)\times(Q_{ij}-Q_{ij}^*) \geq 0, \quad \forall Q_{ij} \geq 0, \tag{15}$$

which, in turn, implies equilibrium conditions (8). Indeed, since if $Q_{ij}^* > 0$, then the term following the multiplication sign in (15) can be either positive, negative, or zero, so for the product to be nonnegative, we must have that the term preceding the multiplication sign in (15) is zero. Hence, the first part of (8) holds true. On the other hand, if $Q_{ij}^* = 0$, then the term after the multiplication sign in (15) is nonnegative and for the product of two terms in (15) to be nonnegative, implies that the first term must be nonnegative, which, in turn, is equivalent to the second part of condition (8) being satisfied.

Similarly, let $Q = Q^*$, and let $\lambda_k = \lambda_k^*$, for all $k \neq j$, and substitute into (10), yielding

$$-(d_j(\lambda^*) - \sum_{i=1}^{m} Q_{ij}^*) \times (\lambda_j - \lambda_j^*) \geq 0, \quad \forall \lambda_j \geq 0, \tag{16}$$

which, in turn, arguing as above, implies equilibrium conditions (9). □

3 Qualitative Properties

In this Section, we provide some qualitative properties of the solution to variational inequality (10). In particular, we derive existence and uniqueness results.We also investigate properties of the function F (see (11)) that enters the variational inequality of interest here.

Since the feasible set $\mathcal{K}$ is not compact we cannot derive existence simply from the assumption of continuity of the supply price, transportation cost and time, and demand functions. Nevertheless, we can impose a rather weak condition to guarantee existence of a solution pattern.

Let $r = (r_1, r_2) \in R^2$ and denote by Ω_r the rectangle in R^{mn+n} such that

$$\Omega_r = \{(Q, \lambda) | 0 \leq Q \leq r_1, 0 \leq \lambda \leq r_2\}, \tag{17}$$

where $Q \leq r_1, \lambda \leq r_2$ means that $Q_{ij} \leq r_1$ and $\lambda_j \leq r_2$ for all ij. Then $\mathcal{K}_r = \mathcal{K} \cap \Omega_r$, the intersection of original feasible set with the rectangle, is a bounded closed convex subset of R^{mn+n}. Thus, the following variational inequality

$$\langle F(X^r), X - X^r \rangle \geq 0, \quad \forall X^r \in \mathcal{K}_r, \tag{18}$$

admits at least one solution $X^r \in \mathcal{K}_r$, from the standard theory of variational inequalities, since $\mathcal{K}_r$ is compact and F is continuous. Following Kinderlehrer and Stampacchia (1980) (see also Theorem 1.5 in Nagurney (1999)), we then have:

Theorem 2
Variational inequality (10) admits a solution if and only if there exist $r_1 > 0, r_2 > 0$, *such that variational inequality (18) admits a solution* $X^r = (Q^r, \lambda^r)$ *in* $\mathcal{K}_r$ *with*

$$Q^r < r_1, \quad \lambda^r < r_2. \tag{19}$$

Proposition 1
Suppose that there exist positive constants M, N,*and* R, *with* $R > 0$, *such that:*

$$\hat{\pi}_i(Q) + w_j^1 c_{ij}(Q_{ij}) + w_j^2 t_{ij}(Q_{ij}) > R, \quad \forall Q, \text{ with } Q_{ij} \geq N, \tag{20}$$

$$d_j(\lambda) \leq N, \quad \forall \lambda \text{ with } \lambda_j > M. \tag{21}$$

Then variational inequality (10) admits at least one solution.

Proof: Follows using analogous arguments as the proof of existence for Proposition 1 in Nagurney and Zhao (1993).

Assumptions (20) and (21) are reasonable from an economics perspective, since when the commodity shipment between a pair of markets is large, we can expect the corresponding supply price or the generalized cost to also be large. Moreover, in the case where the generalized price of the commodity at a demand market is high, we can expect that the demand for the commodity will be low at that market.

We now turn to investigating uniqueness of the equilibrium, that is, a solution to variational inequality (10). We first, however, need to establish the following lemmas.

Lemma 1

Assume that the Jacobian matrices of the transportation cost and transportation time functions are both positive definite, for all $Q \in \mathcal{K}$. Then, the generalized cost function $\hat{c}(Q)$ with component ij given by $w_j^1 c_{ij}(Q_{ij})+ w_j^2 t_{ij}(Q_{ij})$ is strictly monotone for such Q, that is,

$$\langle \hat{c}(Q^1) - \hat{c}(Q^2), Q^1 - Q^2 \rangle > 0, \quad \forall Q^1, Q^2 \in \mathcal{K}, \quad Q^1 \neq Q^2. \tag{22}$$

Proof: Recall, from the standard theory of variational inequalities (see Kinderlehrer and Stampacchia (1980) and Theorem 1.7 in Nagurney (1999)), that if the Jacobian matrix of $\hat{c}(Q)$ is positive definite over $\mathcal{K}$, then $\hat{c}(Q)$ is strictly monotone.

The Jacobian of $\hat{c}(Q)$ can be expressed as:

$$\nabla \hat{c}(Q) = [\nabla c]\left[W^1\right] + [\nabla t]\left[W^2\right], \tag{23}$$

where ∇c is the Jacobian of the transportation cost functions, ∇t is the Jacobian of the transportation time functions, and W^i; $i = 1, 2$ are diagonal $mn \times mn$ matrices with the diagonal components $(w_1^i, w_2^i, \ldots, w_n^i)$ repeating m times. Note that since both the transportation cost and time functions are assumed to be separable, their Jacobian matrices are also diagonal.

Clearly, since the weights are assumed to be positive, both W^1 and W^2 are positive definite matrices. Since the Jacobians of c and t are, by assumption, also positive definite matrices, it follows that $[\nabla c]\,[W^1]$ is positive definite since it is the product of two diagonal and positive definite matrices as is $[\nabla t]\,[W^2]$. Finally, $\nabla \hat{c}$ must be positive definite, since it is the sum of two positive definite matrices. □

Lemma 2

Assume that $\hat{\pi}(Q)$ is strictly monotone increasing, that is, that

$$\langle \hat{\pi}(Q^1) - \hat{\pi}(Q^2), Q^1 - Q^2 \rangle > 0, \quad \forall Q^1, Q^2 \in \mathcal{K}, \quad Q^1 \neq Q^2, \tag{24}$$

$d(\lambda)$ is strictly monotone decreasing, that is,

$$-\langle d(\lambda^1) - d(\lambda^2), \lambda^1 - \lambda^2 \rangle > 0, \quad \forall \lambda^1, \lambda^2 \in \mathcal{K}, \quad \lambda^1 \neq \lambda^2, \tag{25}$$

and that the Jacobian matrices ∇c and ∇t are positive definite over the feasible set. Then $F(X)$ is strictly monotone over $\mathcal{K}$.

Proof: Write

$$\langle F(X') - F(X''), X' - X'' \rangle$$

$$= \sum_{i=1}^{m} \sum_{j=1}^{n} (\hat{\pi}_i(Q') + w_j^1 c_{ij}(Q'_{ij}) + w_j^2 t_{ij}(Q'_{ij}) - \lambda'_j) - (\hat{\pi}_i(Q'')$$

$$+ w_j^1 c_{ij}(Q''_{ij}) + w_j^2 t_{ij}(Q''_{ij}) - \lambda''_j)) \times (Q'_{ij} - Q''_{ij})$$

$$+ \sum_{j=1}^{n} ((\sum_{i=1}^{m} Q'_{ij} - d_j(\lambda')) - (\sum_{i=1}^{m} Q''_{ij} - d_j(\lambda''))) \times (\lambda'_j - \lambda''_j)$$

$$= \sum_{i=1}^{m} \sum_{j=1}^{n} ((\hat{\pi}_i(Q') + w_j^1 c_{ij}(Q'_{ij}) + w_j^2 t_{ij}(Q'_{ij})) - (\hat{\pi}_i(Q'') + w_j^1 c_{ij}(Q''_{ij})$$

$$+ w_j^2 t_{ij}(Q''_{ij})) \times (Q'_{ij} - Q''_{ij}) - \sum_{j=1}^{n} (d_j(\lambda') - d_j(\lambda'')) \times (\lambda'_j - \lambda''_j). \quad (26)$$

We now argue that the right-most term in (26) is strictly greater than zero, and, hence, $F(X)$ is strictly monotone over $\mathcal{K}$. Indeed, by assumption, we have that the demand functions are strictly monotone decreasing, and we have already established in Lemma 1 that the generalized cost functions are also strictly monotone under the assumptions of positive definite Jacobians for the transportation cost and time functions. Finally, it is straightforward to verify that the Jacobian of $\hat{\pi}$ is monotone. Hence, the expression(s) in (26) must be strictly greater than zero and the conclusion follows. □

Theorem 3: Uniqueness

Assume that $\hat{\pi}(Q)$ is strictly monotone increasing and that $d(\lambda)$ is strictly monotone decreasing. Also, assume that ∇c and ∇t are positive definite over $\mathcal{K}$. Then the equilibrium pattern (Q^*, λ^*) satisfying variational inequality (10) is unique.

Proof: Assume that there are two distinct solutions (Q^1, λ^1) and (Q^2, λ^2) to variational inequality (10). Then, we must have that:

$$\sum_{i=1}^{m} \sum_{j=1}^{n} (\hat{\pi}_i(Q^1) + w_j^1 c_{ij}(Q^1_{ij}) + w_j^2 t_{ij}(Q^1_{ij}) - \lambda^1_j) \times (Q_{ij} - Q^1_{ij})$$

$$+ \sum_{j=1}^{n} (\sum_{i=1}^{m} Q^1_{ij} - d_j(\lambda^1)) \times (\lambda_j - \lambda^1_j) \geq 0, \quad \forall (Q, \lambda) \in \mathcal{K}, \quad (27)$$

and

$$\sum_{i=1}^{m}\sum_{j=1}^{n}(\hat{\pi}_i(Q^2) + w_j^1 c_{ij}(Q_{ij}^2) + w_j^2 t_{ij}(Q_{ij}^2) - \lambda_j^2) \times (Q_{ij} - Q_{ij}^2)$$

$$+\sum_{j=1}^{n}(\sum_{i=1}^{m} Q_{ij}^2 - d_j(\lambda^2)) \times (\lambda_j - \lambda_j^2) \geq 0, \quad \forall (Q,\lambda) \in \mathcal{K}. \tag{28}$$

Let $(Q,\lambda) = (Q^1,\lambda^1)$ and substitute into (28) and let $(Q,\lambda) = (Q^2,\lambda^2)$ and substitute into (27). Adding the two resulting inequalities, after algebraic simplification, yields:

$$\sum_{i=1}^{m}\sum_{j=1}^{n}(\hat{\pi}_i(Q^1) + w_j^1 c_{ij}(Q_{ij}^1) + w_j^2 t_{ij}(Q_{ij}^1)) - (w_j^1 c_{ij}(Q_{ij})$$

$$+w_j^2 t_{ij}(Q_{ij}^2)) \times (Q_{ij}^1 - Q_{ij}^2) - \sum_{j=1}^{n}(d_j(\lambda^1) - d_j(\lambda^2)) \times (\lambda_j^1 - \lambda_j^2) \leq 0, \tag{29}$$

but this is in contradiction to the assumptions that the $\hat{\pi}$ functions are strictly monotone increasing, the demand functions d are strictly monotone decreasing, and we know from Lemma 1 that the generalized cost functions are also strictly monotone. Hence, it follows that $(Q^1,\lambda^1) = (Q^2,\lambda^2)$. Thus, uniqueness has been established. □

4 The Dynamics

In this Section, a dynamic counterpart of the multicriteria spatial price network equilibrium model of Section 3 is developed. The set of stationary points coincides with the set of solutions to a variational inequality problem.

The Dynamics of the Commodity Shipments

The dynamic model presented here assumes that the commodity shipments adjust according to the difference between the generalized demand price and the supply price plus the generalized cost associated with a market pair. Mathematically, we have that: For all pairs of markets (i,j) at time τ:

$$\dot{Q}_{ij}(\tau) = \begin{cases} \lambda_j(\tau) - \hat{\pi}_i(Q(\tau)) \\ -w_j^1 c_{ij}(Q_{ij}(\tau)) - w_j^2 t_{ij}(Q_{ij}(\tau)), & \text{when } Q_{ij}(\tau) > 0, \\ \\ \max\{0, \lambda_j(\tau) - \hat{\pi}_i(Q(\tau)) \\ -w_j^1 c_{ij}(Q_{ij}(\tau)) - w_j^2 t_{ij}(Q_{ij}(\tau))\}, & \text{when } Q_{ij}(\tau) = 0. \end{cases} \tag{30}$$

According to (30) the commodity shipment between a pair of supply and demand markets will increase if the generalized demand price at the demand market exceeds the supply price plus the generalized cost associated with shipping the commodity from the supply market. On the other hand, the commodity shipment will decrease if the supply price at the supply market plus the generalized cost exceeds the generalized demand price that the consumers are paying at the demand market. Note also that (30) guarantees that the commodity shipments can never be negative, which would violate feasibility.

The Generalized Demand Price Dynamics

The demand market generalized prices, in turn, evolve at time τ as follows:

$$\dot{\lambda}_j(\tau) \begin{cases} = d_j(\lambda(\tau)) - \sum_{i=1}^{m} Q_{ij}(\tau), & \text{when} \quad \lambda_j(\tau) > 0, \\ = \max\{0, d_j(\lambda(\tau)) - \sum_{i=1}^{m} Q_{ij}(\tau)\}, & \text{when} \quad \lambda_j(\tau) = 0. \end{cases} \tag{31}$$

According to (31), the generalized demand price of the commodity at a demand market will increase if the demand exceeds the supply of the commodity at the demand market; it will decrease if the supply exceeds the demand. Moreover, (31) guarantees that the generalized price at the demand market will not be negative.

Hence, this adjustment process is in concert with those proposed by Nagurney, Takayama, and Zhang (1995a, b) for (single-criteria) spatial price equilibrium problems.

The Projected Dynamical System

The dynamic model described by (30) and (31) can now be rewritten as a *projected dynamical system* (cf. Nagurney and Zhang (1996)). For definiteness, we recall some preliminaries and the definition.

The class of ordinary differential equations of interest here takes the form:

$$\dot{X} = \Pi_{\mathcal{K}}(X, -F(X)), \quad X(0) = X_0 \in \mathcal{K}, \tag{32}$$

where $\mathcal{K}$ is a closed convex set, corresponding to the constraint set in a particular application, $F(X)$ is a vector field defined on $\mathcal{K}$, and $\Pi_{\mathcal{K}}(X, -F(X))$ is given by:

$$\Pi_{\mathcal{K}}(X, -F(X)) = \lim_{\delta \to 0} \frac{(P_{\mathcal{K}}(X - \delta F(X)) - X)}{\delta} \tag{33}$$

and $P_{\mathcal{K}}$ is the projection map:

$$P_{\mathcal{K}}(X) = \arg\min_{z \in \mathcal{K}} \|X - z\|. \tag{34}$$

We refer to the ordinary differential equation in (32) as ODE$(F, \mathcal{K})$.

Observe that the right-hand side of the ordinary differential equation (32) is associated with a projection operator and is, hence, discontinuous on the boundary of $\mathcal{K}$. Therefore, one needs to explicitly state what one means by a solution to an ODE with a discontinuous right-hand side.

Definition 1: Projected Dynamical System
Define the projected dynamical system (PDS) $X_0(t) : \mathcal{K} \times R \mapsto \mathcal{K}$ *as the family of solutions to the Initial Value Problem* (IVP)*(32) for all* $X_0 \in \mathcal{K}$.

It is apparent from the definition that $X_0(0) = X_0$. In the context of the spatial price equilibrium model, $X_0 = (Q_0, \lambda_0)$ is the initial point corresponding to the initial commodity shipment and price pattern. The trajectory of (32) describes the dynamic evolution of and the dynamic interactions between the commodity shipment and the price patterns. A projected dynamical system differs from a classical dynamical system in that the right-hand side in (32) is discontinuous due to the explicit incorporation of the constraint set, where recall that the constraint set here is the nonnegative orthant.

Theorem 4
The set of stationary points of the projected dynamical system (32) coincides with the set of solutions of the variational inequality (11); equivalently, variational inequality (10).

Proof: According to the fundamental theorem of projected dynamical systems (see Nagurney and Zhang (1996)), X^* is a stationary point of the projected dynamical system (32) if and only if it is a solution to variational inequality (11), which is equivalent to (10). □

5 The Discrete-Time Algorithm

In this Section, we propose a discrete-time algorithm, the Euler method, for the computation of the equilibrium pattern. The algorithm provides a discretization of the continuous time adjustment process given in Section 4. Specifically, the Euler method is a special case of the general iterative scheme for projected dynamical systems proposed by Dupuis and Nagurney (1993) (see also Nagurney and Zhang (1996)). The algorithm, hence, computes a solution to variational inequality (10) and also provides a discrete-time approximation to the projected dynamical system (32), the stationary points of which (cf. Theorem 4) coincide with the solutions of variational inequality (10).

Its statement in the general form for the solution of variational inequality (11) and for the time discretization of the corresponding projected dynamical system is given by:

$$X^{k+1} = P_{\mathcal{K}}(X^k - a_k F(X^k)), \tag{35}$$

where k denotes an iteration (or time period) and $\{a_k\}$ is a sequence of positive scalars to be discussed later.

In particular, in the context of the multicriteria spatial price network equilibrium problem formulated as (10), with F as defined following (11) and the feasible set $\mathcal{K}$ being R_+^{mn+n}, the projection operation takes on a very simple form for computational purposes, and, hence, the commodity shipments as well as the demand market prices can be computed at an iteration in closed form as follows:

$$Q_{ij}^{k+1} = \max\{0, a_k(\lambda_j^k - w_j^1 c_{ij}(Q_{ij}^k) - w_j^2 t_{ij}(Q_{ij}^k) - \hat{\pi}_i(Q^k)) + Q_{ij}^k\}, \quad \forall ij, \tag{36}$$

and

$$\lambda_j^{k+1} = \max\{0, a_k(d_j(\lambda^k) - \sum_{i=1}^m Q_{ij}^k) + \lambda_j^k\}, \quad \forall j. \tag{37}$$

We first give the precise conditions for the general convergence theorem, present its statement, and then interpret it for the Euler method applied to the spatial price equilibrium problem. We note that the conditions are given for the general iterative scheme of Dupuis and Nagurney (1993) where an iteration is given by:

$$X^{k+1} = P_{\mathcal{K}}(X^k - a_k F_k(X^k)) \tag{38}$$

with F_k denoting an approximation to F, which, in the case of the Euler method is: $F_k = F$.

Assumption 1

Fix an initial condition $X^0 \in \mathcal{K}$ and define the sequence $\{X^k\}$ by (38). Assume the following conditions.

1. $\sum_{k=1}^{\infty} a_k = \infty, \quad a_k > 0, \quad a_k \to 0$ *as* $k \to \infty$.

2. $d(F_k(X), \bar{F}(X)) \to 0$ *uniformly on compact subsets of* $\mathcal{K}$ *as* $k \to \infty$, *where* $d(X, A) = \inf\{\|X - y\|, y \in A\}$ *and where the bar over the* F *denotes closure.*

3. Define ϕ_y *to be the unique solution to* $\dot{X} = \Pi_{\mathcal{K}}(X, -F(X))$ *that satisfies* $\phi_y(0) = y \in \mathcal{K}$. *The* $\omega-$*limit set*

$$\cup_{y\in\mathcal{K}} \cap_{t\geq 0} \cup_{s\geq t} \{\phi_y(s)\}$$

is contained in the set of stationary points of $\dot{X} = \Pi_{\mathcal{K}}(X, -F(X))$.

4. The sequence $\{X^k\}$ *is bounded.*

5. The solutions to $\dot{X} = \Pi_{\mathcal{K}}(X, -F(X))$ *are stable in the sense that given any compact set* $\mathcal{K}_1$ *there exists a compact set* $\mathcal{K}_2$ *such that* $\cup_{y\in\mathcal{K}\cap\mathcal{K}_1} \cup_{t\geq 0} \{\phi_y(t)\} \subset \mathcal{K}_2$.

The assumptions are phrased as they are because they describe more or less what is needed for convergence, and because there are a number of rather different sets of conditions that imply the assumptions, depending on the application (see, e.g., Nagurney and Zhang (1996)).

Theorem 5 (Dupuis and Nagurney (1993))
Let S denote the solutions to the variational inequality (11), and invoke Assumption 1 and Assumption 2, where

Assumption 2
There exists a $B < \infty$ such that the vector field $-F : R^{mn+n} \mapsto R^{mn+n}$ satisfies the linear growth condition: $\|-F(X)\| \leq B(1+\|X\|)$ for $X \in \mathcal{K}$, and also

$$\langle -F(X) + F(y), X - y \rangle \leq B\|X - y\|^2 \tag{39}$$

for all $X, y \in \mathcal{K}$.

Suppose $\{X^k\}$ is the scheme generated by (38). Then $d(X^k, S) \to 0$ as $k \to \infty$.

Corollary 1 (Dupuis and Nagurney (1993))
Assume the conditions of Theorem 5, and also that S consists of a finite set of points. Then $\lim_{k\to\infty} X^k$ exists and equals a solution to the variational inequality (11).

We now interpret the meaning of Assumptions 1 and 2 in the context of the spatial price equilibrium problem, in order to establish the convergence of the Euler-type method, which is a special case of the general iterative scheme of Dupuis and Nagurney (1993).

In order to establish convergence of the Euler method we need to adopt the following assumption:

Assumption 3
Assume that there exist sufficiently large constants M_d, M_Q, and M_λ, such that

$$d_j(\lambda) \leq M_d, \quad \forall \lambda \in R^n_+, \tag{40}$$

$$\lambda_j \leq \hat{\pi}_i(Q) + w^1_j c_{ij}(Q_{ij}) + w^2_j t_{ij}(Q_{ij}), \quad \text{if } Q_{ij} \geq M_Q, \tag{41}$$

$$d_j(\lambda) \leq \sum_{i=1}^{m} Q_{ij}, \quad \text{if } \lambda_j \geq M_\lambda, \tag{42}$$

for any j and i.

The convergence of the Euler method is stated in the following theorem.

Theorem 6
Suppose that the supply price functions are strictly monotone increasing, the demand functions are strictly monotone decreasing, and the Jacobians

of the transportation cost and time functions are positive definite over the feasible set. Let $\{a_k\}$ be a sequence of positive real numbers that satisfies

$$\lim_{k\to\infty} a_k = 0 \tag{43}$$

$$\sum_{k=1}^{\infty} a_k = \infty. \tag{44}$$

In addition, assume that Assumption 3 holds true. Then the Euler method given by (35) converges to the unique multicriteria spatial price network equilibrium pattern satisfying conditions (8) and (9).

Proof: In view of Theorem 5 above, we need to verify that Assumptions 1 and 2 above are satisfied here.

First, note that, under strict monotonicity, as established in Lemma 2, the vector field $F(X)$ that governs the projected dynamical system (32) satisfies the linear growth condition, namely,

$$\langle -F(X') + F(X''), X' - X'' \rangle \leq 0$$

$$\leq B\|X' - X''\|^2, \tag{45}$$

for any positive B.

The first part of Assumption 1 of Dupuis and Nagurney (1993) is automatically satisfied by the selection of the appropriate a_k sequence and the second part of Assumption 1 automatically holds for the Euler method (see also Nagurney and Zhang (1996)).

The third and fifth parts of Assumption 1 are also satisfied (see Propositions 4.1 and 4.2 in Nagurney and Zhang (1996)) since F is strictly monotone.

All that we need to establish now is the fourth part of Assumption 1, that is, we need to show that the sequence generated by the Euler method is bounded. Assumption 3 guarantees that the sequence generated is bounded following the proof of convergence of the Euler method (see Theorem 7.11 in Nagurney and Zhang (1996)) for the traffic network equilibrium problem with given demand functions.

The proof is complete. □

6 Numerical Examples

In this Section, we present numerical examples for illustrative purposes. Specifically, we consider three spatial price equilibrium problems in which there are two supply markets and two demand markets.

In these examples the supply price functions, the transportation cost and time, and demand functions are identical and are given, respectively, by:

$$\pi_1(s) = 5s_1 + s_2 + 2, \quad \pi_2(s) = 2s_2 + 1.5s_1 + 1.5,$$

so that

$$\hat{\pi}_1(Q) = 5\sum_{j=1}^{2} Q_{1j} + \sum_{j=1}^{2} Q_{2j} + 2, \quad \hat{\pi}_2(Q) = 2\sum_{j=1}^{2} Q_{2j} + 1.5\sum_{j=1}^{2} Q_{1j} + 1.5,$$

$$t_{11}(Q_{11}) = Q_{11}, \quad t_{12}(Q_{ij}) = 2Q_{12} + 3.5,$$

$$t_{21}(Q_{21}) = 3Q_{21} + 16.25, \quad t_{22}(Q_{22}) = 2Q_{22} + 11.5,$$

$$c_{11}(Q_{11}) = 2Q_{11} + 5, \quad c_{12}(Q_{12}) = Q_{12} + 2,$$

$$c_{21}(Q_{21}) = 3Q_{21} + 4, \quad c_{22}(Q_{22}) = 5Q_{22} + 1,$$

$$d_1(\lambda) = -2\lambda_1 - 1.5\lambda_2 + 1128.75, \quad d_2(\lambda) = -4\lambda_2 - \lambda_1 + 1241.$$

The weights w_j^1 and w_j^2 for $j = 1, 2$ differ in each example. The generalized costs were constructed according to (6). Hence, these problems illustrate how the equilibrium pattern changes as the weights change.

The Euler method for all the examples was initialized as follows: the commodity shipment pattern Q^1 was set to zero as were the demand prices λ^1. The $\{a_k\}$ sequence that we utilized was: $.1\times\{1, \frac{1}{2}, \frac{1}{2}, \frac{1}{3}, \frac{1}{3}, \frac{1}{3}, \ldots\}$. The convergence criterion used was: $|Q^{k+1} - Q^k| \leq \epsilon$ and $|\lambda^{k+1} - \lambda^k| \leq \epsilon$ with $\epsilon = .0001$. Hence, the Euler method was considered to have converged when the commodity price and shipment pattern had not changed very much between two iterations and had, effectively, reached a stationary; equivalently, an equilibrium point.

We now report the computed results for the examples. The algorithm was coded in FORTRAN and the system used was an IBM SP2 located at the Computer Science Department at the University of Massachusetts at Amherst. The CPU time is reported exclusive of input and output times.

Example 1

In the first example, we assumed that the consumers in each of the two demand markets weighted travel time and travel cost in the same way, and equally. The weights were: $w_1^1 = w_1^2 = w_2^1 = w_2^2 = 0.5$.

The Euler method converged in 729 iterations and required .01 seconds of CPU time. The computed equilibrium commodity shipment pattern was:

$$Q_{11}^* = 44.194, \quad Q_{12}^* = 0.000, \quad Q_{21}^* = 51.475, \quad Q_{22}^* = 7.914,$$

which induced the equilibrium supply pattern:

$$s_1^* = 44.194, \quad s_2^* = 59.389.$$

The computed demand price pattern was:

$$\lambda_1^* = 351.158, \quad \lambda_2^* = 220.484.$$

The equilibrium conditions (8) and (9) were satisfied with good accuracy. Indeed, only the commodity shipment between supply market 1 and demand market 2 was zero. In this case, the supply price at supply market 1 incurred at the computed equilibrium plus the generalized cost between supply market 1 and demand market 2 exceeded the generalized price of the commodity at demand market 2 by 64.24. For the other pairs of supply and demand markets, which were characterized by positive commodity shipments, the difference between the supply price at a supply market plus the generalized cost between the pair of supply and demand markets and the generalized price at the demand market was 0.0.

Also, in terms of equilibrium condition (9), for the first demand market, the sum of the commodity shipments into it was equal to 95.7, which was the demand incurred at the demand market at the computed equilibrium price pattern. In addition, the sum of the computed equilibrium commodity shipments into the second demand market was equal to 7.91, which was equal to the demand at that demand market evaluated at the computed equilibrium price pattern.

Example 2

In the second example, we now modified the weights as follows: $w_1^1 = 0.0$, $w_1^2 = 1.0$, $w_2^1 = 1.0$, $w_2^2 = 0.0$. Hence, in demand market 1 the consumers are transportation time-sensitive and not transportation cost-sensitive, whereas in demand market 2 the consumers are the opposite.

The Euler method converged in 865 iterations and required .01 seconds of CPU time. The computed equilibrium commodity shipment pattern was:

$$Q_{11}^* = 48.629, \quad Q_{12}^* = 0.000, \quad Q_{21}^* = 49.068, \quad Q_{22}^* = 6.797,$$

which induced the equilibrium supply pattern:

$$s_1^* = 48.629, \quad s_2^* = 55.865.$$

The computed demand price pattern was:

$$\lambda_1^* = 359.647, \quad \lambda_2^* = 221.143.$$

Since the consumers in demand market 1 are now more transportation time-sensitive then they were in Example 1, the commodity shipment on the faster (with the lower transportation time function) "link" (between supply market 1 and demand market 1) increased as compared to the corresponding equilibrium commodity shipment in Example 1. As regards demand market 2, in which consumers are now more transportation cost-sensitive than they were in Example 1, the commodity shipment between supply market 2 and demand market 2 decreased.

As in Example 1, there was no trade between supply market 1 and demand market 2. For this pair of markets, the supply price plus the generalized cost now exceeded the generalized price by 81.865. The analogous difference for the three other market pairs, for which there was trade, i.e., a positive commodity shipment, was 0.0, signifying that equilibrium condition (8) was satisfied to good accuracy, In addition, since the generalized prices were positive at both demand markets, the sum of the commodity shipments into each demand market was (essentially) equal to the computed demand at the respective market at the computed equilibrium price pattern. Indeed, for demand market 1, the sum of commodity shipments into it was 97.7, which was the computed demand. Also, the sum of the commodity shipments into demand market 2 was 6.8 which was equal to the incurred demand at the demand market at the equilibrium price pattern.

Example 3

In the third example, we now modified the weights as follows: $w_1^1 = 1.0$, $w_1^2 = 0.0$, $w_2^1 = 0.0$, $w_2^2 = 1.0$. Hence, in demand market 1 the consumers are now transportation cost-sensitive and not transportation time-sensitive, whereas in demand market 2 the consumers are the opposite.

The Euler method converged in 973 iterations and required .01 seconds of CPU time. The computed equilibrium commodity shipment pattern was:

$$Q_{11}^* = 40.369, \quad Q_{12}^* = 0.000, \quad Q_{21}^* = 53.435, \quad Q_{22}^* = 9.791,$$

which induced the equilibrium supply pattern:

$$s_1^* = 40.360, \quad s_2^* = 63.226.$$

The computed demand price pattern was:

$$\lambda_1^* = 352.798, \quad \lambda_2^* = 219.598.$$

Since the generalized cost now between supply market 2 and demand market 1, which now consists of only the transportation cost has now been reduced either relative to that encountered by the consumers in demand market 1 in Example 1 or Example 2, the commodity shipment Q_{21}^* has increased relative to both equilibrium commodity shipments computed between that pair of markets in the two preceding examples. As regards demand market 2, however, which is now more transportation time-sensitive than in the preceding two examples, the overall commodity shipments into that demand market decrease due to the relatively high transportation times for the commodity.

As in the two preceding examples, there was no trade between supply market 1 and demand market 2, since the supply price at supply market 1 plus the generalized cost between supply market 1 and demand market 2 exceeded the generalized price at demand market 2 by 50.97. The analogous difference for the other market pairs was 0.0, signifying that equilibrium condition (8) held with good accuracy.

In terms of equilibrium condition (9), since the generalized price of the commodity was positive at both demand markets, the computed sum of commodity shipments into each demand market was approximately equal to the incurred demand at the market at the computed equilibrium price pattern. Indeed, for demand market 1, the sum of the commodity shipments was 93.8, which was also the incurred demand for that demand market, whereas for demand market 2, the sum of the commodity shipments into it was 9.8, which was the incurred demand at the computed equilibrium price pattern.

7 Summary and Conclusions

In this paper, we have developed a multicriteria spatial price network equilibrium model and studied it from two perspectives: a static one, with a focus on the equilibrium, and a dynamic one, through a proposed tatonnement process for the evolution of the commodity shipment and price patterns. The model handles consumers who weight the transportation cost and the transportation time associated with the commodity shipment in an individual fashion.

The statics were studied using the finite-dimensional variational inequality formulation of the governing equilibrium conditions whereas the dynamic model was formulated as a projected dynamical system. This is the first time that a multicriteria network equilibrium problem was treated from two such perspectives.

We established that the set of stationary points of the projected dynamical system coincides with the set of solutions of the variational inequality problem. In addition, we proved both existence and uniqueness of the multicriteria spatial price equilibrium pattern under reasonable conditions. These results are sharper than those that have been obtained recently for multicriteria traffic network equilibrium problems.

We provided a discrete-time algorithm, the Euler method, for the approximation of the trajectory, established convergence, and applied the algorithm to several examples for illustrative purposes.

Acknowledgments

The research of the first two authors was supported, in part, by NSF Grant No.: IIS-0002647. The research of the first author was also supported, in part, by NSF Grant No.: CMS-0085720 and by the John F. Smith Memorial Fund at the Isenberg School of Management at the University of Massachusetts at Amherst. The support is gratefully acknowledged.

The authors are grateful to the editor, Patrizia Daniele, for helpful comments and suggestions.

References

[1] S. Dafermos, "A multicriteria route-mode choice traffic equilibrium model," Lefschetz Center for Dynamical Systems, Brown University, Providence, Rhode Island, 1981.

[2] S. Dafermos and A. Nagurney, "Oligopolistic and competitive behavior of spatially separated markets," *Regional Science and Urban Economics* **17** (1987) 245-254.

[3] R. B. Dial, "A model and algorithms for multicriteria route-mode choice," *Transportation Research* **13B** (1979) 311-316.

[4] P. Dupuis and A. Nagurney, "Variational inequalities and dynamical systems," *Annals of Operations Research* **44** (1993) 9-42.

[5] M. Florian and M. Los, "A new look at static spatial price equilibrium models," *Regional Science and Urban Economics* **12** (1982) 579-597.

[6] D. Kinderlehrer and G. Stampacchia, **An Introduction to Variational Inequalities and Their Application**, Academic Press, New York (1980).

[7] F. Leurent, "Modelling elastic, disaggregate demand," in J. C. Moreno Banos, B. Friedrich, M. Papageorgiou, and H. Keller, editors, **Proceedings of the First Meeting of the Euro Working Group on Urban Traffic and Transportation**, Technical University of Munich, Munich, Germany (1993a).

[8] F. Leurent, "Cost versus time equilibrium over a network," *European Journal of Operations Research* **71** (1993b) 205–221.

[9] F. Leurent, "Multicriteria assignment modeling: making explicit the determinants of mode or path choice," in P. Marcotte and S. Nguyen, editors, **Equilibrium and Advanced Transportation Modelling**, Kluwer Academic Publishers, Boston, Massachusetts (1998) 153-174.

[10] P. Marcotte, "Reformulations of a bicriterion equilibrium model," in **Reformulation: Nonsmooth, Piecewise Smooth, Semismooth and Smoothing Methods**, pp. 269-292, M. Fukushima and L. Qi, editors, Kluwer Academic Publishers, Dordrecht, The Netherlands.

[11] A. Nagurney, "Computational comparisons of spatial price equilibrium methods, *Journal of Regional Science* **27** (1987) 55-76.

[12] A. Nagurney, **Network Economics: A Variational Inequality Approach**, second and revised edition, Kluwer Academic Publishers, Dordrecht, The Netherlands (1999).

[13] A. Nagurney, "A multiclass, multicriteria traffic network equilibrium model," *Mathematical and Computer Modelling* **32** (2000), 393-411.

[14] A. Nagurney and J. Dong, "A multiclass, multicriteria traffic network equilibrium model with elastic demand," *Transportation Research B* **36** (2002), 445-469.

[15] A. Nagurney, T. Takayama, and D. Zhang, "Massively parallel computation of spatial price equilibrium problems as dynamical systems," *Journal of Economic Dynamics and Control* **18** (1995a) 3-37.

[16] A. Nagurney, T. Takayama, and D. Zhang, "Projected dynamical systems modeling and computation of spatial price network equilibribria," *Networks* **26** (1995b) 69-85.

[17] A. Nagurney and D. Zhang, **Projected Dynamical Systems and Variational Inequalities with Applications**, Kluwer Academic Publishers, Boston, Massachusetts (1996).

[18] A. Nagurney and L. Zhao, "Networks and variational inequalities in the formulation and computation of market disequilibria: the case of direct demand functions," *Transportation Science* **27** (1993) 4-15.

[19] R. E. Quandt, "A probabilistic abstract mode model," in **Studies in Travel Demand VIII**, Mathematica, Inc., Princeton, New Jersey (1967) 127-149.

[20] P. A. Samuelson, "Spatial price equilibrium and linear programming," *American Economic Review* **42** (1952) 283-303.

[21] M. Schneider, "Access and land development," in **Urban Development Models**, Highway Research Board Special Report **97** (1968) 164-177.

[22] T. Takayama and G. G. Judge, **Spatial and Temporal Price and Allocation Models**, North-Holland, Amsterdam The Netherlands (1971).

NON REGULAR DATA IN UNILATERAL VARIATIONAL PROBLEMS

Pirro Oppezzi
D.I.M.A, Università di Genova
Via Dodecaneso 35 – 16146 Genova, Italy
e-mail: oppezzi@dima.unige.it

1 Introduction

I will describe here some results about "renormalized" (or "entropic") solutions for unilateral nonlinear problems with L^1 or measure data, within the recent literature concerning such variational problems.
These results are contained in several recent joint papers with Anna M. Rossi. ([10], [11], [12], [13], [14])
Let Ω be an open set in $\mathbf{R}^n$, $\psi : \Omega \to \overline{\mathbf{R}}$ measurable.
We consider the unilateral convex set

$$K_\psi = \{v \in \mathrm{H}_o^{1,p}(\Omega) : v \geq \psi\} \quad , \quad 1 < p < n$$

so that the general problem can be formulated as:

$$\text{(I)} \quad \begin{cases} \exists\, u \in K_\psi \quad \text{such that} \\ \langle -\operatorname{div} a(\cdot, u, Du) - f, v - u \rangle \geq 0 \quad \text{for every } v \in K_\psi \end{cases}$$

where for the datum f one may distinguish three levels of assumptions:

- $f \in H^{-1,p'}(\Omega)$ (the dual of $\mathrm{H}_o^{1,p}(\Omega)$)
- $f \in L^1(\Omega)$
- $f \in \mathcal{M}_b(\Omega)$ (the space of real bounded Radon measures)

About the operator $-\operatorname{div} a(\cdot, u, Du)$ two cases are considered

- $a : \Omega \times \mathbf{R} \times \mathbf{R}^n \to \mathbf{R}^n$ is a Carathéodory function
- $a : \Omega \times \mathbf{R} \times \mathbf{R}^n \to 2^{\mathbf{R}^n}$ is measurable and multivalued

In the first case the classical Leray-Lions assumptions are

(i) Coerciveness

$$a(x, s, \xi) \cdot \xi \geq c|\xi|^p + \nu \qquad , \quad c \in \mathbf{R}_+ \,,\ \nu \in L^1(\Omega)$$

(ii) Monotonicity

$$\langle a(x,s,\xi) - a(x,s,\xi'), \xi - \xi' \rangle > 0\,, \quad \xi \neq \xi'\,,\ \xi, \xi' \in \mathbf{R}^n$$

(iii) Growth condition: there exists $\mu \in L^{p'}(\Omega)$ ($p' = \dfrac{p}{p-1}$) such that

$$|a(x,s,\xi)| \leq \mu(x) + c_1|\xi|^{p-1} + c_2|s|^{p-1}\,,\ (s,\xi) \in \mathbf{R} \times \mathbf{R}^n\,,\ \text{a.e. } x \in \Omega$$

In the multivalued case one must simply replace $a(x,s,\xi)$ with each element $\eta \in a(x,s,\xi)$ in coerciveness and growth conditions, while hypothesis (ii) is replaced by

(ii1) $\quad \xi \in \mathbf{R}^n \mapsto a(x,s,\xi) \subset \mathbf{R}^n \quad$ is maximal monotone

(for example $a(x,s,\xi) = \partial_\xi\, g(x,s,\cdot)(\xi)$ with $g(x,s,\cdot)$ convex satisfying a condition as $k_1\,|\xi|^p - k_2 \leq g(x,s,\xi) \leq k_2' + k_1'\,|\xi|^p$.
Moreover a suitable hypothesis

(ii2) $\quad s \in \mathbf{R} \mapsto a(x,s,\xi) \subset \mathbf{R}^n \quad$ is "regular"

will be assumed.
The main difficulty arising from problem (I) is to give a correct formulation in order to obtain existence for general measures f and uniqueness when a strict monotonicity condition is assumed on a with respect to ξ .
Remark that problem (I) contains the equation problem as a particular case when $K_\psi = \mathrm{H}_o^{1,p}(\Omega)$. Let's recall a first answer given by G. Stampacchia (1965) in the linear case.

Theorem 1.1 *There exists a solution and it is unique for the problem*

$$\begin{cases} L\,u = \mu \\ u = 0 \quad \text{on } \partial\Omega \end{cases}, \qquad L\,u = -\,\mathrm{div}\,\Big(\sum_{i=1}^{n} a_{i,j}(x)\,\partial_i\,u\Big)_{j=1,\dots,n}$$

in the sense

$$Lu = \mu \Leftrightarrow \int_\Omega uL^*\varphi = \int \varphi\, d\mu \quad \forall\varphi \in \mathrm{H}_o^{1,2}(\Omega) \cap C(\overline{\Omega}) \text{ such that } uL^*\varphi \in L^1(\Omega)$$

where L^* denotes the adjoint operator of L .

2 The approach by truncation and approximation

Now we describe the approach studied first for the case of the corresponding equation in problem (I) ([6], [9], [1]) and in the inequality case ([10],

[11], [12]) where the operator is multivalued, and in [3] for the single valued case.
In this section we deal with the case where

$$f \in L^1(\Omega)$$

and consider a sequence $(f_h)_{h\in\mathbf{Z}_+}$ in $H^{-1,p'}(\Omega) \cap L^1(\Omega)$ (e.g. $f_h \in C_o^\infty(\Omega)$) such that

$$f_h \to f \qquad \text{strongly in } L^1(\Omega).$$

Let $u_h \in \mathrm{H}_o^{1,p}(\Omega)$ be a solution for the equation

$$\int_\Omega \langle a(x, u_h, Du_h), Dv\rangle \, dx = \int_\Omega f_h \, v \, dx \qquad \forall v \in \mathrm{H}_o^{1,p}(\Omega) \tag{1}$$

or, for the inequality case,

$$\int_\Omega \langle a(x, u_h, Du_h), D(v - u_h)\rangle \, dx \geq \int_\Omega f_h \,(v - u_h) \, dx \qquad \forall v \in K_\psi \tag{2}$$

(see e.g. [8] about existence of u_h in the single valued operator case and see [10] in the multivalued case).
We define the "truncation" $T_k(s) = (s \wedge k) \vee (-k)$ $(k \in \mathbf{R}_+)$.
If we let $v = T_k(u_h)$ in (1), then by coerciveness:

$$\int_\Omega \nu + c\,|DT_k(u_h)|^p \leq \int a(\cdot, u_h, Du_h)\cdot DT_k(u_h) = \int f_h \, T_k(u_h) \leq k\|f_h\|_{L^1(\Omega)}$$

Hence there exist $A, B \in \mathbf{R}$ such that

$$\int_\Omega |DT_k(u_h)|^p \leq k\,A + B \qquad \forall k \in \mathbf{R}_+ ,\ h \in \mathbf{N} \tag{3}$$

and $(T_k(u_h))_{h\in\mathbf{N}}$ is bounded in $\mathrm{H}_o^{1,p}(\Omega)$ for every $k \in \mathbf{R}_+$.
¿From (3) and Sobolev's inequality, it is possible to obtain the uniform Marcinkiewitz estimate

$$|\{x \in \Omega : |u_h(x)| \geq k\}| \leq C\, k^{-p^*/p'} \qquad (p^* = \frac{np}{n-p})$$

Moreover, by a diagonal process, one may find a subsequence, still denoted $(u_h)_{h\in\mathbf{N}}$ and $u : \Omega \to \mathbf{R}$ such that

$$u_h \to u \ \text{ a.e. in } \Omega \quad , \quad T_k(u_h) \rightharpoonup T_k(u) \ \text{ weakly in } \mathrm{H}_o^{1,p}(\Omega)$$

Therefore $T_k(u) \in \mathrm{H}_o^{1,p}(\Omega)$ for each $k \in \mathbf{R}_+$ and satisfies (3).
We show briefly, in the equation case, how the following estimate

$$\lim_{k\to\infty} \int_{\{k \leq |u| \leq k+1\}} |Du|^p = 0 \tag{4}$$

also holds.
If we let $v_h = T_{k+1}(u_h) - T_k(u_h)$ in (1), then as before we obtain

$$\int_{\{k\le|u|\le k+1\}} |Du|^p \le \liminf_h \int |Dv_h|^p \le$$
$$\le C \lim_h \int_{\{|u_h|\ge k\}} |f_h| + |\nu| = C \int_{\{|u|\ge k\}} |f| + |\nu|$$

which gives (4).
By similar but longer computations it is possible, starting with (2), to get (3) and (4) also in the inequality case. Now if $v \in K_\psi$, then $w_h = u_h + T_k(v - u_h) \in K_\psi$. By letting w_h as test function in (2) it follows that:

$$\int_\Omega \langle a(\cdot, u_h, Du_h), DT_k(v - u_h)\rangle \ge \int_\Omega f_h\, T_k(v - u_h) \tag{5}$$

It is easy to check that such an inequality is meaningful also with u in place of u_h if $v \in L^\infty(\Omega)$, so we try to pass to the limit in it.
We observe, from growth condition and, for simplicity, in the case independent on u, that:

$$|a(\cdot, Du_h)|^{p'} \le C\,(|\mu|^{p'} + |Du_h|^p)$$

which gives, thanks to (3), for a subsequence, $(a(\cdot, Du_h)\mathbf{1}_{\{|u_h|\le k\}})_h$ weakly convergent in $L^{p'}(\Omega)$ $\forall k > 0$.
If a verifies the strict monotonicity condition (ii) of Section 1, then we may apply the following:

Lemma 2.1 *([2]) If $v_h \to v_o$ a.e. and weakly in $\mathrm{H}_o^{1,p}(\Omega)$ and*

$$\lim_{h\to\infty} \int \langle a(\cdot, v_h, Dv_h) - a(\cdot, v_h, Dv_o), D(v_h - v_o)\rangle = 0\,,$$

then $Dv_h \to Dv_o$ strongly in $L^p(\Omega)$.

It is possible to verify the condition of the Lemma with $v_h = T_k(u_h)$, $v_o = T_k(u)$. So for a subsequence

$$DT_k(u_h) \to DT_k(u) \quad \text{strongly in } L^p(\Omega) \text{ and a.e. in } \Omega.$$

When $v \in L^\infty(\Omega) \cap K_\psi$ (so that ψ needs to be assumed upper bounded), then

$$\{x \in \Omega : DT_k(v - u_h) \ne 0\} \subset \{x \in \Omega : |u_h| \le k + \|v\|_{L^\infty}\}$$

and

$$a(\cdot, u_h, Du_h)\,\mathbf{1}_{\{|u_h|\le k+\|v\|_{L^\infty}\}} \rightharpoonup a(\cdot, u, Du)\,\mathbf{1}_{\{|u|\le k+\|v\|_{L^\infty}\}}\,,$$

because a is Carathéodory. Therefore the strong convergence

$$DT_k(v-u_h) \to DT_k(v-u) \qquad \text{in } L^p(\Omega)$$

allows to take the limit in (5) and to get, for each $k \in \mathbf{R}_+$, $v \in L^\infty(\Omega) \cap K_\psi$

$$\int_\Omega a(\cdot, u, Du) DT_k(v-u) \geq \int_\Omega f\, T_k(v-u) \tag{6}$$

and (6) is assumed as definition of "entropic solution" for problem (I). We look now to the case where $a(x,s,\cdot)$ is only weakly monotone as in the multivalued case (assumptions (ii1), (ii2) of Section 1).
In [10] it is proved that

$$a(\cdot, u_h, Du_h)\,\mathbf{1}_{\{|u_h|\leq k\}} \underset{L^{p'}}{\rightharpoonup} a(\cdot, u, Du)\,\mathbf{1}_{\{|u|\leq k\}}$$

where $a(\cdot, u_h, Du_h)$, $a(\cdot, u, Du)$ are understood as suitable selections. We shortly describe, in consequence, that:

$$\limsup_h \int a(\cdot, u_h, Du_h)\, DT_k(v-u_h) \leq$$
$$\leq \int a(\cdot, u, Du)\, DT_k(v-u)\,, \quad v \in L^\infty \cap K_\psi \tag{7}$$

By semplicity we consider the case independent on u:

$$\int a(\cdot, Du_h)\, DT_k(v-u_h) = \int_{\{|v-u_h|\leq k\}} a(\cdot, Du_h)\, DT_k(v-u) +$$
$$+ \int_{\{|v-u_h|\leq k\}} \langle a(\cdot, Du_h) - a(\cdot, Du\mathbf{1}_{\{|v-u|\leq k\}}), D(v-u_h) - DT_k(v-u)\rangle +$$
$$+ \int_{\{|v-u_h|\leq k\}} a(\cdot, Du\mathbf{1}_{\{|v-u|\leq k\}}), D(v-u_h) - DT_k(v-u)\rangle\,.$$

By monotonicity the second integral on the right is less or equal to zero; the first goes to $\int a(\cdot, Du)\, DT_k(v-u)$ and the third goes to zero, so (7) is achieved. On the other hand the convergence

$$\int f_h\, T_k(v-u_h) \to \int f\, T_k(v-u)$$

is clear, so (6) follows from (5) and (7) and the following result is proved ([10]).

Theorem 2.2 *If conditions (i), (iii), (ii1), (ii2) of Section 1 are assumed, $f \in L^1(\Omega)$ and ψ is bounded, then there exists an entropic solution of problem (I), i.e. a function $u : \Omega \to \mathbf{R}$ such that $T_k(u) \in \mathrm{H}_o^{1,p}(\Omega)$ for every $k \in \mathbf{R}_+$ and (6) is satisfied for a selection of $a(\cdot, u, Du)$ (still denoted by $a(\cdot, u, Du)$).*

3 Renormalized formulation

Let us consider the function

$$\sigma_k(s) = \begin{cases} 1 & \text{if } |s| \le k \\ 0 & \text{if } |s| \ge k+1 \\ -|s| + k + 1 & \text{if } k < |s| < k+1 \end{cases} .$$

Definition 3.1 We say that $u : \Omega \to \mathbf{R}$ measurable is a renormalized solution of problem (I), when $f \in L^1(\Omega)$, if for each $k \in \mathbf{R}_+$ it verifies

$$\begin{cases} T_k(u) \in \mathrm{H}_o^{1,p}(\Omega) \\ \displaystyle\int a(\cdot, u, Du) D[\theta \sigma_k(u)(v-u)] \ge \int f \theta \sigma_k(u)(v-u) \quad \forall \theta \in \mathrm{H}_o^{1,p} \cap L^\infty, \\ \qquad\qquad \theta \ge 0 \ \forall v \in K_\psi \cap L^\infty(\{|u| \le k+1\}) \\ \displaystyle\lim_{k\to\infty} \int_{\{k \le |u| \le k+1\}} |Du|^p = 0 \end{cases}$$

Remark 3.2 It is not difficult to see that the entropic formulation of Section 2 is equivalent to the one of renormalized, when ψ is upper bounded. In [13], where a theorem of existence of renormalized solutions is given, the above definition was useful to remove the hypothesis of boundedness for ψ.

4 Multivalued operators and more general measures

We describe now more precisely hypothesis (ii2) and existence results in the case where $a : \Omega \times \mathbf{R} \times \mathbf{R}^n \to a(x, s, \xi) \subset \mathbf{R}^n$ is multivalued. We assume that a is measurable, closed-convex valued and satisfies (i), (ii) and (ii1) of Section 1. Moreover

(ii2) If $(u_h)_{h\in\mathbf{N}}$, u are in $\mathrm{H}_o^{1,p}(\Omega)$, $u_h \to u$ a.e. in Ω, $v \in \mathrm{H}_o^{1,p}(\Omega)$ and $\eta(x) \in a(x, u(x), Dv(x))$ is a measurable selection, then, for each $h \in \mathbf{N}$, there exists a selection $\eta_h(x) \in a(x, u_h(x), Dv(x))$ such that $\eta_h \to \eta$ a.e. on Ω.

We observe that assumptions on a imply that $a(x, s, \xi)$ is compact for every $(x, s, \xi) \in \Omega \times \mathbf{R} \times \mathbf{R}^n$.

Remark 4.1 A sufficient condition for (ii2) is the continuity of $s \in \mathbf{R} \mapsto a(x, s, \xi) \in 2^{\mathbf{R}^n}$ with respect to the Hausdorff topology on the metric space of compact subsets of $\mathbf{R}^n$ (see [13], prop. 1.1)

Remark 4.2 Let $\mathcal{A} : \mathrm{H}^{1,p}_o(\Omega) \to 2^{\mathrm{H}^{-1,p'}}$ be defined by

$$\mathcal{A}(u) = \{-\operatorname{div} g : g(x) \in a(x, u(x), Du(x)) \text{ measurable } \},$$

then, when $f \in \mathrm{H}^{-1,p'}$ problem (I) also with a general convex set $K \subset \mathrm{H}^{1,p}_o(\Omega)$ in place of K_ψ, is equivalent to (see [15])

$$\text{(I')} \qquad \text{find } u \in K \text{ such that } f \in (\mathcal{A} + \partial I_K)(u)$$

where ∂I_K is the subdifferential of the indicator function I_K.

The following theorem is proved in [10].

Theorem 4.3 *If* $K \subset \mathrm{H}^{1,p}_o(\Omega)$ *is convex and* $f \in \mathrm{H}^{-1,p'}$*, there exists a solution for (I')*

Remark 4.4 For a general bounded Radon measure μ the following decomposition theorems hold:

(i) ([7]) $\mu = \mu_a + \mu_s$ where μ_a is zero on sets of zero p-capacity and μ_s is concentrated on a set of zero p-capacity.

(ii) ([4]) If μ_a is a bounded Radon measure vanishing on sets of zero p-capacity, then $\quad \mu_a = f - \operatorname{div} F \quad , \quad f \in L^1(\Omega)\ , \ F \in L^{p'}(\Omega)^n$

Define, for $\Omega \subset \mathbf{R}^n$ open,

$$\mathcal{L}^1_p(\Omega) = \{v \in L^{p^*}(\Omega) : \quad Dv \in L^p(\Omega)^n, \exists (\varphi_j)_{j\in\mathbf{N}} \text{ in } C^\infty_o(\Omega), \\ D\varphi_j \to Dv \text{ in } L^p(\Omega)^n \},$$

The following existence theorem is given in [13], where Ω is not necessarly bounded and on $a : \Omega \times \mathbf{R} \times \mathbf{R}^n \to 2^{\mathbf{R}^n}$ the above assumptions are made.

Theorem 4.5 *Let* $f \in L^1(\Omega)$, $F \in L^{p'}(\Omega)^n$ *and* μ_s *be a regular Radon measure concentrated on a set of zero p-capacity. Let* $\psi : \Omega \to [-\infty, \infty]$ *measurable and* $K_\psi = \{v : Dv \in L^p(\Omega),\ v \geq \psi \text{ a.e. in } \Omega\}$.
Then there exists $u : \Omega \to \mathbf{R}$ *such that* $u \geq \psi$ *a.e. in* Ω, $T_k(u) \in \mathcal{L}^1_p(\Omega)$ *for each* $k \in \mathbf{R}_+$ *and there exists* $g : \Omega \to \mathbf{R}$ *measurable such that*

$$g(x) \in a(x, u(x), Du(x)) \ \textit{for a.e.}\ x \in \Omega\,,\ g\,\mathbf{1}_{\{|u|\leq k\}} \in L^{p'}(\Omega) \quad \forall k \in \mathbf{R}_+$$

and u, g *are such that*

$$\int \langle g, D[\,\theta\,\sigma_k(u)\,(v-u)]\rangle \geq \int f\,\theta\,\sigma_k(u)(v-u) + \langle F, D[\,\theta\,\sigma_k(u)\,(v-u)]\,\rangle$$

$$\forall \theta \in \mathrm{H}^{1,p}_{\mathrm{loc}}(\Omega) \cap L^\infty(\Omega) \textit{ such that } D\theta \in L^p(\Omega)^n,\ \theta \geq 0\,;$$
$$\forall v \in \mathcal{L}^1_p(\Omega),\ v \in L^\infty(\{|u| \leq k+1\}),\ v \geq \psi \textit{ on } \{|u| \leq k+1\}\,. \tag{8}$$

Moreover

$$\int \langle g, Dw\rangle \geq \int f\,w + \langle F, Dw\rangle + \int w\,d\mu_s \quad \forall w \in \mathcal{L}^1_r(\Omega),\ r>n,\ w\geq 0 \tag{9}$$

5 Uniqueness and convergence

In [14] we deal with a measure $\mu \in \mathcal{M}_b^p(\Omega)$ = regular Radon bounded measures absolutely continuous with respect to p-capacity. We obtain an existence result for renormalized solutions with a general obstacle ψ such that K_ψ is nonempty.
In such a paper we use estimates on $u - u_0$, $u_0 \in K_\psi$ instead of estimates on u, so the renormalizing limit estimate becomes

$$\lim_{k\to\infty} \int_{\{k\le|u-u_0|\le k+1\}} |D(u-u_0)|^p = 0 \quad , \quad u_0 \in K_\psi\,. \tag{10}$$

Then we require $T_k(u - u_0) \in \mathrm{H}_o^{1,p}(\Omega)$ and inequality (8) is replaced by the following

$$\int \langle g, D[\theta\,\sigma_k(u-u_0)\,(v-u)]\,\rangle \ge \int f\,\theta\,\sigma_k(u-u_0)(v-u) + \\ + \int \langle F, D[\theta\,\sigma_k(u-u_0)\,(v-u)]\,\rangle$$

$$u_0 \in K_\psi\,,\ v \in K_\psi \text{ such that } v-u_0 \text{ is bounded on } \{|u-u_0| \le k+1\}$$
$$\forall\theta \in \mathrm{H}^{1,p}_{\mathrm{loc}}(\Omega) \cap L^\infty(\Omega) \text{ such that } D\theta \in L^p(\Omega)^n,\ \theta \ge 0\,. \tag{11}$$

Moreover we obtain the following convergence result

Theorem 5.1 *If $K_{\psi_h} \to K_\psi$ in the sense of Mosco (in $\mathrm{H}_o^{1,p}(\Omega)$), $\mu_h = f_h - \operatorname{div} F_h \in \mathcal{M}_b^p(\Omega)$ with $f_h \rightharpoonup f$ in $L^1(\Omega)$, $F_h \to F$ strongly in $L^{p'}(\Omega)$ and u_h is a renormalized solution relative to K_{ψ_h}, μ_h, satisfying (10), then for a subsequence and a function $u : \Omega \to \mathbf{R}$ it results $u_h \to u$ a.e. in Ω, $T_k(u_h) \rightharpoonup T_k(u)$ weakly in $\mathrm{H}_o^{1,p}(\Omega)$, u satisfies (10) and u is a renormalized solution relative to $\mu = f - \operatorname{div} F$ and K_ψ, i.e. it satisfies (11).*

With some suitable restrictions on the divergence operator, we obtained in [14] the following uniqueness result

Theorem 5.2 *If $a : \Omega \times \mathbf{R}^n \to 2^{\mathbf{R}^n}$ satisfies conditions (i), (iii) with $c_2 = 0$ and (ii1) of Section 1 and is strictly monotone:*

$$\langle \eta - \eta', \xi - \xi' \rangle > 0 \quad \forall \xi \ne \xi'\,,\ \eta \in a(x,\xi)\,,\ \eta' \in a(x,\xi')$$

then the renormalized solution of problem (I) with $f \in \mathcal{M}_b^p(\Omega)$ is unique.

Remark 5.3 The renormalizing limit condition (10) is not true for a renormalized solution with a general measure data as given by Theorem 4.5. So uniqueness in general case is an open problem except particular cases (the case where $p = 2$ studied by F.Murat and with other restrictions in [5]).

References

[1] P. Bènilan, L. Boccardo, T. Gallouët, R. Gariepy, M. Pierre, J.L. Vasquez, *An L^1-Theory of Existence and Uniqueness of Solutions of Nonlinear Elliptic Equations*, Ann. Sc. Norm. Sup. Pisa (4) **22** (1995) nr. 2, 241–273.

[2] L. Boccardo, F. Murat, J.P. Puel, *Existence of Bounded Solutions for Non Linear Elliptic Unilateral Problems*, Ann. Mat. Pura Appl. (4) **152** (1988), 183–196.

[3] L. Boccardo, G.R. Cirmi, *Existence and Uniqueness of Solution of Unilateral Problems with L^1 Data* , J. Convex Anal. **6** (1999), no. 1, 195–206.

[4] L. Boccardo, T. Gallouët, L. Orsina, *Existence and Uniqueness of Entropy Solutions for Nonlinear Elliptic Equations with Measure Data* - Ann. Inst. H. Poincaré Anal. Non Linéaire **13** (1996), no. 5, 539–551.

[5] G. Dal Maso, F. Murat, L. Orsina, A. Prignet, *Definition and Existence of Renormalized Solutions of Elliptic Equations with General Measure Data* - C. R. Acad. Sci. Paris Sér.I Math. **325** (1997).

[6] R.J. Di Perna, P.L. Lions, *On the Cauchy problem for Boltzman equations: Global existence and weak stability*, Ann. of Math. **130** (1989), 321–366.

[7] M. Fukushima, K. Sato, S. Taniguchi, *On the closable part of pre-Dirichlet forms and the fine support of the underlying measures*, Osaka J. Math. **13** (1996), 539–551.

[8] J.L. Lions, *Quelques méthodes de résolution des problèmes aux limites non linéaires*, Dunod, Paris, 1969.

[9] F. Murat, *Soluciones renormalizadas de EDP elipticas no lineales*, Publications du Laboratoire d'Analyse Numerique n.93023, Université Pierre et Marie Curie, Paris (1993).

[10] P. Oppezzi, A.M. Rossi, *Existence of Solutions for Unilateral Problems with Multivalued Operators*, J. Convex Anal. **2** (1995), 241–261.

[11] P. Oppezzi, A.M. Rossi, *Esistenza di soluzioni per problemi unilateri con dato misura o in L^1*, Ricerche Mat., **XLV**, 2 (1996), 491–513.

[12] P. Oppezzi, A.M. Rossi, *Renormalized Solutions for Divergence Problems with L^1 Data*, Atti Sem. Mat. Fis. Univ. Modena, Suppl. Vol.**46** (1998), 889–914.

[13] P. Oppezzi, A.M. Rossi, *Unilateral Problems with Measure Data*, to appear on Nonlinear Anal.

[14] P. Oppezzi, A.M. Rossi, *Unilateral problems with measure data: links and convergence*, to appear on Diff. Int. Equations.

[15] R.T. Rockafellar, *Convex functions, monotone operators and variational inequalities*, Theory and Applications of Monotone Operators, proc. NATO Institute (Venice, 1968), Oderisi, Gubbio, 1970, 35–65.

EQUILIBRIUM CONCEPTS IN TRANSPORTATION NETWORKS: GENERALIZED WARDROP CONDITIONS AND VARIATIONAL FORMULATIONS

Massimo Pappalardo
Department of Applied Mathematics, University of Pisa
via Bonanno 25/b, 56126 - Pisa, Italy

Mauro Passacantando
Department of Mathematics, University of Pisa
via Buonarroti 2,56127 - Pisa, Italy

Abstract We present two new Wardrop-type definitions of equilibrium for transportation networks and we investigate their relationships with variational formulations of Stampacchia or Minty-type.

Keywords: Transportation network, equilibrium solutions, generalized Wardrop condition, Minty variational inequalities.

2000 Mathematics Subject Classification. 90C30.

1 Introduction

In a transportation network an equilibrium flow is classically defined by Wardrop's principle: each traveler tries to minimize his own travel time irrespective of the other travelers, this principle is known as the *user equilibrium principle.* A lot of analysis has been carried out in order to connect this principle with a variational formulation. In fact, variational inequalities are very useful tools for the study of an equilibrium flow. In this paper we propose new definitions of equilibrium flow, we propose some relationships with variational formulations and we show that this study permits us to select different equilibrium points. Moreover these new definitions aim to analyze the evolution of a traffic network when the flow is not an equilibrium flow.

Now we briefly recall a definition which will be useful in the sequel.

Definition 1.1 *Given a closed and convex set $K \subset \mathbb{R}^n$ and a vector* *function $F : K \to \mathbb{R}^n$, we consider the Variational Inequality, SVI(F,K), which consists in determining a vector $x^* \in K$, such that*

$$\langle F(x^*), x - x^* \rangle \geq 0, \qquad \forall\, x \in K,$$

where $\langle \cdot, \cdot \rangle$ denotes the inner product in $\mathbb{R}^n$. The corresponding Minty Variational Inequality, MVI(F,K), consists in determining a vector $x^* \in K$, such that

$$\langle F(x), x - x^* \rangle \geq 0, \qquad \forall\, x \in K.$$

The following known result establishes relationships between the solutions of SVI(F,K) and those of MVI(F,K).

Theorem 1.1 *(Generalized Minty Lemma) The following statements hold:*

1. *if F is continuous on K, then each solution to MVI(F,K) is a solution of SVI(F,K);*
2. *if F is pseudomonotone on K, that is*

   $$\langle F(y), x - y \rangle \geq 0 \Rightarrow \langle F(x), x - y \rangle \geq 0, \qquad \forall\, x, y \in K,$$

   then each solution to SVI(F,K) is a solution to MVI(F,K).

2 Equilibrium model in a traffic network

Let (N, A, W) be a transportation network where $N = \{N_1, ..., N_p\}$ is the set of nodes, $A = \{A_1, ..., A_n\}$ the set of directed arcs and $W = \{W_1, ..., W_l\}$ the set of OD (origin-destination) pairs. We denote by $\mathcal{R}_j$ the set of those paths R_r, $r = 1, ..., r_j$, that connect the pair $W_j \in W$, and by F_r, $r = 1, ..., r_j$, the path flow on R_r. If we consider the set of all paths $\mathcal{R} = \bigcup_{j=1}^{l} \mathcal{R}_j$ and arrange the path flows into a vector $F \in \mathbb{R}^m$ where $m = r_1 + \cdots + r_l$, we obtain a column vector $F = (F_1, ..., F_m)$, whose components F_r represent the flow on the path R_r, $r = 1, ..., m$ suitably rearranged. A feasible flow has to satisfy demand requirements,

$$\sum_{r=1}^{r_j} F_r = \rho_j, \qquad j = 1, ..., l,$$

where $\rho \geq 0$ is given in R^l. Introducing the pair-path incidence matrix $\phi = (\phi_{jr})$, namely

$$\phi_{jr} = \begin{cases} 1 & \text{if } R_r \in \mathcal{R}_j, \\ 0 & \text{if } R_r \notin \mathcal{R}_j, \end{cases}$$

the demand requirements can be written as

$$\phi F = \rho.$$

Thus, the set of all feasible flows is given by

$$K = \{F \in \mathbb{R}^m : \ F \geq 0, \quad \phi F = \rho\}.$$

The flow on arc A_i is denoted by f_i and f denotes the column vector whose components are f_i, $i = 1, ..., n$. The travel cost on arc A_i is a given function of f which we denote by $c_i(f)$ and the column vector $c(f)$, whose components are $c_i(f)$, denotes the travel cost on all arcs. We denote by $\Delta = (\delta_{i\,r})$ the arc-path incidence matrix, namely

$$\delta_{i\,r} = \begin{cases} 1 & \text{if } A_i \in R_r \quad i = 1, ..., n \quad r = 1, ..., m, \\ 0 & \text{if } A_i \notin R_r \quad i = 1, ..., n \quad r = 1, ..., m, \end{cases}$$

hence it results

$$f_i = \sum_{r=1}^{m} \delta_{i\,r} \, F_r, \qquad i = 1, \ldots, n$$

that is $f = \Delta\, F$, and, denoting by $C_r(F)$ the travel cost on path R_r, we obtain

$$C_r(F) = \sum_{i=1}^{n} \delta_{i\,r} \, c_i(f), \qquad r = 1, \ldots, m,$$

that is

$$C(F) = \Delta^T c(\Delta\, F).$$

Therefore we have a cost function $C : K \to \mathbb{R}^m$ such that $C_r(F)$ gives the marginal cost of sending one additional unit of flow through path r, when the flow F is already present.

Now we may recall the classic Wardrop equilibrium condition.

Definition 2.1 *A feasible flow $H \in K$ is called an* ***equilibrium flow*** *if and only if*

$$\left[\begin{array}{c} \textit{for each OD pair } W_j \in W \textit{ and each } R_q, \ R_s \in \mathcal{R}_j, \\ H_s > 0 \Longrightarrow C_s(H) \leq C_q(H). \end{array} \right] \tag{1}$$

It is classically possible to characterize an equilibrium flow by means of a variational inequality.

Theorem 2.1 *A feasible flow H is an equilibrium flow if and only if H is a solution to $SVI(C,K)$.*

Remark 2.1 *If we consider a network equilibrium flow in the network standard model, namely a model where travel link cost functions are differentiable, positive, strictly increasing and separable in the sense that*

$$c(f) = \sum_{i=1}^{n} c_i(f_i) \qquad i = 1, \ldots, n,$$

we have a clearer situation. In fact in this case the jacobian matrix of the cost function $C(F)$ is symmetric:

$$\frac{\partial C_q}{\partial F_s} = \frac{\partial C_s}{\partial F_q} \qquad \forall\ q, s = 1, \ldots, m.$$

and hence one has $C(F) = \nabla T(\Delta F)$ where

$$T(\Delta F) = T(f) = \sum_{i=1}^{n} \int_0^{f_i} c_i(s)ds.$$

Since functions c_i are strictly increasing, then the following functions

$$f_i \to \int_0^{f_i} c_i(s)ds \qquad i = 1, \ldots, n,$$

are strictly convex, thus $T(\Delta F)$ is also strictly convex and cost function C is strictly monotone.

Therefore in this special case the following are equivalent:

- *Wardrop network equilibrium flow;*
- *minimization points of $T(\Delta F)$;*
- *solutions to SVI(C,K);*
- *solutions to MVI(C,K)*

Since K is a compact set and $T(\Delta F)$ is a strictly convex function, then there exists a unique minimizer for $T(\Delta F)$.

Hence, in such a symmetric network, there is a unique equilibrium flow H; moreover since C is strictly monotone, we obtain

$$\langle C(F), H - F \rangle < 0 \qquad \forall\ F \in K.$$

Wardrop equilibrium condition states that if, at a certain time, the network flow is an equilibrium flow, then no user wants to change his path and therefore the traffic flow will remain constant for all future times. Nevertheless, if the network flows are not equilibrium flows, then Wardrop condition is not useful to know the network traffic evolution.

Moreover we observe that to check whether or not a feasible flow H satisfies Wardrop condition, we do not need to compare H to the other feasible flows; now we are going to introduce a new Wardrop-type condition for a flow $H \in K$ that involves all the network feasible flows. This is done in the following:

Definition 2.2 *A feasible flow $H \in K$ is called a **strong equilibrium flow** if and only if*

$$\left[\begin{array}{c} \text{for each OD pair } W_j \in W, \text{ each } R_q,\ R_s \in \mathcal{R}_j, \\ \text{and each flow } F \in K, \\ C_q(F) < C_s(F) \Longrightarrow F_s > H_s \quad \text{or} \quad F_s = H_s = 0. \end{array}\right] \tag{2}$$

First we remark that a strong equilibrium flow is, in particular, an equilibrium flow.

To understand the above definition better, let us suppose that for some OD pair $W_j \in W$, for some paths $R_q,\ R_s \in W_j$ and for some feasible flow $F \in K$, with $F_s > 0$, we have $C_q(F) < C_s(F)$, then network flow will go down on path R_s; now if a flow H satisfies condition (2), then flow on R_s approaches to H_s. In other words a strong equilibrium flow H takes into account the information $C_q(F) < C_s(F)$ over all the other flows F.

The following examples show a network in which there exists a strong equilibrium flow.

Example 2.1 *Let us consider a network with four nodes $N_1,\ N_2,\ N_3,\ N_4$ and five arcs $A_1,\ A_2,\ A_3,\ A_4,\ A_5$:*

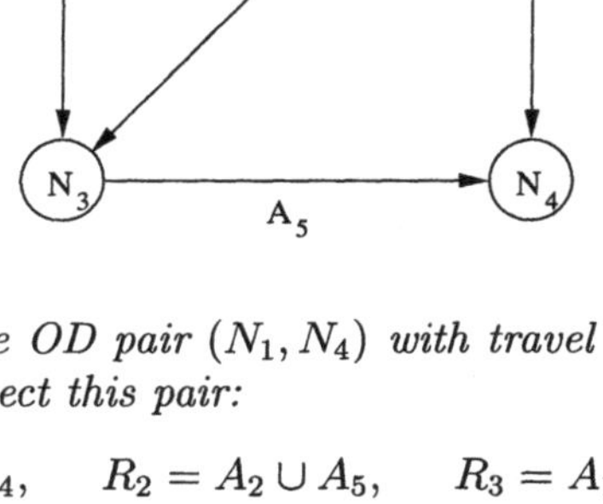

We have only one OD pair (N_1, N_4) with travel demand $\rho = 10$ and three paths that connect this pair:

$$R_1 = A_1 \cup A_4, \qquad R_2 = A_2 \cup A_5, \qquad R_3 = A_1 \cup A_3 \cup A_5,$$

therefore the set of feasible flows is

$$K = \{F \in \mathbb{R}^3 : \; F \geq 0, \; F_1 + F_2 + F_3 = 10\}.$$

We assume that travel cost on all arcs is defined as follows:

$$\begin{cases} c_1(f_1) = f_1 \\ c_2 = 21 \\ c_3(f_3) = f_3 \\ c_4 = 21 \\ c_5(f_5) = f_5, \end{cases}$$

and hence the corresponding travel cost on paths is

$$\begin{cases} C_1(F) = F_1 + \; F_3 + 21 \\ C_2(F) = F_2 + \; F_3 + 21 \\ C_3(F) = F_1 + F_2 + 3\,F_3. \end{cases}$$

We remark that, for all feasible flows $F \in K$, one has:

$$C_1(F) > C_3(F), \qquad \text{and} \qquad C_2(F) > C_3(F),$$

therefore the feasible flow $H = (0,0,10)$ is the unique Wardrop equilibrium flow. Moreover, for an arbitrary flow $F \in K$, the inequality $C_q(F) < C_s(F)$ in the condition (2) is true only if $s = 1$ or $s = 2$, and $H_1 = H_2 = 0$, thus the condition

$$F_s > H_s \quad \text{or} \quad F_s = H_s = 0$$

is equivalent to

$$F_s > 0 \quad \text{or} \quad F_s = 0,$$

which is trivially true.

Example 2.2 *Let us consider the same network as the example 2.1, in which travel demand is $\rho = 10$ and travel cost on arcs is defined as follows:*

$$\begin{cases} c_1(f_1) = f_1 \\ c_2(f_3) = 10 + f_3 \\ c_3(f_3) = f_3 \\ c_4(f_3) = 10 + f_3 \\ c_5(f_5) = f_5, \end{cases}$$

and hence the corresponding travel cost on paths is

$$\begin{cases} C_1(F) = F_1 + 2\,F_3 + 10 \\ C_2(F) = F_2 + 2\,F_3 + 10 \\ C_3(F) = F_1 + F_2 + 3\,F_3. \end{cases}$$

We remark that for all feasible flow $F \in K$ one has:

$$C_1(F) \geq C_3(F), \qquad \text{and} \qquad C_2(F) \geq C_3(F),$$

therefore the feasible flow $H = (0,0,10)$ is the unique equilibrium flow. Moreover, for an arbitrary flow $F \in K$, the inequality $C_q(F) < C_s(F)$ in the condition (2) is true only if $s = 1$ or $s = 2$, and $H_1 = H_2 = 0$, thus the condition

$$F_s > H_s \quad or \quad F_s = H_s = 0$$

is equivalent to

$$F_s > 0 \quad or \quad F_s = 0,$$

which is trivially true.

Besides Wardrop equilibrium, also the condition (2) has connections with a variational formulation, as the following theorem states.

Theorem 2.2 *If H is a strong equilibrium flow, then H is a solution to MVI(C,K).*

Proof. Let $F \in K$ and $W_j \in W$ be arbitrary. We consider a path $R_q \in \mathcal{R}_j$ such that

$$C_q(F) = \min_{R_r \in \mathcal{R}_j} C_r(F).$$

Then from condition (2) one has

$$(C_r(F) - C_q(F))(F_r - H_r) \geq 0 \quad \forall\, R_r \in \mathcal{R}_j.$$

Thus

$$\sum_{r=1}^{r_j} C_r(F)(F_r - H_r) \geq C_q(F) \sum_{r=1}^{r_j} (F_r - H_r) = C_q(F)(\rho_j - \rho_j) = 0.$$

Hence

$$\langle C(F), F - H \rangle = \sum_{j=1}^{l} \sum_{r=1}^{r_j} C_r(F)(F_r - H_r) \geq 0,$$

hence H is a solution of MVI(C,K). □

The vice versa of theorem 2.2 is, in general, false as the following example shows.

Example 2.3 *Let us consider the same network as in the example 2.1, where travel costs on paths are:*

$$\begin{cases} C_1(F) = 3\,F_1 + 2\,F_3 \\ C_2(F) = 5\,F_2 + 2\,F_3 \\ C_3(F) = 2\,F_1 + F_2 + 3\,F_3. \end{cases}$$

We observe that $H = (5, 3, 2)$ *is an equilibrium flow, because*

$$C_1(H) = C_2(H) = C_3(H) = 19,$$

and H *is solution of* $MVI(C, K)$*, in fact for all* $F \in K$ *we can write* $F_3 = 10 - F_1 - F_2$*, and thus we have*

$$\langle C(F), F - H \rangle = 2F_1^2 + 5F_2^2 - F_1F_2 - 17F_1 - 25F_2 + 80 = \phi(F_1, F_2).$$

Since

$$\nabla^2 \phi(F_1, F_2) = \begin{pmatrix} 4 & -1 \\ -1 & 10 \end{pmatrix}$$

has positive eigenvalues $\lambda_{1,2} = 7 \pm \sqrt{10}$, ϕ *is a strictly convex function, his minimum point is* $(5, 3)$ *and* $\phi(5, 3) = 0$*, hence* $\phi(F_1, F_2) \geq 0$ *for all* $(F_1, F_2) \in \mathbb{R}^2$*, that is* H *is solution of* $MVI(C, K)$*.*

However H *is not a global equilibrium flow, because, if we consider* $F = (6, 3, 1) \in K$*, one has*

$$17 = C_2(F) < C_3(F) = 18 \qquad \text{and} \qquad 1 = \mathrm{F}_3 < \mathrm{H}_3 = 2.$$

For a feasible flow $H \in K$ we now introduce another Wardrop-type condition giving the "stability" of H with respect to some perturbations of flow H on a arbitrary couple of paths connecting an OD-pair.

Definition 2.3 *A feasible flow* $H \in K$ *is called a* ***stable equilibrium flow*** *if and only if*

$$\left[\begin{array}{c} \text{for each OD pair } W_j \in W, \text{ each } R_q,\ R_s \in \mathcal{R}_j, \\ \text{and each flow } F \in K \text{ such that } F_i = H_i, \forall\ i \neq q, s \text{ and } F_s < H_s \\ H_s > 0 \Longrightarrow C_s(F) \leq C_q(F) \end{array} \right] \tag{3}$$

To understand the above definition better, let us suppose that a feasible flow H be a stable equilibrium flow, R_q, R_s be two paths connecting a certain OD pair and that the flow H_s be positive; if we perturb H only on the paths R_q, R_s pushing flow from R_s to R_q, then the path R_q does not cost less than the path R_s.

In the following theorem we check the relationships between stable equilibrium flow and solutions of SVI(C,K) and MVI(C,K).

Theorem 2.3

1. *If a flow $H \in K$ solves MVI(C,K), then H is a stable equilibrium flow;*

2. *if $H \in K$ is a stable equilibrium flow and the cost function C is continuous, then H is an equilibrium flow.*

Proof.

1. We consider an arbitrary OD pair W_j and two fixed paths $R_q, R_s \in \mathcal{R}_j$. Since $H \in K$ is a solution to MVI(C,K) then for all feasible flow $F \in K$ one has
$$\langle C(F), F - H \rangle = \sum_{j=1}^{l} \sum_{r=1}^{r_j} C_r(F)(F_r - H_r) \geq 0;$$
in particular, if we choose a feasible flow F such that $F_i = H_i, \forall\, i \neq q, s$ and $F_s < H_s$, then
$$(C_s(F) - C_q(F))(F_s - H_s) \geq 0,$$
and hence $C_s(F) \leq C_q(F)$.

2. For each OD pair W_j and for each $R_q, R_s \in \mathcal{R}_j$, with $H_s > 0$, we have $C_s(H) \leq C_q(H)$ by continuity of the cost function C. □

From the above theorem it follows that the stable equilibrium flows are selected network equilibrium flows, with the following property: if a feasible flow H is a network equilibrium flow and R_q, R_s are two paths connecting a certain OD pair W_j, with $H_s > 0$, then we know that $C_s(H) \leq C_q(H)$; moreover, if H is a stable equilibrium flow then $C_s(F) \leq C_q(F)$ for all perturbation flows F which shift flow from R_s to R_q.

We remark that a stable equilibrium flow does not coincide with all the network equilibrium flows as the following example shows.

Example 2.4 *Let us consider the same network as the example 2.1, in which travel cost on arcs is:*
$$\begin{cases} c_1(f_1) = f_1 \\ c_2(f_2, f_3) = f_2 + 4\, f_3 \\ c_3(f_2, f_3) = f_2 + 2\, f_3 \\ c_4(f_3, f_4) = 2\, f_3 + 2\, f_4 \\ c_5(f_5) = f_5, \end{cases}$$
and therefore the travel cost on paths is
$$\begin{cases} C_1(F) = 3\, F_1 + 3\, F_3 \\ C_2(F) = 2\, F_2 + 5\, F_3 \\ C_3(F) = F_1 + 2\, F_2 + 4\, F_3. \end{cases}$$

We assume that travel demand is $\rho = 10$. *An equilibrium flow is* $H = (4, 2, 4)$, *in fact*

$$C_1(H) = C_2(H) = C_3(H) = 24,$$

but H *is not a stable equilibrium flow because if we perturb* H *on paths* R_2 *and* R_3 *so that* $F_1 = 4$, $F_2 = 2 + x$ *and* $F_3 = 4 - x$, *with* $0 < x < 4$, *one has*

$$C_3(F) = 24 - 2\,x > C_2(F) = 24 - 3\,x.$$

Moreover we can remark, by the following example, that all the stable equilibrium flows are not solution to MVI(C,K).

Example 2.5 *We consider the same network as the example 2.1, with the following travel cost on paths:*

$$\begin{cases} C_1(F) = 5\,F_1 + 4\,F_3 \\ C_2(F) = 5\,F_2 + 4\,F_3 \\ C_3(F) = 3\,F_1 + 3\,F_2 + 3\,F_3 \end{cases}$$

and travel demand is $\rho = 9$.

An equilibrium flow is $H = (3, 3, 3)$, *in fact*

$$C_1(H) = C_2(H) = C_3(H) = 27,$$

moreover H *is a stable equilibrium flow. In fact if we choose a feasible flow* F *such that* $F_1 = 3$, $F_2 = 3 - x$ *and* $F_3 = 3 + x$, *with* $0 < x < 3$, *one has*

$$C_2(F) = 27 - x < 27 = C_3(F),$$

if $F_2 = 3$, $F_1 = 3 - x$ *and* $F_3 = 3 + x$ *then*

$$C_1(F) = 27 - x < 27 = C_3(F),$$

if $F_3 = 3$, $F_1 = 3 - x$ *and* $F_2 = 3 + x$ *then*

$$C_1(F) = 27 - 5\,x < 27 + 5\,x = C_2(F).$$

If we choose the perturbation $-3 < x < 0$ *we will obtain analogous inequalities.*

However H *is not solution to MVI(C,K) because for* $F = (0, 0, 9)$ *we have*

$$\langle C(F), F - H\rangle = \langle (36, 36, 27), (-3, -3, 6)\rangle = -54.$$

Finally we show an example that gives a global picture:

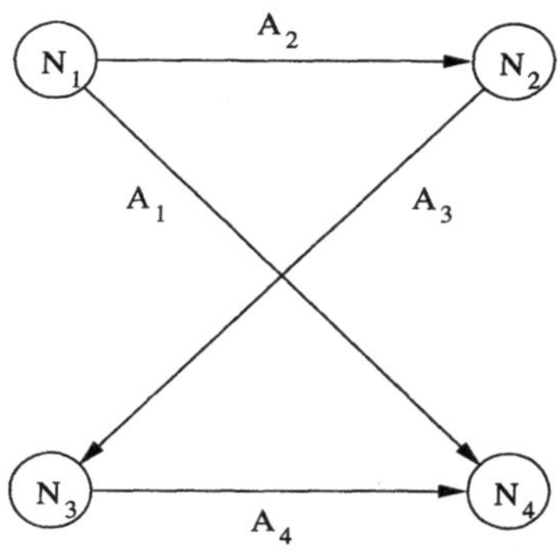

Example 2.6 *Let us consider the following network with four nodes N_1, N_2, N_3, N_4 and four arcs A_1, A_2, A_3, A_4:*

We have only one OD pair (N_1, N_4) with travel demand $\rho = 10$ and only two paths connecting this pair:

$$R_1 = A_1, \qquad R_2 = A_2 \cup A_3 \cup A_4,$$

therefore the set of feasible flows is

$$K = \{F \in \mathbb{R}^2 : \ F \geq 0, \ F_1 + F_2 = 10\}.$$

We assume that travel cost on paths is defined as follows:

$$\begin{cases} C_1(F) = 2\,F_1 + 5 \\ C_2(F) = \frac{1}{2}\,F_1^2 - 4\,F_1 + 15 \end{cases}$$

From the figure it is easy to argue that there are two equilibrium flows: $H' = (2, 8)$ and $H'' = (10, 0)$.

We remark that H' is a strong equilibrium flow because: if $C_2(F) > C_1(F)$ then $F_1 < H_1'$, that is $F_2 > H_2'$, and if $C_1(F) > C_2(F)$, then $F_1 > H_1'$.

Moreover H'' is an equilibrium flow but it is not a stable equilibrium: indeed $H_1'' > 0$ but $C_1(F) > C_2(F)$ for each feasible flow F such that $2 < F_1 < 10$.

Finally, we sum up in the following figure the relationships between Wardrop-type equilibria and solutions to SVI(C,K) and MVI(C,K) when the cost function C is assumed to be continuous on K.

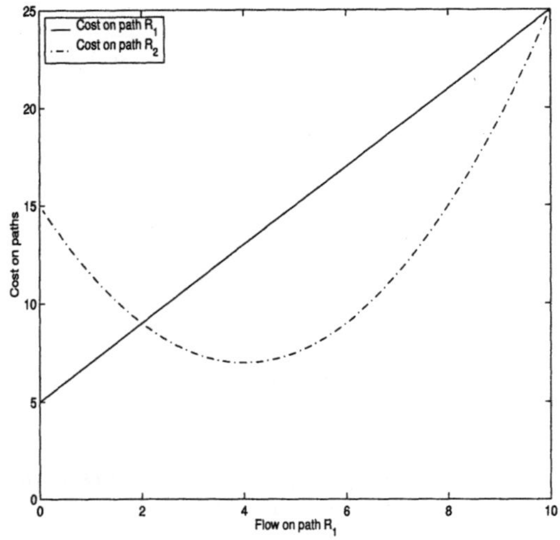

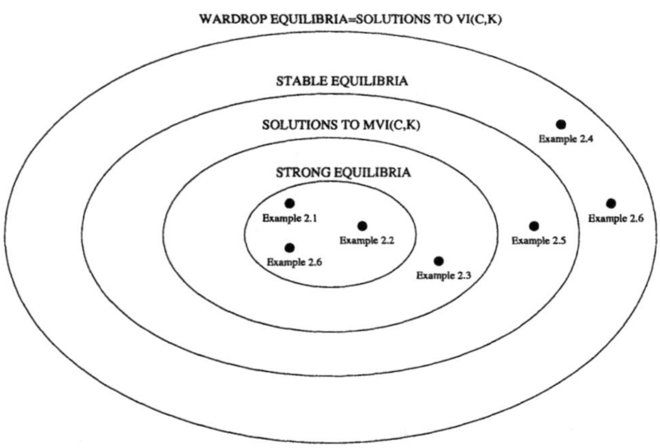

References

[1] P. Daniele, A. Maugeri and W. Oettli, *Time-Dependent Traffic Equilibria*, Journal of Optimization Theory and Applications, **103**, (1999), 543-555.

[2] P. Dupuis, *Large deviations analysis of reflected diffusions and constrained stochastic approximation algorithms in convex sets*, Stochastic **21** (1987), 63-96.

[3] P. Dupuis and A. Nagurney, *Dynamical systems and variational inequalities*, Annals of Operations Research, **44**, (1993), 9-42.

[4] P. Ferrari, *Equilibrium in asymmetric multimodal transport networks with capacity constraints*, Le Matematiche, vol. XLIX, (1994), 223-241.

[5] T.L. Friesz, D.H. Bernstein, N.J. Mehta, R.L. Tobin and S. Ganjlizadeh, *Day-to-day dynamic network disequilibria and idealized traveler information systems*, Operations Research **42** (1994), 1120-1136.

[6] P.T. Harker and J.-S. Pang, *Finite-dimensional variational inequality and nonlinear complementarity problems: a survey of theory, algorithms and applications*, Mathematical Programming **48** (1990), 161-220.

[7] S. Karamardian and S. Schaible, *Seven kinds of monotone maps*, Journal of Optimization Theory and Applications, **66** (1990), 37-46.

[8] D. Kinderlehrer and G. Stampacchia, *An introduction to variational inequality and their applications*, Academic Press, New York, 1980.

[9] S. Komlósi, *On the Stampacchia and Minty Variational Inequalities*, in Generalized convexity and optimization for economic and financial decisions, edited by Giorgi G. and Rossi F. A., Pitagora Editrice, Bologna, 1999, 231-261.

[10] T. Larsson and M. Patriksson, *On side constrained models of traffic equilibria*, in Variational inequalities and network equilibrium problems, edited by F. Giannessi and A. Maugeri, Plenum Press, New York, 1995.

[11] T. Larsson and M. Patriksson, *Equilibrium characterizations of solutions to side constrained asymmetric traffic assignment models*, Le Matematiche, vol. XLIX, (1994), 249-280.

[12] A. Maugeri, *Optimization problems with side constraints and generalized equilibrium principles*, Le Matematiche, vol. XLIX, (1994), 305-312.

[13] A. Maugeri, *Variational and quasi-variationalinequalities in network flow models. Recent developments in theory and algorithms*, in Variational inequalities and network equilibrium problems, edited by F. Giannessi and A. Maugeri, Plenum Press, New York, 1995.

[14] A. Maugeri, W. Oettli and D. Schläger, *A flexible form of Wardrop's principle for traffic equilibria with side constraints*, Rendiconti del Circolo Matematico di Palermo, serie II, suppl. 48, (1997), 185-193.

[15] A. Nagurney and D. Zhang, *Formulation, stability, and computation of traffic network equilibria as projected dynamical systems*, Journal of Optimization Theory and Applications, **93**, (1997), 417-444.

[16] R.T. Rockafellar, *Monotone relations and network equilibrium*, in Variational Inequalities and Network Equilibrium Problems, edited by Giannessi F. and Maugeri A., Plenum Press, New York, 1995, 271-288.

VARIATIONAL GEOMETRY AND EQUILIBRIUM

Michael Patriksson
Department of Mathematics
Chalmers Univeristy of Technology
SE-412 96 Gothenburg, Sweden
e-mail: `mipat@math.chalmers.se`

R. Tyrrell Rockafellar
Department of Mathematics
University of Washington
Seattle, WA 98195-4350, USA
e-mail: `rtr@math.washington.edu`

Abstract Variational inequalities and even quasi-variational inequalities, as means of expressing constrained equilibrium, have utilized geometric properties of convex sets, but the theory of tangent cones and normal cones has yet to be fully exploited. Much progress has been made in that theory in recent years in understanding the variational geometry of nonconvex as well as convex sets and applying it to optimization problems. Parallel applications to equilibrium problems could be pursued now as well.

This article explains how normal cone mappings and their calculus offer an attractive framework for many purposes, and how the variational geometry of the graphs of such mappings, as nonconvex sets of a special nature, furnishes powerful tools for use in ascertaining how an equilibrium is affected by perturbations. An application to aggregated equilibrium models, and in particular multi-commodity traffic equilibrium, is presented as an example.

Keywords: variational geometry, normal cones, tangent cones, variational inequalities, quasi-variational inequalities, equilibrium, traffic, multi-commodity flow, variational analysis, perturbations.

1 Introduction

The concept of equilibrium has long had a close connection with optimization. Traditionally, an equilibrium point $\bar{x}$ has often been construed as a solution to a vector equation $F(x) = 0$. In that case, if F is the gradient ∇f of a function f, the equilibrium satisfies $\nabla f(\bar{x}) = 0$ and is said to be described by a "variational principle." The function f might have a local minimum or maximum at $\bar{x}$, and indeed variational principles have often centered on minimum energy, but of course $\bar{x}$ could merely be a "stationary point" of f.

The connection between equilibrium and optimization has really always been on this level. Equilibrium conditions typically are properties that resemble *first-order* conditions for optimality. They do correspond to local optimality in many cases, or at least to a competitive balance among various optimizing agents, but not in every case.

The question of how to account for constraints in formulations of equilibrium has gained in importance in modern times. Some constraints can be handled through the introduction of Lagrange multipliers. Others, though, fit poorly with classical techniques and demand new mathematics in order to achieve satisfactory treatment. Anyway, Lagrange multipliers are something "secondary." The basic issue is how to relativize an equilibrium to a set C, but there is great diversity in the ways that C might be specified. Such diversity should somehow be accommodated in a "primary" manner.

A major advance came with the notion of a variational inequality over a convex set C, whatever the structure of that set. This innovation, in the era when convex analysis was starting up, provided a broad approach to generalized constraints of convex type. Moreover, through restatements in a Lagrangian setting, it supported treatment of some constraints of nonconvex type as well—provided that those constraints could be represented by multipliers which in turn would be incorporated among the equilibrium variables.

Since then, work on variational inequalities has led to many successes. Extensions such as to quasi-variational inequalities have carried this further. But equilibrium theory has yet to take advantage of some of the recent progress in optimization, where much has been learned about how to handle very general classes of sets C directly, without necessarily introducing multipliers or relying on excursions through convex analysis.

The key to those accomplishments in optimization theory has come from *variational geometry*, i.e., the study of tangent cones and normal cones to a general set C, their properties, relationships, and calculus. Variational geometry has proved to be valuable not only in characterizing optimality but also in understanding how solutions are affected by parametric perturbations. In the latter role it has required abandoning the preconception from convex analysis that tangent cones and normal cones, once they have "rightly" been defined, should themselves always be convex.

The aim of this article is to indicate how ideas of variational geometry can more fully be put to use in understanding equilibrium, thereby perhaps opening new avenues of development as well as consolidating some of the theory that already exists.

We begin by reviewing the way that a variational inequality reflects the concept of a normal cone in convex analysis. We go on then to discuss normal cones to nonconvex sets as they are now understood, demonstrating that quasi-variational inequalities correspond, at least sometimes, to normal cones in that wider framework. The crucial object in both cases is a set-valued normal cone *mapping* N_C, yet such mappings are often quite out of sight when people speak of variational or quasi-variational inequalities.

Properties of a normal cone mapping can be very powerful, especially in sensitivity analysis of solutions. In fact the variational geometry of the *graphs* of normal cone mappings is the mainstay for results in that direction.

We illustrate how that operates by means of a formula for perturbations of equilibrium in a traffic model where the set C is convex, but instead of being specified directly by equations or inequalities, is expressed as a *sum* of sets with those specifications. We show also, in a different example, how even nonconvex sets C can be amenable to perturbation treatment through their associated normal cone mappings N_C.

2 Variational Inequalities and Normals to Convex Sets

For simplicity, we focus in this article on equilibrium models in the space $\mathbb{R}^n$ and, in other respects too, forgo possibilities for greater generality in order to concentrate on the main points of our discussion.

Let $F : \mathbb{R}^n \to \mathbb{R}^n$ be a continuous mapping (single-valued) and let $C \subset \mathbb{R}^n$ be a nonempty, closed, convex set. The *variational inequality* for F over C, with solution $\bar{x}$, is customarily posed as the condition

$$\bar{x} \in C, \qquad \langle F(\bar{x}), x - \bar{x}\rangle \geq 0 \;\text{ for all } x \in C. \tag{1}$$

In this formulation, a variational inequality is comprised of an infinite collection of linear inequalities which must be satisfied by $\bar{x}$ in addition to the condition $\bar{x} \in C$, which itself could of course in turn be represented by an infinite collection of linear inequalities, inasmuch as any closed convex set is the intersection of a collection of closed half-spaces. Nobody insists on *always* expressing a convex set that way, however, since that would not be convenient and the picture of the set and its geometry could get lost. For the same reason, (1) can fall short of being the best way to think about a variational inequality, even though it is the source of the "inequality" part of the name.

A better approach is to make use of the normal cone concept in convex analysis [8], [10], which captures the inequality aspect of (1) more conducive to geometric thinking and open-ended calculus. A vector v is said to be *normal* to the convex set C at a point $\bar{x}$ if $\bar{x} \in C$ and

$$\langle v,\, x - \bar{x}\rangle \geq 0 \quad \text{for all } x \in C. \tag{2}$$

The set of all such vectors is denoted by $N_C(\bar{x})$ and is called the *normal cone* to C at $\bar{x}$. It is indeed a *cone* (a set containing the origin and including for each of its elements $v \neq 0$, if any, the ray $\{\lambda v \mid \lambda \geq 0\}$). Moreover it is closed and convex. For points $\bar{x} \notin C$, it is expedient to take $N_C(\bar{x}) = \emptyset$ so as to get a fully defined set-valued *normal cone mapping* $N_C : I\!R^n \rightrightarrows I\!R^n$. This mapping can be described directly by the optimization rule

$$v \in N_C(\bar{x}) \iff \bar{x} \in \operatorname*{argmax}_{x \in C} \langle v, x\rangle. \tag{3}$$

For $v \neq 0$, the relation $v \in N_C(\bar{x})$ can also seen pictorially as meaning that v is an outward normal to a supporting half-space to C at $\bar{x}$.

In terms of the normal cone mapping N_C, the variational inequality condition in (1) can be recast in the form

$$-F(\bar{x}) \in N_C(\bar{x}), \quad \text{or} \quad F(\bar{x}) + N_C(\bar{x}) \ni 0, \tag{4}$$

with (3) then providing the popular interpretation of a variational inequality as corresponding to optimization of a linear form.

Although the re-expression in (4) might, at first, seem to provide nothing much beyond (1), it shifts the perspective from a system of inequalities to finding a "zero" of a set-valued mapping $F + N_C$. It turns attention to the nature of that mapping and the geometry of its graph. Furthermore, it provides guidance to generalization by suggesting that, when the need arises to go beyond the case of a convex set C, the central issue ought to be what definition of normality to adopt in the absence of convexity.

The graph of $F + N_C$ depends heavily on the graph of N_C. One has $y \in (F + N_C)(x)$ if and only if $y - F(x) \in N_C(x)$, so that

$$\operatorname{gph}(F + N_C) = M^{-1}(\operatorname{gph} N_C) \quad \text{for } M : (x, y) \mapsto (x,\, y - F(x)). \tag{5}$$

The mapping M is a homeomorphism of $I\!R^n$ with itself, since F has been assumed to be continuous, and it is actually a diffeomorphism when F is smooth (i.e., continuously differentiable). Therefore, the geometry of the graph of N_C holds critical information about the variational inequality.

What can be said about that geometry? First and foremost is the fact that $\operatorname{gph} N_C$ is a *graphically Lipschitzian manifold* in a global sense. A set is said to be such a manifold of dimension d at one of its points if there is a smooth change of coordinates that transforms it, locally around that point, into the graph of a Lipschitz continuous mapping from d coordinates into the remaining coordinates. In the case of N_C, a change of coordinates that is well known to have this effect is

$$(x, v) \longleftrightarrow (z, w) \quad \text{with } z = x + v, \ w = x - v.$$

It sets up $\operatorname{gph} N_C$ as the graph of the mapping $J : z \mapsto (P_C(z),\ z - P_C(z))$, where P_C is the nearest-point projection mapping onto C:

$$\operatorname{gph} N_C = \{(P_C(z),\ z - P_C(z)) \,|\, z \in \mathbb{R}^n\}. \tag{6}$$

Indeed, the indicated mapping J is one-to-one between $z \in \mathbb{R}^n$ and $(x, v) \in \operatorname{gph} N_C$ and is globally Lipschitz continuous in both directions. Thus, $\operatorname{gph} N_C$ is globally a graphically Lipschitzian manifold of dimension n within $\mathbb{R}^{2n}$.

The representation in (6) is the *Minty parameterization* of $\operatorname{gph} N_C$. It stems from N_C being a *maximal monotone* mapping. Recall that a set-valued mapping $S : \mathbb{R}^n \rightrightarrows \mathbb{R}^n$ is called monotone if $\langle v' - v,\ x' - x \rangle \geq 0$ whenever $v \in S(x)$ and $v' \in S(x')$, and is said to be maximal in this respect if there is no monotone mapping $S' : \mathbb{R}^n \rightrightarrows \mathbb{R}^n$ such that $\operatorname{gph} S' \supset \operatorname{gph} S$, $\operatorname{gph} S' \neq \operatorname{gph} S$. Minty showed in 1962 [5] that the graph of any maximal monotone mapping has a Lipschitz continuous parameterization like (6), except with a certain other mapping P in place of the projection P_C. The maximal monotonicity of N_C is a property shared with the subgradient mappings associated with lower semicontinuous, proper, convex functions in general and also enjoyed in many other situations, but we will not go into that here. (See [10, Chapter 12], for instance.)

On the basis of (5), the graphically Lipschitzian property of N_C carries over locally to $F + N_C$ when F is smooth. We will extract more from this later when we turn to the sensitivity analysis of solutions to a variational inequality. Obviously, for any property of N_C to be usable in practice there must be machinery for working out in detail how the property is manifested in terms of the specific structure of C. We will come back to this once we have passed beyond convexity to broader versions of variational geometry.

Another set-valued mapping that deserves mention here in connection with N_C is the tangent cone mapping T_C of convex analysis. The *tangent cone* to C at a point $\bar{x} \in C$ is

$$T_C(\bar{x}) = \operatorname{cl} \{w = \lambda(x - \bar{x}) \,|\, x \in C,\ \lambda \geq 0\}. \tag{7}$$

Like $N_C(\bar{x})$, it too is a closed convex cone, and moreover these two cones are polar to each other:

$$\begin{aligned} N_C(\bar{x}) &= \{v \,|\, \langle v, w\rangle \le 0,\ \forall w \in T_C(\bar{x})\}, \\ T_C(\bar{x}) &= \{w \,|\, \langle v, w\rangle \le 0,\ \forall v \in N_C(\bar{x})\}. \end{aligned} \tag{8}$$

The question of how far this polarity persists when normal cones and tangent cones are generalized will occupy us as we proceed.

3 Quasi-Variational Inequalities and Normals to General Sets

Suppose now that the set $C \subset \mathbb{R}^n$, although still nonempty and closed, is not necessarily convex. What conditions on C and F might be suitable candidates as replacements for a variational inequality? One idea has been a quasi-variational inequality. It depends on the intermediary of a mapping D that assigns to each $x \in C$ a set $D(x)$ containing x. With respect to such a mapping D, for which different choices may be admitted, the *quasi-variational inequality* for F over C, with solution $\bar{x}$, is the condition

$$\bar{x} \in C, \qquad \langle F(\bar{x}),\, x - \bar{x}\rangle \ge 0 \ \text{ for all } x \in D(\bar{x}). \tag{9}$$

Usually the requirement $\bar{x} \in D(\bar{x})$ is added to the statement (9), but *we will assume here that*

$$x \in D(x) \ \text{ for all } x \in C. \tag{10}$$

In principle *no generality is lost by this assumption*—for conceptual purposes—since we can arrange for it be satisfied by replacing C in that situation by its subset $C' = \{x \in C \,|\, x \in D(x)\}$. Although possible fixed-point aspects of a quasi-variational inequality in the more usual formulation are suppressed from view by this device, we can hope then to focus more clearly on the equilibrium aspects related to optimization. The theory of existence of solutions to a quasi-variational, posed in our reduced manner, could well require a careful utilization of fixed-point technology through representing our C as the truncation C' of some larger, possibly convex set C to which the mapping D can be extended, but that is a separate matter which need not distract us from our present goals.

In the special case where $D(x) = C$ for all $x \in C$, the quasi-variational inequality reverts to a variational inequality. Beyond that, as long as the mapping D is closed-convex-valued, (9) can aptly be viewed as a roving variational inequality for F over a set that shifts with the solution candidate. This is appealing especially when $D(x)$ is envisioned as a local approximation to C at x.

For other insights, it will be helpful instead to think of a quasi-variational inequality as maybe involving a "proposal for generalized normality," at least in some cases. At an arbitrary $\bar{x} \in C$, let

$$N(\bar{x}) = \{v \mid \langle v, x - \bar{x} \rangle \leq 0, \ \forall x \in D(\bar{x})\}, \tag{11}$$

noting that $N(\bar{x})$ is a certain closed, convex cone. For $\bar{x} \notin C$, take $N(\bar{x}) = \emptyset$. A set-valued mapping $N : I\!R^n \rightrightarrows I\!R^n$ is thereby defined for which, in parallel to (4), the quasi-variational inequality (9) comes out as equivalent to

$$-F(\bar{x}) \in N(\bar{x}), \quad \text{or} \quad F(\bar{x}) + N(\bar{x}) \ni 0. \tag{12}$$

May the vectors $v \in N(\bar{x})$ rightly be regarded as "normals" to C at $\bar{x}$, in some reasonable sense? That has to depend on the choice of the mapping D; without a filtering of possibilities, the results could be too bizarre. Nonetheless, choices of D consistent with "normality" do exist, as will be explained next, even though not every useful example of a quasi-variational would have to conform to such an interpretation.

Three versions of normal cone now dominate theory in the finite-dimensional context we are operating in. In describing them, we follow the patterns of notation and terminology in the recent book or Rockafellar and Wets [10, Chapter 6]. The *regular* normal cone to C at a point $\bar{x} \in C$, consisting of the *regular* normal vectors v, is

$$\hat{N}_C(\bar{x}) = \{v \mid \langle v, x - \bar{x} \rangle \leq o(|x - \bar{x}|) \text{ for } x \in C\}, \tag{13}$$

where $|\cdot|$ denotes the Euclidean norm and the "o" inequality stands for the property that

$$\limsup_{\substack{x \to \bar{x} \\ x \in C, \, x \neq \bar{x}}} \frac{\langle v, x - \bar{x} \rangle}{|x - \bar{x}|} \leq 0.$$

The *general* normal cone to C at $\bar{x}$, the elements v of which are simply called the normal vectors to C at $\bar{x}$, is defined from this by

$$N_C(\bar{x}) = \{v \mid \exists x^\nu \in C, \ v^\nu \in \hat{N}_C(x^\nu), \text{ with } (x^\nu, v^\nu) \to (\bar{x}, v)\}. \tag{14}$$

(We systematically use superscript $\nu = 1, 2, \ldots$ to express sequences.) The third object,

$$\bar{N}_C(\bar{x}) = \ \text{closed convex hull of } N_C(\bar{x}), \tag{15}$$

is the *Clarke* normal cone (see [10, Chapter 6] for more on this cone and its history). Although $\bar{N}_C(\bar{x})$ and $\hat{N}_C(\bar{x})$ are closed convex cones, $N_C(\bar{x})$ is a closed cone that need not be convex. Obviously

$$\hat{N}_C(\bar{x}) \subset N_C(\bar{x}) \subset \bar{N}_C(\bar{x}).$$

The set C is said to be *Clarke regular* at $\bar{x}$ if every normal vector is a regular normal vector, i.e., if the limit process in (14) generates no additional vectors v at $\bar{x}$. In that case, a very important one for many—but not all—applications, all three cones coincide. Such regularity prevails in particular when C is convex; then the "o" term in (13) can be replaced by 0, and normality reduces to the single concept of convex analysis that we were dealing with previously.

To understand how these normal cones might be connected with quasi-variational inequalities, ideas of tangency need to brought in. The *general* tangent cone to C at $\bar{x}$, the elements of which are simply called tangent vectors, is defined through set limits as

$$T_C(\bar{x}) = \limsup_{\lambda \searrow 0} \frac{1}{\lambda}(C - \bar{x}), \tag{16}$$

whereas the *regular* tangent cone, consisting of the *regular* tangent vectors to C at $\bar{x}$, is

$$\hat{T}_C(\bar{x}) = \liminf_{\substack{\lambda \searrow 0 \\ x \to \bar{x},\, x \in C}} \frac{1}{\lambda}(C - x). \tag{17}$$

Both $T_C(\bar{x})$ and $\hat{T}_C(\bar{x})$ are closed cones, the first also being called the *contingent* cone and the second the *Clarke* tangent cone. Evidently

$$T_C(\bar{x}) \supset \hat{T}_C(\bar{x}).$$

When C is convex, $T_C(\bar{x})$ and $\hat{T}_C(\bar{x})$ agree with the tangent cone of convex analysis that was defined in (7).

Although $T_C(\bar{x})$ can be nonconvex, $\hat{T}_C(\bar{x})$ is always convex. This surprising property goes hand in hand with another remarkable facts of basic variational geometry:

$$T_C(\bar{x}) = \hat{T}_C(\bar{x}) \iff N_C(\bar{x}) = \hat{N}_C(\bar{x}). \tag{18}$$

In other words, *the case where every normal vector v to C at $\bar{x}$ is a regular normal vector, which was taken above as the definition of Clarke regularity, can equally well be portrayed as the case where every tangent vector w to C at $\bar{x}$ is a regular tangent vector.* (This is why the term "regular" is employed systematically as above.) Besides, one has

$$\hat{T}_C(\bar{x}) = \liminf_{\substack{x \to \bar{x} \\ x \in C}} T_C(x), \qquad N_C(\bar{x}) = \limsup_{\substack{x \to \bar{x} \\ x \in C}} \hat{N}_C(x), \tag{19}$$

where the second limit merely restates the definition in (14) but the first is a rather deep result. Through (18) and (19), Clark regularity can be identified with a semicontinuity property of T_C as well as one of $\hat{N}_C$.

The reciprocity goes further still with the fact that the following polarity relationships always hold:

$$\begin{aligned} \hat{N}_C(\bar{x}) &= \{v \,|\, \langle v, w\rangle \leq 0,\ \forall w \in T_C(\bar{x})\}, \\ \hat{T}_C(\bar{x}) &= \{w \,|\, \langle v, w\rangle \leq 0,\ \forall v \in N_C(\bar{x})\}. \end{aligned} \tag{20}$$

Therefore, *C is Clarke regular at $\bar{x}$ if and only if $T_C(\bar{x})$ and $N_C(\bar{x})$ are closed convex cones polar to each other.* On the other hand,

$$\begin{aligned} \bar{N}_C(\bar{x}) &= \{v \,|\, \langle v, w\rangle \leq 0,\ \forall w \in \hat{T}_C(\bar{x})\}, \\ \hat{T}_C(\bar{x}) &= \{w \,|\, \langle v, w\rangle \leq 0,\ \forall v \in \bar{N}_C(\bar{x})\}, \end{aligned} \tag{21}$$

so that $\hat{T}_C(\bar{x})$ and $\bar{N}_C(\bar{x})$ always form a pair of closed convex cones polar to each other, regardless of Clarke regularity.

We can return now to quasi-variational inequalities. Let us observe that the normal cone schemes generated in that setting by (11), which translates the quasi-variational inequality condition (9) into the mapping formulation in (12), are perhaps not as varied as they might appear. They can really be seen as coming from *the choice of a mapping T that assigns to any point $\bar{x} \in C$ a cone $T(\bar{x})$.* Specifically, given D let

$$T(\bar{x}) = \{w = \lambda(x - \bar{x}) \,|\, x \in K(\bar{x}),\ \lambda \geq 0\}. \tag{22}$$

The formula for $N(\bar{x})$ in (11) can then be rewritten equivalently as a polarity relation:

$$N(\bar{x}) = \{v \,|\, \langle v, w\rangle \leq 0,\ \forall w \in T(\bar{x})\}. \tag{23}$$

Therefore, instead of speaking at all about a mapping D that assigns to each $x \in C$ a set $D(x)$ containing x, one could speak directly, from the start, about a cone-valued mapping T. No generality is lost in this maneuver because, given a choice of T, one can return to a D formulation by taking $D(x) = T(x) + x$. *The putative normal cones $N(\bar{x})$ that underlie quasi-variational inequalities can thus be interpreted as arising by duality from the introduction of putative tangent cones $T(\bar{x})$.*

Two specializations are now immediate. The choice $T(x) = T_C(x)$ turns the quasi-variational inequality into the case of (12) in which $N(x) = \hat{N}_C(x)$, whereas the choice $T(x) = \hat{T}_C(x)$ corresponds in (12) to $N(x) = \bar{N}_C(x)$. *When C is Clarke regular, these cases coincide and the quasi-variational inequality comes out as*

$$-F(\bar{x}) \in N_C(\bar{x}), \quad \text{or} \quad F(\bar{x}) + N_C(\bar{x}) \ni 0, \tag{24}$$

which exactly mirrors the variational inequality in (4), except that N_C is no longer merely the normal cone mapping of convex analysis. When

C lacks Clarke regularity, however, the problem of finding a solution $\bar{x}$ to (24) does *not* amount to a quasi-variational inequality, since $N_C(\bar{x})$ can fail then to be convex, whereas any cone $N(\bar{x})$ coming from a quasi-variational inequality must be convex by (23).

The condition in (24) in the case of a gradient mapping $F = \nabla f$ is recognized now as the generally best expression of first-order optimality in minimizing f over C, irrespective of Clarke regularity. Research has shown that N_C enjoys a more robust calculus than $\hat{N}$ or $\bar{N}$, and in addition has deep ties to certain Lipschitz-type properties in geometry and analysis (see Theorem 9.41 of [10], for example). This argues strongly that (24) should perhaps serve broadly as *the fundamental model for constrained equilibrium of F relative to C*. From that perspective, many quasi-variational inequalities would, in practice, emerge as examples of (24) associated especially with Clarke regular classes of sets C.

Other cones than $T_C(\bar{x})$ or $\hat{T}_C(\bar{x})$ are sometimes encountered as choices of $T(\bar{x})$ in the paradigm of quasi-variational inequalities. For instance, when C is expressed by a system of equations and inequalities, or even beyond that, one can consider as $T(\bar{x})$ the set of vectors w for which there is a smooth curve $x : [0, \epsilon] \to C$ with $x(0) = \bar{x}$ and $x'(0) = w$. For most purposes, though, this cone concept, familiar from the Kuhn-Tucker approach to Lagrange multipliers in nonlinear programming, is too feeble to provide much mathematical traction unless some kind of "constraint qualification" is fulfilled. Constraint qualifications typically guarantee, however, that C is Clarke regular at $\bar{x}$ with $\operatorname{cl} T(\bar{x}) = T_C(\bar{x}) = \hat{T}_C(\bar{x})$. This version of $T(\bar{x})$ fits squarely then with the tangent cones already discussed and does not offer anything significantly different.

A weaker property than Clarke regularity in this context is the *derivability* of C at $\bar{x}$. It is said to hold when the "lim sup" in definition (16) coincides with the corresponding "lim inf" (with respect to λ), or in other words, when the sets $[C - \bar{x}]/\lambda$ actually converge to something as $\lambda \searrow 0$. The elements of $T_C(\bar{x})$ are then the vectors w such that one can choose $x(\lambda) \in C$, for λ in an interval $[0, \epsilon]$, so as to have $x(0) = \bar{x}$ and $x'_+(0) = w$. Here $x'_+(0)$ is the limit of $[x(\lambda) - x(0)]/\lambda$ as $\lambda \searrow 0$; only the existence of that one-sided derivative at $\lambda = 0$ is required, and at other $\lambda \in [0, \epsilon]$, the "curve" need not even be continuous. This is distinctly less restrictive than the curve property of the vectors w in the Kuhn-Tucker cone above. In comparison, definition (16) itself only requires of a vector $w \in T_C(\bar{x})$ that there be sequences $\lambda^\nu \searrow 0$ and $x^\nu \to \bar{x}$ with $[x^\nu - \bar{x}]/\lambda^\nu \to w$.

4 Calculus and Solution Perturbations

No general formulation of equilibrium for a mapping F relative to a set C would help much unless there were ways of bringing the abstract condition down to the particular structure of C. For equilibrium models in the form (4), or (24), that we have been emphasizing, this means having a good calculus of normal cone mappings N_C.

Many results are available in this calculus and can be found in [10, Chapter 6], but here we will state only two of the most fundamental. The first result concerns sets that are inverse images of other sets under smooth mappings:

$$C = A^{-1}(K) = \{x \mid A(x) \in K\} \tag{25}$$

where $A : \mathbb{R}^n \to \mathbb{R}^m$ is continuously differentiable and $K \subset \mathbb{R}^m$ is closed and nonempty. We denote by $\nabla A(\bar{x})$ the $m \times n$ Jacobian matrix of A at $\bar{x}$ and by $\nabla A(\bar{x})^*$ its transpose. *If a point $\bar{x} \in C$ satisfies the constraint qualification*

$$y \in N_K(A(\bar{x})),\ \ \nabla A(\bar{x})^* y = 0 \quad \Longrightarrow \quad y = 0, \tag{26}$$

then the (general) normal cone $N_C(\bar{x})$ at that point satisfies the inclusion

$$N_C(\bar{x}) \subset \{\nabla A(\bar{x})^* y \mid y \in N_K(A(\bar{x}))\}. \tag{27}$$

Moreover if K is Clarke regular at $A(\bar{x})$ (as for instance when K is convex), then C is Clarke regular at $\bar{x}$ and the inclusion holds as an equation.

For illustration, suppose K is the cone $\mathbb{R}^m_-$, which corresponds to C being specified by a system of m smooth inequality constraints. The constraint qualification is equivalent then to the standard one of Mangasarian and Fromovitz, and the elements $y = (y_1, \ldots, y_m)$ of $N_K(A(\bar{x}))$ give the *Lagrange multipliers* associated with the constraints at $\bar{x}$. An equilibrium expressed by (24) would therefore involve such multipliers. Equations or mixtures of equations and inequalities can be handled similarly by other choices of K as a cone, but the stated result covers more than just traditional constraint systems and indeed supports an effective *calculus of regularity.*

Apart from the Clarke regularity case where all three types of normal cone coincide anyway, there is no comparable result for regular normal cones. For Clarke normal cones, the same calculus rule does stay valid; i.e., $\bar{N}_C(\bar{x})$ and $\bar{N}_K(A(\bar{x}))$ can validly replace $N_C(\bar{x})$ and $N_K(A(\bar{x}))$ in (26) and (27). But the corresponding constraint qualification,

$$y \in \bar{N}_K(A(\bar{x})),\ \ \nabla A(\bar{x})^* y = 0 \quad \Longrightarrow \quad y = 0, \tag{28}$$

is much more restrictive than (26). Through polarity, (28) is equivalent to requiring that the (convex) regular tangent cone $\hat{T}_C(\bar{x})$ cannot be separated from the range of the linear transformation $w \mapsto \nabla A(\bar{x})w$, which is a subspace expressible as $\nabla A(\bar{x})\mathbb{R}^n$, and this stipulation can be written in turn as

$$\hat{T}_C(\bar{x}) + \nabla A(\bar{x})\mathbb{R}^n = \mathbb{R}^m. \tag{29}$$

When $\hat{T}_C(\bar{x}) = \{0\}$, for instance, which is an all too frequent occurrence in working with regular tangents in the absence of Clarke regularity, (29) insists on $\nabla A(\bar{x})$ actually having full row rank m.

In contrast, the more versatile constraint qualification (26) is *not* equivalent to a condition in terms of tangent vectors (apart from the case of Clarke regularity). *Tangency conditions are thus distinctly weaker and less far-reaching than normality conditions in variational geometry.* This is counter to popular thinking that tangent vectors ought to be "primary" and normal vectors "secondary."

The second of the fundamental rules in the calculus of normal cone mappings that we will look at here concerns images instead of inverse images. Suppose that

$$C = A(K) = \{A(u) \mid u \in K\} \tag{30}$$

where $A : \mathbb{R}^m \to \mathbb{R}^n$ is continuously differentiable and $K \subset \mathbb{R}^m$ is closed and nonempty. *Under the assumption that $A^{-1}(B) \cap K$ is bounded in $\mathbb{R}^m$ for every bounded set $B \subset \mathbb{R}^n$, one has for any $\bar{x} \in C$ that*

$$\begin{aligned} \hat{N}_C(\bar{x}) &\subset \{v \mid \forall u \in A^{-1}(\bar{x}) \cap K:\ A(u)^*v \in \hat{N}_K(u)\}, \\ N_C(\bar{x}) &\subset \{v \mid \exists u \in A^{-1}(\bar{x}) \cap K:\ A(u)^*v \in N_K(u)\}. \end{aligned} \tag{31}$$

If A is affine and K is convex, so C is convex (hence Clarke regular), one has

$$N_C(\bar{x}) = \{v \mid A(u)^*v \in N_K(u)\} \quad \textit{for any single } u \in A^{-1}(\bar{x}) \cap K. \tag{32}$$

The boundedness assumption is only needed for the second inclusion in (31); it is superfluous for the validity of the first inclusion in (31) or for the convex case in (32), provided that C is closed.

Note that this second rule provides no normal cone *equation* or criterion for Clarke regularity in a nonconvex setting, and in that way it contrasts with the first rule. Both rules have many consequences, obtained through special choices of A and K.

Next we take up the topic of solution perturbations. We adopt for this purpose the equilibrium model in (24), which we know covers variational inequalities and a major class of quasi-variational inequalities, but we now consider F to be parameterized by an element w belonging to an open set $W \subset \mathbb{R}^d$. The object of study is the (generally set-valued) *solution mapping*

$$S : w \in W \mapsto \{x \mid F(w,x) + N_C(x) \ni 0\}. \tag{33}$$

Our analysis centers on a fixed pair $(\bar{w}, \bar{x})$ in the graph of S, i.e., with $\bar{x} \in S(\bar{w})$, and the issue of what may happen to $\bar{x}$ under perturbations of $\bar{w}$.

We suppose that F is continuously differentiable on $W \times \mathbb{R}^n$ and denote its Jacobians in the w and x arguments by $\nabla_w F(w,x)$ and $\nabla_x F(w,x)$. We make the following *assumption of ample parameterization*:

$$\nabla_w F(\bar{w}, \bar{x}) \text{ has full rank } n. \tag{34}$$

This assumption is relatively unrestrictive, in the sense that the introduction of additional "canonical" parameters can always force it to be satisfied. More on this matter and the results quoted below can be found in the paper of Dontchev and Rockafellar [3], in complement to the book of Rockafellar and Wets [10].

Ample parameterization guarantees in particular that *S is graphically Lipschitzian of dimension d around $(\bar{w}, \bar{x})$ when N_C is graphically Lipschitzian of dimension n around $(\bar{x}, \bar{v})$, where $\bar{v} = -F(\bar{w}, \bar{x})$*. As we know from earlier, N_C meets that provision when C is convex, but it also does when C is a nonconvex set expressible in the form (25) with K convex and the constraint qualification (26) fulfilled at $\bar{x}$. Through various choices of K, that covers cases where C is specified by smooth equations and inequalities under the Mangasarian-Fromovitz constraint qualification.

We will be occupied by a concept of differentiation for set-valued mappings that is based on the variational geometry of the graph of S in $\mathbb{R}^d \times \mathbb{R}^n$. The *graphical derivative* of S at $\bar{w}$ for $\bar{x}$ is the mapping $DS(\bar{w} \mid \bar{x}) : \mathbb{R}^d \rightrightarrows \mathbb{R}^n$ defined by

$$x' \in DS(\bar{w} \mid \bar{x})(w') \iff (w', x') \in T_{\text{gph}\, S}(\bar{w}, \bar{x}). \tag{35}$$

When $\bar{x}$ is the only element of $S(\bar{w})$, the notation simplifies to $DS(\bar{w})$ and $D^*S(\bar{w})$. If S were actually differentiable at $\bar{w}$, these would be the linear mappings associated with the Jacobian matrix and its transpose, but of course we cannot count on that special case and have to proceed more generally.

The mapping S is called *proto-differentiable* at $\bar{w}$ for $\bar{x}$ when $\text{gph}\, S$ is derivable at $(\bar{w}, \bar{x})$. This property is of particular interest when S is graphically Lipschitzian, as we can commonly expect from of the observations above. If S were single-valued and Lipschitz continuous on an neighborhood of $\bar{w}$ (for which criteria are available in some cases—see Dontchev and Rockafellar [3] for an overview), proto-differentiability would reduce to *semi-differentiability*: $DS(\bar{w})$ would be a single-valued, positively homogeneous, Lipschitz continuous mapping such that

$$S(w) = S(\bar{w}) + DS(\bar{w})(w - \bar{w}) + o(|w|). \tag{36}$$

This is the same as classical differentiability, except for $DS(\bar{w})$ not having to be a linear mapping. An expansion like (36) is not characteristic of proto-differentiability in general, but proto-differentiability nonetheless corresponds to a strong kind of approximation when S is graphically Lipschitzian. Indeed, with respect to a coordinate change as in the definition of the latter property, which identifies gph S locally—from a different angle—as the graph of a single-valued Lipschitz continuous mapping, proto-differentiability of S turns into semi-differentiability of that mapping. Conditions guaranteeing the proto-differentiability of S therefore have some importance.

These concepts of graphical differentiation are applicable also to the mapping N_C, and this will be the key. *The graphical derivatives of S are given by the formula*

$$\begin{array}{l} DS(\bar{w}\,|\,\bar{x})(w') = \{x' \mid -G(w',x') \in DN_C(\bar{x}\,|\,\bar{v})(x')\}, \ \text{where} \\ G(w',x') = \nabla_w F(\bar{w},\bar{x})w' + \nabla_x F(\bar{w},\bar{x})x', \quad \bar{v} = -F(\bar{w},\bar{x}). \end{array} \tag{37}$$

Through this, S is proto-differentiable at $\bar{w}$ for $\bar{x}$ if and only if N_C is proto-differentiable at $\bar{x}$ for $\bar{v}$. Moreover, that is true for instance when C has a general constraint representation as in (25) with K polyhedral convex and the constraint qualification (26) satisfied at $\bar{x}$; in particular, it is true when C itself is polyhedral convex.

The formula for $DN_C(\bar{x}\,|\,\bar{v})$ in the general case of a constraint representation of C as in (25) will not be presented; it is available in the book of Rockafellar and Wets [10]. We concentrate rather on the case where C itself is polyhedral convex. The formula then is appealingly simple:

$$DN_C(\bar{x}\,|\,\bar{v}) = N_T \quad \text{for the cone } T = T_C(\bar{x}) \cap \bar{v}^{\perp}, \tag{38}$$

where $\bar{v}^{\perp}$ denotes the subspace orthogonal to $\bar{v}$. The polyhedral cone T in (38) is the *critical cone* to C at $\bar{x}$ for $\bar{v}$ and can be expressed equivalently by

$$T = \underset{x' \in T_C(\bar{x})}{\operatorname{argmax}} \langle \bar{v}, x' \rangle = T_D(\bar{x}) \quad \text{for } D = \underset{x \in C}{\operatorname{argmax}}\, \langle \bar{v}, x \rangle. \tag{39}$$

It is revealed now by (37) that *when C is polyhedral convex, the vectors $x' \in DS(\bar{w}\,|\,\bar{x})(w')$, describing the differential perturbations of $\bar{x}$ associated with a differential perturbation w' of $\bar{w}$, are then the solutions to* *an auxiliary variational inequality over the critical cone T:*

$$\begin{aligned} DS(\bar{w}\,|\,\bar{x})(w') &= \{x' \mid -G(w',x') \in N_T(x')\}, \text{ where} \\ G(w',x') &= \nabla_w F(\bar{w},\bar{x})w' + \nabla_x F(\bar{w},\bar{x})x', \\ T &= T_C(\bar{x}) \cap F(\bar{w},\bar{x})^{\perp} \\ &= T_D(\bar{x}) \text{ for } D = \operatorname{argmin}_{x\in C} \langle F(\bar{w},\bar{x}), x\rangle. \end{aligned} \tag{40}$$

An alternative description of how the sensitivity analysis of a parameterized variational inequality over a polyhedral set can be carried out has been presented by Robinson [7]. It is likewise based in effect on (38) but in this case in terms of "normal maps" that express the Minty parameterization of the graphs of N_C and N_T. For nonpolyhedral C, a framework of normal maps is less attractive, but formulas for $DN_C(\bar{x}\,|\,\bar{v})$ exist still in some major situations, as mentioned.

5 Application to an Equilibrium Model with Aggregation

These calculus results, culminating for the polyhedral case in the perturbation formula (40), have been elaborated by Patriksson and Rockafellar [6] in the framework of *aggregation*, which underlies traffic equilibrium. Consider the solution mapping S in (33) for the case of a parameterized variational inequality, or equilibrium model, having

$$C = C_1 + \cdots + C_r \quad \text{with each } C_k \text{ polyhedral convex.} \tag{41}$$

Here C is the image $A(K)$ of the set $K = C_1 \times \cdots \times C_r$ under the linear transformation $A : (x_1, \ldots, x_r) \mapsto x_1 + \cdots + x_r$. The normal cone rule in (32) applies and says that, for any $\bar{x} \in C$ and any choice of vectors $\bar{x}_k \in C_k$ with $\bar{x} = \bar{x}_1 + \cdots + \bar{x}_r$, one has

$$N_C(\bar{x}) = N_{C_1}(\bar{x}_1) \cap \cdots \cap N_{C_r}(\bar{x}_r). \tag{42}$$

The normal cones in this formula are polyhedral convex and have the corresponding tangent cones as their polars, so by taking polars on both sides of (42) one gets the dual formula

$$T_C(\bar{x}) = T_{C_1}(\bar{x}_1) + \cdots + T_{C_r}(\bar{x}_r), \tag{43}$$

again with all cones polyhedral convex. The especially interesting thing now is the form of the critical cone T, as described by (38), (39) or equivalently (40) with $\bar{v} = -F(\bar{w},\bar{x})$:

$$T = T_1 + \cdots + T_r \quad \text{with each } T_k \text{ polyhedral convex} \tag{44}$$

for the cones

$$\begin{aligned} T_k &= T_{C_k}(\bar{x}_k) \cap F(\bar{w}, \bar{x})^\perp \\ &= T_{D_k}(\bar{x}_k) \quad \text{for } D_k = \operatorname{argmin}_{x_k \in C_k} \langle F(\bar{w}, \bar{x}), x_k \rangle. \end{aligned} \tag{45}$$

The auxiliary variational inequality in (40) thus exhibits in this case an aggregation structure mirroring that in the given variational inequality under (41). We see further that the normal cone $N_T(x')$ in (40) has the expression

$$\begin{aligned} N_T(x') = N_{T_1}(x'_1) \cap \cdots \cap N_{T_r}(x'_r) \\ \text{for any } x'_k \in T_k \text{ with } x' = x'_1 + \cdots + x'_r. \end{aligned} \tag{46}$$

Therefore, the differential perturbations x' of $\bar{x}$ associated with a differential perturbation w' of $\bar{w}$ are the vectors of the form

$$\begin{aligned} x' = x'_1 + \cdots + x'_r \text{ in which} \\ x'_k \text{ minimizes } \langle G(w', x'), \cdot \rangle \text{ over } T_k. \end{aligned} \tag{47}$$

In our paper [6], we have worked out in detail the implications of this for solution perturbations to network models of traffic equilibrium with origin-destination pairs. Here we apply it to a simpler yet broader model of multi-commodity flow.

Let $\mathcal{G} = (\mathcal{N}, \mathcal{A})$ be a transportation network, where $\mathcal{N}$ and $\mathcal{A}$ are the sets of nodes and arcs (directed links). For $k = 1, \ldots, r$, let $x_k(j)$ denote the quantity of flow of commodity type k in the arc $j \in \mathcal{A}$. Let $b_k(i)$ be the given supply of commodity type k at node $i \in \mathcal{N}$, with negative supply corresponding to demand and zero supply expressing a conservation requirement. The flows $x_k : \mathcal{A} \to \mathbb{R}$ of type k that we admit are the ones belonging to the polyhedral convex set

$$\begin{aligned} C_k = \Big\{ x_k \,\Big|\, & x_k(j) \in I_k(j) \; \forall j \in \mathcal{A}, \\ & \textstyle\sum_{j \in \mathcal{A}} e(i,j) x_k(j) = b_k(i) \; \forall i \in \mathcal{N} \Big\}, \end{aligned} \tag{48}$$

where $e(i,j)$ is the incidence of node i with arc j (this being $+1$ if i is the initial node of j, but -1 if i is the terminal node of j, and 0 otherwise), and $I_k(j)$ is a nonempty, closed interval constraining the flow values allowed for commodity k in arc j. (As a special case, one could have $I_k(j) = [0, 0]$ in certain arcs where commodity k is not permitted.)

Equilibrium in the multi-commodity setting revolves around minimizing the travel costs for the individual commodities while coping with the fact that those costs depend on the aggregate flow contributed by these commodities and thus reflect an interdependence among the different kinds of traffic. The travel cost in arc j is a function $t_j(x)$ of the aggregate flow $x = x_1 + \cdots + x_r$. An *equilibrium* consists, by definition, of a collection of commodity flows $\bar{x}_k \in C_k$ for $k = 1, \ldots, r$ such that

$$\bar{x}_k \in \operatorname*{argmin}_{x_k \in C_k} \sum_{j \in \mathcal{A}} \bar{t}(j) x_k(j), \quad \text{with } \bar{t}(j) = t_j(\bar{x}), \quad \bar{x} = \bar{x}_1 + \cdots + \bar{x}_r. \quad (49)$$

Traffic equilibrium, so defined, can be translated into a variational inequality in two ways, "extensive" or "aggregate." Let

$$t : (w, x) \mapsto (\ldots, t_j(w, x), \ldots)$$

be the mapping that, for a given pair (w, x), assigns to the arcs $j \in \mathcal{A}$ the corresponding travel costs $t_j(w, x)$. The sum in (49) can be construed then as $\langle t(\bar{x}, \bar{w}), x_k \rangle$. In the *extensive* formulation of equilibrium, the focus is on elements

$$(x_1, \ldots, x_r) \in \Gamma = C_1 \times \cdots \times C_r$$

and the mapping

$$\Phi : (w, x_1, \ldots, x_r) \mapsto (t(w, x), \ldots, t(w, x)) \qquad (r \text{ copies}).$$

Normal cones to Γ are given by

$$N_\Gamma(\bar{x}_1, \ldots, \bar{x}_r) = N_{C_1}(\bar{x}_1) \times \cdots \times N_{C_r}(\bar{x}_r),$$

so the condition in (49) comes out as

$$-\Phi(\bar{w}, \bar{x}_1, \ldots, \bar{x}_r) \in N_\Gamma(\bar{x}_1, \ldots, \bar{x}_r), \quad (50)$$

or in other words, the parameterized variational inequality over Γ for Φ.

In the *aggregate* formulation of equilibrium, on the other hand, the condition in (49) is rendered as

$$-t(\bar{w}, \bar{x}) \in N_C(\bar{x}) \quad \text{for } C = C_1 + \cdots + C_r. \quad (51)$$

It purely and simply refers to aggregate flows. *The extensive and aggregate formulations of equilibrium are equivalent on the basis of the normal cone formula (42) that holds under (41).* To pass from (51) to (50), an *arbitrary* choice can be made of commodity flows $\bar{x}_k$ such that $\bar{x}_1 + \cdots + \bar{x}_r = \bar{x}$. (Such flows $\bar{x}_k$ can be determined computationally by solving a system of linear equations and inequalities.) Obviously the aggregate version involves vastly fewer variables, but the aggregate polyhedron C does not come with a direct specification in terms of linear constraints.

To see next what happens with perturbations of equilibrium in this traffic situation, with emphasis on the aggregate model, suppose that the travel costs depend on a parameter vector $w \in W \subset I\!R^d$, so that we have $t_j(w,x)$ in arc j and altogether a mapping $t : (w,x) \mapsto t(w,x)$. Consider the associated solution mapping

$$S : w \mapsto \{x \in C \,|\, -t(w,x) \in N_C(x)\} \text{ with } C = C_1 + \cdots + C_r \tag{52}$$

along with a particular $\bar{w} \in W$ and a corresponding aggregate flow $\bar{x} \in S(\bar{w})$. *Assume that the mapping t is continuously differentiable and, as the F in (34), satisfies our ample parameterization condition*:

$$\nabla_w t(\bar{w}, \bar{x}) \text{ has full rank } n = |\mathcal{A}|, \tag{53}$$

where $|\mathcal{A}|$ is the cardinality of $\mathcal{A}$; in other words, for any choice of values $t'(j)$ for $j \in \mathcal{A}$, there exists w' such that $\nabla_w t_j(\bar{w}, \bar{x})w' = t'(j)$ for all j.

We wish to specialize the perturbation formula (40) to this framework with $F = t$. For this purpose we make an *arbitrary* choice of flows $\bar{x}_k \in C_k$ yielding $\bar{x}_1 + \cdots + \bar{x}_r = \bar{x}$ and let

$$\begin{aligned} &T = T_1 + \cdots + T_r \text{ with} \\ &\quad T_k = T_{D_k}(\bar{x}_k) \text{ for } D_k = \operatorname{argmin}_{x_k \in C_k} \langle t(\bar{w}, \bar{x}), x_k \rangle, \end{aligned} \tag{54}$$

noting that D_k is again polyhedral. The aggregate cone T, likewise polyhedral, is the critical cone on which the formula in (40) will operate, as we already have established. The mapping G in this formula will be given by

$$\begin{aligned} G(w', x') = \nabla_w t(\bar{w}, \bar{x})w' + \nabla_x t(\bar{w}, \bar{x})x' \\ \text{with } x' = x'_1 + \cdots + x'_r,\ x'_k \in T_k. \end{aligned} \tag{55}$$

Therefore, in the network context the differential perturbations x' of $\bar{x}$ associated with a differential perturbation w' of $\bar{w}$ are the vectors of the form

$$\begin{aligned} &x' = x'_1 + \cdots + x'_r \text{ in which } x'_k \text{ minimizes} \\ &\langle \nabla_w t(\bar{w}, \bar{x})w' + \nabla_x t(\bar{w}, \bar{x})x', \cdot \rangle \text{ over } T_k. \end{aligned} \tag{56}$$

The question, though, is how to understand through (54) the special nature of the subcones T_k in this network context. In particular, we wish to know whether the calculation of perturbations by way of (56) comes down to solving another *traffic equilibrium* problem of reduced type.

In investigating that, we can invoke the known optimality conditions for a linear min cost network flow problem as applied to the subproblems in (56) that define the sets D_k. It will be demonstrated now constructively in this way that *the cones T_k in (54) have the form*

$$\begin{aligned} T_k = \Big\{ x'_k \,\Big|\, & x'_k(j) \in I'_k(j) \;\forall j \in \mathcal{A}, \\ & \textstyle\sum_{j \in \mathcal{A}} e(i,j) x_k(j) = 0 \;\forall i \in \mathcal{N} \Big\}, \end{aligned} \tag{57}$$

where each $I'_k(j)$ *is an interval of the type* $[0,0]$, $[0,\infty)$, $(-\infty,0]$, *or* $(-\infty,\infty)$. Since this form is just like that of the sets C_k in (48), except that the intervals $I'_k(j)$ are much more special and the quantities $b_k(j)$ have become 0, it will follow that the perturbation formula (56) does indeed amount to solving a special traffic equilibrium *subproblem.*

Optimality conditions for the problem of minimizing the linear function $\langle t(\bar{w},\bar{x}),\cdot\rangle$ over C_k, with C_k having the general form in (48), are available in [9, Chapter 7]. They involve the notion of a potential $u:\mathcal{N}\to I\!R$ and its differential $\Delta u:\mathcal{A}\to I\!R$, where

$$\Delta u(j) = -\sum_{i\in\mathcal{N}} u(i)e(i,j) = u(\text{final node of } j) - u(\text{initial node of } j).$$

According to these results, a flow x_k belongs to the argmin set D_k in this problem if and only if $x_k \in C_k$ and there is a potential u_k such that

$$\Delta u_k(j) - t_j(\bar{w},\bar{x}) \in N_{I_k(j)}(x_k(j)) \quad \text{for every arc } j\in\mathcal{A}. \tag{58}$$

Moreover the potentials u_k that fill this role for a particular $x_k \in D_k$ are precisely the solutions to a certain dual problem (as explained in [9]. Therefore, we can arbitrarily select one such $\bar{u}_k$ (obtained for instance as a by-product of using algorithms such as in [9] to solve the cost minimization problem in question), and the flows $x_k \in D_k$ will be characterized then as the ones for which the values $x_k(j)$ satisfy the condition in (58). But this is equivalent to saying that in terms of the intervals

$$\bar{I}_k(j) = \begin{cases} I_k(j) & \text{if } \Delta\bar{u}_k(j) - t_j(\bar{w},\bar{x}) = 0,\\ \{\text{right endpoint of } I_k(j)\} & \text{if } \Delta\bar{u}_k(j) - t_j(\bar{w},\bar{x}) > 0,\\ \{\text{left endpoint of } I_k(j)\} & \text{if } \Delta\bar{u}_k(j) - t_j(\bar{w},\bar{x}) < 0, \end{cases} \tag{59}$$

one has

$$D_k = \Big\{x_k \,\Big|\, x_k(j)\in \bar{I}_k(j)\ \forall j\in\mathcal{A},\ \textstyle\sum_{j\in\mathcal{A}} e(i,j)x_k(j) = b_k(i)\ \forall i\in\mathcal{N}\Big\}. \tag{60}$$

In other words, D_k is, like C_k, a flow polyhedron, but with respect to certain smaller intervals $\bar{I}_k(j)$ dictated by the optimality.

Tangent cones to polyhedral convex sets specified by linear constraint systems are readily determined. We immediately get from (60) that, for any particular $\bar{x}_k \in D_k$, we have

$$T_{D_k}(\bar{x}_k) = \Big\{x'_k \,\Big|\, x'_k(j)\in T_{\bar{I}_k(j)}(\bar{x}_k(j))\ \forall j\in\mathcal{A},\ \textstyle\sum_{j\in\mathcal{A}} e(i,j)x'_k(j) = 0\ \forall i\in\mathcal{N}\Big\}. \tag{61}$$

It remains then only to set $I'_k(j) = T_{\bar{I}_k(j)}(\bar{x}_k(j))$ for all $j \in \mathcal{A}$, so that

$$I'_k(j) = \begin{cases} [0,0] & \text{if } \bar{x}_k(j) \text{ is an interior point of } \bar{I}_k(j), \\ [0,\infty) & \text{if } \bar{x}_k(j) \text{ is the left endpoint (only) of } \bar{I}_k(j), \\ (-\infty,0] & \text{if } \bar{x}_k(j) \text{ is the right endpoint (only) of } \bar{I}_k(j), \\ (-\infty,\infty) & \text{if } \bar{I}_k(j) \text{ is the one-point interval } \{\bar{x}_k(j)\}. \end{cases} \tag{62}$$

In summary, *perturbations in this network setting can be calculated by resolving the equilibrium condition (56) for a "differential" multicommodity flow problem having linear costs and constraints as in (57) and (62).*

This equilibrium subproblem can itself be articulated further now in either extensive or aggregate form and in that manner tackled as a variational inequality. In extensive form, there are many variables once more, in principle, but it can be seen from the derivation of the intervals $I'_k(j)$ in (62) from the intervals $\bar{I}_k(j)$ in (59) that $I'_k(j) = [0,0]$ whenever $\bar{I}_k(j)$ is a one-point interval. In fact, that can be anticipated to most arcs j, in which case the flow variables $x_k(j)$ for those arcs can be suppressed by setting them equal to 0. Thus, the number of variables involved in the subproblem is likely really to be very much smaller than in the original problem.

In aggregate form, there are other possibilities for the equilibrium subproblem that calculates perturbations. It has the form

$$-\nabla_w t(\bar{w},\bar{x})w' - \nabla_x t(\bar{w},\bar{x})x' \in N_T(x') \quad \text{for } T = T_1 + \cdots + T_r, \tag{63}$$

the cones T_k being specified by the flow constraints in (57) and (62). The usual difficulty in this picture would be that such constraints do not carry over to a direct constraint representation of T, but because we are dealing with flows, there are additional tools available. The flows in T_k are "circulations" without source or sink, and the intervals $I'_k(j)$ merely impose "sign restrictions" on what can pass through the various arcs j; when $I'_k(j)$ reduces to 0, the arc can effectively be deleted in consideration of commodity k. The theory of *conformal realization* of flows in [9, Chapter 4B] can be utilized to obtain—in a constructive manner—a representation

$$T_k = \left\{ \sum\nolimits_{l=1}^{L_k} \lambda_{kl} e_{kl} \,\middle|\, \lambda_{kl} \in \Lambda_{kl} \right\}$$

in which each e_{kl} is a so-called *elementary* flow (nonzero only on the arcs of an elementary closed path) and each Λ_{kl} is either the interval $[0,\infty)$ or the interval $(-\infty,\infty)$. Having determined such representations for the cones T_k, one could combine them into one for T and in that way arrive through a change of variables at a version of the aggregate variational inequality in terms of the variables λ_{kl}, of which there might not be very many.

Finally, something should be said about the role of quasi-variational inequalities in traffic equilibrium. Our discussion has centered on variational inequalities only, but research on quasi-variational inequality models has also been carried out in this setting; see for example De Luca and Maugeri [1, 4, 2]. Such efforts have been directed toward more general models than ours that treat *elastic* supplies and demands; they can be construed as models in which the values $b_k(i)$ are not fixed but rather can respond to the costs achieved at optimality. In passing to that kind of framework, would the quasi-variational inequalities so obtained, once "truncated" in the manner we adopted in Section 3, fit with pattern of generalized tangent and normal cones that we have suggested?

That could well be the case, but there might be a simpler alternative. Elastic supplies and demands can also be handled in many cases by expanding the network to allow for additional arcs which connect supply nodes and demand nodes to the "outside world" and then introducing further traffic costs on those arcs. In this way, a quasi-variational inequality of such type can, at least in typical cases like those with simple origin-destination pairs, be reformulated as an ordinary variational inequality in the expanded network. The fact that our model allows general intervals $I_k(j)$ may help in that respect, since an arc introduced for only one of the commodities can be assigned the trivial interval $[0, 0]$ for the other commodities.

References

[1] M. De Luca and A. Maugeri, *Quasi-variational inequalities and applications to equilibrium problems with elastic demand*, in *Nonsmooth Optimization and Related Topics* (F. H. Clarke, V.F. Dem'yanov, and F. Giannessi, editors), Plenum Press, 61–77, 1988.

[2] M. De Luca, *Generalized quasi-variational inequalities and traffic equilibrium problems*, in *Variational Inequalities and Network Equilibrium Problems* (F. Giannessi and A. Maugeri, editors), Plenum Press, 45–54, 1995.

[3] A. L. Dontchev and R. T. Rockafellar, *Ample parameterization of variational inclusions*, SIAM Journal on Optimization, to appear, 2001.

[4] A. Maugeri, *Variational and quasi-variational inequalities in network flow models. Recent developments in theory and algorithms*,

in *Variational Inequalities and Network Equilibrium Problems* (F. Giannessi and A. Maugeri, editors), Plenum Press, 195–211, 1995.

[5] G. J. Minty, *Monotone (nonlinear) operators in Hilbert space*, Duke Mathematical Journal 29, 341–346, 1962.

[6] M. Patriksson and R. T. Rockafellar, *Sensitivity analysis of variational inequalities over aggregated polyhedra, with application to traffic equilibria*, preprint, 2001.

[7] S. M. Robinson, *Sensitivity analysis of variational inequalities by normal-map techniques*, in *Variational Inequalities and Network Equilibrium Problems* (F. Giannessi and A. Maugeri, editors), Plenum Press, 257–269, 1995.

[8] R.T. Rockafellar, *Convex Analysis*, Princeton University Press, Princeton, New Jersey (available from 1997 also in paperback in the series Princeton Landmarks in Mathematics), 1970.

[9] R. T. Rockafellar, *Network Flows and Monotropic Optimization*, Wiley; republished in 1998 by Athena Scientific, 1984.

[10] R.T. Rockafellar and R.J-B Wets, *Variational Analysis*, Springer Verlag, Berlin, 1998.

ON THE CALCULATION OF EQUILIBRIUM IN TIME DEPENDENT TRAFFIC NETWORKS

Fabio Raciti
Dipartimento di Matematica e Informatica- Università di Catania,
Viale A. Doria, 6 - 95125 CATANIA (ITALIA)
e-mail: fraciti@dmi.unict.it

Abstract We analyse a method which has been recently proposed to give global approximation to time dependent equilibrium in traffic networks, but has not yet been applied to any concrete traffic problem. Moreover, we discuss in detail a class of problems for which the exact solution is available in order to compare the exact solution with the approximate one.

Keywords: Traffic Networks, Variational Inequalities, subgradient.

1 Introduction

In a recent paper [2] time dependent traffic equilibria have been studied in the general framework of variational inequalities. Let us briefly rewiew the time dependent model proposed in [2]. (For general references on the traffic assignement problem, and on the functional methods based on variational inequalities see also references in [2]).

A traffic network consists of a set W of origin-destination pairs and a set $\mathcal{R}$ of routes. The set of all $r \in \mathcal{R}$ which link a given $w \in W$ is denoted by $\mathcal{R}(w)$. The traffic network is considered at all times $t \in \tau$, where $\tau = [0, T]$. For each time $t \in \tau$ there is a route-flow vector $F(t) \in \Re^{\mathcal{R}}$. Feasible flows are flows which satisfy the time dependent capacity constraints and demands, i.e., which belongs to the set:

$$K := \{F \in \mathcal{L} \mid \lambda(t) \leq F(t) \leq \mu(t), \quad \phi F(t) = \rho(t) \text{ a.e. on } \tau\}$$

where $\lambda(t) \leq \mu(t)$ and $\rho(t)$ are given, and ϕ is the well known pair-route incidence matrix whose elements $(\phi)_{w,r}$ are set equal 1 if route r connects the pair w, 0 else. $F : \tau \mapsto \Re^{\mathcal{R}}$ is the flow trajectory over time. Flows trajectories are supposed to be elements of $\mathcal{L} := L^p(\tau, \Re^{\mathcal{R}})$ with $p > 1$. The dual space, $L^q(\tau, \Re^{\mathcal{R}})$ with $\dfrac{1}{p} + \dfrac{1}{q} = 1$, will be denoted by

$\mathcal{L}^*$. A Mapping $C := K \mapsto \mathcal{L}^*$ is then given which assigns to each flow trajectory $F(t) \in K$ the cost trajectory $C[F(t)] \in \mathcal{L}^*$.

On $\mathcal{L}^* \times \mathcal{L}$ the canonical bilinear form is then introduced :

$$\langle\langle G, F\rangle\rangle := \int_{\mathcal{T}} \langle G(t), F(t)\rangle dt \quad G \in \mathcal{L}^*, F \in \mathcal{L}$$

According to [2] $H \in \mathcal{L}$ is an *equilibrium flow* iff:

$$H \in K \text{ and } \langle\langle C(H), F - H\rangle\rangle \geq 0 \ \ \forall F \in K \tag{1}$$

which is equivalent to the following:

$$\forall w \in W, \forall q, s \in \mathcal{R}(w), \text{and a.e on } \tau \text{ there holds :}$$

$$C_q[H(t)] < C_s[H(t)] \Longrightarrow H_q(t) = \mu_q(t) \ \text{ or } \ H_s(t) = \lambda_s(t)$$

From the user point of view, however, one can prefer the last condition to define an equilibrium flow and use the variational inequality as a characterization. Starting from this framework the authors state existence theorems for equilibria. Moreover, in [2], a procedure is given for the approximate calculation of equilibria by using the subgradient method and exploiting the fact that under standard assumption [1] (1) is equivalent to Minty variational inequality:

$$H \in K \text{ and } \langle\langle C(F), H - F\rangle\rangle \geq 0 \ \ \forall F \in K \tag{2}$$

2 Calculation of Equilibria

Given an abstract problem of the form:

$$H \in K, \ \varphi(H, F) \geq 0 \ \ \forall F \in K \tag{3}$$

where $K \subset B$ and $\varphi : B \times B \mapsto \Re$, let us recall that a gap function for problem (3) is a function $\Psi : B \mapsto \Re \bigcup \{\infty\}$ such that, for all $H \in B$, $\Psi(H) \geq 0$ and moreover:

$$\psi(H) = 0 \text{ if and only if } H \text{solves (3)}$$

Then $H \in B$ solves (3) if and only if $\psi(H) \leq 0$.

In connection with our equilibrium problem we choose:

$$B := \{F \in \mathcal{L} | \lambda(t) \leq F(t) \leq \mu(t) \text{ a.e.on}[0, T]\}$$

In order to carry out calculation choose $\mathcal{L} = L^2([0, T]), \Re^{\mathcal{R}}$

and make the following assumptions on the operator C:
(C1) C in pseudomonotone;
(C2) C is hemicontinuous with respect to the weak topology on K
(C3) the set $\{C(F)|F \in K\}$ is norm-bounded
Let, for all $H \in B$,

$$\psi_1(H) = \max_{F \in K} \langle\langle C(F), H - F \rangle\rangle, \quad \psi_2(H) = \int_{[0,T]} ||\phi H(t) - \rho(t)||_2^2 dt$$

The real valued function $\Psi(H) = max\{\psi_1(H), \psi_2(H)\}$ is a gap function for our problem and is convex and weakly lower semicontinuous. Thanks to $(C2)$, the subdifferential:

$$\partial\Psi(H) := \{\tau \in \mathcal{L} | \Psi(F) - \Psi(H) \geq \langle\langle \tau, F - H \rangle\rangle, \ \forall F \in B\}$$

is nonempty for all $H \in B$.
Let then:

$$\Gamma = \{H \in B | \Psi(H) \leq 0\}$$

Because of the equivalence of (1) and (2), $H \in \mathcal{L}$ is a solution of (1) if and only if $H \in \Gamma$.

3 The algorithm

The subgradient method for finding an element of Γ runs as follows:
Choose $H^0 \in B$. Given $H^n \in B \setminus \Gamma$, let:

$$H^{n+1} = Proj_B(H^n - \rho_n \tau_n)$$

where $\tau_n \in \partial\Psi(H^n)$ and $\rho_n = \Psi(H^n)/||\tau_n||^2$. $Proj_B$ denotes the euclidean projection onto B. Under assumption $(C3)$ one can show that the τ_n can be chosen in such a way that $||\tau_n||$ remain bounded. Then, if $H^n \notin \Gamma \ \forall n$ one can prove that $\Psi(H^n) \mapsto 0$, that the sequence $\{H^n\}$ has weak cluster points, and every cluster point is in Γ. In addition, if the sequence $\{H^n\}$ has a strong cluster point $\overline{H}$, then $\overline{H}$ is unique and $\{H^n\}$ converges strongly to $\overline{H}$.

Now we want to make some remarks about the application of this algorithm to a concrete traffic problem. The main difficulty is the calculation of elements of $\partial\Psi_1(H)$. In order to find elements of $\partial\Psi_1(H)$, for each $H \in B$, one has to evaluate the cost operator in a point $\overline{F}$ where the functional: $\langle\langle C(F), H - F \rangle\rangle$ reaches the maximum. For a linear cost operator this functional can be linear or quadratic. In the linear case the point $\overline{F}$ lies in the boundary of the convex K and is independent of the point $H \in B$. In the quadratic case one has to solve a quadratic optimization problem on K which can be carried out easly when the functional is concave (this happens, for instance, for diagonal coercive operators). However, for a general cost operator this maxmization problem is non-concave and the technical difficulties which arise can render the algorithm impratical. Another problem which arises in applications is the calculation of the projection onto B. It turns out, thanks to the particular structure of B, that the projection onto B of $(H^n - \rho_n \tau_n)$ is the vector whose components are the positive parts of the components of $(H^n - \rho_n \tau_n)$. As our purpose is to apply the proposed algorithm whithout any discretization method, we have to calculate, at each step, the positive parts of $\mathcal{R}$ functions, constrained to the interval $[0, T]$, and this will increase the computational effort of our method. In order to test the algorithm we apply it to a class of problems for which the exact solution is known.

4 Applications and Examples

Let us consider a network with only one O-D pair $w_1 := w$ and suppose that there are N routes which connect w: $R_1, R_2, ..., R_N$. For our purposes we are not interested in the detailed structure of the network (i.e. the link structure of each route). Let the cost operator be given by the following law:

$$C_j[F(t)] = b_j F_j(t)$$

where $b_j > 0 \; \forall j = 1...N$. Thus, C is a coercive operator with coercivity constant $b = min\{b_1, b_2, \ldots, b_N\}$.

Then let the convex K be:

$$K := \{F \in L^2([0,T], \Re^N) \mid 0 \leq F(t),$$

$$F_1(t) + F_2(t) + ... + F_N(t) = \rho(t) \quad \text{a.e. } [0,T]\}$$

For this problem one can calculate the exact equilibrium H:

$$H_j(t) = \rho(t) \frac{b_1 b_2 ... b_{j-1} b_{j+1} ... b_N}{\sum\limits_{h>k} b_h b_k}$$

In order to find elements of $\partial\psi_1(H)$ we have, first of all, solve ($\forall H \in B$) the problem:

$$\max_{F\in K}\int_0^T (b_1F_1(t),...,b_NF_N(t))^T(H_1(t)-F_1(t),...,H_N(t)-F_N(t))dt$$

$$=\max_{0\leq F_2+F_3+...+F_N\leq\rho}\int_0^T I[F_2(t),F_3(t),...,F_N(t)]dt$$

where

$$I(F_2,F_3,...,F_N)=-\sum_{j=2}^{N}(b_1+b_j)F_j^2-2b_1\sum_{j>k}F_jF_k+$$

$$+\sum_{j=2}^{N}[b_1(2\rho-H_1)+b_jH_j]F_j+b_1\rho H_1-b_1\rho^2$$

Because of the particular structure of $I(F_2,F_3,...,F_N)$ the maximization problem can be solved exactly for each $H\in B$ and the algorithm can start. In the following we consider in detail the case where $N=2$. In spite of its simple structure, this example will shed some light on the general properties of the proposed algorithm.

Let the convex K be:

$$K:=\{F\in L^2([0,1],\Re^2)\mid 0\leq F(t),\quad F_1(t)+F_2(t)=\rho(t)\quad \text{a.e. }[0,1]\}$$

Now we assign the cost operator:

$$C:\begin{cases} C_1[F(t)]=b_1F_1(t) & \forall t\in[0,1]\\ C_2[F(t)]=b_2F_2(t) & \forall t\in[0,1]\end{cases}$$

with these data, one can show that the exact equilibrium is reached in:

$$H:\begin{cases} H_1(t)=\rho(t)\dfrac{b_2}{b_1+b_2} & \forall t\in[0,1]\\ H_2(t)=\rho(t)\dfrac{b_1}{b_1+b_2} & \forall t\in[0,1]\end{cases}$$

In order to find elements of $\partial\psi_1(H)$ we have, first of all, solve $\forall H\in B$ the problem:

$$\max_{F\in K}\int_0^1 (b_1F_1(t),b_2F_2(t))^T(H_1(t)-F_1(t),H_2(t)-F_2(t))dt$$

$$= \max_{0\le F_2\le\rho} \int_0^1 I[F_2(t)]dt$$

where

$$I[F_2] = -(b_1+b_2)F_2^2 - (b_1H_1 - b_2H_2 - 2b_1\rho)F_2 + b_1\rho H_1 - b_1\rho^2$$

The solution is:

$$\overline{F}(t): \begin{cases} \overline{F}_2(t) = \dfrac{2b_1\rho + b_2H_2 - b_1H_1}{2(b_1+b_2)} & \forall t \in [0,1] \\ \overline{F}_1(t) = \rho(t) - \overline{F}_2(t) & \forall t \in [0,1] \end{cases}$$

if the r.h.s. satisfies the constraints, otherwise it will be $\overline{F}_2 = \rho$ or $\overline{F}_2 = 0$.

Thus, $C(\overline{F}_1, \overline{F}_2) = (b_1\overline{F}_1, b_2\overline{F}_2) \in \partial\psi_1(H)$
and

$$\psi_1(H) = \int_0^1 [(b_1+b_2)\overline{F}_2^2 + b_1\rho H_1 - b_1\rho^2]dt$$

Now choose $b_1 = 1, b_2 = \frac{1}{2}$ and $\rho(t) = t \quad \forall t \in [0,1]$. The exact equilibrium is then: $H(t) = (t/3, 2t/3)$. For the calculation of elements of $\partial\psi_1(H)$ let us observe that:

$$I(F_2) = \left[-\frac{3}{2}F_2^2 - F_2\left(-2t + H_1 - \frac{H_2}{2}\right) + tH_1 - t^2\right]$$

which yields: $\overline{F}_2 = \dfrac{2t - H_1 + \frac{H_2}{2}}{3}$ if $0 \le \overline{F}_2 \le t$. Otherwise $\overline{F}_2 = 0$ or $\overline{F}_2 = t$. Then $C(\overline{F}) = \left(\overline{F}_1, \frac{1}{2}\,\overline{F}_2\right) = \left(\dfrac{t + H_1 - \frac{H_2}{2}}{3}, \dfrac{t - \frac{H_1}{2} + \frac{H_2}{4}}{3}\right) \in \partial\psi_1(H)$. The calculation of elements of $\partial\psi_2(H)$ is straightforward because an explicit formula is known:

$$2\phi(\phi H(t) - \rho(t)) = 2[H_1(t) + H_2(t) - t, H_1(t) + H_2(t) - t] \in \partial\psi_2(H)$$

The other quantities which we need before starting the algorithm are:

$$\psi_1(H) = \int_0^1 \left[\frac{1}{2}\overline{F}_2^2 + tH_1 - t^2\right] dt$$

$$\psi_2(H) = \int_0^1 |H_1(t) + H_2(t) - t|^2 dt$$

$$||\tau_1||^2 = \frac{1}{9}\int_0^1 [(t + H_1 - H_2/2)^2 + (t - H_2/2 + H_2/4)^2]dt$$

$$||\tau_2||^2 = 8\int_0^1 (H_1 + H_2 - t)^2 dt$$

Let now, for example, $H^0 = [0, t] \in B \setminus \Gamma$ and define the error in the natural way:

$$E^n = ||H - H^{(n)}||_{L^2}$$

Then, after a stopping rule has been fixed, we can start our calculations, which we have performed with the help of MAPLEV [4]. In Tab.1, a few iterations are shown, from which one can deduce that convergence is very slow.

n	$H_x^{(n)}$	$H_y^{(n)}$	$E^{(n)}$
1	0	0.9137 t	0.2395
2	0.0215t	0.9352 t	0.2376
3	0.0042 t	0.8970 t	0.2319
4	0.0289 t	0.9217 t	0.2292
5	0.0150 t	0.8920 t	0.2251
6	0.0382 t	0.9152 t	0.2227
7	0.0248 t	0.8874 t	0.2190
8	0.0467 t	0.9093 t	0.2168
9	0.0338 t	0.8830 t	0.2133
10	0.0546 t	0.9038 t	0.2112
100	0.1769 t	0.7983 t	0.1180
1000	0.2790 t	0.7192 t	0.0436

Table 1. Iterations with coercivity constant $b = 1/2$.

Things go much better, however, if we consider an operator whose coercivity constant is smaller than the previous one:

$$C : \begin{cases} C_1[F(t)] = F_1(t) & \forall t \in [0,1] \\ C_2[F(t)] = (1/200)F_2(t) & \forall t \in [0,1] \end{cases}$$

In this case, the exact equilibrium is:

$$H(t) = \left(t\frac{1}{201}, t\frac{200}{201}\right)$$

while in Tab.2 a few iterations are shown (where we have chosen $H^0 = [0, t] \in B \setminus \Gamma$). This time one can see that after 100 iterations the percentual error is about 1 %.

n	$H_x^{(n)}$	$H_y^{(n)}$	$E^{(n)}$
1	02562 t	0.4963 t	0.3223
2	0.3180t	0.5582 t	0.3102
3	0.1691 t	0.5543 t	0.2714
4	0.2382 t	0.6243 t	0.2532
5	0.2728 t	0.6580 t	0.2485
6	0.1454 t	0.6541 t	0.2128
7	0.1955 t	0.7042 t	0.2007
8	0.2206 t	0.7292 t	0.1976
9	0.1192 t	0.7252 t	0.1691
10	0.1581 t	0.7641 t	0.1599
100	0.0124 t	0.9846 t	0.0073

Table 2. Iterations with coercivity constant $b = 1/200$.

5 Conclusions

We have applied the algorithm proposed in [2] in concrete traffic networks. We have not carried out any discretization of the problem, because our aim was to get a *global* approximation of the equilibrium, in the interval $[0, T]$. Because of this fact, implementation is not straightforward and we have preferred to use the MAPLEV programming language in order to use symbolic manipulation as far as possible. By means of particular examples we have realized that without some coercivity property the speed of convergence is quite slow. A deeper theoritical analysis is required in order to improve convergence, at least for some other classes of cost operators. However, the most serious technical problem is the calculation of elements of $\partial\psi_1(H)$. Thus, in our opinion, the criterium that one should use in order to decide wether it is convenient to use this method to solve a certain traffic problem should be based on a preliminar analysis of the maximization problem above mentioned.

References

[1] A. Maugeri, W. Oettli, D. Schläger, *A Flexible Form of Wardrop's Principle for Traffic Equilibria with Side Constraints*, Rendiconti del Circolo Matematico di Palermo Serie II, Supplemento, Vol.48, pp.185-193,1997.

[2] P. Daniele, A. Maugeri and W. Oettli, *Time-Dependent Traffic Equilibria*, Jota,Vol. 103, No. 3 December 1999. (See also references therein).

[3] M.J. Smith,*A new Dinamic Traffic Model And the Existence and calculation of Dynamic User Equilibra on Congested Capacity-Constrained Road Networks*, Transportation Research, Vol. 27B, pp.49-63,1993.

[4] A tutorial introduction to MAPLE V, B.W. Char et al, Springer Verlag New York, 1992.

MECHANICAL EQUILIBRIUM AND EQUILIBRIUM SYSTEMS *

Tamás Rapcsák
Computer and Automation Institute,
Hungarian Academy of Sciences,
1111 Budapest, Kende utca 13-17
e-mail: rapcsaksztaki.hu

Abstract In the paper, it is shown that the principle of virtual work considered an axiom of mechanics by Lagrange (1788) and Farkas (1906) can be embedded in a general equilibrium system, the quasi-variational inequalities introduced by Bensoussan and Lions in 1973, assuming force fields and holonomic-scleronomic constraints. Then, the dual form of the principle of virtual work is formulated in this case, the procedure for solving mechanical equilibrium problems, the existence of solutions and some examples are discussed.

1 Introduction

Gyula Farkas (1847-1930) was one of the greatest Hungarian scientists in mathematics and physics. In 1893, he participated in a celebration in Padova where the 300th year anniversary of Galilei's starting his activity there was held. Farkas discussed the use of the virtual work in the papers of Galilei. The principle of virtual work was studied by Farkas in his whole life, written 9 papers on the principle and 9 papers related to its mathematical problems. His scientific activity is very closely related to the trend based on experience, and he was one of the most consistent advocates of this approach. He was convinced that a great deal of unsuccessfulness in the theories was due to the fact that too strict restrictions are introduced in the basic concepts [27].

Farkas dealt with the foundation of force equilibrium in mechanics and thermodynamics when creating the famous theorem of homogeneous linear inequality systems. In 1898, Farkas published his theorem in Hungarian, then in 1901, in German. Nowadays, this result is referred to by a great number of authors, and since then, this statement has been the starting point of several equilibrium theories (e.g., mechanics, eco-

*Work partially sponsored by OTKA T029572.

nomics, finance, networks) and mathematical disciplines (e.g., optimization theory, system of linear inequalities). It seems, however, that the point where Farkas started from and the connection of this result with mechanical force equilibrium are not known in details either in mathematics or in physics. This is probably one of the reasons why the overall systematization and discussion of the mathematical problems relevant to mechanical force equilibrium have not got prominence till now and the various approaches focused on special cases only (see details in [4], [15], [25]).

Papers [29] and [30] show that Farkas' results related to mechanical equilibrium imply, practically, the first-order necessary optimality conditions of inequality constrained nonlinear optimization problems (the regularity condition is missing). This statement has been a basic result in nonlinear optimization since the publication of Karush in 1939 as well as Kuhn and Tucker in 1951, thus, within this theory, its connection with classical mechanics and the clarification of the right "place" of Farkas are of great importance.

During the recent years, within optimization theory, equilibrium systems have got into the centre of theoretical approach. The main aim of the paper is to show that the principle of virtual work is in close connection with one of the most general equilibrium systems, i.e., the quasi-variational inequalities, introduced in 1973. It can be stated that the discovery of Lagrange (1788) and Farkas (1901, 1906) - having had influence up to our days - according to which the principle of virtual work, which is a general equilibrium system equal to the experience, can be considered one of the axioms of mechanics. The discovery allows to classify the mathematical problems related to the equilibrium of mechanical systems with finite number of degrees of freedom.

In Section 2, some physical motivation can be found, in Section 3, the statement of the mechanical force equilibrium problem, in Section 4, the principle of virtual work, in Section 5, the characterization of the constraints related to motions, in Section 6, the formulation of quasi-variational inequalities, in Section 7, the principle of virtual work in force fields under scleronomic and holonomic constraints, in Section 8, the dual form of the principle of virtual work in the discussed case, in Section 9, the procedure for solving mechanical equilibrium problems and in Section 10, the question of the existence of solutions is dealt with.

2 Physical motivation

In 1687, Newton laid down the solid foundation of mechanics by describing the motion of a mass point in the 3-dimensional Euclidean space, based on Newton's second law: "mass times acceleration equals moving force". It was deduced from the motion of a mass point in the field of gravity on the earth, and was then applied to the motion of planets under the action of the sun.

The analytical form of mechanics was introduced by Euler and Lagrange considering a system of mass points and taking constraints for the motion as well into account. A nice synthesis of these efforts is Lagrangian mechanics the purpose of which is the description of the motion and equilibrium of a system of mass points subject to ideal holonomic constraints, i.e., a differentiable manifold, and in which the motion is governed by the Lagrangian function (see, e.g., Arnold, 1978). In the Newtonian case, the differentiable manifold is equal to the 3-dimensional Euclidean space and the Lagrangian function to the difference of the kinetic energy depending on the velocities and the potential energy depending on the position only. An important property of Lagrangian mechanics is that, by using the D'Alembert principle, the description of the motion of a system of mass points is equivalent to that of equilibrium positions. Several generalizations of Lagrangian mechanics were published, e.g., Lagrangian mechanics with generalized Lagrangian functions [19] and time-dependent Lagrangian mechanics [1].

Since neither in optimization theory (e.g., variational and quasi-variational inequalities) nor in physics (e.g., the acting forces are not derivable from a potential function) important equilibrium systems are governed by real-valued functions but rather by more general mappings, our aim is to follow the original idea of Newton by characterizing mechanical equilibrium and taking different kind of constraints into consideration based on the force vector function and general equilibrium systems.

3 Statement of the mechanical force equilibrium problem

One of the most general settings of the problem in Newtonian mechanics to be discussed is as follows. Let n mass points, on which active forces $P_1, \dots, P_n$ have an effect, be given in R^3 (in the 3-dimensional Euclidean space). Besides, let the constraints which put the only restrictions on the motion of system of mass points be given. The position and motion of system of n mass points can be given by the help of the time functions of the space coordinates in R^3. In the paper, the case will be discussed when a possible motion of the point system in an interval of time is given by a twice continuously differentiable vector function $\mathbf{x} : [t_0, t_1] \subseteq R \to R^{3n}$ satisfying the constraints where $t_0, t_1 \in R$ are given values. The vector functions $\dot{\mathbf{x}}(t)$, $\ddot{\mathbf{x}}(t)$, $t \in [t_0, t_1] \subseteq R$, denote the velocity and acceleration vector at time t, respectively.

We assume that the active forces affecting the system are continuous functions of the position, velocity and time. If the forces depend on the position only, we mean force fields (e.g., the central force fields and the gravitational force fields). A very important case is the conservative force field where there exists a potential energy function $V : R^{3n} \to R$ depending on the position only, for which the equation

$$\mathbf{P}(\mathbf{x})^T = -\nabla V(\mathbf{x}), \qquad \mathbf{x} \in R^{3n}, \tag{1}$$

holds. (By agreement, the gradient of a function is a row vector.) Note that both the central force fields and the gravitational force fields are conservative force fields.

The mechanical system is considered to be in force equilibrium if on every mass point, the sum of active forces and reaction forces is zero so that reaction forces cause the points to maintain constraints. The main problem is to determine the mechanical states of system of mass points or, in a special case, the characterization of the equilibrium positions of the system.

4 The principle of virtual work

The principle of virtual work is one of the oldest laws of classical mechanics stated for the case of equality type constraints, first, by Bernoulli in 1717. In the famous work of Lagrange (1788), this principle appeared as an axiom of mechanics. For the case of inequality constraints, the principle was stated by Fourier in 1798, thereupon by Gauss in 1829. But as early as in 1827, Cournot and in 1838, Ostrogradsky, putting the principle of virtual work into use, met the difficulty that virtual displacements are given implicitly. Therefore, the virtual displacements were eliminated in the case of equality and inequality type constraints, i.e., the statement of Farkas theorem was used but not proved. It was first proved by Farkas in 1898. Some historical relations of the principle of virtual work can be read in details in [4], [29], [30].

The use of virtual displacements requires being defined. The classical interpretation can be found, e.g., in Banach's book (1951).

Definition 4.1 *The displacement of the given mass point system from point $\mathbf{x}_1$ to point $\mathbf{x}_2$ is said to be possible if the vector function describing the displacement satisfies the constraints. The velocities belonging to them are called possible velocities. Virtual displacements mean the directions of possible velocities.*

By the principle of virtual work, if a system of mass points is in an equilibrium position, then the work of the active forces in the directions of the virtual displacements are less than or equal to zero, i.e., if the active forces are denoted by the vector $\mathbf{P} \in R^{3n}$ in an equilibrium position, the virtual displacements at this point by the vector $\mathbf{v} \in R^{3n}$, then by the principle of virtual work, the inequality

$$\mathbf{P}^T \mathbf{v} \leq 0 \tag{2}$$

fulfils for every virtual displacement $\mathbf{v} \in R^{3n}$.

5 Characterization of the constraints

The characterization of the mechanical state of a system of mass points needs the mathematical classification of the physical constraints which are based on [31], [32].

I. According to variables in the constraints, the following cases are distinguished:

$$\begin{array}{ll} \text{I. a.: } \mathbf{x}; & \text{I. c.: } \mathbf{x}, t; \\ \text{I. b.: } \mathbf{x},\ \dot{\mathbf{x}}; & \text{I. d.: } \mathbf{x},\ \dot{\mathbf{x}}, t. \end{array} \tag{3}$$

II. According to relations defining the constraints:

II. a.: the constraint is defined by an equality;

II. b.: the constraint is defined by an inequality;

II. c.: the constraint is defined by a surface or manifold; (4)

II. d.: the constraint is defined by a non-finite form.

Thereafter, as a system of constraints, any meaningful combination of the above cases can be considered, as well.

III. According to the configuration space, the following two cases are distinguished:

III. a.: the configuration space is Euclidean;
III. b.: the configuration space is not Euclidean, but some other differentiable manifolds (e.g., Arnold, 1978). (5)

Following the special terms used in physics, if among variables entering the constraints, time t does not occure (I. a., I. b.), then the system is called scleronomic, otherwise, rheonomic (I. c., I. d.). According to another classification (Gantmacher, 1970), in the cases I. a. and I. c., the constraints are finite or geometric, in cases I. b. and I. d., differential or kinetic. According to relations providing the constraints, in the cases II. a., II. b., II. c. and in the combinations of the same, the system is holonomic if, however, among the constraints there is also one of those of type II. d., then it is non-holonomic or anholonomic. Note that in this latter issue, the terminology is not uniform, for the definition of the holonomic system differs in the books of Appell (1909), Banach (1951) and Gantmacher (1970). Here, we have accepted the most prevalent definition.

Case II. d. (when the constraint is not given in a finite form) means that the fulfilment of a linear equality or inequality is directly required for the virtual displacements or possible velocities. Naturally, conditions of such type may differ in the states of the mass point system. Discussions accordingly are detailed in Appell (1909), Farkas (1906) and Gantmacher (1970). An interesting fact is that the linear anholonomic constraints appearing in the paper of Farkas published in 1906 seem to be the most general even in our days [14].

6 Quasi-variational inequalities (QVI)

Let H be a real Hilbert space, $A \subseteq H$ a subset, $F : A \to H$, $X : A \Rightarrow H$, $g : H \times A \to R^m$ mappings (X set-valued). Quasi-variational inequalities (in short, QVI) are equilibrium systems where the aim is to find a point $x \in A$ such that $x \in K(x)$ and

$$\langle F(x), y - x \rangle \geq 0, \quad \forall y \in K(x) = \{y \in X(x) |\ g(y; x) \in \mathcal{C}\} \neq \emptyset, \qquad (6)$$

where $\mathcal{C} \subseteq R^m$ is a closed and convex cone with apex at the origin, R^m the m-dimensional Euclidean space and the symbol $\langle , \rangle$ denotes the scalar product in the Hilbert space. This form can be interpreted as a choice among the fixed-points of the point-to-set map K following the requirements of formula (6). The present problem contains the classical QVI if $\forall x \in X(x)$ and the conditions $g(y; x) \in \mathcal{C}$ hold identically in the variable y belonging to the set $X(x)$. Equilibrium systems (6) were first introduced by Bensoussan and Lions in 1973. If X and g do not depend on the variable x, then formula (6) determines variational inequalities (VI), due to Stampacchia in 1964. The first systematic study of VI can be found in [22].

If the sets $K(x) = X(x)$, $x \in A$, determine convex cones, then equilibrium systems (6) are called quasi-complementarity systems (QCS). A special case of QCS can be deduced if the sets $K(x) = X(x)$, $x \in A$, determine closed and convex cones with apex at the origin. An equivalent setting is as follows:

$$x \in K(x), \quad F(x) \in K^*(x), \quad \langle F(x), x \rangle = 0, \tag{7}$$

where K^* is the positive polar of the cone K. If the cone K does not depend on the variable x, then the classical complementarity systems (CS) are obtained.

We remark that the formulation of QVI in (6) is not the most general one. The existence theorems of a solution for a QVI is based on the convexity and compactness of the set A, the continuity of the mappings and a fixed-point theorem or an equivalent statement [6],[8]. The existence of a solution in (6) and duality are investigated in [17]. Some application areas of QVI are related to traffic networks [9],[24]. Recent results can be found in [20], [26].

7 Principle of virtual work in force fields under scleronomic and holonomic constraints

In this part, the relation between the principle of virtual work and QVI is studied in force fields under scleronomic and holonomic constraints. These classical mechanical systems seem to be from among the simplest ones. The constraints of the mechanical systems are the combinations of the cases I.a, II.a, II.b, II.c, III.a, given in the form of

$$h_j(\mathbf{x}) = 0, \qquad j = 1, \dots, p, g_i(\mathbf{x}) \geq 0, \qquad i = 1, \dots, m, \mathbf{x} \in R^{3n}, \tag{8}$$

where the functions h_j, $j = 1, \dots, p$; and g_j, $j = 1, \dots, m$, are twice continuously differentiable; p and m are positive integers. Let us introduce the notation

$$A = \{\mathbf{x} \in R^{3n} | \; h_j(\mathbf{x}) = 0, \;\; j = 1, \dots, p; \;\; g_i(\mathbf{x}) \geq 0, \;\; i = 1, \dots, m\}.$$

In other words, the states of the investigated mechanical systems are represented by the vectors of the set $A \subseteq R^{3n}$. In nonlinear optimization, the sets $I(\mathbf{x}) = \{i | \; g_i(\mathbf{x}) = 0, \; i \in \{1, \dots, m\}\}, \mathbf{x} \in A$, denote the active indices of the inequality constraints.

Definition 7.1 *In the case of constraints (8), the cones of attainable directions are as follows:*

$$C_{[\mathbf{h},\mathbf{g}]}(\mathbf{x}) = \{\mathbf{v} \in R^{3n} | \; \exists \mathbf{x}(t) : [0, \varepsilon] \subseteq R \rightarrow R^{3n}, \quad \varepsilon > 0,$$

$$h_j(\mathbf{x}(t)) = 0,\ j = 1,\dots,p; \quad g_i(\mathbf{x}(t)) \geq 0,\ i \in I(\mathbf{x}),\ t \in [0,\varepsilon],$$
$$\mathbf{x}(0) = \mathbf{x}, \quad d\mathbf{x}(t)/dt_{|t=0} = \mathbf{v}\}, \quad \mathbf{x} \in A.$$

It is emphasized that the values ε appearing in Definition 7.1 depend on the position and velocity. Curves satisfying the constraints are called feasible. Let

$$D_{[\mathbf{h},\mathbf{g}]}(\mathbf{x}) = \{\mathbf{v} \in R^{3n} | \ \nabla h_j(\mathbf{x})\mathbf{v} = 0, \quad j = 1,\dots,p; \\ \nabla g_i(\mathbf{x})\mathbf{v} \geq 0,\ i \in I(\mathbf{x})\}, \quad \mathbf{x} \in A, \tag{9}$$

and

$$I(\mathbf{x},\mathbf{v}) = \{i \in I(\mathbf{x}) \mid \nabla g_i(\mathbf{x})\mathbf{v} = 0\}, \quad \mathbf{x} \in A, \quad \mathbf{v} \in D_{[\mathbf{h},\mathbf{g}]}(\mathbf{x}). \tag{10}$$

Definition 7.2 *The gradient-type Karush-Kuhn-Tucker (GKKT) constraint qualification holds at a point $\mathbf{x} \in A$ if for every $\mathbf{v} \in D_{[\mathbf{h},\mathbf{g}]}(\mathbf{x})$ the gradients*

$$\nabla g_i(\mathbf{x}), \quad i \in I(\mathbf{x},\mathbf{v}); \quad \nabla h_j(\mathbf{x}), \quad j = 1,\dots,p, \tag{11}$$

are linearly independent. The GKKT constraint qualification fulfills on the set A if it fulfills at every point of A.

Remark 7.1 *The GKKT constraint qualification is weaker than the linear independency of the active gradients. Examples, showing the difference between the two conditions, are the polyedric cones determined by the intersection of subspaces of a greater number than that of the dimension of the space.*

Theorem 7.1 *Consider the mechanical equilibrium problems under constraints (8) in the force field defined by $\mathbf{P} : A \rightarrow R^{3n}$. If the GKKT constraint qualification fulfils on A, then the determination of an equilibrium position is equivalent to the solution of a QCS where a point $\mathbf{x} \in A$ has to be found for which*

$$\mathbf{P}(\mathbf{x})^T\mathbf{v} \leq 0, \qquad \mathbf{v} \in D_{[\mathbf{h},\mathbf{g}]}(\mathbf{x}), \quad \mathbf{x} \in A. \tag{12}$$

Proof. By the principle of virtual work, the inequalities

$$\mathbf{P}(\mathbf{x}^*)^T\mathbf{v} \leq 0 \tag{13}$$

hold for all the virtual displacements in an equilibrium state $\mathbf{x}^* \in A$. A direct consequence of Definition 4.1 is that the virtual displacements multiplied by any real number $\lambda > 0$ are virtual displacements as well, i.e., the virtual displacements with respect to constraints (8) form a cone at every point $\mathbf{x} \in A$, moreover, these cones are equal to $C_{[\mathbf{h},\mathbf{g}]}(\mathbf{x})$, $\mathbf{x} \in A$. By the principle of virtual work, it follows that a system of n mass points at a position $\mathbf{x}^* \in A$ is in mechanical equilibrium iff

$$\mathbf{P}(\mathbf{x}^*)^T\mathbf{v} \leq 0, \qquad \mathbf{v} \in C_{[\mathbf{h},\mathbf{g}]}(\mathbf{x}^*). \tag{14}$$

In the next step, it will be shown that if constraints (8) satisfy the GKKT condition, then the virtual displacements with respect to constraints (8) form the cones $D_{[\mathbf{h},\mathbf{g}]}(\mathbf{x})$, $\mathbf{x} \in A$. As

$$C_{[\mathbf{h},\mathbf{g}]}(\mathbf{x}) \subseteq D_{[\mathbf{h},\mathbf{g}]}(\mathbf{x}), \qquad \mathbf{x} \in A, \tag{15}$$

it is sufficient to prove that

$$D_{[\mathbf{h},\mathbf{g}]}(\mathbf{x}) \subseteq C_{[\mathbf{h},\mathbf{g}]}(\mathbf{x}), \qquad \mathbf{x} \in A. \tag{16}$$

Let us consider an arbitrary $\mathbf{v} \in D_{[\mathbf{h},\mathbf{g}]}(\mathbf{x})$, $\mathbf{x} \in A$. By the fulfilment of GKKT, $\mathbf{v}$ belongs to the intersection of a finite number of hyperspaces with the linearly independent normal vectors $\nabla h_j, j = 1, \ldots, p$; ∇g_i, $i \in I(\mathbf{x}, \mathbf{v})$. The corresponding constraints are active, thus the intersection of these constraints determines a Riemannian manifold endowed with the induced Euclidean metric (see, e.g., [33]). Since the vector $\mathbf{v}$ belongs to the tangent space of this manifold at the investigated point, it follows that a uniquely determined geodesic of the Riemannian manifold starts from the point $\mathbf{x}$ in the direction $\mathbf{v}$ in a neighbourhood of the point $\mathbf{x}$. On the other hand, all the functions g_i, $i \in I(\mathbf{x}) \setminus I(\mathbf{x}, \mathbf{v})$ must increase along this curve because

$$\nabla g_i(\mathbf{x})\mathbf{v} > 0, \qquad i \in I(\mathbf{x}) \setminus I(\mathbf{x}, \mathbf{v}).$$

Thus, this geodesic is a feasible curve, it fulfils the requirements of a possible displacement and the vector $\mathbf{v}$ belongs to the virtual displacements. This fact proves that the virtual displacements with respect to constraints (8) form the cones $D_{[\mathbf{h},\mathbf{g}]}(\mathbf{x})$,$\mathbf{x} \in A$.

Now, it will be shown that system (14) can be originated from a QCS. An equivalent formulation of (14) is as follows:

$$\mathbf{P}(\mathbf{x}^*)^T\mathbf{v} \leq 0, \qquad \mathbf{x}^* + \mathbf{v} \in \mathbf{x}^* + D_{[\mathbf{h},\mathbf{g}]}(\mathbf{x}^*), \tag{17}$$

where the sum denotes the Minkowski sum. If

$$H = R^{3n}, \; F(\mathbf{x}) = -\mathbf{P}(\mathbf{x}), \; K(\mathbf{x}) = X(\mathbf{x}) = \mathbf{x} + D_{[\mathbf{h},\mathbf{g}]}(\mathbf{x}), \; \mathbf{x} \in A, \tag{18}$$

$$\mathbf{y} = \mathbf{x} + \mathbf{v}, \quad \mathbf{v} \in D_{[\mathbf{h},\mathbf{g}]}(\mathbf{x}), \qquad \mathbf{x} \in A,$$

then, $\mathbf{x} \in K(\mathbf{x}) \neq \emptyset$, $\mathbf{x} \in A$, and QVI (6) gives QCS (12), which is the statement. □

The cones of virtual displacements are $C_{[\mathbf{h},\mathbf{g}]}(\mathbf{x})$, $\mathbf{x} \in A$, where the virtual displacements are given only implicitly. In order to make use of the principle of virtual work, an explicit characterization of the virtual displacements given by the cones $D_{[\mathbf{h},\mathbf{g}]}(\mathbf{x})$, $\mathbf{x} \in A$, seems to be necessary.

In optimization theory, the KKT constraint qualification fulfils at an arbitrary point $\mathbf{x}_0 \in A$ if

$$C_{[\mathbf{h},\mathbf{g}]}(\mathbf{x}_0) = D_{[\mathbf{h},\mathbf{g}]}(\mathbf{x}_0), \tag{19}$$

and this equality is generally ensured by assuming the linear independency of the active gradients. The GKKT condition fits better to the structure of the constraints, and it provides a constructive proof for the fulfilment of equality (19) also in cases where the linear independency of the active gradients does not hold.

The principle of virtual work characterizes the mechanical equilibrium at a given point. Theorem 7.1 states that the determination of an equilibrium position by checking the condition of the equilibrium at given points is equivalent to the solution of QCS (12). This approach makes the embedding of the principle of virtual work into general equilibrium systems possible. Since general equilibrium systems are in the focus of theoretical and numerical investigations in optimization theory, this approach should mean an essential development of this principle.

A classical example for the motion of a mass point under an equality constraint is the planar pendulum where a mass point hung on a thread of a given length is moving on the effect of the gravitational force. Here, the equality constraint derives from the fact that the length of the thread is constant.

8 Dual form of the principle of virtual work in force field under scleronomic and holonomic constraints

In this part, the dual problem of QCS (12) is formulated based on the Farkas theorem. Let the symbols *span*{ } and *coneconv*{ } denote the subspace and convex cone spanned by the vectors between parenthesis, respectively, furthermore, let the sum be the Minkowski sum.

Theorem 8.1 *Consider the mechanical equilibrium problems under constraint (8) in the force field defined by* $\mathbf{P} : A \to R^{3n}$. *If the GKKT constraint qualification fulfils on A, then the dual form of QCS (12) consists* *in finding a point* $\mathbf{x} \in A$ *for which*

$$\mathbf{P}(\mathbf{x}) \in span\{\nabla h_j(\mathbf{x})^T,\ j = 1, \dots, p\} + coneconv\{\nabla g_i(\mathbf{x})^T,\ i \in I(\mathbf{x})\}. \tag{20}$$

Proof. By Theorem 7.1, if $x^* \in A$ is an equilibrium position, then

$$\mathbf{P}(\mathbf{x}^*)^T \mathbf{v} \leq 0, \qquad \forall \mathbf{v} \in D_{[\mathbf{h},\mathbf{g}]}(\mathbf{x}^*). \tag{21}$$

Since the cone $D_{[\mathbf{h},\mathbf{g}]}(\mathbf{x}^*)$ is given by a linear system of equalities and inequalities, the inequality $\mathbf{P}(\mathbf{x}^*)^T \mathbf{v} \leq 0$ is a consequence of this system. By the Farkas theorem, relation (21) holds iff there exist multipliers λ_j, $j = 1, \dots, p$, and nonnegative multipliers μ_i, $i \in I(\mathbf{x}^*)$, such that

$$\mathbf{P}(\mathbf{x}^*)^T = \sum_{j=1}^{p} \lambda_j \nabla h_j(\mathbf{x}^*) + \sum_{i \in I(\mathbf{x}^*)} \mu_i \nabla g_i(\mathbf{x}^*). \tag{22}$$

Equality (22), containing elements from the dual space only, can be considered the dual form of (21), because the linear optimization problem

$$\max\ \mathbf{P}(\mathbf{x}^*)^T \mathbf{v}, \qquad \mathbf{v} \in D_{[\mathbf{h},\mathbf{g}]}(\mathbf{x}^*), \tag{23}$$

has a feasible solution and a finite optimum value, so by the duality theorem in linear optimization, equality (22) is deduced. If the dual characterization (22) of an equilibrium position is combined with the search for such a position, then equilibrium problems (20) are obtained. □

The dual form of the principle of virtual work does not contain the virtual displacements, thus, in concrete problems, this form seems to be more advantageous than the original one. Duality questions in QVI were considered by Giannessi (1995).

The conservative force field is an important special case of the force fields where a potential function $V : R^{3n} \to R$ exits depending only on the position for which

$$\mathbf{P}(\mathbf{x}) = -\nabla V(\mathbf{x})^T, \qquad \mathbf{x} \in R^{3n}. \tag{24}$$

We remark that both the central force fields and the gravitational force fields are conservative force fields. Let us consider a conservative force field and assume that the GKKT condition fulfils in an equilibrium position $\mathbf{x}^* \in A$ of the system of mass points. Then, by Theorem 8.1, there exist multipliers $\lambda_1, \dots, \lambda_p$ and nonnegative multipliers μ_i, $i \in I(\mathbf{x}^*)$, such that

$$-\nabla V(\mathbf{x}^*) = \sum_{j=1}^{p} \lambda_j \nabla h_j(\mathbf{x}^*) + \sum_{i \in I(\mathbf{x}^*)} \mu_i \nabla g_i(\mathbf{x}^*). \tag{25}$$

It is known well from optimization theory that if a regularity condition holds at $\mathbf{x}^*$, then condition (25) is the first-order necessary optimality condition at the point $\mathbf{x}^*$ of the nonlinear optimization problem

$$\begin{aligned} &\min\ V(\mathbf{x}) \\ &h_j(\mathbf{x}) = 0, \qquad j = 1, \ldots, p, \\ &g_i(\mathbf{x}) \geq 0, \qquad i = 1, \ldots, m, \\ &\mathbf{x} \in R^{3n}. \end{aligned} \tag{26}$$

This is a special case of the principle of virtual work, the Courtivron principle by which, in a conservative force field, a mechanical equilibrium position is a KKT point of the potential function V subject to the given constraints. The clarification of the relation among the Courtivron principle, results of Farkas and nonlinear optimization is one of the main results of papers [29], [30].

The following type of examples was studied by Farkas. Here, it is shown how to determine the mechanical equilibrium position of a mass point in a polyhedral cone.

Example 8.1. Let a polyhedral cone in R^3 be given by the following inequalities:

$$\begin{array}{ll} x_1 - x_3 & \geq 0, \\ -x_1 - x_2 & \geq 0, \\ x_2 - x_3 & \geq 0, \\ (x_1, x_2, x_3) & \in R^3, \end{array}$$

or by matrix formulation, let

$$A = \begin{bmatrix} \mathbf{a}_1^T \\ \mathbf{a}_2^T \\ \mathbf{a}_3^T \end{bmatrix} = \begin{bmatrix} 1 & & -1 \\ -1 & -1 & \\ & 1 & -1 \end{bmatrix},$$

then, an equivalent formulation of the inequality system is

$$A\mathbf{x} \geq 0, \qquad \mathbf{x} \in R^3.$$

If the force vector is given by

$$\mathbf{P}(\mathbf{x}) = (0, 0, -1)^T, \quad \mathbf{x} \in M_1 = \{\mathbf{x} \in R^3 \mid A\mathbf{x} \geq 0\},$$

then,

$$K(\mathbf{x}) = X(\mathbf{x}) = \mathbf{x} + D_{[A]}(\mathbf{x}) = \{\mathbf{v} \in R^3 \mid \mathbf{a}_i^T\mathbf{v} \geq 0, \quad i \in I(\mathbf{x})\},$$

where $I(\mathbf{x})$, $\mathbf{x} \in M_1$, are the sets of active indices.

The QSC to determine the equilibrium positions is as follows:

$$v_3 \geq 0, \qquad \forall \mathbf{v} \in D_{[A]}(\mathbf{x}), \quad \mathbf{x} \in M_1.$$

By using the dual form, we obtain that

$$\mathbf{P}(0) = (0, 0, -1)^T = \frac{1}{2} \sum_{i=1}^{3} \mathbf{a}_i,$$

thus, the point zero is a mechanical equilibrium position of the mass point.

9 Procedure for solving mechanical equilibrium problems

In order to solve a mechanical equilibrium problem based on this approach, the following steps should be taken:

1. determination of the configuration space;
2. determination of the virtual displacements;
3. solution of the resulted QCS:
4. "**a.**" investigation of the existence and unicity of solutions;
5. "**b.**" elaboration methods for solving the QCS problem numerically.

Consider a classic and a very old example, a planar pendulum (see Figure 1).

Its configuration space, a circle, is a compact differentiable manifold and the virtual displacements span the tangent line of the circle at every point. Let us determine the equilibrium positions of the planar pendulum by solving the corresponding QCS.

Example 9.1. The configuration space of the planar pendulum is the circle which can be given in the form of

$$M_2 = \{(x_1, x_2) \in R^2 \mid h(\mathbf{x}) = x_1^2 + x_2^2 - 1 = 0\}.$$

The gravitational force is given by

$$\mathbf{P}(\mathbf{x}) = (0, -1)^T, \qquad \mathbf{x} \in M_2,$$

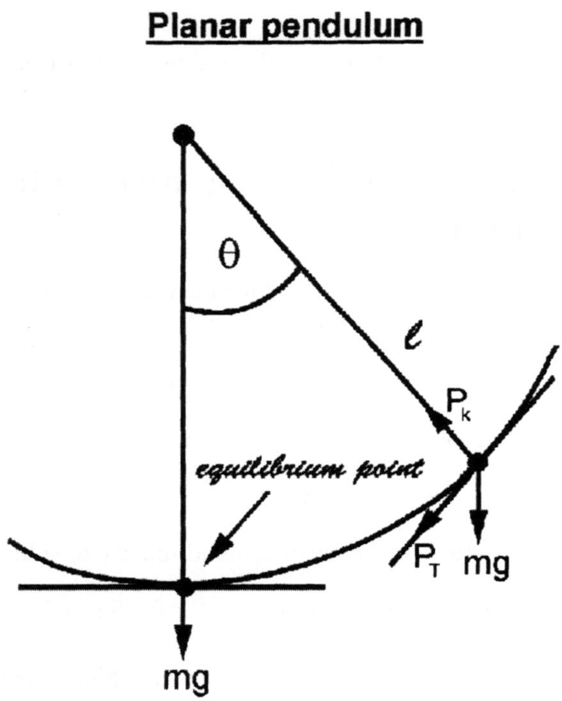

Figure 1

where the gravitational constant is not taken into account, because it does not alter the solution. Since,

$$D_{[h]}(\mathbf{x}) = TM_2(\mathbf{x}), \qquad \mathbf{x} \in M_2,$$

where TM_2 denotes the tangent space of the manifold M_2, by the principle of virtual work, the determination of the equilibrium positions of the planar pendulum is equivalent to the QCS in which a point $\mathbf{x} \in M_2$ has to be found such that

$$v_2 \geq 0, \qquad \forall \mathbf{v} \in TM_2(\mathbf{x}), \quad \mathbf{x} \in M_2.$$

From Figure 1 it is obvious that the equilibrium positions are $(0,1)$ and $(0,-1)$.

Let us determine the equilibrium positions of a double planar pendulum by solving the corresponding QCS.

Example 9.2. In the case of a double planar pendulum, the motion of a mass point is constrained by two equalities in such a way that the mass point is at one end of a rod with one joint, the other end of the rod is fixed and the mass point is moving on the effect of the gravitational force. Thus, the configuration space of a planar double pendulum is the direct product of two circles, i.e., the two torus and the virtual displacements determine the tangent spaces at every point.

Let us assume that the rod is fixed at the point $(0,0)$, the coordinates of the joint and the mass point are (x_1, x_2), and (y_1, y_2), respectively. Then, the configuration space of the double planar pendulum can be given in the form of

$$h_1(\mathbf{x}, \mathbf{y}) = \frac{1}{2}x_1^2 + \frac{1}{2}x_2^2 - \frac{1}{2} = 0, \qquad (\mathbf{x}, \mathbf{y}) \in R^4,$$

$$h_2(\mathbf{x}, \mathbf{y}) = \frac{1}{2}(x_1 - y_1)^2 + \frac{1}{2}(x_2 - y_2)^2 - \frac{1}{2} = 0, \qquad (\mathbf{x}, \mathbf{y}) \in R^4.$$

Let

$$M_3 = \{(\mathbf{x}, \mathbf{y}) \in R^4 | \; h_1(\mathbf{x}, \mathbf{y}) = 0, \; h_2(\mathbf{x}, \mathbf{y}) = 0\}$$

denote the configuration space. Since the gradients

$$\nabla h_1(\mathbf{x}, \mathbf{y}) = (x_1, x_2, 0, 0), \qquad (\mathbf{x}, \mathbf{y}) \in M_3,$$

$$\nabla h_2(\mathbf{x}, \mathbf{y}) = \left(x_1 - y_1, x_2 - y_2, -(x_1 - y_1), -(x_2 - y_2)\right), \qquad (\mathbf{x}, \mathbf{y}) \in M_3,$$

are linearly independent, the set M_3 is a differentiable manifold. The gravitational force can be given by

$$\mathbf{P}(\mathbf{x}, \mathbf{y}) = (0, 0, 0, -1)^T, \qquad (\mathbf{x}, \mathbf{y}) \in M_3.$$

Then, the determination of the equilibrium positions of the double planar pendulum is equivalent to the solution of the following QCS: find a point $(\mathbf{x}, \mathbf{y}) \in M_3$ such that

$$\mathbf{P}(\mathbf{x}, \mathbf{y})^T \mathbf{v} \le 0, \qquad \forall \mathbf{v} \in TM_3(\mathbf{x}, \mathbf{y}), \quad \text{i.e.,}$$

$$v_4 \ge 0, \qquad \forall \mathbf{v} \in TM_3(\mathbf{x}, \mathbf{y}),$$

where

$$TM_3(\mathbf{x}, \mathbf{y}) = \{\mathbf{v} \in R^4 \mid x_1 v_1 + x_2 v_2 = 0,$$

$$(x_1 - y_1)v_1 + (x_2 - y_2)v_2 - (x_1 - y_1)v_3 - (x_2 - y_2)v_4 = 0\}, \quad (\mathbf{x}, \mathbf{y}) \in M_3.$$

By using the dual form, the QCS is equivalent to the solution of the equality system

$$\begin{pmatrix} 0 \\ 0 \\ 0 \\ -1 \end{pmatrix} = \alpha_1 \begin{pmatrix} x_1 \\ x_2 \\ 0 \\ 0 \end{pmatrix} + \alpha_2 \begin{pmatrix} x_1 - y_1 \\ x_2 - y_2 \\ -(x_1 - y_1) \\ -(x_2 - y_2) \end{pmatrix} = \begin{pmatrix} \alpha_1 x_1 + \alpha_2 (x_1 - y_1) \\ \alpha_1 x_2 + \alpha_2 (x_2 - y_2) \\ -\alpha_2 (x_1 - y_1) \\ -\alpha_2 (x_2 - y_2) \end{pmatrix}$$

for $(\mathbf{x}, \mathbf{y}) \in M_3$ and $(\alpha_1, \alpha_2) \in R^2$. From the last two equations, we have that

$$\alpha_2 (x_1 - y_1) = 0,$$

$$\alpha_2 (x_2 - y_2) = 1,$$

from which it follows that

$$\alpha_2 \ne 0, \; x_1 - y_1 = 0, \; \alpha_2 (x_2 - y_2) = 1, \quad x_2 - y_2 \ne 0.$$

From the first two equations, we obtain that

$$\alpha_1 x_1 = 0,$$

$$\alpha_1 x_2 + 1 = 0,$$

from which it follows that

$$\alpha_1 \ne 0, \; x_1 = 0, \; \alpha_1 x_2 = -1.$$

Summarizing the consequences, we have that in an equilibrium position, the following relations fulfill:

$$\alpha_1 \ne 0, \; \alpha_2 \ne 0, \; x_1 = 0, \; y_1 = 0, \; \alpha_2 (x_2 - y_2) = 1 \text{ and } \alpha_1 x_2 = -1.$$

Due to the feasibility of the point $(\mathbf{x}, \mathbf{y})$, we have that $x_2 = \pm 1$, and from the second equation that $(\pm 1 - y_2)^2 = 1$, thus,

$$\begin{array}{lllll} x_2 = +1 & \text{and} & y_2 = 0 & \text{or} & y_2 = 2; \\ x_2 = -1 & \text{and} & y_2 = 0 & \text{or} & y_2 = -2. \end{array}$$

So, if $x_2 = +1$, then, $\alpha_1 = -1$ and $\alpha_2 = +1$ or $\alpha_2 = -1$, and if $x_2 = -1$, thus, $\alpha_1 = 1$ and $\alpha_2 = -1$, or $\alpha_2 = +1$. We can conclude that the equilibrium positions of the double planar pendulum are the points

$$\begin{array}{l}(0,\ 1,\ \ 0,\ 0);\\(0,\ 1,\ \ 0,\ 2);\\(0,-1,\ 0,\ 0);\\(0,-1,\ 0,-2).\end{array}$$

10 Existence of solutions

In this part, an existence theorem based on the basic one is stated. Then, it will be shown that the existence of solutions for a subclass of QCS leads to the existence of zeros of given vector fields on differentiable manifolds which is a difficult topological question. From methodological point of view, this latter approach allows to replace the original problem with a smooth global optimization one on a Riemannian manifold studied in details in the book [33].

Theorem 10.1 *Consider the mechanical equilibrium problems under constraints (8) in the force field defined by* $\mathbf{P} : A \rightarrow R^{3n}$. *If the configuration space* A *is convex and compact, both* $\mathbf{P}$ *and* $D_{[\mathbf{h},\mathbf{g}]}$ *are continuous, then QCS (12) has at least one solution.*

Proof. Since the set A is convex and compact, both $\mathbf{P}$ and $D_{[\mathbf{h},\mathbf{g}]}$ are continuous, all the sets $D_{[\mathbf{h},\mathbf{g}]}(\mathbf{x})$, $\mathbf{x} \in A$ are nonempty, closed and convex, thus by Theorem 6.1 in [6], QCS (12) has at least one solution. □

Remark 10.1 *This statement can be extended to the case of a convex and closed set* A *with discontinuous functions* $\mathbf{P}$ *and* $D_{[\mathbf{h},\mathbf{g}]}$ *[8].*

Theorem 10.2 *Consider the mechanical equilibrium problems under constraints (8) in the force field defined by* $\mathbf{P} : A \rightarrow R^{3n}$. *If the configuration space* A *is a differentiable manifold, then a point of* A *is an equilibrium position iff the force vector is orthogonal to the tangent space at this point.*

Proof. Since the cones of the virtual displacements coincide with the tangent spaces at every point, the statement directly follows from Theorems 7.1 and 8.1. □

This result was stated in the case of natural systems in Lagrangian mechanics, e.g., in Arnold's book (p.94, 1989).

Definition 10.1 *Let* A *be a subset of a differentiable manifold. Then, a vector field is defined on* A *if a continuous mapping assigns a tangent vector to every point of* A.

Corollary 10.1 *Let* $Pr\mathbf{P}(\mathbf{x})$, $\mathbf{x} \in A$, *denote the orthogonal projection of the vector function* $\mathbf{P}$ *with respect to the Euclidean metric to the corresponding tangent spaces of* A. *If the assumptions of Theorem 10.2 hold, then a point of* A *is an equilibrium position iff the global optimum of the problem*

$$\min \; (Pr\mathbf{P}(\mathbf{x}))^2 \mathbf{x} \in A, \tag{27}$$

is equal to zero.

We remark that the Euclidean metric of R^{3n} induces a Riemannian metric on A, thus (27) is a smooth global optimization problem on a subset of a Riemannian manifold if the vector function P is a smooth one. Let us construct this smooth optimization problem in the case of the planar pendulum.

Example 10.1. It will be shown how to transform the QCS related to the planar pendulum into a smooth optimization problem. Let us project the force vector orthogonally into the tangent spaces of M_2. In this case, the projection matrix is

$$I - \mathbf{x}(\mathbf{x}^T\mathbf{x})^{-1}\mathbf{x}^T = I - \mathbf{x}\mathbf{x}^T =$$

$$= \begin{pmatrix} 1 & 0 \\ 0 & 1 \end{pmatrix} - \begin{pmatrix} x_1^2 & x_1x_2 \\ x_1x_2 & x_2^2 \end{pmatrix} = \begin{pmatrix} x_2^2 & -x_1x_2 \\ -x_1x_2 & x_1^2 \end{pmatrix},$$

thus, the smooth vector field on M_2 is equal to

$$\begin{pmatrix} x_2^2 & -x_1x_2 \\ -x_1x_2 & x_1^2 \end{pmatrix} \begin{pmatrix} 0 \\ -1 \end{pmatrix} = \begin{pmatrix} x_1x_2 \\ -x_1^2 \end{pmatrix} = x_1 \begin{pmatrix} x_2 \\ -x_1 \end{pmatrix}, \qquad \mathbf{x} \in M_2,$$

and the smooth optimization problem is as follows

$$\begin{gathered} \min \; x_1^2 \\ \mathbf{x} \in M_2, \end{gathered}$$

whose solutions are $(0, 1)$ and $(0, -1)$. Because the solutions are the zeros of the projected vector field $Pr\mathbf{P}(\mathbf{x})$, $\mathbf{x} \in M_2$, we can conclude that the solutions are the equilibrium positions of the planar pendulum.

A famous result in topology, the hairy ball theorem, states that every vector field on the even-dimensional spheres S^{2n}, where n is an arbitrary positive integer, is zero somewhere. More picturesquely, a hairy ball cannot be combed. This result is of topological character and is based on the Euler characteristic of a compact and connected differentiable manifold. The characterization of the zeros of vector fields on the odd-dimensional spheres and on subsets of differentiable manifolds related to the existence of solutions in QCS seems to be an open question.

References

[1] M. Anastasiei and H. Kawaguchi, *A geometrical theory of time-dependent Lagrangians I. Non-linear connections*, Tensor 48 (1989) 273-282.

[2] Appell, *Mécanique Rationelle*, Gauthier-Villars, Paris, 1909.

[3] V.I. Arnold, *Mathematical methods of classical mechanics*, Springer-Verlag, New York, Heidelberg, Berlin, 1978, 1989.

[4] S. Banach, *Mechanics*, Nauki, Warszawa, Wroclaw, 1951.

[5] A. Bensoussan and J.L. Lions, *Nouvelle formulation des problèmes de controle impulsionnel et applications*, Comptes Rendus de l'Académie des Sciences, Paris 276 (1973) A1189-A1192.

[6] D. Chan and J.S. Pang, *The generalized quasi-variational inequality problem*, Mathematics of Operations Research 7 (1982) 211-222.

[7] A. Cournot, *Extension du principle des vitesses virtuelles au cas ou les conditions de liaison du système sont exprimées par des inégalités*, Bulletin des Sciences Mathématiques (Première Section du Bulletin Universel des Sciences et de l'Industrie publié sous la direction de Férussac) 8 (1827) 165-170.

[8] P. Cubiotti, *Finite-dimensional quasi-variational inequalities associated with discontinuous functions*, Journal of Optimization Theory and Applications 72 (1992) 577-582.

[9] M. De Luca and A. Maugeri, *Quasi-variational inequalities and application to equilibrium problems with elastic demand*, in: Nonsmooth optimization and related topics, F.H. Clarke et al. (eds.), Plenum Press, New York (1989) 61-77.

[10] Gy. Farkas, *Algebraic basis of the application of Fourier principle in mechanics*, Mathematikai és Természettudományi Értesítő 16 (1898) 361-364. (in Hungarian)

[11] J. Farkas, *Theorie der einfachen Ungleichungen*, Journal für die Reine und Angewandte Mathematik 124 (1901) 1-27.

[12] J. Farkas, *Beiträge zu den Grundlagen der analytischen Mechanik*, Journal für die Reine und Angewandte Mathematik 13 (1906) 165-201.

[13] J. Fourier, *Mémoire sur le statique*, Journal de l'École Politechnique 5 (1798).

[14] N.A. Fufaev, *About an example of a system with a nonholonomic constraint of second order*, Zeitschrift für Angewandte Mathematik und Mechanik 70 (1990) 593-594.

[15] F. Gantmacher, *Lectures in analytical mechanics*, Mir Publishers, Moscow, 1970.

[16] C.F. Gauss, *Über ein neues allgemeines Grundgesetz der Mechanik*, Journal für die Reine und Angewandte Mathematik 4 (1829) 232-235.

[17] F. Giannessi, *Separation of sets and gap functions for quasi-variational inequalities, in: Variational inequalities and network equilibrium problems*, F. Giannessi and A. Maugeri (eds.), Plenum Press, New York and London (1995) 101-121.

[18] W. Karush, *Minima of function of several variables with inequalities as slide conditions*, Master's Thesis, Department of Mathematics, University of Chicago, 1939.

[19] T. Kawaguchi and R. Miron, *On the generalized Lagrange spaces with the metric* $\gamma_{ij}(x) + (1/c^2)y_i y_j$, Tensor 48 (1989) 52-63.

[20] W.K. Kim and K.-K. Tan, *On generalized vector quasi-variational inequalities*, Optimization 46 (1999) 185-198.

[21] H.W. Kuhn and A.W. Tucker, *Nonlinear programming, in: Proceedings of the Second Berkeley Symposium on Mathematical Statistics and Probability*, University of California Press, Berkeley, California (1951) 481-492.

[22] D. Kinderlehrer and G. Stampacchia, *An introduction to variational inequalities and their applications*, Academic Press, New York, London, Toronto, 1980.

[23] J.L. Lagrange, *Mécanique analytique I-II*, Paris, 1788.

[24] A. Maugeri, *Variational and quasi-variational inequalities in network flow models. Recent developments in theory and algorithms*, in: Variational inequalities and network equilibrium problems, F. Giannessi and A. Maugeri (eds.), Plenum Press, New York and London (1995) 195-211.

[25] K. Nagy, *Theoretical physics*, Tankönyvkiadó, 1989. (in Hungarian)

[26] M.A. Noor, *Generalized quasi-variational inequalities and implicit Wiener-Hopf equations*, Optimization 45 (1999) 197-222.

[27] R. Ortvay, *To the memory of Gyula Farkas academician*, Memorial talks on the late academicians of the Hungarian Academy of Sciences 21 (1933) 15. (in Hungarian)

[28] M. Ostrogradsky, *Mémoire sur les déplacements instantanés des systèmes assujettis à des conditions variables*, Mémoires de l'Académie Impériale des Sciences de Saint-Pétersbourg, Sixième Série 1 (1838) 565-600.

[29] A. Prékopa, *On the development of optimization theory*, Alkalmazott Matematikai Lapok 4 (1978) 165-191. (in Hungarian)

[30] A. Prékopa, *On the development of optimization theory*, American Mathematical Monthly 87 (1980) 527-542.

[31] T. Rapcsák and J. Szenthe, *On the connection between mechanical equilibrium and nonlinear programming*, Alkalmazott Matematikai Lapok 12 (1986) 161-174. (in Hungarian)

[32] T. Rapcsák and J. Szenthe, *On the connection between mechanical force equilibrium and nonlinear programming*, Zeitschrift für Angewandte Mathematik und Mechanik 70 (1990) 557-564.

[33] T. Rapcsák, *Smooth nonlinear optimization in R^n*, Kluwer Academic Publishers, 1997.

[34] G. Stampacchia, *Formes bilinéaires coercitives sur les ensembles convex*, Comptes Rendus de l'Académie des Sciences, Paris (1964) 4413-4416.

FALSE NUMERICAL CONVERGENCE IN SOME GENERALIZED NEWTON METHODS

Stephen M. Robinson
Department of Industrial Engineering
University of Wisconsin-Madison
1513 University Avenue
Madison, WI 53706-1572, USA
e-mail: smrobins@facstaff.wisc.edu

Abstract This paper deals with generalized Newton methods for solving nonsmooth equations. In particular, these methods can be used to solve variational inequalities by reformulating the latter as equations involving the associated normal map. We ask whether, if the method appears to converge nicely to a point, one is safe in supposing that point to be at least close to a zero of the function. We show that the answer is no, so that these methods can produce what one might term false numerical convergence. We then demonstrate a general method for constructing normal maps for which this false convergence phenomenon can occur, and we construct such an example in $\mathbb{R}^5$ for the method of Kojima and Shindo. Finally, we look at some ways in which a user might avoid being deceived by false numerical convergence.

Keywords: Newton methods, generalized equations, variational inequalities, nonsmooth equations.

2000 Mathematics Subject Classification. 49J, 90C, 65K.

1 Introduction

In this paper we are concerned with several generalized Newton methods that can be used to solve nonsmooth equations. In particular, these methods can be used to solve variational inequalities by reformulating the latter as equations involving the associated normal map [17, 19]. There are quite a few such methods, but the ones of interest for our purposes fall into two broad classifications: those using what we have elsewhere termed a "point-based approximation" (PBA) [18] as a generalized derivative, and those using an element selected from some set of derivative-like objects such as the generalized Jacobian [12], or simply the union of all derivatives of active functions at the point in question [9].

One can prove point-of-attraction results for both classes of methods. These results state that if the method is started sufficiently close to a zero of the function in question, and if in addition certain regularity properties hold near that zero, then the method will converge quadratically to the zero. In that respect the methods behave in similar ways.

The question we address here is in a sense the converse of such a result. We ask whether, if the method appears to converge nicely to a point, one is safe in supposing that point to be a zero of the function. We show that the answer is no, so that these methods can produce what one might term false numerical convergence.

By false numerical convergence we mean that the iteration enters and remains in a very small region, within which the function values are so small that a user may assume they differ from zero only by roundoff error. Thus, a user confronted by this phenomenon might erroneously conclude that the method had found a zero, whereas in fact the region in question may contain no zero and, in fact, the function itself may have no zero at all.

This kind of distinction is not just theoretical; false numerical convergence does in fact deceive people, even in the case of equations involving smooth functions. For an example, consider the numerical solution of a maximum-likelihood factor analysis problem reported by Rao in [14]. That solution used a simple iteration procedure that, unbeknownst to Rao, could not be guaranteed to converge because the iterating function was not strongly contractive near the zero in question. As shown later in [5], Rao's reported solution was far from the true solution, so that in this case false numerical convergence resulted in the reporting of incorrect results in the literature.

Four sections follow this introduction. In Section 2 we describe several generalized Newton methods that we shall consider in the rest of the paper. Then in Section 3 we introduce the false numerical convergence phenomenon and exhibit a method for constructing examples of such false convergence. In Section 4 we use this method to construct such an example for the method of Kojima and Shindo. Finally, in Section 5 we consider ways of avoiding the phenomenon of false convergence.

2 Some generalized Newton methods

In this section we give a brief introduction to some generalized Newton methods that exemplify the two classes mentioned above. We present only a selection of methods, and do not pretend to provide a survey or a comprehensive discussion. Our purpose is simply to give enough background so that the subsequent discussion is comprehensible.

The first class of method uses a "point-based approximation" as a proxy for the derivative used in the conventional Newton method. We describe first the method given in [18], and then the well known Josephy-Newton method, which can be regarded as a special case of it. We then discuss briefly the PATH solver [3], a very powerful and effective software package in wide use for solving mixed complementarity problems.

Suppose f is a function from an open subset Ω of $\mathbb{R}^n$ to $\mathbb{R}^m$. We say that a function $D : \Omega \times \Omega \to \mathbb{R}^m$ is a *point-based approximation (PBA)* to f on Ω if there is a scalar κ, called the *modulus* of the approximation, such that for each u and v in Ω (1) the function $D(u, \cdot) - D(v, \cdot)$ is Lipschitzian on Ω with modulus $\kappa\|u - v\|$, and (2) one has

$$\|f(v) - D(u, v)\| \leq .5\kappa\|v - u\|^2.$$

The PBA is an attempt to abstract to a more general setting some of the useful properties of the linearization that we could obtain by taking $D(u, v) = f(u) + df(u)(v - u)$ in the case in which f has a Fréchet derivative df on Ω.

Having the idea of a PBA, we can describe the generalized Newton method of [18] for finding a zero of f. It is very simple: one starts with a point $x_0 \in \Omega$ and defines a sequence $\{x_k\}$ by requiring that $D(x_k, x_{k+1}) = 0$ for $k = 0, 1, \ldots$. Of course some hypotheses are needed to ensure that such a sequence is well defined, and that it converges to a zero of f. The following theorem, a specialization to the present finite-dimensional case of Theorem 3.2 of [18], details these. In it we use the notation $B(x, \rho)$ for the closed ball of radius ρ around x, and for a function $g : \Omega \to \mathbb{R}^m$ we write

$$\beta(g, \Omega) = \inf\{\|u - v\|^{-1}\|g(u) - g(v)\| \mid u \neq v,\ u, v \in \Omega\}.$$

This β gives a quantitative measure of the extent to which g separates points.

Theorem 1 *Let Ω be an open convex subset of $\mathbb{R}^n$ and f a continuous function from Ω to $\mathbb{R}^n$. Suppose that f has a point-based approximation D on Ω with modulus κ. Assume that for some $x_0 \in \Omega$ and some positive real number r_0 we have the following:*

a. $\beta(D(x_0, \cdot), \Omega) \geq d_0 > 0$.

b. $0 < h := d_0^{-1}\kappa r_0 \leq .5$.

c. For each $y \in B(0, d_0(c - r_0))$ the equation $D(x_0, x) = y$ has a solution x, where

$$c := (r_0/h)(1 - (1 - 2h)^{1/2}).$$

d. The solution $G(x_0)$ *of* $D(x_0, G(x_0)) = 0$ *satisfies* $\|G(x_0) - x_0\| \le r_0$.
e. $\Omega \supset B(x_0, c)$.
Then the Newton iteration defining x_{k+1} *by*

$$D(x_k, x_{k+1}) = 0, \qquad k = 0, 1, \ldots,$$

is well defined and the sequence $\{x_k\}$ *converges to* $x_* \in B(x_0, c)$ *such that* $f(x_*) = 0$. *Further, if* $h < 1/2$ *then for each* k *we have*

$$\|x_k - x_*\| \le 2\alpha\theta^{(2^k)}/[1 - \theta^{(2^k)}],$$

where

$$a = (r_0/h)(1 - 2h)^{1/2} = (r_0/h) - c,$$
$$\theta = (1 - (1 - 2h)^{1/2})/(1 + (1 - 2h)^{1/2}),$$

whereas if $h = 1/2$ *then for each* k,

$$\|x_k - x_*\| \le 2^{1-k} r_0.$$

This theorem establishes a convergence result of Kantorovich type (that is, similar to the Newton-Kantorovich result for the classical Newton method given in Theorem 12.6.2 of [11], with the optimal error bounds of Gragg and Tapia [4]) for the generalized Newton iteration it describes. It is shown in Section 4 of [18] that for normal maps f_C constructed from differentiable functions f, the corresponding normal map constructed from the linearization of f will be a PBA for f_C. This particular case subsumes the Josephy-Newton method, for which a Newton-Kantorovich theorem had already been proved by Josephy in Theorem 2 of [6] (see also [7] and [15]).

Dirkse and Ferris developed the PATH solver to implement a stabilized version of the above Newton method for mixed complementarity problems [3]. The stabilization uses the path search technique developed by Ralph [13]. The PATH solver has since been bundled into the GAMS algebraic modeling language [1]. The result, with continued evolution and improvement of the solver, is a highly effective and easy to use software package. We shall employ it later in this paper to deal with an example that we construct below.

The second class of method is much simpler than those just described. The basic idea is to deal with a nonsmooth function by using a multivalued derivative substitute, instead of the single-valued construction provided by a PBA. Perhaps the simplest method of this class, which we shall use below, is that of Kojima and Shindo [9] for piecewise C^1 functions. Suppose f is such a function defined from $\mathbb{R}^n$ to $\mathbb{R}^n$ using an underlying family f_i of C^1 functions associated with a subdivision of the space, such that f is continuous and for each $x \in \mathbb{R}^n$ the function $f(x)$ agrees with (at least) one of the f_i. For precise definitions and conditions on the subdivision, we refer to [9]. For a given x, we call the indices i for which $f(x) = f_i(x)$ *active.*

The Kojima-Shindo method defines a sequence $\{x_k\}$ by starting with some $x_0 \in \mathbb{R}^n$ and defining x_{k+1} by

$$f(x_k) + D_i(x_k)(x_{k+1} - x_k) = 0,$$

where $D_i(x_k)$ is the derivative at x_k of any of the functions f_i such that i is active at x_k. Thus, the method involves only the solution of a system of linear equations at each step, which is generally much simpler than trying to solve an equation involving a point-based approximation.

Kojima and Shindo proved a point of attraction result for their method (Theorem 1 of [9]). This result provides that if $f(x_*) = 0$, if all of the derivatives D_i associated with indices i active at x_* are nonsingular at x_* and Lipschitzian near x_*, and if the method is started sufficiently close to x_*, then the resulting sequence exists and converges quadratically to x_*. This result requires one to assume that a zero exists, whereas theorems of the Newton-Kantorovich type prove that it exists along with proving convergence of the Newton sequence.

One can generalize the Kojima-Shindo procedure by selecting D_i from a more general class of derivative substitutes. One example of such a method is the nonsmooth Newton method of Qi and Sun [12], in which the function f is only assumed to be semismooth instead of being piecewise C^1 as in the work of Kojima and Shindo. Qi and Sun use the generalized Jacobian ∂f defined by Clarke [2, Section 2.6]. They proved a point of attraction result for their method, together with results about the rate of convergence (which depends on the degree of semismoothness). Other such generalizations exist, and further references to methods of this type can be found in [12].

This section has briefly introduced several generalized Newton methods, and has given an overview of their convergence properties. In the next section we turn to the question of false numerical convergence.

3 False numerical convergence

This section introduces the false numerical convergence phenomenon and exhibits a method of constructing examples of such false convergence. In particular, an example is demonstrated for the method of Kojima and Shindo. The term "numerical convergence" is borrowed from a paper of Nickel and Ritter [10], although the phenomenon we study here is somewhat different from that considered in their work.

To describe the false-convergence phenomenon it will be useful to introduce the idea of a region of attraction. We follow Kelley [8] in saying that a sequence $\{x_k\}$ is *eventually* in a set X if there is some K such that for each $k \geq K$ one has $x_k \in X$.

Now let Ω be a subset of $\mathbb{R}^n$ and Φ be a function from Ω to Ω. Let A be a connected subset of Ω. Let us say that A is a *region of attraction* for the iteration $x_{k+1} = \Phi(x_k)$ if there is a neighborhood Q of A, contained in Ω, such that for each x_0 in Q the sequence $\{x_k\}$ generated by the iteration is eventually in A.

By itself, this definition is not very interesting (for example, if Ω is open then it is always a region of attraction). However, it is helpful in making somewhat more precise what we mean by false numerical convergence. In terms of the above definition we say that such convergence occurs for an iteration seeking a zero of a function $f : \Omega \to \mathbb{R}^m$ if there is a region of attraction A such that (1) A is small, and (2) f takes only small values on A.

Of course, this description of false numerical convergence is still imprecise because it uses the word "small" twice, and we have not said what we mean by "small." It is impossible to give a precise value for this smallness, because what we are trying to capture is a qualitative phenomenon: namely, that the region A and the set of function values $f(A) := \cup\{f(x) \mid x \in A\}$ will look to a user like, respectively, a point of $\mathbb{R}^n$ and the origin of $\mathbb{R}^m$, in both cases contaminated by roundoff error. Thus what "small" should be will depend on, among other things, the type of computing machine being used and the user's judgment.

We can illustrate this idea by introducing a simple one-dimensional variational inequality that we shall later use as a building block to construct a more complicated example. Let γ, δ and ϵ be real numbers and let p be the vector $(\gamma, \delta, \epsilon)$. Define a function f depending on p and a real variable x by

$$f(p, x) = .5\gamma x^2 + \delta x - \epsilon. \tag{1}$$

Now for fixed p consider the problem of finding x such that for each $c \in \mathbb{R}_+$, $\langle c - x, f(p, x)\rangle \geq 0$. Here $\mathbb{R}_+$ is the nonnegative halfline in $\mathbb{R}$; later we also write $\mathbb{R}_-$ for the nonpositive halfline.

We can write this in the form of a generalized equation as

$$0 \in f(p, x) + N_+(x), \tag{2}$$

where $N_+(x)$ denotes the normal cone of $\mathbb{R}_+$ at x. It is not very hard to see that if $\gamma = 0$, $\delta < 0$ and $\epsilon > 0$, then (2) cannot have a solution because whenever x belongs to $\mathbb{R}_+$ the function $\delta x - \epsilon$ is strictly negative.

However, it is interesting to see what happens if we reformulate (2) as a normal map and attempt to solve it using the generalized Newton method of Kojima and Shindo. Reformulation of a generalized equation $0 \in h(x) + N_C(x)$ yields the normal map $h_C(z)$ defined by

$$h_C(z) = h[P_C(z)] + z - P_C(z),$$

where $P_C(z)$ is the projection of z on C. For our problem $C = \mathbb{R}_+$ and $h(x) = f(p, x)$, so we obtain

$$f(p, \cdot)_+(z) = \begin{cases} z - \epsilon & \text{if } z < 0, \\ .5\gamma z^2 + \delta z - \epsilon & \text{if } z \geq 0. \end{cases}$$

If we temporarily set $\gamma = 0$ then the graph of $f(p, \cdot)_+$ consists of the union of two halflines, each terminating at the point $(0, -\epsilon)$, one having slope $+1$ and the other having slope δ.

For this problem the generalized Newton method of Kojima and Shindo yields a very simple sequence: if $x_k < 0$ then $x_{k+1} = \epsilon$, if $x_k > 0$ then $x_{k+1} = \delta^{-1}\epsilon$ unless $\delta = 0$ (when x_{k+1} is undefined), while if $x_k = 0$ then if $\delta \neq 0$ the point x_{k+1} can be either $\delta^{-1}\epsilon$ or ϵ, depending upon which of the two active linear functions one selects. Accordingly, if we keep $\delta < 0$ and $\epsilon > 0$, the interval $A = [\delta^{-1}\epsilon, \epsilon]$ is a region of attraction. If we set $\delta = -1$, for example, then the diameter of A is 2ϵ, while the function value at each point in A has absolute value no greater than 2ϵ. Accordingly, if ϵ is small then the method can appear to a user to have converged to a zero of the function at the origin.

At this point it is important to note that we have been assuming computation with infinite precision. That is, we assume no error at all in computing the successive iterates. The error occurs in the user's interpretation of what he or she sees: apparent stability of the iterates, and small function values. The user is quite likely to conclude that the method has stabilized close to a zero of the normal map, or equivalently a solution of the variational inequality.

One might object at this point that this is nothing new: it is well known that the ordinary Newton method for smooth functions can cycle, and for a sufficiently unpleasant function one ought to be able to produce this kind of apparent convergence. This is quite true, and we can illustrate a way to do it. Let $\gamma = 0$, $\delta = -1$, and $\epsilon > 0$; take the function $f(p, x)$ used above and, for some $\alpha \in (0, \epsilon)$ smooth $f(p, \cdot)$ on $[-\alpha, \alpha]$ to replace the sharp point by a C^1 function. To see one way to do this, consider the derivative d of $f(p, \cdot)$, which is $+1$ for $x < 0$ and -1 for $x > 0$. Now define a function $e(x)$ by requiring e to equal d on $(-\infty, -\alpha]$ and on $[\alpha, +\infty)$, but in $[-\alpha, \alpha]$ to be $-\alpha^{-1}x$. Thus, on that interval the graph of e is the straight line segment joining $(-\alpha, 1)$ and $(\alpha, -1)$. The integrals of d and of e over the interval $[-\alpha, \alpha]$ are the same. Therefore the function g defined for any x by

$$g(x) = f(p, -\epsilon) + \int_{-\epsilon}^{x} e(s)ds$$

will agree with f off $[-\alpha, \alpha]$, while it is everywhere C^1 (in fact the derivative is Lipschitzian). Now if one applies Newton's method starting at any point outside $[-\alpha, \alpha]$, after one step the process will alternate between the points $-\epsilon$ and ϵ.

The answer to this objection is that of course false numerical convergence is possible for differentiable functions, but we do not generally encounter it in practical problems. Indeed, to produce it in the example of the last paragraph we had to construct a rather strange function, which would be unlikely to arise in practice unless one began, as we did, with a nonsmooth function and then tried to smooth it. On the other hand, the kind of corner on the graph that produces this phenomenon occurs very frequently in dealing with nonsmooth functions: indeed, in the one-dimensional case if one substituted for $f(p, x)$ in (2) *any* C^1 function $h(x)$ defined on $\mathbb{R}_+$ such that $h(0) < 0$, $h'(0) \leq -1$, and h' is nonincreasing, then the resulting normal map has such a corner.

Up to now we have just dealt with simple one-dimensional examples so that the underlying process would be easy to visualize. In the next section we build a more realistic model and show how an inexperienced user might easily be deceived by false convergence.

4 An example

Here we use the simple function f introduced above to construct a problem that appears more realistic. To that end, for $i = 1, \ldots, 5$ we define $p_i := (\gamma_i, \delta_i, \epsilon_i)$ and points y and u in $\mathbb{R}^5$ with coordinates y_i and u_i by the data in Table 1.

Table 1: Parameter data

i	*1*	*2*	*3*	*4*	5
γ_i	-14	-56	-84	-24	-65
δ_i	1.2	4.5	-1.5	3.7	2.5
ϵ_i	0	0	10^{-5}	0	0
y_i	1	4	22/7	5	6
u_i	.5	.9	.6	.4	.8

Define a function $F(x)$ on $\mathbb{R}^5$ by

$$F_i(x) = f(p_i, x_i - y_i), \qquad i = 1, \ldots, 5, \tag{3}$$

and a subset C of $\mathbb{R}^5$ by $C = y + \mathbb{R}^5_+$. Then C is the product of five intervals of the form $[y_i, +\infty)$, so that the generalized equation

$$0 \in F(x) + N_C(x) \tag{4}$$

is a mixed complementarity problem (MCP). This problem is of exceedingly simple structure (its Jacobian is diagonal), and it has no solution because if it did then the generalized equation

$$0 \in f(p_3, \xi) + N_+(\xi) \tag{5}$$

would have to be solvable. But $\gamma_3 < 0$ and $\delta_3 < 0 < \epsilon_3$, so (5) cannot have a solution.

We wrote a simple MATLAB program to implement the Kojima-Shindo method for solving the normal map $F_C(z) = 0$ obtained from (4), starting at the point $z_0 := y + u$. Table 2 shows the iterates $z_0, \ldots, z_8$ and Table 3 the values of $F_C(z_i)$, both as MATLAB displayed them using the default "format short" setting. An unsuspecting user might very well interpret this as quick, trouble-free convergence of Newton's method to a solution. However, as we have already shown, this problem has no solution.

Table 2: Iterates z^i

i	z_1^i	z_2^i	z_3^i	z_4^i	z_5^i
0	1.5000	4.9000	3.7429	5.4000	6.8000
1	1.3017	4.4941	3.4342	5.3254	6.4202
2	1.2107	4.2950	3.2801	5.3092	6.2313
3	1.1776	4.2027	3.2036	5.3083	6.1387
4	1.1716	4.1679	3.1663	5.3083	6.0960
5	1.1714	4.1610	3.1495	5.3083	6.0801
6	1.1714	4.1607	3.1438	5.3083	6.0770
7	1.1714	4.1607	3.1429	5.3083	6.0769
8	1.1714	4.1607	3.1429	5.3083	6.0769

At this point we might consider a possible objection to this kind of experiment. That objection would run as follows: "It's true that in this case the Kojima-Shindo method appeared to converge to a solution when in fact there was none. But the point in question was almost a solution, because if the function had been very slightly perturbed then a solution would have existed." The short answer to this is that the function wasn't perturbed, and a solution doesn't exist, and it's not good when methods make non-solutions appear to be solutions. However, a more thoughtful reply could take into account the fact that in using the conventional

Table 3: Function values $f(z^i)$

i	$f_1(z^i)$	$f_2(z^i)$	$f_3(z^i)$	$f_4(z^i)$	$f_5(z^i)$
0	-1.1500	-18.6300	-16.0200	-0.4400	-18.8000
1	-0.2752	-4.6127	-4.0017	-0.0667	-4.6880
2	-0.0580	-1.1097	-0.9971	-0.0032	-1.1601
3	-0.0077	-0.2386	-0.2460	-0.0000	-0.2785
4	-0.0002	-0.0339	-0.0583	-0.0000	-0.0594
5	-0.0000	-0.0013	-0.0119	-0.0000	-0.0082
6	-0.0000	-0.0000	-0.0014	-0.0000	-0.0003
7	-0.0000	-0.0000	-0.0000	-0.0000	-0.0000
8	-0.0000	0.0000	-0.0000	-0.0000	-0.0000

Newton method we always accept some error in the computed solution, so that the point we end up with is in fact not a solution. One might argue that the objection cited above simply extends this line of thinking to the nonsmooth case, so it is not unreasonable.

However, there is a crucial difference between these two cases. In the conventional case (except for pathological functions like the smoothed function we constructed above) the derivative of the function whose zero we are seeking is nonsingular at that solution, so that the function maps a neighborhood of the solution homeomorphically onto a neighborhood of the origin. Moreover, unless the problem is very ill conditioned the homeomorphism is Lipschitzian with a reasonable modulus. Accordingly, within those neighborhoods small errors in the solution correspond to small deviations from a zero value of the function, and vice versa. This is not at all the case here, as the function F_C is not a local homeomorphism if $z_3 = 22/7$.

If a smooth function were not a homeomorphism at a point then its derivative would be singular there, and an attempt to solve the system of linear equations required by the Newton method would uncover this singularity. Even if we were not quite at the solution, the derivative would be almost singular and the Newton method would typically display extreme instability (e.g., very large distance between successive iterates). These phenomena do not occur in the nonsmooth case, as is illustrated by the example above, or even by the simple one-dimensional function that we constructed in Section 3. Therefore the conventional thinking does not work in this case, and the objection raised above is invalid. However, we shall return to this homeomorphism question in the next section, where we show that in some cases we can justify using the local homeomorphism reasoning even in the nonsmooth case.

In this section we used the simple one-dimensional function of Section

3 as a building block to construct a mixed complementarity problem in $\mathbb{R}^5$ for which the Kojima-Shindo method displays false numerical convergence. We exhibited output from an unsophisticated MATLAB program, such as an inexperienced user might have written, that could lead the user to think that the program had found a solution when in fact no solution exists. In the next section we turn to the question of how a user might guard against this kind of false numerical convergence.

5 Avoiding false numerical convergence

This section addresses a question motivated by the example in the last section: namely, suppose one wants to use a generalized Newton method to solve variational inequalities that have been reformulated as nonsmooth equations via the normal map. What steps can a user take to guard against the phenomenon of false numerical convergence?

Unfortunately, we have no panacea that will keep a user completely safe. However, there are precautionary measures that can help, and we discuss two of these here. We describe them for the mixed complementarity problem, as that is probably the most common form in which such problems are now solved, and is also currently the form for which the best software exists.

The first suggestion, applicable in many situations and particularly here, is to use sophisticated off-the-shelf software whenever it is available. Such software is likely to be more carefully written, and to employ more checks and tests of the model being solved, than would be a one-shot program written by a user whose primary interest was not developing software.

Fortunately, for the mixed complementarity problem the PATH solver is available (see the discussion in Section 2. When we ran the model discussed above with GAMS/PATH (Version 4.4b with default settings), the solver terminated immediately, reporting that the complementarity problem was infeasible. Therefore with this method of solution false convergence was not an issue.

To use methods of the first class described in Section 2 one has to have fairly well developed software, because those methods require the solution of subproblems that are typically of the same class as the problem being solved but of simpler structure. For example, the Josephy-Newton method (hence also the PBA method) when applied to a MCP like that discussed here requires the solution of other MCPs, for each of which the underlying set is the same as that in the original problem but the function is a linearization of the original function. One therefore solves a sequence of linear MCPs. The need to carry out these fairly sophisticated computations tends to push the user in the direction of using a prepared package, and therefore helps to guard against false convergence.

That is not the case with methods of the second class, because they are so simple: they require only the computation of a derivative element and then the solution of a system of linear equations. That being the case, is there any way in which a user can guard against the problem we have exhibited above?

We can suggest one method that in the case of nonsmooth equations derived from normal maps (which includes the example given above) should protect users from at least some of the hazards of this problem. That is to check the orientation of the Jacobian at an alleged solution. In order to explain how to do this and why it is worthwhile, we need to make a detour into theory.

We pointed out already that a crucial difference between the solution of a conventional system of C^1 equations by Newton's method and the present situation is that in the C^1 case one generally has a nonsingular Jacobian at the solution and therefore the function is locally a Lipschitzian homeomorphism. This means that small computational errors, such as the function's not being quite equal to zero, are likely to result in only small errors in the computed solution point. This is not at all the case in the example we gave. Therefore, if there were an easy way to determine whether the function was a local homeomorphism one might be able to distinguish cases in which small computational errors are likely to be innocuous from those in which they may cause great trouble. Fortunately there is a way to determine this, and in the case of the MCP it is not difficult to apply. We explain next how to do so.

Suppose we start with a generalized equation $0 \in G(x) + N_C(x)$ in which C is a polyhedral convex subset of $\mathbb{R}^n$ and G is a C^1 function from an open subset Ω of $\mathbb{R}^n$ that contains C to $\mathbb{R}^n$. We transform this generalized equation into the equivalent normal-map equation

$$0 = G_C(z) := G[P_C(z)] + z - P_C(z), \tag{6}$$

where P_C is the Euclidean projector on the set C. The relations

$$x_0 = P_C(z_0), \qquad z_0 = x_0 - G(x_0)$$

place solutions z_0 of (6) in one-to-one correspondence with solutions x_0 of the generalized equation. We are interested in the behavior of G_C near such a solution z_0; in particular we want to know when it will be a local homeomorphism.

By applying results in [16] and [17] it is possible to show that G_C will be a local homeomorphism at z_0 provided that it satisfies a certain orientation condition there. This condition is phrased in terms of a subdivision of $\mathbb{R}^n$ called the *normal manifold*, and we now briefly explain that. Because it is much simpler in the case of the MCP and because that case is the one in which we are primarily interested here, from here on we assume that C is the Cartesian product of n intervals I_i, each being of one of three forms: $[L_i, +\infty)$, $[L_i, U_i]$, or $(-\infty, U_i]$, where for $i = 1, \ldots, n$ the L_i and U_i are fixed real numbers. Some indices i may have only one such number (e.g., if the interval I_i is of the form $[L_i, +\infty)$ then for that i there is no U_i), but if L_i and U_i both exist then we assume that $L_i < U_i$.

We now define closed regions of $\mathbb{R}^n$ by a certain construction. First, for each i consider intervals defined as follows:

- If the interval I_i is of the form $[L_i, +\infty)$, we use $(-\infty, L_i]$ and $[L_i, +\infty)$.
- If the interval I_i is of the form $[L_i, U_i]$, we use $(-\infty, L_i]$, $[L_i, U_i]$, and $[U_i, +\infty)$.
- If the interval I_i is of the form $[U_i, +\infty)$, we use $(-\infty, U_i]$ and $[U_i, +\infty)$.

Now for each i we select one of the intervals just defined, using the definition corresponding to the particular form of I_i, and then form the Cartesian product of these. Clearly there will be multiple ways of doing this: for example, in the case of the ordinary complementarity problem (in which each L_i is zero and there are no U_i) we obtain 2^n regions, while in the case of a bounded C, in which each i has both L_i and U_i, there will be 3^n. Let us denote the set of such regions (for the C we are using) by Σ and a generic element of Σ by σ. The collection Σ is the *normal manifold* of C; it is a subdivided piecewise linear manifold whose union is all of $\mathbb{R}^n$. The sets σ are called its *cells.* For more on its properties, see [17].

For the local homeomorphism property in which we are interested, we need one more concept. For a polyhedral convex subset S of $\mathbb{R}^n$ we define the *critical cone* $K_S(x, y)$ of S at x with respect to a point $y \in \mathbb{R}^n$ to be the intersection of the tangent cone $T_S(x)$ of S at x with the subspace $\{z \in \mathbb{R}^n \mid \langle y, z \rangle = 0\}$. One can verify that for a set like C that is a product of closed intervals, the critical cone is also a product of such intervals. However, the critical cone often has a substantially simpler structure. For example, suppose that C is the subset of $\mathbb{R}^3$ defined by $C = [-1, 1] \times [0, +\infty) \times (-\infty, 1]$. If $x = (0, 1, 0)$ and $y = 0$ then $K_C(x, y) = \mathbb{R}^3$. If x is the point $(0, 1, 1)$ and $y = (0, 0, 1)$ then $K_C(x, y) = \mathbb{R}^2 \times \mathbb{R}_-$. If x is the point $(1, 1, 1)$ and $y = (0, 0, 1)$ then $K_C(x, y) = \mathbb{R}_- \times \mathbb{R} \times \mathbb{R}_-$.

With these definitions, the local homeomorphism property can be very simply stated as follows. Suppose that $z_0 \in \mathbb{R}^n$ and that we write

x_0 for $P_C(z_0)$. Let K be the critical cone $K_C(x_0, G(x_0))$. Then G_C is a local homeomorphism at $z_0 \in \mathbb{R}^n$ if the normal map $dG(x_0)_K$, formed from the derivative $dG(x_0)$ of G at x_0 and the critical cone K, is coherently oriented (this means that the determinant of the affine map representing $dG(x_0)_K$ in each cell of the normal manifold of K has the same nonzero sign). For a proof, see Theorem 3 of [19], using there the simple perturbation structure $f(x, y) = f(x) - y$ to derive an inverse-function result that yields the desired homeomorphism property.

Now let us apply this knowledge to the example given above. We suppose for the moment that we do not know about the bad properties of the function F constructed in (3). We shall try to determine, using the point z_0 and the function values $G_C(z_0)$ as given in Tables 2 and 3, whether the normal map we are using is in fact a local homeomorphism at the candidate solution z_0. If it is, then we might feel comfortable appealing to the argument used above for the conventional version of Newton's method: namely, that even though the normal map may not be quite equal to zero at the candidate solution, it is reasonable to suppose that very near that point is an actual zero of the map, so that we can use z_0 as an approximation to the solution we are seeking. On the other hand, if we cannot verify that G_C is a local homeomorphism at z_0 then we should certainly be on our guard. As we are solving a mixed complementarity problem, we can take advantage of the special structure of C in computing the critical cone and in verifying coherent orientation.

We note that our set C is the product of five intervals of the form $[y_i, +\infty)$, where the y_i are given in Table 1. Moreover, for $i = 1, 2, 4, 5$ we see that $(z_0)_i > y_i$, whereas $(z_0)_3 = y_3$. It follows that $z_0 \in C$ and hence $z_0 = x_0$ in this case, and that the critical cone K is

$$K = \mathbb{R}^2 \times \mathbb{R}_+ \times \mathbb{R}^2.$$

The normal manifold associated with this K has just two cells, given by

$$\sigma_1 = \mathbb{R}^2 \times \mathbb{R}_+ \times \mathbb{R}^2, \qquad \sigma_2 = \mathbb{R}^2 \times \mathbb{R}_- \times \mathbb{R}^2.$$

In σ_1 the matrix of the affine map representing $dF(x_0)_K$ is just $dF(x_0)$, whereas in σ_2 it is the matrix obtained from $dF(x_0)$ by replacing the third column by the third column of the 5-dimensional identity matrix. Consulting the definition of F and remembering that $dF(x_0)$ is diagonal, we see that coherent orientation holds if and only if $df(p_3, 0) > 0$. Using the data in Table 1 we find that $f(p_3, x) = -42x^2 - 1.5x - 10^{-5}$, so $df(p_3, x) = -84x - 1.5$. This function is negative at the origin, so we see that coherent orientation fails. This should serve as a danger signal comparable to finding a singular Jacobian in the smooth case, and should motivate us to investigate much more closely the apparently nice solution shown in Tables 2 and 3.

We have now found two ways in which one might guard against false numerical convergence: the first is to use good professional software, and the second is to check the orientation of the normal map $dF(x_0)_K$. We demonstrated how to carry out the latter procedure for the case of a mixed complementarity problem, and we showed how in our example it could alert us to the possibility of a badly behaved problem.

Finally, we pointed out that if coherent orientation holds then this guarantees that the normal map being used is a local homeomorphism at the solution, so small errors in computation are less likely to be disastrous. On the other hand, if coherent orientation fails then this should indicate that we may be dealing with a very bad function, so that we should check our alleged solution much more closely. We said at the beginning of this section that neither of these methods would be a panacea, but they can help to keep users from being deceived by false numerical convergence.

Acknowledgments This material is based upon work supported by the U. S. Army Research Laboratory and the U. S. Army Research Office under grant number DAAG55-97-1-0324.

References

[1] A. Brooke, D. Kendrick and A. Meeraus, *GAMS, A User's Guide.* The Scientific Press, South San Francisco, CA, USA, 1988.

[2] F.H. Clarke, *Optimization and Nonsmooth Analysis.* Wiley, New York, 1983.

[3] S. Dirkse and M.C. Ferris, The PATH solver: a non-monotone stabilization scheme for mixed complementarity problems, *Optimization Methods and Software*, 5:123–156, 1995.

[4] W.B. Gragg and R.A. Tapia, Optimal error bounds for the Newton-Kantorovich theorem. *SIAM Journal on Numerical Analysis*, 11:10–13, 1974.

[5] R.I. Jennrich and S.M. Robinson, A Newton-Raphson algorithm for maximum-likelihood factor analysis, *Psychometrika*, 34:111–123, 1969.

[6] N.H. Josephy, Newton's method for generalized equations. Technical Summary Report 1965, Mathematics Research Center, University of Wisconsin-Madison, Madison, WI, USA, 1979a.

[7] N.H. Josephy, *Newton's Method for Generalized Equations and the PIES Energy Model*, Ph.D. Dissertation, Department of Industrial Engineering, University of Wisconsin–Madison, Madison, WI, 1979b.

[8] J.L. Kelley, *Geneeral Topology.* Graduate Texts in Mathematics, Springer-Verlag, New York. Originally published 1955 by Van Nostrand, 1975.

[9] M. Kojima and S. Shindo, Extension of Newton and quasi-Newton methods to systems of PC^1 equations, *Journal of the Operations Research Society of Japan*, 29:352–374, 1986.

[10] K. Nickel and K. Ritter, Termination criterion and numerical convergence. Technical Summary Report 1068, Mathematics Research Center, University of Wisconsin, Madison, WI 53706, USA, 1970.

[11] J.M. Ortega and W.C. Rheinboldt, *Iterative Solution of Nonlinear Equations in Several Variables.* Academic Press, New York, 1970.

[12] L. Qi and J. Sun, A nonsmooth version of Newton's method. *Mathematical Programming*, 58:353–367, 1993.

[13] D. Ralph, Global convergence of damped Newton method for nonsmooth equations, via the path search, *Mathematics of Operations Research*, 19:352–389, 1994.

[14] C.R. Rao, Estimation and tests of significance in factor analysis, *Psychometrika*, 20:93–111, 1955.

[15] S.M. Robinson, Generalized equations, In Bachem, A., Grötschel, M., and Korte, B., editors, *Mathematical Programming: The State of the Art, Bonn 1982*, Berlin, Springer-Verlag, 1983.

[16] S.M. Robinson, An implicit-function theorem for a class of nonsmooth functions, *Mathematics of Operations Research*, 16:292–309, 1991.

[17] S.M. Robinson, Normal maps induced by linear transformations. *Mathematics of Operations Research*, 17:691–714, 1992.

[18] S.M. Robinson, Newton's method for a class of nonsmooth functions. *Set-Valued Analysis*, 2:291–305, 1994.

[19] S.M. Robinson, Sensitivity analysis of variational inequalities by normal-map techniques, In Giannessi, F. and Maugeri, A., editors, *Variational Inequalities and Network Equilibrium Problems*, pages 257–269, New York, Plenum Press, 1995.

DISTANCE TO THE SOLUTION SET OF AN INEQUALITY WITH AN INCREASING FUNCTION

Alex M. Rubinov
SITMS, University of Ballarat, Australia

Abstract We show that the distance (in max-norm) from a point x^0 to the solution set of the inequality $\{x : f(x) \leq c\}$, with an increasing function f defined on either $\mathbb{R}^n_+$ or $\mathbb{R}^n$, can be expressed as the upper envelope of a certain family of functions $\psi_J(x^0, y) = \min_{i \in J(y)} (x_i^0 - y_i)^+$, where $J(y) \subset \{1, \ldots, n\}$.

1 Introduction

Evaluation of the distance from a point to a set arises in many areas. This problem is very important from the point of view of metric regularity and its applications (see [4]). Estimation of error bounds requires to evaluate the distance from a point to the set of solutions of the system of inequalities $f_j(x) \leq 0$, $(j \in J)$ where J is an index set. This question has been studied in detail for convex functions f_j (see for example, [1]). Convexity of f_j allows one to apply some results and methods from convex analysis. Methods of monotonic analysis (see, for example,[9]) have been used in [8] for the calculation of the distance with respect to the norm $\|\cdot\|_\infty$ to the solution set of the inequality

$$f(x) \leq c, \qquad x \in \mathbb{R}^n_+, \tag{1}$$

where f is a so-called increasing convex-along-rays (ICAR) function defined on the cone $\mathbb{R}^n_+$ of all n-vectors with nonnegative coordinates; see Section 4 for the definition of ICAR functions. (Unfortunately there was an error in calculation in [8]; here we present a corrected result.) For a convex function $f : \mathbb{R}^n \to \mathbb{R}$ the solution set of the inequality $f(x) \leq 0$ is convex; for an increasing function $f : \mathbb{R}^n_+ \to \mathbb{R}$ the solution set of this inequality is *normal*. (The set $S \subset \mathbb{R}^n_+$ is called normal if $(x \in S, x' \in \mathbb{R}^n_+, x' \leq x) \implies x' \in S$.) The following two results play the main role in the study of the distance to the solution set of the inequality (1) where f is an ICAR function:

Theorem 1.1 *([10], see also [9]) A function $f : \mathbb{R}^n_+ \to \mathbb{R}$ is ICAR if and only if there exists a set $U \subset (\mathbb{R}^n_+ \setminus \{0\}) \times \mathbb{R}_+$ such that* $f(x) = \max_{(l,c) \in U} \min_{i : l_i > 0} (l_i x_i - c)$.

Theorem 1.2 *([8], see also [9]) Let* $\mathbb{R}^n$ *be equipped with max-norm, that is* $\|x\| = \max_i |x_i|$ *for all* $x \in X$*. Let* Γ *be a set of indices and let* $(G_\gamma)_{\gamma\in\Gamma}$ *be a totality of normal closed sets. Let* $G = \bigcap_{\gamma\in\Gamma} G_\gamma$*. Then*

$$\operatorname{dist}(x, G) = \sup_{\gamma\in\Gamma} \operatorname{dist}(x, G_\gamma), \qquad (x \in \mathbb{R}^n_+).$$

In the first part of the present paper we consider the inequality (1) with an increasing function f defined on $\mathbb{R}^n_+$. Applying Theorem 1.2, we express the distance to the solution set $S(c)$ of this inequality through the weak Pareto boundary of $S(c)$. In particular, if the function f is strictly increasing, this boundary coincides with the set $\{x \in \mathbb{R}^n_+ : f(x) = c\}$. We also examine the system

$$f_j(x) \le c, \ (j \in J), \qquad x \in \mathbb{R}^n_+ \tag{2}$$

where J is a finite index set and f_j $(j \in J)$ is an increasing function. In order to compute the distance to the solution set of this system we should replace (2) by the single inequality $f(x) \le 0$, where $f(x) = \max_{j\in J} f_j(x)$. In the case of ICAR functions we show that the technique developed in [8] allows one to express the distance to the solution set of (2) in terms of the function f_j itself, without exploiting the maximum function f.

In the second part of the paper we study the inequality

$$f(x) \le c, \qquad (x \in \mathbb{R}^n)) \tag{3}$$

where f is an increasing function defined on the entire space $\mathbb{R}^n$. The solution set $T(c)$ of (3) is downward. (The set $T \subset \mathbb{R}^n$ is called downward if $(x \in T, x' \le x) \implies x' \in T$.) We express the distance to the solution set $T(c)$ through the weak Pareto boundary of $T(c)$, however, the definition of weak Pareto points for a downward set is different from that for normal sets.

We show that for both inequalities (1) and (3) the distance to the solution sets can be expressed in terms of the function $\psi_J(x,y) = \min_{i\in J}(x_i - y_i)^+$, where $J \subset I = \{1,\dots,n\}$. However there is a different choice J for (1) and (3). We have $J = I$ for inequality (3) and $J = \{i \in I : y_i > 0\}$, hence J depends on y, for inequality (1).

2 Preliminaries

Let n be a positive integer and $I = \{1,\dots,n\}$. Consider the space $\mathbb{R}^n$ of all vectors $(x_i)_{i\in I}$. We shall use the following notations:

- if $x \in \mathbb{R}^n$, then x_i is the i-th coordinate of x;

- if $x, y \in \mathbb{R}^n$ then $x \geq y \iff x_i \geq y_i$ for all $i \in I$;
- if $x, y \in \mathbb{R}^n$ then $x > y \iff x \geq y \ \& \ x \neq y$;
- if $x, y \in \mathbb{R}^n$ then $x \gg y \iff x_i > y_i$ for all $i \in I$;
- $\mathbb{R}^n_+ = \{x = (x_i)_{i \in I} \in \mathbb{R}^n : x_i \geq 0 \text{ for all } i \in I\}$;
- $x^+ = \max(x, 0)$ (in other words, $x_i^+ = \max(x_i, 0)$ for all $i \in I$).

We assume that $\mathbb{R}^n$ is equipped with max-norm: $\|x\| = \max_{i \in I} |x_i|$ ($x \in \mathbb{R}^n$).

In the first part of this paper we consider some functions defined on the cone $\mathbb{R}^n_+$ and some subsets of this cone. Assume that $\mathbb{R}^n_+$ is equipped with the following non-symmetric coupling function:

$$\langle l, x \rangle = \begin{cases} \min_{i \in I(l)} l_i x_i, & l \in \mathbb{R}^n_+ \setminus \{0\}, x \in \mathbb{R}^n_+ \\ 0 & l = 0, x \in \mathbb{R}^n_+, \end{cases}$$

where

$$I(l) = \{i \in I : l_i > 0\}. \tag{4}$$

Definition 2.1 A subset G of the cone $\mathbb{R}^n_+$ is called normal if $(x \in G, x' \in \mathbb{R}^n_+, x' \leq x) \implies x' \in G$.

Definition 2.2 Let G be a normal set. A point $x \in G$ is called a weak Pareto point (briefly w.P. point) of G, if $\lambda x \notin G$ for all $\lambda > 1$. The totality of all w.P. points of G is called the weak Pareto boundary of G and denoted by $\operatorname{wb} G$.

Remark 2.1 This definition is slightly different from the definition of a weak Pareto point, which is used in multi-objective optimization. Recall that a point x of a set $X \subset \mathbb{R}^n$ is called a weak Pareto point of X in the sense of multi-objective optimization, if $x + \operatorname{int} \mathbb{R}^n_+ \cap X = \emptyset$. Let $G \subset \mathbb{R}^n_+$ be a normal set. It is easy to check that each weak Pareto point in the sense of Definition 2.2 is a weak Pareto point in the sense of multi-objective optimization . For points $x \in G \cap \operatorname{int} \mathbb{R}^n_+$ both definitions of weak Pareto points coincides. However, this is not longer true if x is a boundary point of G. The following example confirms this assertion:

Example 2.1 Let $G \subset \mathbb{R}^2$ be the union of two segments $G = G_1 \cup G_2$ where G_1 is the convex hull of points $(0, 1)$ and $(0, 0)$ and G_2 is the convex hull of points $(1, 0)$ and $(0, 0)$. It is easy to check that G is a normal set and this set has only two weak Pareto points in the sense of Definition 2.2: $(1, 0)$ and $(0, 1)$. At the same time the set of weak Pareto points in the sense of multi-objective optimization coincides with G.

For a vector $l \in \mathbb{R}^n_+$ consider the vector l^{-1} with coordinates

$$l_i^{-1} = \begin{cases} \frac{1}{l_i} & i \in I(l); \\ 0 & i \notin I(l). \end{cases}$$

We need the following results which were established in [7, 8].

Proposition 2.1 *(see the proof of Proposition 2.2 in [7]). Let G be a normal subset of $\mathbb{R}^n_+$ and $x \notin G$. Then $\langle x^{-1}, g \rangle < 1$ for each $g \in G$.*

Proposition 2.2 *([7]) Let $l \in \mathbb{R}^n_+$ and $S := \{x \in \mathbb{R}^n_+ : \langle l, x \rangle \leq 1\}$. Then for each $x^0 \in \mathbb{R}^n_+$ we have*

$$\text{dist}\,(x^0, S) = \min_{i \in I(l)} \left(x_i^0 - \frac{1}{l_i} \right)^+ .$$

where $a^+ = \max(a, 0)$.

We shall express the distance from a point to a set in terms of the following functions ψ_J, $(J \subset I)$:

$$\psi_J(x, y) = \min_{i \in J} (x_i - y_i)^+ . \tag{5}$$

Let x^0 and S be the point and the set as in Proposition 2.2. We can present the conclusion of this proposition in the following form

$$\text{dist}\,(x^0, S) = \psi_{I(l)}(x^0, l^{-1}).$$

3 Distance to the solution set of the inequality with an arbitrary increasing function

We say that a certain property (P) of a function $f : \mathbb{R}^n_+ \to \mathbb{R}$ holds along-rays if the restriction of f to the ray $R_x = \{\lambda x : \lambda \geq 0\}$ starting from zero and passing through x enjoys property (P) for each $x \in \mathbb{R}^n_+$. In other words, (P) holds along-rays if the function of one variable f_x defined by

$$f_x(\alpha) = f(\alpha x), \qquad (\alpha \geq 0) \tag{6}$$

possesses the property (P). In particular, we consider the following classes of functions:

1) lower semicontinuous-along-rays functions f: the function f_x is lower semicontinuous for all $x \in \mathbb{R}^n_+$;

2) continuous along rays functions f: the function f_x is continuous for all $x \in \mathbb{R}^n_+$;

3) convex-along-rays functions f: the function f_x is convex for all $x \in \mathbb{R}^n_+$;

4) positively homogeneous functions. A function p is called positively homogeneous (of degree one) if $p(\lambda x) = \lambda p(x)$ for all $\lambda \geq 0$ (that is, p linear-along-rays);

A function $f : \mathbb{R}^n_+ \to \mathbb{R}$ is called increasing if $x \geq y$ implies $f(x) \geq f(y)$. In this paper we shall study increasing functions with a certain property-along-rays. First, we show that the lower semicontinuity-along-rays implies the lower semicontinuity for increasing functions. (Note that the class of increasing continuous-along-rays functions is broader than the class of increasing continuous functions.) It follows directly from the definition that level sets $S(c) = \{x \in \mathbb{R}^n_+ : f(x) \leq c\}$ of an increasing function f are normal for all $c \geq f(0)$. A normal set G is called closed-along-rays if

$$\lambda_n > 0, \ \lambda_n g \in G \ (n = 1, 2, \ldots) \text{ and } \lambda_n \to \lambda > 0 \implies \lambda g \in G.$$

It is known ([9], Remark 2.16) that a normal closed-along-rays set is closed.

Proposition 3.1 *Let f be an increasing lower semicontinuous along-rays function. Then f is lower semicontinuous.*

Proof: We need to prove that all the level sets $S(c)$ of the function f with $c \geq f(0)$ are closed. Let $c \geq f(0)$. Let $\lambda_n > 0$ and $\lambda_n \to \lambda > 0$, let also $\lambda_n x \in S(c)$, that is $f(\lambda_n x) \leq c$. Since f is lower semicontinuous along-rays, it follows that $f(\lambda x) \leq \liminf_n f(\lambda_n x) \leq c$, so $\lambda x \in S(c)$. Thus the normal set $S(c)$ is closed-along-rays, hence closed. □

We now study the solution set of the inequality $f(x) \leq c$ with an increasing continuous-along rays function.

Proposition 3.2 *Let f be an increasing continuous-along rays function and $S(c) = \{x : f(x) \leq c\}$ with $c \geq f(0)$ be the level set of the function f. Then*

$$S(c) = \bigcap \{S_y : y \in wb\, S(c)\} \tag{7}$$

where

$$S_y := \{x \in \mathbb{R}^n_+ : \langle y^{-1}, x\rangle \leq 1\} \tag{8}$$

and $wb\, S(c)$ is the weak Pareto boundary (in the sense of Definition 2.2) of the normal set $S(c)$. It is assumed here that the intersection over the empty set coincides with $\mathbb{R}^n_+$.

Proof: First we assume that $\text{wb}\, S(c)$ is empty. Let $y \in \mathbb{R}^n_+ \setminus \{0\}$. Note that the set $A_y = \{\alpha : \alpha y \in S(c)\}$ is nonempty, since $0 \in A_y$. Let $\alpha \in A_y$ and $0 \leq \alpha' < \alpha$. Then $\alpha' y < \alpha y$, so $f(\alpha' y) \leq f(\alpha y) \leq c$. We have $\alpha' \in A_y$. Thus A_y is a segment. Let $a = \sup A_y$. If $a < +\infty$, then continuity-along-rays implies $ay \in S(c)$. At the same time $\alpha y \notin S(c)$ for all $\alpha > a$, so ay is a w.P.point of $S(c)$. Since by our assumption the set $S(c)$ has no w.P. points, it follows that $a = +\infty$. Hence, $A_y = \mathbb{R}_+$, so the set $S(c)$ contains the entire ray $R_y = \{\alpha y : \alpha \geq 0\}$. Since y is an arbitrary nonzero element of $\mathbb{R}^n_+$, it follows that $S(c) = \mathbb{R}^n_+$. Thus (7) holds in the case under consideration.

Now let us assume that the set $\text{wb}\, S(c)$ is nonempty and let $y \in \text{wb}\, S(c)$. Then $\lambda y \notin S(c)$ for all $\lambda > 1$. It follows from Proposition 2.1 that

$$\langle (\lambda y)^{-1}, x\rangle < 1 \text{ for all } x \in S(c) \text{ and } \lambda > 1,$$

so

$$\langle y^{-1}, x\rangle < \lambda \text{ for all } x \in S(c) \text{ and all } \lambda > 1.$$

Hence $\langle y^{-1}, x\rangle \leq 1$ for all $x \in S(c)$; in other words $x \in S_y$ for an arbitrary $y \in \text{wb}\, S(c)$. We have proved that $S(c) \subset \bigcap\{S_y(c) : y \in \text{wb}\, S(c)\}$. Assume now that $x \notin S(c)$, that is $f(x) > c$. Consider the function of one variable f_x defined by (6). This function is continuous and $f_x(1) = f(x) > c$, $f_x(0) = f(0) \leq c$, so the set $\{\alpha : \alpha x = c\}$ is nonempty. Since f_x is increasing and continuous it follows that this set coincides with a closed segment, which we denote by $[\mu', \mu]$. Note that $\mu < 1$. Indeed, since $x \notin S(c)$ and f is increasing it follows that $\nu x \notin S(c)$ for all $\nu \geq 1$. Let $\bar{y} = \mu x$. Due to the definition of μ, we have $\bar{y} \in S(c)$. ¿From this definition, it also follows that $\lambda \bar{y} \notin S(c)$ for arbitrary $\lambda > 1$, so $\bar{y} \in \text{wb}\, S(c)$. We have $\langle \bar{y}^{-1}, x\rangle = 1/\mu > 1$, hence $x \notin S_{\bar{y}}$ and $x \notin \bigcap\{S_y : y \in \text{wb}\, S(c)\}$. We have proved that $\bigcap\{S_y : y \in \text{wb}\, S(c)\} \subset S(c)$. □

Theorem 3.1 *Let $f : \mathbb{R}^n_+ \to \mathbb{R}$ be an increasing continuous-along-rays function and $c \geq f(0)$. Let $S(c) = \{x : f(x) \leq c\}$ be the level set of this function, which is also the solution set of the inequality $f(x) \leq c$. Then for each $x^0 \in \mathbb{R}^n_+ \setminus S(c)$ there holds*

$$\text{dist}\,(x^0, S(c)) = \sup\{\psi_{I(y)}(x^0, y) : y \in wb\, S(c)\}, \tag{9}$$

where ψ_J is the function defined by (5), so $\psi_{I(y)}(x^0, y) = \min_{i \in I(y)}(x_i^0 - y_i)^+$.

Proof: It follows from Proposition 3.2 and Theorem 1.2 that

$$\text{dist}\,(x^0, S(c)) = \sup\{\text{dist}\,(x_0, S_y) : y \in \text{wb}\, S(c)\},$$

where S_y defined by (8). Since $I(y) = I(y^{-1})$ we have, due to Proposition 2.2,

$$\text{dist}\,(x^0, S_y) = \min_{i \in I(y)} \left(x_i^0 - \frac{1}{(y^{-1})_i} \right)^+ = \min_{i \in I(y)} (x_i^0 - y_i)^+.$$

Thus

$$\text{dist}\,(x^0, S) = \sup\{ \min_{i \in I(y)} (x_i^0 - y_i)^+ : y \in \text{wb}\, S(c)\} =$$

$$= \sup\{\psi_{I(y)}(x^0, y) : y \in \text{wb}\, S(c)\}.$$

□

Proposition 3.3 *Let f be a continuous-along-rays and strictly increasing-along rays function. (The latter means that $f(\alpha x) < f(\beta x)$ for each $x \in \mathbb{R}^n \setminus \{0\}$ and $\alpha < \beta$.) Then*

$$\text{dist}\,(x^0, S(c)) = \sup_{f(y)=c} \psi_{I(y)}(x^0, y)).$$

Proof: It is sufficient to show that

$$\text{wb}\, S(c) = \{x \in \mathbb{R}_+^n : f(x) = c\}. \tag{10}$$

and make use of Theorem 3.1. We now check that (10) holds. Let $x \in \text{wb}\, S(c)$, then $f(\lambda x) > c$ for all $\lambda > 1$ so, due to continuity-along-rays, $f(x) \geq c$. On the other hand, $f(x) \leq c$, since $x \in S(c)$. Thus $f(x) = c$. Assume now that $f(x) = c$. Since f is strictly increasing-along-rays, it follows that $f(\lambda x) > f(x)$ for all $\lambda > 1$ so $\lambda x \notin S(c)$. Thus $x \in \text{wb}\, S(c)$. □

4 Distance to the solution set of the inequality with an ICAR function

A function $f : \mathbb{R}_+^n \to \mathbb{R}$ is called increasing convex-along-rays (briefly ICAR) if f is increasing and the function f_x defined by (6) is convex for each $x \in \mathbb{R}_+^n$. Since f is increasing it follows that $f_x(0) \leq f_x(\alpha)$ for all $\alpha > 0$, so the convex function f_x is continuous on $[0, +\infty)$. Thus, an ICAR function is increasing continuous-along-rays, so all results of the previous section are valid for ICAR function. In this section we present an expression for the distance to the solution set $S(c)$ of the inequality $f(x) \leq c$ with an ICAR function f, where we do not exploit the structure of the level set $S(c)$.

Let $f : \mathbb{R}^n_+ \to \mathbb{R}$ be an ICAR function and $y \in \mathbb{R}^n_+, y \neq 0$. Since the function f_y is convex, it follows that there exists the directional derivative

$$f'(y,y) := \lim_{\alpha \to +0} \frac{1}{\alpha}(f(y+\alpha y) - f(y))$$

and this derivative is equal to the right derivative $(f_y)'_+(1)$ of the function f_y at the point $\alpha = 1$. Since f is increasing it follows that $f'(y,y) \geq 0$ for all $y \in \mathbb{R}^n_+$. In the sequel we shall use the excess function E_f of the function f (see [9]). By definition,

$$E_f(y) = f(y) - f'(y,y), \qquad y \in \mathbb{R}^n_+.$$

Consider an ICAR function f such that $f(0) \leq 0$. Then the solution set $S(0)$ of the inequality $f(x) \leq 0$ is not empty The following result holds ([8], Lemma 4.1).

Lemma 4.1 *Let $f : \mathbb{R}^n_+ \to \mathbb{R}$ be an ICAR function, such that $f(0) \leq 0$. Then*
a) $E_f(y) \leq 0$ for all $y \in \mathbb{R}^n_+$, $y \neq 0$;
b) for each $x^0 \in \mathbb{R}^n_+ \setminus S(0)$ there holds

$$\text{dist}\,(x^0, S(0)) = \sup_{f'(y,y)>0} \text{dist}\,(x^0, P^y),$$

where

$$P^y := \{x \in \mathbb{R}^n_+ : \langle l^y, x\rangle \leq -E_f(y)\} \tag{11}$$

and $l^y := f'(y,y)/y$.

Due to item $b)$ of Lemma 4.1, we shall consider only points y such that $f'(y,y) > 0$. Let

$$D(f) = \{y \in \mathbb{R}^n_+ : f'(y,y) > 0\}. \tag{12}$$

It follows from the item $a)$ of Lemma 4.1 that $f'(y,y) - f(y) \geq 0$ for all $y \in \mathbb{R}^n_+ \setminus \{0\}$. We consider points y such that $f'(y,y) > f(y)$ (hence $f'(y,y) > f(y)^+$) and points y such that $f'(y,y) = f(y) > 0$ separately. Let

$$\begin{aligned} D_1(f) &= \{y \in \mathbb{R}^n_+ : f'(y,y) > f(y)^+\}, \\ D_2(f) &= \{y \in \mathbb{R}^n_+ : f'(y,y) = f(y) > 0\} \end{aligned} \tag{13}$$

Clearly $D_1(f) \cap D_2(f) = \emptyset$. Due to Lemma 4.1 we have $D(f) = D_1(f) \cup D_2(f)$.

Applying Lemma 4.1 and Proposition 2.2 we can establish the following result:

Proposition 4.1 *Let $f : \mathbb{R}^n_+ \to \mathbb{R}$ be an ICAR function such that $f(0) \le 0$ and $S(0)$ be the solution set of the inequality $f(x) \le 0$. Then for each $x^0 \in \mathbb{R}^n_+ \setminus S(0)$ there holds:*

$$\mathrm{dist}\,(x^0, S(0)) = \max\left(\sup_{y\in D_1(f)} \psi_{I(y)}\left(x^0, \frac{-E_f(y)}{f'(y,y)}y\right), \max_{y\in D_2(f)} \psi_{I(y)}(x^0,0)\right).$$

Remark 4.1 Note that $\psi_{I(y)}(x^0,0) = \min_{i\in I(y)} x^0_i$.

Actually Proposition 4.1 has been proved in [8]. Unfortunately a wrong version of Proposition 2.2 has been used in [8], so the result there was not correct. However the scheme of the proof presented in [8] is completely true and can be used for the proof of Proposition 4.1. For the sake of completeness we present a scheme of this proof:
Scheme of a proof of Proposition 4.1: Since $D(f) = D_1(f) \cup D_2(f)$ we have, by applying Lemma 4.1:

$$\begin{aligned}\mathrm{dist}\,(x^0, S(0)) &= \sup_{y\in D(f)} \mathrm{dist}\,(x^0, P^y)\\ &= \max(\sup_{y\in D_1(f)} \mathrm{dist}\,(x^0, P^y), \sup_{y\in D_2(f)} \mathrm{dist}\,(x^0, P^y)).\end{aligned} \tag{14}$$

The second supremum in (14) was calculated in [8], Theorem 4.1, (see the second part of the proof of this theorem)). It was shown there that

$$\sup_{y\in D_2(f)} \mathrm{dist}\,(x^0, P^y) = \max_{y\in D_2(f)} \min_{i\in I(y)} x^0_i. \tag{15}$$

Let us calculate $\mathrm{dist}\,(x^0, P^y)$ for $y \in D_1(f)$. Since $E_f(y) < 0$ for such y, we have, by applying (11) and the equality $l^y = f'(y,y)/y$

$$\begin{aligned}P^y &= \{x : \langle l^y, x\rangle \le -E_f(y)\} = \left\{x : \left\langle \frac{l^y}{-E_f(y)}, x\right\rangle \le 1\right\}\\ &= \left\{x : \left\langle \frac{f'(y,y)}{-E_f(y)y}, x\right\rangle \le 1\right\}.\end{aligned}$$

Keeping in mind that $I(l^y) = I(y)$ and applying Proposition 2.2, we conclude that

$$\mathrm{dist}\,(x^0, P^y) = \min_{i\in I(y)} \left(x^0_i + \frac{E_f(y)}{f'(y,y)}y_i\right)^+ = \psi_{I(y)}\left(x^0, \frac{-E_f(y)}{f'(y,y)}y\right). \tag{16}$$

The desired result follows from (14), (16) and (17). □

Remark 4.2 Let $\mathcal{I}_2(f) = \{I(y) : y \in D_2(f)\}$. Note that $\mathcal{I}_2(f)$ is a finite set. Then

$$\sup_{y \in D_2(f)} \operatorname{dist}(x^0, P^y) = \max_{I \in \mathcal{I}_2(f)} \min_{i \in I} x_i^0, \tag{17}$$

so

$$\operatorname{dist}(x^0, S(0)) = \max\left(\sup_{y \in D_1(f)} \psi_{I(y)}\left(x^0, \frac{-E_f(y)}{f'(y,y)}y\right), \max_{I \in \mathcal{I}_2(f)} \psi_I(x^0, 0)\right).$$

We now describe the distance for the solution set of the inequality $f(x) \le c$, where f is an ICAR function and $c \ge f(0)$.

Theorem 4.1 *Let $f : \mathbb{R}^n_+ \to \mathbb{R}$ be an ICAR function and $c \ge f(0)$. Let $x^0 \in \mathbb{R}^n_+ \setminus S(c)$, where $S(c)$ is the solution set of the inequality $f(x) \le c$. Then*

$$\operatorname{dist}(x^0, S(c)) = \max\left(\sup_{D_1^c(f)} \psi_{I(y)}\left(x^0, \frac{c - E_f(y)}{f'(y,y)}y\right), \max_{I \in \mathcal{I}_2^c(f)} \min_{i \in I} x_i^0\right),$$

where $D_1^c(f) = \{y : f'(y,y) > (f(y) - c)^+\}$, $D_2^c(f) = \{y : f'(y,y) = f(y) - c > 0\}$, $\mathcal{I}_2^c(f) = \{I(y) : y \in D_2^c(f)\}$.

Proof: Consider the function $f_c(x) = f(x) - c$. Then the set $S(c)$ coincides with the set $\{x : f_c(x) \le 0\}$. Since $(f_c)'(y,y) = f'(y,y)$ it follows that $E_{f_c}(y) = E_f(y) - c$, so the desired result follows directly from Remark 4.2. □

Now let us consider an increasing positively homogeneous of degree one (briefly, IPH) function. Clearly each IPH function is ICAR. Let f be an IPH function. It is well known (and easy to check) that $f'(y,y) = f(y)$, so $E_f(y) = 0$ for all $y \in \mathbb{R}^n_+$. Let $c > 0$. Then the relation $f'(y,y) > (f(y) - c)^+$ is equivalent to $f(y) > 0$. The equality $f'(y,y) = f(y) - c$ does not hold for all $y \in \mathbb{R}^n_+$, so the set $\mathcal{I}_2^c(f)$ is empty. It follows from Remark 4.2 that

$$\operatorname{dist}(x_0, S(c)) = \sup_{f(y) > 0} \psi_{I(y)}\left(x^0, \frac{c}{f(y)}y\right). \tag{18}$$

Since f is positively homogeneous, we can present (18) in the following form:

$$\operatorname{dist}(x^0, S(c)) = \sup_{f(y) = c} \psi_{I(y)}(x^0, y) \tag{19}$$

This result was established in [8].

Remark 4.3 Note that (19) follows also from Proposition 3.3.

We now calculate the distance to the solution set of a system

$$f_j(x) \leq 0, \qquad (j \in J) \tag{20}$$

where J is a finite set of indices and $f_j (j \in J)$ is an ICAR function. For each $x \in \mathbb{R}^n_+$ consider sets:

$$R(x) = \{k \in J : f_k(x) = \max_{j \in J} f_j(x)\},$$

$$Q(x) = \{q \in R(x) : f'_q(x,x) = \max_{j \in R(x)} f'_j(x,x)\}.$$

Theorem 4.2 *Let J be a finite set of indices and $f_j : \mathbb{R}^n_+ \to \mathbb{R}$ $(j \in J)$ be an ICAR function. Assume that $f_j(0) \leq 0$ for all $j \in J$ and let S be a solution set of the inequality (20). Then for each $x^0 \in \mathbb{R}^n_+ \setminus S(0)$ there holds*

$$\text{dist}\,(x^0, S) = \max\left(\sup_{y \in D_1(f_q)} \psi_{I(y)}\left(x^0, \frac{-E_{f_q}(y)}{f'_q(y,y)} y\right), \max_{I \in \mathcal{I}_2(f_q)} \psi_I(x^0, 0)\right),$$

where q is an arbitrary index from the set $Q(y)$.

Proof: Let $f(x) = \max_{j \in J} f_(x)$. Clearly the system (20) is equivalent to the inequality $f(x) \leq 0$, so

$$\text{dist}\,(x^0, S(0)) = \max\left(\sup_{y \in D_1(f)} \psi_{I(y)}\left(x^0, \frac{-E_f(y)}{f'(y,y)} y\right), \max_{I \in \mathcal{I}_2(f)} \psi_i(x^0, 0)\right). \tag{21}$$

It is well-known (see, for example, [3]) that $f'(y,y) = \max_{j \in R(x)} f'_j(y,y) = f_q(y,y)$. Since $q \in R(x)$, we have $f(y) = f_q(y)$ and $E_f(y) = E_{f_q}(y)$. We also have $D_1(f) = D_1(f_q)$ and $\mathcal{I}_2(f) = \mathcal{I}_2(f_q)$, so the desired result follows directly from (21). □

5 Inequalities with an increasing function defined on the entire space

In this section we consider the distance to the set

$$T(c) = \{x \in \mathbb{R}^n : f(x) \leq c\},$$

where f is an increasing function defined on $\mathbb{R}^n$, that is, $x \geq y$ implies $f(x) \geq f(y)$. It is assumed that $T(c)$ is nonempty. It is easy to check that the set $T(c)$ is *downward*, that is $x \in T(c), y \leq x \implies y \in T(c)$. We need some properties of downward sets.

Theorem 5.1 *([5], Theorem 4) Let Γ be a set of indices, $(A_\gamma)_{\gamma\in\Gamma}$ be a family of closed downward sets and the set $A := \bigcap_{\gamma\in\Gamma} A_\gamma$ is not empty. Then*

$$\text{dist}\,(x, A) = \sup_{\gamma\in\Gamma} \text{dist}\,(x, A_\gamma), \qquad (x \in \mathbb{R}^n).$$

Theorem 5.2 *([5], Proposition 4 and Theorem 1) Let $l \in \mathbb{R}^n$ and $T_l = \{x \in \mathbb{R}^n : \min(x_i - l_i) \le 0\}$, let further $x^0 \in \mathbb{R}^n$. Then*

$$\text{dist}\,(x^0, T_l) = \psi_I(x^0, l))$$

We also need the following simple assertion.

Proposition 5.1 *Let A be a downward set and $y \notin A$. Then $\min_{i\in I}(x_i - y_i) < 0$ for all $x \in A$.*

Proof: Assume that $\min_i(x_i - y_i) \ge 0$ for $x \in A$. Then $x \ge y$ and, since A is downward, it follows that $y \in A$, which contradicts our assumption. □

Let us denote by $\mathbf{1}$ the vector $(1, 1, \dots, 1)$.

Proposition 5.2 *Let A be a downward set. A point $x \in A$ is a boundary point of A if and only if $x + \lambda\mathbf{1} \notin A$ for all $\lambda > 0$.*

Proof: If $x+\lambda\mathbf{1} \notin A$ then x is a boundary point. Assume now that $x \in A$ is a point such that $x + \lambda\mathbf{1} \in A$ for some $\lambda > 0$. Since A is downward it follows that the ball $\{y : x - \lambda\mathbf{1} \le y \le x + \lambda\mathbf{1}\}$ is contained in A, hence x is not a boundary points. □

Remark 5.1 It is easy to check that the boundary of A coincides with the set of all the weak Pareto point in the sense of multi-objective optimization. (See Remark 2.1 for the definition of this set.)

The boundary of a downward set A will be denoted by $\text{bd}\, A$.

A function f is called *continuous along diagonal lines* if its restriction to each line of the form $(x + \lambda\mathbf{1})_{\lambda\in\mathbb{R}}$ is continuous. For $y \in \mathbb{R}^n$ consider the set

$$T_y = \{x \in \mathbb{R}^n : \min_{i\in I}(x_i - y_i) \le 0\} \tag{22}$$

We now prove the following assertion.

Proposition 5.3 *Let $f : \mathbb{R}^n \to \mathbb{R}$ be an increasing continuous along diagonal lines function and let the set $T(c) = \{x \in \mathbb{R}^n : f(x) \le c\}$ be nonempty. Then*

$$T(c) = \bigcap\{T_y : y \in bdT(c)\}, \tag{23}$$

where T_y is defined by (22). (It is assumed that the intersection over the empty set coincides with $\mathbb{R}^n$.)

Proof: First we assume that $\operatorname{bd} T(c)$ is empty. Let $y \in \mathbb{R}^n$. Then the set $B_y = \{\lambda \in \mathbb{R} : y + \lambda \mathbf{1} \in T(c)\}$ is nonempty. Indeed, for each $z \in T(c)$ there exists $\lambda < 0$ such that $y + \lambda \mathbf{1} \leq z$. Since $T(c)$ is downward it follows that $y + \lambda \mathbf{1} \in T(c)$. Clearly $(\lambda \in B_y, \lambda' \leq \lambda) \implies \lambda' \in B_y$. Thus B_y is a segment. Let $b = \sup B_y$. If $b < +\infty$, then $y + b\mathbf{1} \in T(c)$ (it follows from continuity along diagonal lines) and $y + \lambda \mathbf{1} \notin T(c)$, so $y + b\mathbf{1}$ is a weak Pareto point of $T(c)$, which contradicts our assumption. Thus $b = +\infty$. Since $T(c)$ is downward, it follows that $B_y = (-\infty, +\infty)$. We proved that $T(c)$ contains lines $\{y + \lambda \mathbf{1} : \lambda \in \mathbb{R}\}$ for all $y \in \mathbb{R}^n$, so $T(c) = \mathbb{R}^n$. Thus (23) is valid in the case under consideration.

Now let us assume that $\operatorname{bd} T(c)$ is nonempty and let $y \in \operatorname{bd} T(c)$. Then $y + \varepsilon \mathbf{1} \notin T(c)$ for all $\varepsilon > 0$. Due to Proposition 5.1, we have $\min_i (x_i - y_i - \varepsilon) < 0$ for all $x \in T(c)$ and $\varepsilon > 0$, so $\min_i(x_i - y_i) \leq 0$ for all $x \in T(c)$. Thus each $x \in T(c)$ belongs to the set T_y with arbitrary $y \in \operatorname{bd} T(c)$.

Let $x \notin T(c)$, that is, $f(x) > c$. Since f is increasing , it follows that $f(x + \alpha \mathbf{1}) > c$ for all $\alpha > 0$. Since $T(c)$ is nonempty and downward, it follows that there exists a number $\alpha < 0$ such that $f(x + \alpha \mathbf{1}) \leq c$. Since the function $x \mapsto f(x + \alpha \mathbf{1})$ is increasing and continuous, it follows that the set $\{\alpha : f(x + \alpha \mathbf{1})\} = c$ is nonempty and coincides either with a segment $[\nu, \mu]$ or with a segment $(-\infty, \mu]$ with $\mu < 0$. Let $\bar{y} = x + \mu \mathbf{1}$. It follows from the definition of μ that $\bar{y}$ is a weak Pareto point of $T(c)$. We have $\min_i(x_i - \bar{y}_i) = -\mu > 0$, hence $x \notin T_{\bar{y}}$ and $x \notin \bigcap \{T_y : y \in \operatorname{bd} T(c)\}$.
□

Theorem 5.3 *Let $f : \mathbb{R}^n \to \mathbb{R}$ be an increasing continuous along diagonal lines function and let the set $T(c) = \{x \in \mathbb{R}^n : f(x) \leq c)\}$ be nonempty. Let $x^0 \notin T(c)$. Then*

$$\operatorname{dist}(x^0, T(c)) = \sup_{y \in bd T(c)} \psi_I(x^0, y).$$

Proof: It follows from Proposition 5.3, Theorem 5.1 and Theorem 5.2 that

$$\operatorname{dist}(x^0, T(c)) = \sup_{y \in \operatorname{bd} T(c)} \operatorname{dist}(x^0, D_y) = \sup\{\psi_I(x^0, y) : \ y \in \operatorname{bd} T(c)\}.$$

□

6 Inequalities with a topical function

An increasing function $f : \mathbb{R}^n \to \mathbb{R}$ is called topical if $f(x + \lambda \mathbf{1}) = f(x) + \lambda$. (For properties and examples of topical functions see [2, 6].) The following assertion directly follows from results of the paper [6].

Theorem 6.1 *A function $f : \mathbb{R}^n \to \mathbb{R}$ is topical if and only*

$$f(x) = \max_{y \in \mathbb{R}^n} \min_{i \in I} (f(y) - y_i + x_i)$$

Corollary 6.1 *Let $z^y = f(y)\mathbf{1} - y$, $(y \in \mathbb{R}^n)$. Consider the set T_{z^y} defined by (22), that is*

$$T_{z^y} = \{x \in \mathbb{R}^n : \min_{i \in I}(f(y) - y_i + x_i) \leq 0\}. \tag{24}$$

Then $T(0) = \bigcap_{y \in \mathbb{R}^n} T_{z^y}$.

We now prove the following statement.

Theorem 6.2 . *Let f be a topical function such the set $T(0) = \{x \in \mathbb{R}^n : f(x) \leq 0\}$ is nonempty. Let $x^0 \notin T(0)$. Then*

$$\text{dist}\,(x^0, T(0)) = \sup_{y \in \mathbb{R}^n} \psi_I(f(x^0, f(y)\mathbf{1} - y).$$

Proof: Consider the set T_{z^y} defined by (24). Due to Theorem 5.2 we have

$$\text{dist}\,(x^0, T_{z^y}) = \min(x_i^0 - y_i + f(y))^+.$$

It follows from Corollary 6.1 and Theorem 5.1 that

$$\text{dist}\,(x^0, T(0)) = \sup_{y \in \mathbb{R}^n} \text{dist}\,(x^0, T_{z^y}) = \sup_{y \in \mathbb{R}^n} \min_{i \in I}(f(y) - y_i + x_i^0)^+.$$

□

Acknowledgement

The author very grateful to Prof. I. Singer and Dr. J. Dutta for valuable comments and to anonymous referee for for constructive comments on the first version of the paper.

References

[1] H. H. Bauschke and J. M. Borwein, On projection algorithms for solving convex feasibility problems. SIAM Rev. 38 (1996), 367-426.

[2] J. Gunawardena, An Introduction to idempotency. In: Idempotency (J. Gunawardena,ed.), Cambridge University Press, pp.1-49, Cambridge, 1998.

[3] V.F. Demyanov and A.M. Rubinov, Constructive Nonsmooth Analysis, Approximation and Optimization, No 7, Peter Lang, Frankfurt am Main, 1995.

[4] A.D. Ioffe, Metric regularity and subdifferential calculus, Uspehi Mat. Nauk 55:3 (2000), 103-162, (in Russian). English translation: Russian Math Survey 55:3 (2000).

[5] J-E. Martinez-Legaz, A.M. Rubinov and I. Singer, Downward sets and their separation and approximation properties, Research Report No 0024, University of Ballarat, 2000.

[6] A.M. Rubinov and I. Singer, Topical and sub-topical functions, downward sets and abstract convexity, Research Report, No 0023, University of Ballarat, 2000.

[7] A. M. Rubinov and Singer, Best approximation by normal and conormal sets, J. Approximation Theory, to appear

[8] A.M. Rubinov and I. Singer, Distance to the intersection of normal sets and applications, *Numerical Functional Analysis and Optimization*, **21** (2000), pp. 521-536.

[9] A.M. Rubinov, Abstract Convexity and Global Optimization, Kluwer Academic Publishers, Dorderecht, 2000.

[10] A.M. Rubinov and B. M. Glover, Increasing convex-along-rays functions with applications to global optimization, *Journal of Optimization Theory and Applications.* **102** (1999), 615-642.

TRANSPORTATION NETWORKS WITH CAPACITY CONSTRAINTS

Laura Scrimali*
Dipartimento di Matematica e Informatica - Università di Catania,
Viale A. Doria, 6 - 95125 CATANIA (ITALIA)
e-mail: scrimali@dmi.unict.it

Abstract In this paper we deal with transportation networks when some capacity constraints are introduced on route flows. Our purpose is to show how the equilibrium solution, according to a generalized form of Wardrop's principle (see [10, 11]), can still be found. We shall also provide an example of a network in which, paradoxically, the increase in the travel demands reduces the total costs.

1 Introduction

Our aim is to examine transportation networks in the presence of capacity constraints on route flows. The introduction of capacity constraints is meant to make models more realistic and reliable, but, generally, only cases of restrictions on link flows are studied. Indeed, link flows can be directly observed and the corresponding models are manageable. Nevertheless, when the traffic manager wants to take route flows under control, the extension to the case of route constraints is necessary. Therefore, in this framework, we shall discuss the existence of the equilibrium solution as well as its characterization in terms of Wardrop's generalized principle (see [10, 11]).
Some authors (see [3, 4, 5, 7, 8]) share the opinion that the presence of capacity constraints affects the equilibrium in such a way that it does not always exist. It is assumed that the user classic equilibrium condition holds and it is observed that the optimal flow does not coincide with the solution of the related variational inequality, which describes the traffic problem. The equivalence is admitted solely if some additional costs, which are represented by the linear combination of multipliers of capacity constraints, are introduced. Hence, the existence of the equilibrium is bound to the augmented travel costs, for which Wardrop's classic equilibrium principle holds. Nevertheless, it is not considered that in the

*Ph.D. Student in Applied Mathematics and Computer Science at Department of Mathematics, University of Catania, Italy

presence of capacity constraints Wardrop's principle assumes a generalized form and, as a consequence, the equilibrium flow is always equivalent to the solution of the associated variational inequality. The main result is that it is still possible to study the traffic problem in terms of variational inequalities. We shall provide some examples of transportation networks with capacity constraints on the routes, whose equilibrium solutions will be obtained applying the computational procedure shown by A. Maugeri in [9], which well adapts to our purposes.
Even the interpretation of constrained models deserves attention: we suggest that route-flow constraints can have a meaningful role as control measures. The traffic manager may need to impose some restrictions on route flows to make the users choose certain routes instead of others, so as to guarantee an optimal traffic distribution, in the sense that the total travel cost is minimized. Moreover, we remark that, in some cases, it is possible to increase the travel demand and unexpectedly to have a decrease in total travel costs, as Fisk's famous paradox shows (see [1, 2, 6]).

2 Wardrop's generalized equilibrium condition

We want to remark not only that the capacity-constraint traffic equilibrium problems can always be studied in terms of variational inequalities, but also that they always have solutions. Our instruments will be some theoretical results previously acquired in this field and the computational method shown in [9].
Let us consider a traffic network where $\mathcal{W}$ is the set of Origin Destination (O/D) pairs and $\mathcal{R}$ is the set of routes. We note that each route $r \in \mathcal{R}$ connects only one pair $w \in \mathcal{W}$; we denote by $\mathcal{R}(w)$ the set of all routes connecting a given $w \in \mathcal{W}$. We assume that the feasible flows have to satisfy some capacity restrictions

$$\lambda_r \leq F_r \leq \mu_r \qquad \forall r \in \mathcal{R},$$

and demand requirements

$$\sum_{r \in \mathcal{R}(w)} F_r = \rho_w \qquad \text{for all } w \in \mathcal{W}, \tag{1}$$

where $\lambda \leq \mu \in \mathbb{R}_+^{\mathcal{R}}$ and $\rho \in \mathbb{R}^{\mathcal{W}}$ is non negative. In order to express the demand requirements in a matrix-vector notation, we shall introduce the pair-route incidence matrix $\Phi = (\Phi_{wr})$, with $w \in \mathcal{W}$ and $r \in \mathcal{R}$, where:

$$\Phi_{wr} = \begin{cases} 1 & \text{if } r \in \mathcal{R}(w) \\ 0 & \text{otherwise.} \end{cases}$$

Thus (1) can be written as:

$$\Phi F = \rho$$

and the set of all feasible flows is given by:

$$\mathcal{K} := \{F \in \mathbb{R}_+^{\mathcal{R}} : \lambda \leq F \leq \mu, \Phi F = \rho\} \tag{2}$$

where $\Phi\lambda \leq \Phi F \leq \Phi\mu$. Let $C : \mathcal{K} \to \mathbb{R}^{\mathcal{R}}$ be a given cost function on routes, thus a cost vector $C(F) \in \mathbb{R}^{\mathcal{R}}$ corresponds to every feasible flow $F \in \mathcal{K}$.

Definition 2.1 (Wardrop's Generalized Principle ([10, 11])) *A flow $H \in \mathcal{K}$ is an equilibrium flow if and only if*

$$\text{for all } w \in \mathcal{W} \text{ and all } q, s \in \mathcal{R}(w),$$
$$C_q(H) < C_s(H) \Rightarrow H_q = \mu_q \quad \text{or} \quad H_s = \lambda_s. \tag{3}$$

It is possible to characterize equilibrium flows by means of variational inequalities and to this end we recall the following theorem (see [13, 12]):

Theorem 2.1 *Let $\mathcal{K}$ be given by (2), a flow $H \in \mathbb{R}^{\mathcal{R}}$ is an equilibrium flow if and only if:*

$$H \in \mathcal{K} \text{ and } \langle C(H), F - H \rangle \geq 0, \quad \forall F \in \mathcal{K}. \tag{4}$$

Thus, we are entitled to conclude that the research of an equilibrium flow in the sense of (3) is reduced to solving a variational inequality. Hence, we can refer to the theory of variational inequalities and, as our hypotheses are in agreement with those required by existence theorems, we can count on the presence of solutions. In the next section, we shall propose a significant example as application of the above results. We shall apply the computational procedure shown in [9] which allows us to see clearly how it is possible to calculate explicitly the solution, which either lies inside the set of feasible flows, thereby making the costs equal or else lies on a face of the set. We refer to the concrete case of Fisk's model, however the procedure holds for every kind of network.

3 A triangular network

Let us consider the network in Figure 1, where the set of nodes is $N = \{P_1, P_2, P_3\}$ and the set of links, which coincides with the set of origin-destination pairs, is $L = \{(P_3, P_2), (P_3, P_1), (P_2, P_1)\}$. The set of routes connecting the O/D pairs is $R = \{R_1, R_2, R_3, R_4\}$, where $R_1 = P_3P_2$, $R_2 = P_3P_1$, $R_3 = P_2P_1$, $R_4 = P_3P_2 \cup P_2P_1$. Let F_1, F_2, F_3, F_4 be the flows on the routes; ρ_1, ρ_2, ρ_3 the travel demands, and f_1, f_2, f_3 the flows between the pairs $(P_3, P_2), (P_3, P_1), (P_2, P_1)$.

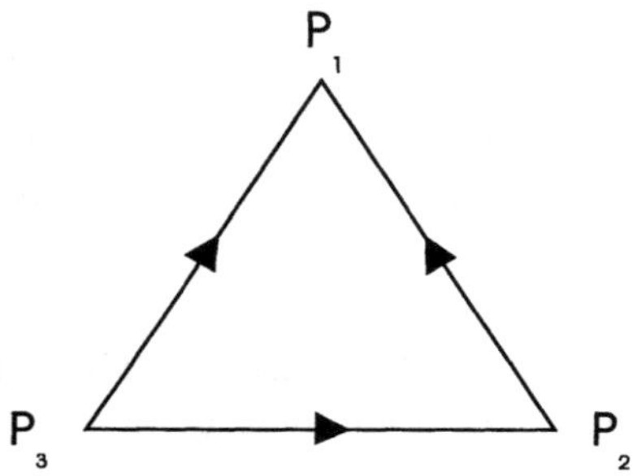

Figure 1

We suppose that link costs are:

$$\begin{aligned} c_1(f) &= f_1 \\ c_2(f) &= f_2 + \alpha \\ c_3(f) &= f_3 \end{aligned}$$

where $f = (f_1, f_2, f_3)$ and α is a non negative constant.
It results that:

$$\begin{aligned} f_1 &= F_1 + F_4 \\ f_2 &= F_2 \\ f_3 &= F_3 + F_4, \end{aligned}$$

the route costs $C_1(F), C_2(F), C_3(F), C_4(F)$ with $F = (F_1, F_2, F_3, F_4)$ are given by:

$$\begin{aligned} C_1(F) &= F_1 + F_4 \\ C_2(F) &= F_2 + \alpha \\ C_3(F) &= F_3 + F_4 \\ C_4(F) &= F_1 + F_3 + 2F_4. \end{aligned}$$

We assume that the route flows have to satisfy some capacity constraints so that the set of all feasible flows is given by:

$$\mathcal{K} = \{F \in \mathbb{R}^4_+ : \lambda_r \leq F_r \leq \mu_r \quad r = 1, 2, 3; F_1 = \rho_1, F_2 + F_4 = \rho_2, F_3 = \rho_3\}$$

where $\lambda \leq \mu \in \mathbb{R}^3_+$, $\lambda_1 \leq \rho_1 \leq \mu_1$, $\rho_2 > \lambda_2$ and $\lambda_3 \leq \rho_3 \leq \mu_3$.
The variational inequality which expresses the equilibrium problem has the form:

$$\text{find } H \in \mathcal{K} \text{ such that } C(H)(F-H) \geq 0, \qquad \forall F \in \mathcal{K}, \tag{5}$$

where $C(H)(F-H) = \sum_{i=1}^{4} C_i(H)(F_i - H_i)$.
We apply the computational procedure shown in [9] and confine our study only to the case $\mu_2 < \rho_2$.
Deducing F_4 from $F_2 + F_4 = \rho_2$, the variational inequality (5) becomes:

$$\text{find } \widetilde{H} \in \widetilde{\mathcal{K}} \text{ such that } \Gamma(\widetilde{H})(\widetilde{F} - \widetilde{H}) \geq 0, \qquad \forall \widetilde{F} \in \widetilde{\mathcal{K}}$$

where

$$\widetilde{\mathcal{K}} = \{\widetilde{F} \equiv F_2 \in \mathbb{R}_+ : \lambda_2 \leq F_2 \leq \mu_2\},$$

and

$$\Gamma(\widetilde{H}) = C_2(\widetilde{H}) - C_4(\widetilde{H}).$$

We find that if the following condition holds:

$$3\lambda_2 + \alpha - 2\rho_2 \leq \rho_1 + \rho_3 \leq 3\mu_2 + \alpha - 2\rho_2 \tag{6}$$

the solution of the system:

$$\begin{cases} C_2(\widetilde{H}) = C_4(\widetilde{H}) \\ H_1 = \rho_1 \\ H_3 = \rho_3 \\ H_2 + H_4 = \rho_2 \end{cases}$$

is given by:

$$\begin{cases} H_1 = \rho_1 \\ H_2 = \dfrac{\rho_1 + 2\rho_2 + \rho_3 - \alpha}{3} \\ H_3 = \rho_3 \\ H_4 = \dfrac{-\rho_1 + \rho_2 - \rho_3 + \alpha}{3}. \end{cases}$$

If

$$\rho_1 + \rho_3 < 3\lambda_2 + \alpha - 2\rho_2 \tag{7}$$

then the solution, obtained intersecting $\widetilde{\mathcal{K}}$ with $\{F_2 = \lambda_2\}$, is:

$$\begin{cases} H_1 = \rho_1 \\ H_2 = \lambda_2 \\ H_3 = \rho_3 \\ H_4 = \rho_2 - \lambda_2; \end{cases}$$

while if

$$\rho_1 + \rho_3 > 3\mu_2 + \alpha - 2\rho_2 \tag{8}$$

the solution, which lies in $\widetilde{\mathcal{K}} \cap \{F_2 = \mu_2\}$, is given by:

$$\begin{cases} H_1 = \rho_1 \\ H_2 = \mu_2 \\ H_3 = \rho_3 \\ H_4 = \rho_2 - \mu_2. \end{cases}$$

We can observe how, in any case, the computational procedure leads to the solution, which either makes the costs equal or lies on a face of the convex set $\mathcal{K}$.

4 More about generalized equilibrium principle

Wardrop's generalized principle, as defined in (3), can also be expressed in another form, where some additional costs are introduced.
After presenting the new formulation, we provide an example where the equilibrium solution satisfies the generalized principle, but it obeys Wardrop's classic principle if the augmented costs are considered instead of the initial ones.
Let H be an equilibrium flow according to (3) and let us consider:

$$C^w(H) = \max_{r \in B_w} C_r(H),$$

where

$$B_w = \{r : r \in \mathcal{R}(w), H_r > \lambda_r\}.$$

Moreover we set:

$$L_r^w = C^w(H) - C_r(H) \qquad r \in B_w, \text{ and}$$

$$\widetilde{C}_r(H) = \begin{cases} C_r(H) + L_r^w = C^w(H) & \text{if } r \in B_w \\ C_r(H) & \text{if } r \notin B_w,\ r \in \mathcal{R}(w) \end{cases}$$

where $\widetilde{C}_r(H)$, $r \in \mathcal{R}(w)$ represents the *generalized* or *modified* cost. Hence, the generalized equilibrium principle assumes the following form:

Definition 4.1 *$H \in \mathcal{K}$ is an equilibrium vector if*

$$\text{for all } w \in \mathcal{W} \text{ and all } q, s \in \mathcal{R}(w),$$
$$\widetilde{C}_s(H) > \widetilde{C}_q(H) \Rightarrow H_s = \lambda_s. \tag{9}$$

Now we want to illustrate the role of generalized costs by means of a concrete example.
Let us consider the network as in Figure 2.

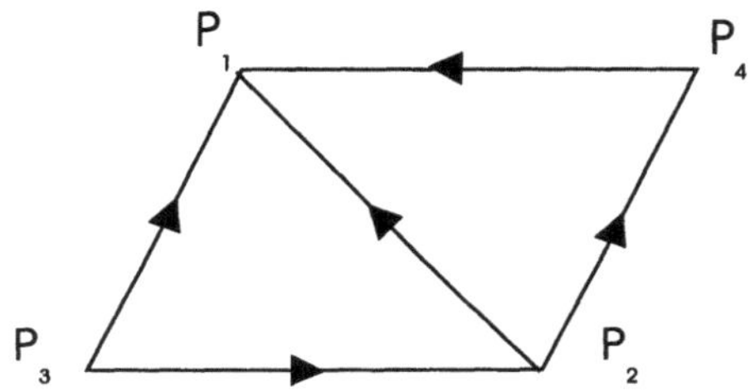

Figure 2

The set of routes connecting the O/D pairs P_3P_2, P_3P_1 is $R = \{R_1, R_2, R_3, R_4\}$, where $R_1 = P_3P_2$, $R_2 = P_3P_1$, $, R_3 = P_3P_2 \cup P_2P_1$, $R_4 = P_3P_2 \cup P_2P_4 \cup P_4P_1$. Let F_1, F_2, F_3, F_4 be the route flows; ρ_1, ρ_2 the travel demands, and f_1, f_2, f_3, f_4, f_5 the link flows.
Let us set:

$$\begin{array}{ll} c_1(f) = f_1 & f_1 = F_1 + F_3 + F_4 \\ c_2(f) = f_2 + \alpha & f_2 = F_2 \\ c_3(f) = f_3 & f_3 = F_3 \\ c_4(f) = f_4 & f_4 = F_4 \\ c_5(f) = f_5 + \beta & f_5 = F_4 \end{array}$$

where f is the link flow vector and α, β are non negative constants. Therefore, the route costs $C_1(F), C_2(F), C_3(F), C_4(F)$, where F is the route flow vector, are given by:

$$\begin{aligned} C_1(F) &= F_1 + F_3 + F_4 \\ C_2(F) &= F_2 + \alpha \\ C_3(F) &= F_1 + 2\,F_3 + F_4 \\ C_4(F) &= F_1 + F_3 + 3F_4 + \beta. \end{aligned}$$

We assume that the set of all feasible flows be:

$$\begin{aligned} \mathcal{K} = \quad & \{F \in \mathbb{R}^4_+ : \lambda_r \le F_r \le \mu_r;\ r = 1, 2, 3;\ F_1 = \rho_1, \\ & F_2 + F_3 + F_4 = \rho_2, F_3 = \rho_3\} \end{aligned}$$

where $\lambda \leq \mu \in \mathbb{R}^3_+$, $\lambda_1 \leq \rho_1 \leq \mu_1$, $\rho_2 > \lambda_2 + \lambda_3$.
The variational inequality associated to the problem is then:

$$\text{find } H \in \mathcal{K} \text{ such that } C(H)(F-H) \geq 0, \qquad \forall F \in \mathcal{K}, \tag{10}$$

where $C(H)(F-H) = \sum_{i=1}^4 C_i(H)(F_i - H_i)$.
We confine our study only to the case $\rho_2 > \mu_2 + \mu_3$.
Deducing F_4 from $F_2 + F_3 + F_4 = \rho_2$, the variational inequality (10) becomes:

$$\text{find } \widetilde{H} \in \widetilde{\mathcal{K}} \text{ such that } \Gamma_2(\widetilde{H})(F_2-H_2)+\Gamma_3(\widetilde{H})(F_3-H_3) \geq 0, \qquad \forall \widetilde{F} \in \widetilde{\mathcal{K}}$$

with

$$\widetilde{\mathcal{K}} = \{\widetilde{F} \in \mathbb{R}^2_+ : \lambda_r \leq F_r \leq \mu_r, r = 2,3,\ F_2 + F_3 \leq \rho_2\},$$

and

$$\Gamma_2(\widetilde{H}) = C_2(\widetilde{H}) - C_4(\widetilde{H})$$
$$\Gamma_3(\widetilde{H}) = C_3(\widetilde{H}) - C_4(\widetilde{H}).$$

We find that if the following conditions hold:

$$8\lambda_2 + 3\alpha - \beta \leq 3\rho_1 + 5\rho_2 \leq 8\mu_2 + 3\alpha - \beta \tag{11}$$
$$4\lambda_3 - \alpha - \beta \leq -\rho_1 + \rho_2 \leq 4\mu_3 - \alpha - \beta \tag{12}$$

the solution of the system

$$\begin{cases} \Gamma_2(\widetilde{H}) = 0 \\ \Gamma_3(\widetilde{H}) = 0 \\ \widetilde{H} \in \widetilde{\mathcal{K}} \end{cases}$$

is:

$$\begin{cases} H_1 = \rho_1 \\ H_2 = \dfrac{\rho_1 + 5\rho_2 - 3\alpha + \beta}{8} \\ H_3 = \dfrac{-\rho_1 + \rho_2 + \alpha + \beta}{4} \\ H_4 = \dfrac{-\rho_1 + \rho_2 + \alpha - 3\beta}{8}. \end{cases}$$

If

$$8\lambda_2 + 3\alpha - \beta > 3\rho_1 + 5\rho_2 \tag{13}$$

then the solution, obtained intersecting $\widetilde{\mathcal{K}}$ with $\{F_2 = \lambda_2\}$, is:

$$\begin{cases} H_1 = \rho_1 \\ H_2 = \lambda_2 \\ H_3 = \dfrac{2\rho_2 - 2\lambda_2 + \beta}{3} \\ H_4 = \dfrac{\rho_2 - \lambda_2 - \beta}{3}. \end{cases}$$

If

$$8\mu_2 + 3\alpha - \beta < 3\rho_1 + 5\rho_2 \tag{14}$$

the solution, which lies in $\widetilde{\mathcal{K}} \cap \{F_2 = \mu_2\}$, is given by:

$$\begin{cases} H_1 = \rho_1 \\ H_2 = \mu_2 \\ H_3 = \dfrac{2\rho_2 - 2\mu_2 + \beta}{3} \\ H_4 = \dfrac{\rho_2 - \mu_2 - \beta}{3}. \end{cases}$$

If

$$4\lambda_3 - \alpha - \beta > -\rho_1 + \rho_2 \tag{15}$$

the solution, which is obtained intersecting $\widetilde{\mathcal{K}}$ with $\{F_3 = \lambda_3\}$, is:

$$\begin{cases} H_1 = \rho_1 \\ H_2 = \dfrac{\rho_1 + 3\rho_2 - 2\lambda_3 - \alpha + \beta}{4} \\ H_3 = \lambda_3 \\ H_4 = \dfrac{-\rho_1 + \rho_2 - 2\lambda_3 + \alpha - \beta}{4}. \end{cases}$$

If

$$4\mu_3 - \alpha - \beta < -\rho_1 + \rho_2 \tag{16}$$

the solution lies in $\widetilde{\mathcal{K}} \cap \{F_3 = \mu_3\}$ and it is represented by:

$$\begin{cases} H_1 = \rho_1 \\ H_2 = \dfrac{\rho_1 + 3\rho_2 - 2\mu_3 - \alpha + \beta}{4} \\ H_3 = \mu_3 \\ H_4 = \dfrac{-\rho_1 + \rho_2 - 2\mu_3 + \alpha + \beta}{4}. \end{cases}$$

It is noteworthy that Wardrop's generalized principle (3) is verified, but, if the modified costs are introduced, the classic formulation holds. In fact, for the pair $w = (P_3, P_1)$, if (14) holds, we find that:

$$
\begin{aligned}
C^w(H) = \max_{r \in B_w} C_r(H) = C_3(H) \\
B_w = \{2, 3, 4\} \\
L_3 = L_4 = 0 \\
L_2 = C_3(H) - C_2(H)
\end{aligned}
$$

If we calculate the modified costs, we have that:

$$
\begin{cases}
\widetilde{C}_r(H) = C_r(H),\ r = 3, 4 \\
\widetilde{C}_2(H) = C_2(H) + L_2
\end{cases}
$$

and since all the generalized costs are equal, Wardrop's classic condition is obviously satisfied. We draw similar conclusions when (16) holds.
The possibility of introducing the generalized costs is then appealing. In fact, they are equal on all the routes whose flows are strictly grater than the minimum, while they coincide with the initial costs on the other routes. Therefore if lower bounds are never reached, the generalized costs, which all have the same value, fulfil the classic principle and we are entitled to proceed as no restrictions were imposed. In the other cases, we apply the second generalized formulation (9), which, however, can always be reduced to the classic one, if we shift the flows (setting $\widehat{H}_s = H_s - \lambda_s$) and transform the corresponding travel demand.

5 Capacity constraints and paradox

The sensitivity of travel costs to changes in input data can be successfully exploited in order to have an optimal distribution of traffic, namely a reduction of congestion and total costs, see [1, 2, 6]. It can be observed that, even in the presence of capacity constraints, the change of the initial data, *i.e.* the travel demands, can produce interesting results. In fact, the following paradox can be shown: in some cases, an increase in travel demands leads to a decrease in total costs. Thus, the introduction of capacity constraints is seen as a means of traffic control. We shall refer to the above example, which generalizes Fisk's paradox (see [6]).
Let us increase the travel demand for the pair (P_2, P_3), adding the positive quantity d^*. A new problem has to be solved, in which the travel demands are given by $(\rho_1 + d^*, \rho_2, \rho_3)$. If

$$3\lambda_2 + \alpha - 2\rho_2 \le \rho_1 + d^* + \rho_3 \le 3\mu_2 + \alpha - 2\rho_2$$

holds, then the solution $H^* = (H_1^*, H_2^*, H_3^*, H_4^*)$ is given by:

$$\begin{cases} H_1^* = H_1 + d^* \\ H_2^* = H_2 + \dfrac{d^*}{3} \\ H_3^* = H_3 \\ H_4^* = H_4 - \dfrac{d^*}{3}. \end{cases}$$

The new cost functions are:

$$\begin{cases} C_1^*(H^*) = C_1(H) + \dfrac{2d^*}{3} \\ C_2^*(H^*) = C_2(H) + \dfrac{d^*}{3} \\ C_3^*(H^*) = C_3(H) - \dfrac{d^*}{3} \\ C_4^*(H^*) = C_4(H) + \dfrac{d^*}{3}. \end{cases}$$

It is easy to show that under convenient conditions the increase in the travel demand can make the total cost decrease.
Let $C^*(H^*)H^*$ denote the total cost, it results that:

$$C^*(H^*)H^* = C(H)H + \frac{d^*}{3}(2(2\rho_1 + \rho_2 - \rho_3) + \alpha) + \frac{2d^{*2}}{3}.$$

We have a decrease in total costs if

$$C^*(H^*)H^* < C(H)H,$$

namely:

$$\rho_3 > 2\rho_1 + \rho_2 + \frac{\alpha}{2} + d^*.$$

A numerical example is obtained choosing: $\rho_1 = 2$, $\rho_2 = 10$, $\rho_3 = 45$, $\lambda_2 = 5$, $\mu_2 = 8$, $d^* = 4$, $\alpha = 50$.
Regarding the cases (7) and (8) we observe that when flows reach the upper or lower bound there are no parameter value which can reduce the amount of the costs.

References

[1] M. De Luca and A. Maugeri, Quasi-Variational Inequalities and Applications to the Traffic Equilibrium Problem; Discussion of a Paradox, *Journal of Computational and Applied Mathematics* **28** (1989) 163-171.

[2] M. De Luca and A. Maugeri, Variational Inequalities Applied to the Study of Paradoxes in Equilibrium Problems, *Optimization* **25** (1992) 249-259.

[3] P. Ferrari, Equilibrium in Transport Networks with Capacity Constraints, in: F. Giannessi, A. Maugeri, eds., *Proceedings of the International School of Mathematics "G. Stampacchia" 19th Course on "Variational Inequalities and Network Equilibrium Problems", Erice, June 19-25, 1994* (Plenum Press, New York, 1995) 85-100.

[4] P. Ferrari, The Meaning of Capacity Constraints Multipliers in the Theory of Road Network Equilibrium, *Supplemento Rendiconti del Circolo Matematico di Palermo, Serie 2* **48** (1997) 107-120.

[5] P. Ferrari, Optimal Flow Pattern in Road Networks, in: F. Giannessi, A. Maugeri, P. Pardalos eds., *Equilibrium Problem: non Smooth Optimalization and Variational Inequalities* (Kluwer Academic Publisher, 2000).

[6] C. Fisk, More Paradoxes in the Equilibrium Assignment Problem, *Transportation Research Part B* **13** (1979) 305-309.

[7] T. Larsson and M. Patriksson, Equilibrium Characterizations of Solutions to Side Constrained Asymmetric Traffic Assignment Models, *Le Matematiche* **49** (1994) 249-280.

[8] T. Larsson and M. Patriksson, On Side Constrained Models of Traffic Equilibria, in: F. Giannessi, A. Maugeri, eds., *Proceedings of the International School of Mathematics "G. Stampacchia" 19th Course on "Variational Inequalities and Network Equilibrium Problems", Erice, June 19-25, 1994* (Plenum Press, New York, 1995) 169-178.

[9] A. Maugeri, Convex Programming, Variational Inequalities and Applications to the Traffic Equilibrium Problems, *Applied Mathematics and Optimization* **16** (1987) 169-185.

[10] A. Maugeri, Optimization Problems with Side Constraints and Generalized Equilibrium Principles, *Le Matematiche* **49** (1994) 305-312.

[11] A. Maugeri and W. Oettli and D. Schläger, A Flexible Form of Wardrop's Principle for Traffic Equilibria with Side Constraints, *Supplemento Rendiconti del Circolo Matematico di Palermo, Serie 2* **48** (1997) 185-193.

[12] M.J. Smith, The Existence, Uniqueness and Stability of Traffic Equilibrium, *Transportation Research* **138** (1979) 295-304.

[13] J.G. Wardrop, Some Theoretical Aspects of Road Traffic Research, in: *Proceedings of the Institute of Civil Engineers, Part II* (1952) 325-378.

MIX
Papier aus verantwortungsvollen Quellen
Paper from responsible sources
FSC® C105338

If you have any concerns about our products,
you can contact us on
ProductSafety@springernature.com

In case Publisher is established outside the EU,
the EU authorized representative is:
Springer Nature Customer Service Center GmbH
Europaplatz 3, 69115 Heidelberg, Germany

Printed by Libri Plureos GmbH
in Hamburg, Germany